Dyneins

STRUCTURE, BIOLOGY AND DISEASE

Dynein Mechanics, Dysfunction, and Disease

Volume 2

Second Edition

Edited by

Stephen M. King
University of Connecticut Health Center
Farmington, CT, United States

ELSEVIER

ACADEMIC PRESS
An imprint of Elsevier

Academic Press is an imprint of Elsevier
125 London Wall, London EC2Y 5AS, United Kingdom
525 B Street, Suite 1800, San Diego, CA 92101-4495, United States
50 Hampshire Street, 5th Floor, Cambridge, MA 02139, United States
The Boulevard, Langford Lane, Kidlington, Oxford OX5 1GB, United Kingdom

Library of Congress Cataloging-in-Publication Data
A catalog record for this book is available from the Library of Congress

British Library Cataloguing-in-Publication Data
A catalogue record for this book is available from the British Library

ISBN: 978-0-12-809470-9

For Information on all Academic Press publications visit our website at
https://www.elsevier.com/books-and-journals

Working together
to grow libraries in
developing countries

www.elsevier.com • www.bookaid.org

Publisher: Sara Tenny
Acquisition Editor: Sara Tenny
Editorial Project Manager: Fenton Coulthurst
Production Project Manager: Kiruthika Govindaraju
Designer: Harris Greg

Cover Art Legend: A model of the structure of cytoplasmic dynein (grey) bound to its essential cofactor
dynactin (multicolor) via the Golgi vesicle cargo adaptor Bicaudal-D2 (orange). The dynein is reaching down
to contact the microtubule track (dark and light green) that it will drag the cargo along. In the background
are examples of electron density from the cryo-electron microscopy structure of dynactin from Urnavicius
et al (2015) Science 347, 1441-1446. The image was created by Janet Iwasa (University of Utah) and was
provided courtesy of Andrew Carter's group (MRC Laboratory of Molecular Biology, Cambridge).

Typeset by TNQ Books and Journals

Contents

Volume 1

Part I   History and Evolution

1.   Discovery of dynein and its properties: a personal account   5

Ian R. Gibbons

2.   Origins of cytoplasmic dynein   89

Richard B. Vallee

3.   The evolutionary biology of dyneins   101

Bill Wickstead

Contents

Contents

Contents

Contents

Contents

Contents

Volume 2

Part I Structure and Mechanics of Dynein Motors

Contents

Contents

Contents

Contents

Contents

List of Contributors

Volume 1

Adam W. Avery
University of Minnesota, Minneapolis, MN, United States

Gordon K. Chan
University of Alberta, Edmonton, AB, Canada

Anudariya B. Dean
SUNY Upstate Medical University, Syracuse, NY, United States

Paurav B. Desai
SUNY Upstate Medical University, Syracuse, NY, United States;
UMASS Medical School, Worcester, MA, United States

Ian R. Gibbons
University of California Berkeley, Berkeley, CA, United States

Thomas S. Hays
University of Minnesota, Minneapolis, MN, United States

Edward H. Hinchcliffe
University of Minnesota, Austin, MN, United States

Yuqing Hou
University of Massachusetts Medical School, Worcester, MA,
United States

Emily L. Hunter
Emory University, Atlanta, GA, United States

Juyeon Hwang
Emory University, Atlanta, GA, United States

Takashi Ishikawa
Paul Scherrer Institute (PSI), Villigen, Switzerland; ETH Zurich,
Zurich, Switzerland

Rupam Jha
The Francis Crick Institute, London, United Kingdom

Ritsu Kamiya
Gakushuin University, Tokyo, Japan

Stephanie A. Ketcham
Johns Hopkins University, Baltimore, MD, United States

K. Kevin Pfister
University of Virginia, Charlottesville, VA, United States

Stephen M. King
University of Connecticut Health Center, Farmington, CT, United States

Cody W. Lewis
University of Alberta, Edmonton, AB, Canada

Min-Gang Li
University of Minnesota, Minneapolis, MN, United States

Richard J. McKenney
University of California – Davis, Davis, CA, United States

David R. Mitchell
SUNY Upstate Medical University, Syracuse, NY, United States

Amanda L. Neisch
University of Minnesota, Minneapolis, MN, United States

Mary E. Porter
University of Minnesota, Minneapolis, MN, United States

Winfield S. Sale
Emory University, Atlanta, GA, United States

Trina A. Schroer
Johns Hopkins University, Baltimore, MD, United States

Chikako Shingyoji
The University of Tokyo, Tokyo, Japan

Thomas Surrey
The Francis Crick Institute, London, United Kingdom

Noriko Ueki
Tokyo Institute of Technology, Yokohama, Japan

Richard B. Vallee
Columbia University, New York, NY, United States

Kevin T. Vaughan
University of Notre Dame, Notre Dame, IN, United States

Ken-ichi Wakabayashi
Tokyo Institute of Technology, Yokohama, Japan

Bill Wickstead
University of Nottingham, Nottingham, United Kingdom

Maureen Wirschell
University of Mississippi Medical Center, Jackson, MS, United States

George B. Witman
University of Massachusetts Medical School, Worcester, MA, United States

Xin Xiang
Uniformed Services University of the Health Sciences, Bethesda, MD, United States

Toshiki Yagi
Prefectural University of Hiroshima, Hiroshima, Japan

Volume 2

Lionel Berthoux
Université du Québec à Trois-Rivières, QC, Canada

Andrew P. Carter
Medical Research Council Laboratory of Molecular Biology, Cambridge, United Kingdom

Amrita Dawn
Beckman Research Institute at the City of Hope, Duarte, CA, United States

David J. Doobin
Columbia University, New York, NY, United States

Mike Fainzilber
Weizmann Institute of Science, Rehovot, Israel

Elizabeth M.C. Fisher
University College London, London, United Kingdom

Ken'ya Furuta
National Institute of Information and Communications Technology, Kobe, Japan

Veikko F. Geyer
Yale University, New Haven, CT, United States

Steven P. Gross
University of California, Irvine, CA, United States

William O. Hancock
Pennsylvania State University, University Park, PA, United States

Kent L. Hill
University of California, Los Angeles, CA, United States

Erika L.F. Holzbaur
University of Pennsylvania Perelman School of Medicine, Philadelphia, PA, United States

Jonathon Howard
Yale University, New Haven, CT, United States

Simon Imhof
University of California, Los Angeles, CA, United States

Kazuo Inaba
University of Tsukuba, Shimoda, Japan

Frank Jülicher
Max Planck Institute for the Physics of Complex Systems, Dresden, Germany

Tarun M. Kapoor
The Rockefeller University, New York, NY, United States

Sandip Koley
Weizmann Institute of Science, Rehovot, Israel

Christine M. Lightcap
Beckman Research Institute at the City of Hope, Duarte, CA, United States

Niki T. Loges
University Hospital Muenster, Muenster, Germany

Miroslav P. Milev
Concordia University, Montréal, QC, Canada

Hannah M. Mitchison
University College London, UCL Great Ormond Street Institute of Child Health, London, United Kingdom

Armen J. Moughamian
University of Pennsylvania Perelman School of Medicine, Philadelphia, PA, United States; University of California, San Francisco, CA, United States

Andrew J. Mouland
Lady Davis Institute at the Jewish General Hospital, Montréal, QC, Canada; McGill University, Montréal, QC, Canada

Kazuhiro Oiwa
National Institute of Information and Communications Technology, Kobe, Japan

Heymut Omran
University Hospital Muenster, Muenster, Germany

Stephen H. Pilder
Temple University School of Medicine, Philadelphia, PA, United States

Michael Price
University of Nebraska Medical Center, Omaha, NE, United States

Babu J.N. Reddy
University of California, Irvine, CA, United States

Hitoshi Sakakibara
National Institute of Information and Communications Technology, Kobe, Japan

Pablo Sartori
Institute for Advanced Study, Princeton, NJ, United States

Helgo Schmidt
Institut de Génétique et de Biologie Moléculaire et Cellulaire (IGBMC), Illkirch, France

Miriam Schmidts
Radboud University Medical Center, Nijmegen, The Netherlands; University Hospital Freiburg, Freiburg, Germany

Amanda E. Siglin
Beckman Research Institute at the City of Hope, Duarte, CA, United States

Joseph H. Sisson
University of Nebraska Medical Center, Omaha, NE, United States

Jonathan B. Steinman
The Rockefeller University, New York, NY, United States

George T. Shubeita
New York University Abu Dhabi, Abu Dhabi, United Arab Emirates

Marco Terenzio
Weizmann Institute of Science, Rehovot, Israel

Richard B. Vallee
Columbia University, New York, NY, United States

John B. Vincent
Campbell Family Mental Health Research Institute, Toronto, ON, Canada; University of Toronto, Toronto, ON, Canada

John C. Williams
Beckman Research Institute at the City of Hope, Duarte, CA, United States

Maureen Wirschell
University of Mississippi Medical Center, Jackson, MS, United States

Fan Yang
University of Mississippi Medical Center, Jackson, MS, United States

Xaojian Yao
University of Manitoba, Winnipeg, MB, Canada

Ahmet Yildiz
University of California, Berkeley, CA, United States

Biography

Stephen King Biosketch

Stephen M. King is a professor of Molecular Biology and Biophysics at the University of Connecticut School of Medicine and director of the electron microscopy facility. Dr. King received his BSc (Hons) from the University of Kent, Canterbury, UK, and his PhD from University College London. He performed postdoctoral studies at the Worcester Foundation for Experimental Biology in Shrewsbury, Massachusetts, before moving to Connecticut in 1993 following several months at Palmer Station Antarctica studying cold adaption of dyneins. He has worked on the structure, function, and regulation of dyneins for over 30 years, mainly using *Chlamydomonas* and Planaria as model systems. His current research employs a broad array of structural, biochemical, genetic, and cell biological approaches to address fundamental questions concerning dynein assembly, function, and regulation.

Preface

Molecular motors are required for a vast array of cellular functions and consequently have attracted the attention of many biologists and clinicians. Over the last several decades, much has been learned about the mechanisms of action of molecular motors and the consequences of their dysfunction; for example, how members of the myosin and kinesin superfamilies convert ATP hydrolysis to mechanical work is now understood in exquisite detail. However, one class of motor—the dyneins—which translocate along microtubules has stood out in terms of their complexity and the difficulties encountered in trying to understand their structural and functional attributes. Not only are the motor units themselves enormous (over 500 kDa each), which in and of itself imparts a whole host of technical challenges, but these motor subunits are associated with a complicated array of additional components that play key roles in stable formation of the holoenzymes, binding appropriate cargos and regulating motor function in response to a wide variety of ever-changing regulatory inputs. Likewise, the array of activities that dyneins participate in and the consequences of their malfunction are myriad and complex.

The first edition of *Dyneins: Structure, Biology and Disease* was published in 2012 as a single volume containing 24 chapters. Since then, enormous progress has been made in many areas of dynein biology including, for example, understanding the structural basis of dynein motor function, the mechanisms of cytoplasmic preassembly of axonemal dyneins, and the consequences of dynein dysfunction for human disease. In this second edition, the scope has been greatly expanded, necessitating two volumes with a total of 40 chapters, to try and capture the major advances that have recently been made. Only two chapters (on the initial discovery of dynein and on the role of dynein in non-Mendelian transmission of mouse *t* haplotypes) are reprinted directly from the first edition with only minor alterations; all the others are either completely new or highly updated and modified from the originals. In addition, a website (http://booksite. elsevier.com/9780128094716) accompanies these volumes and includes videos of dynein-based motility and other materials that complement specific chapters. Each chapter is written by true experts in their respective fields, and it is my hope that this compendium will prove useful both to seasoned dynein researchers and to those thinking about entering the field for the first time and studying this fascinating and enormously complex family of molecular motors.

This two-volume work is divided into five broad segments. The first volume is devoted to the *Biology of Dynein Motors* and starts with a section concerning the history of dynein research and the evolutionary biology of dyneins. The initial discovery of dynein in cilia is described by Gibbons, and the subsequent identification of cytoplasmic dynein is detailed by Vallee. This is followed by a chapter from Wickstead who describes dynein phylogenetic bioinformatics and plots the evolutionary biology of these motors, including discussion of how these unusual AAA+ motors might have arisen prior to the last eukaryotic common ancestor. The second section examines dyneins in the context of ciliary biology. Desai et al. detail the current understanding of the complex mechanisms by which these massive multicomponent axonemal motors are initially assembled in cytoplasm, and King covers their detailed biochemical composition. The arrangement of dyneins within the axonemal superstructure is described by Ishikawa, and Yagi and Kamiya discuss the genetics of axonemal dyneins and how studies in a broad array of model systems have resulted in major conceptual advances. The multiple mechanisms of outer- and inner-arm axonemal dynein regulation are detailed in chapters by King and Hwang et al., and the role of the dynein regulatory complex is described by Porter. Shingyoji analyzes dynein-driven ciliary/flagellar motility from a more physiological perspective, while Ueki and Wakabayashi discuss dynein-mediated photoresponses, including waveform conversion and redox-based control of dynein function in response to altered light conditions. The final chapter in this section by Witman and Hou describes our current understanding of the specialized dynein that powers retrograde intraflagellar transport and which is required for assembly of both motile and nonmotile (primary) cilia. The final section in Volume 1 covers the biology of cytoplasmic dynein. The subunit composition of this complex motor is described by Pfister, while McKenney details the regulatory mechanisms controlling motor activity. Insights into dynein function derived from fungal genetics are reviewed by Xiang, and details of how dynactin modifies dynein function are described by Ketcham and Schroer. The last four chapters in this section examine dyneins role in mitotic checkpoint silencing (Lewis and Chan), its functions at the nuclear envelope, cortex, and kinetochore during mitosis (Hinchcliffe and Vaughan), the mechanisms of microtubule plus end tracking (Jha and Surrey), and how dynein impacts gametogenesis, embryogenesis, and neuronal function in *Drosophila* (Neisch et al.).

The second volume subtitled *Dynein Structure, Mechanics, and Disease* is built around two sections. The first is focused on dynein structure and mechanics—an area that has seen massive strides in the last few years. Oiwa and colleagues describe the analysis of dynein structure and mechanism as determined from electron microscopic studies. This is followed by Schmidt and Carter who describe the dynein heavy chain structure and power stroke mechanism based on crystallographic and cryo electron microscopy studies. This is followed by a chapter by Williams et al. detailing our current understanding of the high-resolution structure of the intermediate and light chains that are needed for assembly of the dynein particle. Biochemical methods to purify native and recombinant dyneins are described by Inaba. The next three chapters cover motor mechanics: Yildiz details single molecule mechanics in vitro, Shubeita et al. consider dynein biophysics in vivo, and Hancock describes the mechanics of bidirectional transport. Steinman and Kapoor detail dynein inhibitors ranging from the phosphate analogue vanadate, originally identified as a dynein inhibitory factor in commercial preparations of ATP, to the more recently identified ciliobrevins. The final chapter in this section by Geyer et al. provides a practical guide to computational modeling of dynein activity and how this generates flagellar waveforms; a Matlab code enabling the reader to perform their own analyses can be downloaded from the website (http://booksite. elsevier.com/9780128094716) that accompanies these volumes. The last section describes the consequences of dynein dysfunction and how in many situations it leads to human disease. Milev et al. describe how viruses manipulate host dynein motors. Terenzio et al. discuss the use of mouse genetics to probe the consequences of cytoplasmic dynein dysfunction, and Doobin and Vallee detail how the regulation of this motor impacts brain development. This is followed by chapters examining the role of dyneins as causative factors in primary ciliary dyskinesia and other ciliopathies (Loges and Omran), and how disrupting dynein-mediated retrograde intraflagellar transport leads to severe skeletal abnormalities (Schmidts and Mitchison). Price et al. then describe the effects of chronic alcohol exposure on dynein function, while Imhof and Hill discuss dynein-based motility of parasitic protozoa, and Vincent details an intriguing connection between a dynein light chain and congenital mirror movement disorder. The final chapter by Pilder examines the potential role of dyneins in non-Mendelian chromosomal inheritance in mice.

Finally, I would like to extend my thanks to all the authors, who spent so much time and effort in preparing chapters for these volumes, and to the staff at Elsevier for all their assistance in making this second edition a reality.

Stephen M. King
West Hartford, Connecticut

A Cautionary Note About Dynein Nomenclature

Dynein was first identified in the early 1960s (see Chapter 1 by Gibbons), and the detailed composition of purified axonemal dynein motor complexes was initially determined in several organisms (including sea urchins [1], *Chlamydomonas* [2–4], and *Tetrahymena* [5]) during the late 1970s and early 1980s, many years before comprehensive sequence analysis was possible. Thus, the original names for various dynein components were derived from biochemical analysis of their migration patterns in denaturing SDS and/or urea polyacrylamide gels. This approach to naming components continued as dyneins from other organisms such as the ascidian *Ciona* [6], trout [7], bull [8], and sea anemone [9] were isolated. Consequently, the complexity imparted by the sheer number of dynein subunits, combined with variations in their relative migration due to sequence or compositional differences, resulted in closely related components being given completely distinct names. Furthermore, the same or very similar names were often assigned in various organisms to subunits, now known to be totally different. This history engendered a level of confusion that has in some cases caused components to be misidentified in the literature and still today provides a very significant barrier to students entering the field. For example, although the β heavy chains of sea urchin and *Chlamydomonas* outer-arm dyneins are orthologous, the sea urchin α heavy chain is actually equivalent to the γ heavy chain in *Chlamydomonas*, while the *Chlamydomonas* α heavy chain is a variant form of the β heavy chain. Confusion surrounding dynein nomenclature was made even worse during annotation of the human and mouse genomes when some components were named after their sea urchin orthologues while others followed the *Chlamydomonas* system. For example, the leucine-rich repeat protein termed DNAL1 in mammals was named after the *Chlamydomonas* outer-arm LC1 light chain; however, in sea urchins, the orthologous component is termed LC2. In contrast, the mammalian DNAL4 light chain was designated based on the sea urchin light chain numbering scheme (where it is LC4), while its *Chlamydomonas* equivalent is LC10; in *Chlamydomonas*, LC4 refers to a calmodulin-like Ca^{2+}-binding protein. A similar level of confusion applies to many other components as well.

In this volume, a brief guide to the differing nomenclatures used for axonemal outer- and inner-arm dyneins is presented in Tables 5.2 and 5.4 of Chapter 5 by King, while the naming scheme employed for the subunits of the dynein that powers retrograde intraflagellar transport is defined in Table 13.1 of Chapter 13 by Witman and Hou, and canonical cytoplasmic dynein nomenclature is detailed in Chapter 15 by Pfister. Furthermore, the detailed phylogenetic classification of dynein components is described in Chapter 3 by Wickstead. If there is any doubt about the identity of a dynein subunit, the reader is also urged to consult the agreed nomenclature for cytoplasmic dyneins [10] and/or the unified taxonomic guide to ciliary dyneins [11], which provides a comprehensive analysis delineating orthologous components across phylogeny.

Stephen M. King
West Hartford, Connecticut

References

[1] C.W. Bell, E. Fronk, I.R. Gibbons, Polypeptide subunits of dynein 1 from sea urchin sperm flagella, J. Supramol. Struct. 11 (3) (1979) 311–317.

[2] K.K. Pfister, R.B. Fay, G.B. Witman, Purification and polypeptide composition of dynein ATPases from *Chlamydomonas* flagella, Cell Motil. 2 (6) (1982) 525–547.

[3] G. Piperno, D.J. Luck, Inner arm dyneins from flagella of *Chlamydomonas reinhardtii*, Cell 27 (2 Pt 1) (1981) 331–340.

[4] G. Piperno, D.J. Luck, Axonemal adenosine triphosphatases from flagella of *Chlamydomonas reinhardtii*. Purification of two dyneins, J. Biol. Chem. 254 (8) (1979) 3084–3090.

[5] M.E. Porter, K.A. Johnson, Characterization of the ATP-sensitive binding of *Tetrahymena* 30 S dynein to bovine brain microtubules, J. Biol. Chem. 258 (10) (1983) 6575–6581.

[6] P. Padma, A. Hozumi, K. Ogawa, K. Inaba, Molecular cloning and characterization of a thioredoxin/nucleoside diphosphate kinase related dynein intermediate chain from the ascidian, *Ciona intestinalis*, Gene 275 (1) (2001) 177–183.

[7] J.L. Gatti, S.M. King, A.G. Moss, G.B. Witman, Outer arm dynein from trout spermatozoa. Purification, polypeptide composition, and enzymatic properties, J. Biol. Chem. 264 (19) (1989) 11450–11457.

[8] S.P. Marchese-Ragona, C. Gagnon, D. White, M.B. Isles, K.A. Johnson, Structure and mass analysis of 12S and 19S dynein obtained from bull sperm flagella, Cell Motil. Cytoskelet. 8 (4) (1987) 368–374.

[9] H. Mohri, K. Inaba, M. Kubo-Irie, H. Takai, Y. Yano-Toyoshima, Characterization of outer arm dynein in sea anemone, *Anthopleura midori*, Cell Motil. Cytoskelet. 44 (3) (1999) 202–208.

[10] K.K. Pfister, E.M.C. Fisher, I.R. Gibbons, T.S. Hays, E.L.F. Holzbaur, J.R. McIntosh, M.E. Porter, T.A. Schroer, K.T. Vaughan, G.B. Witman, S.M. King, R.B. Vallee, Cytoplasmic dynein nomenclature, J. Cell Biol. 171 (3) (2005) 411–413.

[11] E. Hom, G.B. Witman, E.H. Harris, S.K. Dutcher, R. Kamiya, D.R. Mitchell, G.J. Pazour, M.E. Porter, W.S. Sale, M. Wirschell, T. Yagi, S.M. King, A unified taxonomy for ciliary dyneins, Cytoskeleton 68 (2011) 555–565.

I

Structure and Mechanics of Dynein Motors

In this chapter

Electron microscopy of isolated dynein complexes and the power stroke mechanism

Kazuhiro Oiwa, Hitoshi Sakakibara, Ken'ya Furuta
National Institute of Information and Communications Technology, Kobe, Japan

1.1 Introduction

Dynein is the force generating ATPase that has been identified in ciliary axonemes of *Tetrahymena* cilia as a high molecular weight protein with Mg ATPase activity [1]. Structural studies of dynein with an atomic resolution have recently been performed using X-ray crystallography [2–6] with expression systems of cytoplasmic dyneins in human cells [4], yeast cells [7], and *Dictyostelium* cells [8,9]. Before the early 2010s, no structural information on the atomic resolution existed for the entire motor domain of dynein despite approximately 50 years of investigation since the discovery of dynein. This is because the size of the dynein is too large for nuclear magnetic resonance investigations and the flexibility of the molecule, including the linker, stalk and tail, has hampered the crystallographic studies. The lack of expression systems providing a "biochemical amount" of homogeneous dynein molecules was also an obstacle needing to be overcome for the progress in dynein crystallography. Under these circumstances, a substantial amount of knowledge for determining the structure and molecular architecture of dynein molecules has arisen from the findings obtained by electron microscopy on dynein arms in axonemes and isolated dynein molecules. In this chapter, we try to provide a guide into the future by scrutinizing past achievements.

1.1.1 Overview of dynein molecular architecture

Electron microscopy has provided insight on the molecular organization of the entire dynein molecule. Despite their distinct roles in cells, cytoplasmic and axonemal dyneins, which form large protein complexes, are constructed with the similar basic components: the complexes contain heavy chains, several intermediate chains with WD repeats involved in cargo attachment, and at least three distinct classes of light chains, namely the highly conserved LC8, and members of the roadblock/LC7, and T-complex testis-specific protein 1 (Tctex1) protein families.

Dynein belongs to the AAA+ protein superfamily (ATPases associated with diverse activities) [10] and like other AAA+ proteins, it has a ring of six AAA+ modules at its motor core. In contrast to the conventional AAA+ proteins, six AAA+ modules are concatenated in a heavy chain with three unique appendages (Fig. 1.1). A dynein heavy chain has a molecular mass of typically 500–540 kDa, consisting of approximately 4500 amino acid residues. It contains a fundamental motor domain in the C-terminal, an approximately 320–380 kDa fragment, which incorporates sites for both ATP hydrolysis and microtubule binding (Fig. 1.1B and C). The architecture of the motor domain has been revealed by sequence analysis [10], two-dimensional [11–13] and three-dimensional electron microscopy [14,15] and, most recently, X-ray crystallography [2–6].

The motor domain is composed of six AAA+ modules each of 35–40 kDa (numbered as 1 to 6 from the N-terminus), and three appendages: the stalk with the microtubule-binding domain (MTBD), the buttress/strut, and the linker. The linker connects the AAA1 domain and the tail and is normally docked onto the head ring (Fig. 1.1D). This linker is thought to deliver the power stroke. A tail is formed on the N-terminal third of the heavy chain and constitutes an extended structural domain, which mediates dimerization/trimerization of the heavy chains and provides a scaffold for the intermediate and light intermediate chains, and offers a fixed attachment of dynein to the A tubule of an axonemal doublet.

The stalk emanating from the AAA4 is composed of an antiparallel coiled coil and the MTBD lies at the tip of the coiled coil. Near its base, the stalk interacts with the buttress/strut that emerges from AAA5. Although the detailed structure of the buttress/strut was first described in the crystal analyses [2–6], it was observed by the bifurcation of the stalk near its base by negative staining electron microscopy [13,14]. Such a modular architecture of the dynein motor domain suggests that the MTBD is segregated by ~25 nm from the ATPase site, which allows us to engineer the MTBD without impairing the ATPase activity [16].

The topic of this chapter is the electron microscopy studies on "isolated" dynein molecules. As is well known, the development and improvement of electron microscopy instruments, computational methods, and careful consideration of some experimental factors have enabled us to approach achieving atomic resolution for entirely asymmetric protein molecules (for a review, see Ref. [17]). Before providing the details of electron microscopy on isolated dynein molecules, however, we shall begin with a brief review of the morphology of the outer dynein arms on doublet microtubules in an axoneme since what we have known about the molecular configurations of dynein has long been what was understood about the dynein arm structures. In addition, this historical information would be helpful for determining how to conduct your own investigations on dynein research. One of the ultimate goals of dynein research is to provide an understanding of the structure and dynamics of dynein molecules under steric constraints (i.e., integrated into the axonemal lattice) and under interactions with microtubules. Thus, understanding the dynein arm configuration in situ on the basis of the structures of individual components is still of great importance.

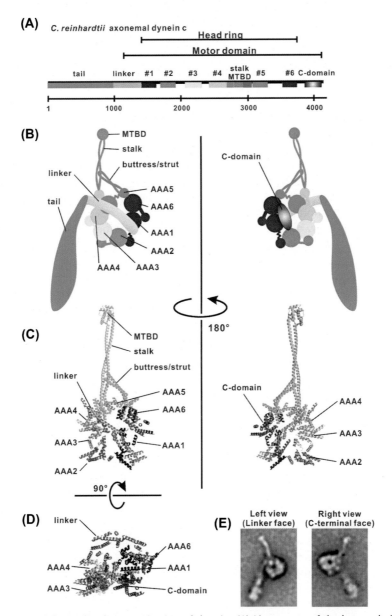

Figure 1.1 Overview of the molecular organization of dynein. (A) Linear map of the heavy chain of *Chlamdomonas* axonemal inner-arm dynein c (BAE19786, *C. reinhardtii*), showing the domain structure: tail (light blue), linker (pink), AAA modules, stalk, and the microtubule-binding domain (MTBD). (B) Schematic drawings of the dynein heavy chain in the APO state. The six AAA+ domains are arranged in a ring, which are indicated in red, orange, yellow, cyan, green, and magenta for AAA1 through AAA6, respectively. Each module is composed of an N-terminal large domain (the *larger circles*) and a C-terminal small domain (the *smaller circles*). Dynein has two distinct faces. The face on which the linker docks are the linker face, corresponding to the face seen in the left view (E), and the face on which the C-terminal localizes is the C-terminal face, corresponding to that seen in the right view in the electron microscopy class averages (E). (C) Crystal structure of the *Dictyostelium* cytoplasmic dynein (PDB ID: 3AY1). The six individual AAA modules are highlighted in same colors used in (A). The left and right panels show the linker face and the C-terminal face, respectively. The linker (pink) is mainly composed of a series of helices. (D) Side view of the AAA ring. The linker is observed arching over the linker face of the ring and having contact only at the AAA1 and AAA4 at its terminal portion. (E) Negative staining sample of dynein-c in the absence of ATP. Single particle analysis shows the details of the two faces of the head ring. The head has several lobes of density forming an asymmetric ring. Two elongated structures extend from the head ring. *(C) The figure was prepared using PyMOL provided by DeLano Scientific LLC* (http://www.pymol.org).

1.2 Historical background of dyneins

As described in Chapter 1 (vol. 1 of this book), dynein was first identified as the microtubule-associated ATPase, extracted from axonemes by exposure to a low-ionic-strength buffer lacking divalent cations or by brief exposure to buffers containing a high-salt concentration (either 0.6 M KCl or NaCl) [1,18]. The ATPase was named after the cgs unit of force, dyne [18]. Dynein-extracted axonemes do not possess arm structures extending from A tubules of doublet microtubules and are unable to generate bending waves in the presence of ATP. However, if the extract from the axonemes is added back to the extracted axonemes, the normal morphology of the dynein arms and motility is restored. This unique property of restoration was used in many later studies on the functional behavior and structure of axonemal dyneins [19–26]. Since it was, thus, demonstrated that dynein ATPase was localized in the arm structure [1], the shape of the arms (dynein arms) and the change in shape were extensively studied as the targets of electron microscopy.

Non-axonemal dynein ATPase activity was first described in extracts of unfertilized sea urchin eggs [27]. A significant fraction of egg dyneins were, however, derived from a precursor pool for ciliogenesis [28]. Although dynein-like polypeptides were investigated in cells and tissues that did not form the ciliary precursor, there was limited success in early studies [29]. Cytoplasmic dynein was, however, identified as a microtubule translocator in bovine brain [30], thanks to the development of in vitro motility assays, which could directly evaluate the mechanochemical properties (i.e., motility) of dynein-like polypeptides. Since then, cytoplasmic dyneins in various cells and tissues have been purified by means of the affinity to microtubules in the absence of ATP and they have been well characterized. The molecular configuration of cytoplasmic dyneins was examined by STEM (see below [31]). The full-length sequences of cytoplasmic dynein heavy chains were obtained in the early 1990s from the analysis of cDNA clones: *Dictyostelium* [8]; rat [32]; *Saccharomyces cerevisiae* [7]. The success in this sequencing led to the development of the expression systems of cytoplasmic dynein.

1.2.1 Electron microscopy of the axonemes

Eukaryotic flagella and cilia have a conserved common structure, named the 9+2 structure, in which nine doublet microtubules surround a pair of central singlet microtubules. Several axonemal components are found in the axoneme; radial spokes, central apparatus, and inner/outer dynein arms, which are the motors that drive the movement of cilia and flagella (Fig. 1.2). Dynein arms were first visualized as two short dense protrusions extending from one side of most of the doublets in the sea urchin sperm flagellar axoneme fixed with

Dyneins

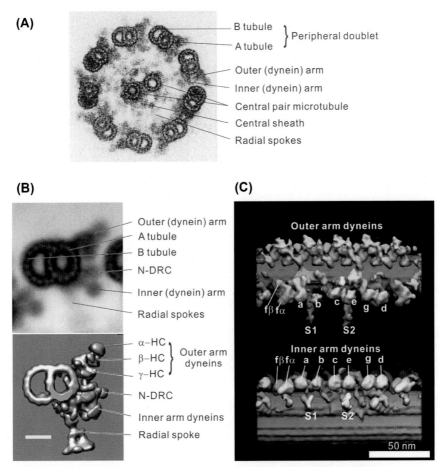

Figure 1.2 (A) Electron micrograph showing axonemal structures of a wild-type flagellum of *Chlamydomonas reinhardtii* in the cross section. The axoneme was demembranated and viewed from the base of the flagellum. Nine peripheral doublet microtubules surround a pair of central singlet microtubules. Two rows of dynein arms are observed on the surface of the peripheral double microtubule and extend toward the adjacent doublet. (B) An average of the structural unit of an axoneme, taken from eight micrographs, consisting of a doublet microtubule, inner and outer dynein arms and a radial spoke. The lower panel shows surface rendered representations of dynein arms arrayed on the microtubule doublets from the cryo–electron tomograms. Individual dynein molecules are identified. N-DRC indicates the nexin–dynein regulatory complex. View from the base of the flagellum. (C) The cryo–electron tomogram of a 96 nm-long axonemal unit. The unit contains four outer dynein arms, six single headed inner arm dyneins (a, b, c, d, e, and g), and one double-headed I1 (dynein-f). Scale bar 50 nm. *(B) Modified from K.H. Bui, H. Sakakibara, T. Movassagh, K. Oiwa, T. Ishikawa, Molecular architecture of inner dynein arms in situ in* Chlamydomonas reinhardtii *flagella, J. Cell Biol. 183 (5) (2008) 923–932.*

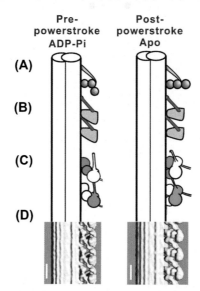

Pre-
powerstroke
ADP-Pi

Post-
powerstroke
Apo

(A)

(B)

(C)

(D)

Figure 1.3 Outer dynein arms described in previous electron microscopy studies. Schematic diagram to describe how the outer dynein arms were observed in conventional electron microscopy. Dynein arms in the absence of ATP are on the right microtubule. On the left microtubule, dynein arms in the presence of ADP and vanadate are shown. From the top to the bottom, the outer dynein arm described by (A) Avolio et al. [35], (B) Witman and Minerveni [40], and (C) Goodenough and Heuser [53]. (D) Cryo–electron tomography shows the dynein arms viewed from the external side of the axoneme [103]. Three plates correspond to the head rings and all three head rings are stacked. In this Fig. 1.3, the base of the flagellum is toward the bottom and the tip is on the top. All electron microscopic observations suggest that the dynein arms pull the adjacent doublet microtubule toward the tip of the flagellum.

osmium tetroxide [33]. In *Chlamydomonas* flagella or *Tetrahymena* cilia, the outer arm in the axoneme shows 24 nm structural repeats along a doublet and consists of three large globular domains (heads) joined to the A tubule [34,35]; for cryo–electron tomography, see Refs. [36,37]. The previous electron micrographs of axonemes showed that the outer-arm dynein was observed as electron dense blobs with a hook [38] in the cross section, stacked three sphere [35,39] or hummer [40] (Fig. 1.3) in the longitudinal view. However, the limited spatial resolution of electron microscopy in those days and low accessibility to dynein arms in situ resulted in the confusion of the dynein arm configuration. For further details of the progress in the early development of axonemal microstructures, interested readers should consult the review by Gibbons [41] or Chapter 1 (vol. 1 of this book).

Conformational changes of dynein arms coupled with nucleotide states were thus described by electron microscopy of axonemes under the various nucleotide conditions. For instance, when ATP is abruptly removed from demembranated axonemes with beating in the presence of ATP, these axonemes adopt the rigor state, in which the axonemes become rigid, freeze their waveforms, and form so-called "rigor waves" [42]. In the rigor wave, dynein arms are in the rigor state, in which dynein arms bind to adjacent doublet microtubules. In the presence of adenylyl–imidodiphosphate (AMP–PNP), a nonhydrolyzable ATP analogue [43], AMP–PNP competes with ATP for binding to the active site of dynein and results in dynein being in the state where the nucleotide is bound but not yet hydrolyzed [44]. In the presence of the phosphate analogue, orthovanadate (VO_3^{4-}), which is known as a dynein ATPase inhibitor [45] and acts as a phosphate analogue, axonemal dynein, which is trapped in the ADP-Pi state [46], and the axonemes are relaxed [47,48].

Witman and Minervini [40] showed that in the presence of ATP, the outer dynein arms on a doublet microtubule, which were negatively stained with uranyl acetate, were tilted proximally at an angle of 70 degrees relative to the A tubule of the doublet. In contrast, the arms in the absence of ATP or in the presence of AMP–PNP showed distinct appearances: the arms showed an upright-configuration with a tilt angle of 82 degrees. According to these observations, they calculated a relative sliding between the two doublets produced by an arm movement of 4.5 nm (Fig. 1.2). Although a higher spatial resolution was required for a more precise description of the outer-arm morphology, using the current level of knowledge, these descriptions were sufficient to reflect the real features of the dynein arm (See also Chapter 6 (vol. 1 of this book) regarding cryo–electron tomography of axonemes).

On the basis of these observations and the analogy of the actomyosin cross-bridge cycle [49], dynein arms were proposed to undergo a cycle of conformational changes coupled with the nucleotide conditions: dynein dissociates from microtubules by the binding of ATP, becomes primed for subsequent structural changes coupled with hydrolysis, then rebinds to the microtubule, accelerates product release, and undergoes the power stroke. The determination of the kinetics of axonemal dynein was extensively carried out using an outer-arm dynein purified from *Tetrahymena* cilia and revealed its unique properties as ATPase [50]. Owing to the strict substrate specificity of dynein ATPase and the poor availability of probes monitoring dynein ATPase states, however, the determination of the kinetics of dynein ATPase has not made much progress since the series of initial investigations. A breakthrough was achieved by the establishment of an expression system and the introduction of Förster resonance energy transfer (FRET) probes into the dynein construct [51,52].

1.2.2 From two-dimensional to three-dimensional information of dynein arm structures

Quick-freeze deep-etch replica techniques have had a great impact on dynein research and showed the dynein arms in situ with unprecedented detail [53–58]. In these techniques, biological specimens are frozen, fractured, and replicated by the evaporation of carbon/platinum onto the specimens. Thin carbon/platinum replication makes freeze-etched samples observable at high resolution using transmission electron microscopy. Since only the outermost features of the shadowed surface are visible, this technique limits observation to the exterior of the molecule. However, the shadowing materials depositing on the structure make fine structures visible and provide three-dimensional information of the molecule.

The technique applied to flagellar axonemes revealed that an outer dynein arm in situ has an elliptical head connected to the A tubule by a thin connection through two spherical feet (Fig. 1.3). The most striking feature the technique revealed is the fine structure, the stalk (also named B-link) that binds to the B tubule of the adjacent doublet. The observations showed that the configuration of the outer arm and the interaction between the stalk head and the B tubule varied according to the nucleotide conditions. The outer arms in rigor were roughly perpendicular to the long axis of the axonemes but ATP induced the arms to undergo a distinct basal tilt [53]. These observations are consistent with those made by Witman and Minervini [40]. The dynein arms showed ~12 nm displacement relative to the base [53,54,58–60] (Fig. 1.3). However, each visible outer arm in the studies was composed of two or three heavy chains. For a conformational change of the individual heavy chains, structure analysis on isolated dyneins or tomographic techniques with high spatial resolution were necessary (see Section 1.5).

1.2.3 Electron microscopy on isolated dynein arms from axonemes

As described above, the dynein arm configuration in situ had been the cause of much confusion owing to the limited spatial resolution of electron microscopy at that time and the difficulty to access the interior of an axoneme. However, the confusion regarding the configuration of the dynein arms bound on a doublet microtubule was partly resolved by electron microscopy on isolated dynein molecules although a more precise understanding of the configuration was only possible with the emergence of cryo–electron tomography (Chapter 6 (vol. 1 of this book) by Ishikawa, or Refs. [25,36,37,61,62]).

For the isolation of dynein, it could be said that the high-salt extract from the axoneme has long been the only source of axonemal dyneins. By using the extract, not only structural studies but also biochemical and kinetic studies on axonemal dyneins have been carried out, as described above (for a review, see Ref. [50]). The availability of a biochemical amount of dynein restricts such sources of axonemes, such as cilia of *Tetrahymena* and *Chlamydomonas*, and sea urchin sperm flagella.

STEM was used for isolated dyneins [63]. STEM allows an unfixed, unstained material to be examined using a low electron dose while it measures the mass of individual particles by integration of the electron scattering intensities. Analysis of dynein molecules by STEM demonstrated that the outer-arm dynein isolated from *Tetrahymena* [63] and *Chlamydomonas* [64] consists of three globular heads attached by three flexible strands to a common globular base and was described as a three-headed bouquet. The three-headed bouquet appearance was also visualized by the rapid-freeze, deep-etch method [54]. The method was also applied to the dynein arms of sea urchin sperm flagella and showed that the outer-arm dynein of sea urchin sperm flagella had two heads [57]. The molecular organization of dynein was thus characterized using these techniques and showed a dynein molecule is composed of a tail (originally termed "stem"), a globular head and a stalk domain.

The flower bouquet configuration with heads splayed apart is now thought to be an artifact since the molecule exhibits a strong interaction with the surface and suffers from surface tension during dryness. To reduce the strong interaction between dynein molecules and the surface, the mica flakes for dynein adsorption were coated with small peptides or proteins, such as cytochrome *c*. The outer dynein arms on such a surface showed a compact form resembling toadstools. The toadstool configuration represents a closer approximation to the native configuration of the outer-arm protein in solution [54]. Cryo–electron tomography finally showed the manner of packing of the three flower bouquet into the dynein arm structure in situ [36,37]. The flower bouquet configuration of the dynein arm was, of course, invaluable for determining the anatomy of the dynein arm complex. In addition, it has also been recently shown that the head-to-head interactions play important roles in the regulation of cytoplasmic dynein activity [65].

1.2.4 Cytoplasmic dyneins isolated from various types of cells

Cytoplasmic dynein was first described as a microtubule-binding protein [66]. It was found that the brain cytosolic microtubule-associated protein 1C (MAP 1C) is a microtubule-activated ATPase, capable of translocating microtubules in vitro in the direction corresponding to retrograde transport [30,66]. Biochemical analysis of this protein as well as STEM revealed that MAP 1C is a brain cytoplasmic form of dynein ATPase [31]. The molecular configuration of cytoplasmic dynein was also described by negative staining electron microscopy. Amos and her colleagues characterized the cytoplasmic dynein purified from porcine brains and showed a phi-shaped configuration of a whole dynein molecule and a small deposit of the stain in the center of globular head domains [67,68].

The establishment of a recombinant expression system of cytoplasmic dynein has taken a long time. Koonce et al. [8] first established the expression system of the cytoplasmic dynein of *Dictyostelium discoideum* and provided the foundation for numerous structural and functional analyses. They characterized the configuration of the motor domain and showed the ringlike configuration with a central hole of the motor domain by using negative staining electron microscopy [11]. However, the expressed dynein showed a poor motility of the microtubules in vitro. Nishiura et al. [9] succeeded in the expression of active cytoplasmic dynein from *Dictyostelium* and this paved the way for the determination of the kinetics of dynein ATPase, which had not experienced significant progress following the initial studies [50], as well as a description of how the mechanochemical cycle was related to force generation and movement [52]. Another important expression system of cytoplasmic dynein was established in *S. cerevisiae* [7]. The system provided insight into the requirements for the processive movement of a dynein molecule and stable active dimers for single molecule experiments [7].

In the following sections, we shall provide a description of the advanced electron microscopy techniques for analyzing the molecular configurations and mechanical properties and summarize the advantages and advances of these techniques in dynein research.

1.3 Electron microscopic techniques used in recent dynein research

As described above, electron microscopy on axonemes and dynein molecules has enabled great progress for understanding their structure and functions. Furthermore, electron microscopy combined with image processing has become a powerful technique to study protein structures covering a wide range of resolutions. Recently, atomic or near-atomic resolution structures of several viruses and protein assemblies have been determined by cryo–electron microscopy coupled with single particle analysis [69,70]; for a review by Zhou, see Ref. [17]. This technique now allows us to build an ab initio atomic model by following the amino acid side chains in cryo–electron microscopy density maps.

1.3.1 Issues for electron microscopy observation on biological materials

Because of the low electron density of biological molecules, the contrast of the image is usually very low without metal staining. Especially in cryo–electron microscopy, this low contrast leads to noisy micrographs of biological specimens. Since the signal-to-noise ratio of electron micrographs can be improved by image processing, it is now an integral part of electron microscopy that large numbers of individual particles are categorized, aligned, and averaged.

In addition, averaging techniques can reduce the requirement of a high electron dose, thereby reducing the radiation damage of the samples. As the recent electron microscopes are equipped with a charge-coupled device (CCD) camera or direct detection camera [71] and data processing elements are pipelined, the speed and efficiency of the electron micrograph analysis has been substantially improved; electron micrographs are immediately transferred to a computer and the particles in the micrographs can be isolated, classified, aligned, and averaged.

1.3.2 Advantages of electron microscopy over X-ray crystallography

The progress in cryo–electron microscopy clearly illustrates the advantages of electron microscopy compared with X-ray crystallographic structural analysis. X-ray crystallography provides only intensity profiles of reflections in the Fourier space. Since diffraction patterns lack phase information, the real space images cannot be restored with a simple method. This phase problem is well known in X-ray crystallography, but it does not exist in electron microscopy. Although the resolution of most electron microscopy analyses is not sufficient for a discussion concerning the structures at atomic resolutions, electron microscopy provides real space images as primary data and can yield structural information for macromolecular complexes that are often very difficult to be obtained using X-ray crystallography.

The ability of electron microscopy to access individual molecules as real space images permits us to discuss the properties, which are masked by ensemble averaging (i.e., in X-ray crystallography). Protein structures are not rigid entities and undergo a range of motions over various time scales. Owing to the difficulty in crystallization or the smear of the electron density over a large space, structural determination of flexible proteins is difficult using X-ray crystallography. However, since electron microscopy can provide the various shapes of individual molecules, we can approach the issue of the flexibility of the proteins. There are some risks in recording artifactual images of the protein molecules owing to interactions with the surface and the surface tension, as described above. Nevertheless, with appropriate attention to these potential artifacts, electron microscopy techniques, such as the negative staining technique, can provide reliable structural information. In fact, by using negative staining and single particle imaging processing techniques, the structure and flexibility of single molecules of axonemal dyneins have been determined, as described below [13,14,72,73].

Even if the sample is heterogeneous, it can be analyzed with electron microscopy. This ability allows dynamic and unstable complexes to be studied. For example, dynein molecules have been analyzed while they change their configuration during mechanochemical cycles [13,14] and during the regulation of the motor activity [65]. The ability to observe individual molecules and the ensemble provide the equilibrium information of the molecules under different chemical states. The number of particles we observe, of course, must be large enough not to be influenced by artifacts. Coupled with single molecule experiments, valuable information about the structure-to-function relationship of proteins can be obtained [65].

Furthermore, electron microscopy does not require crystals of the target proteins and does not have an upper limit as to the size or complexity of the macromolecular complex. This feature enables us to study the structure of a whole dynein molecule of axonemal inner-arm dynein c [13,14], dynein–microtubule complexes [73,74], a large protein complex like dynein–dynactin complex [75,76], and a whole axoneme being a target of cyroelectron tomography [36,37,61,62]. This feature also provides the molecular configuration of the target molecules without steric constraints. In crystals, the molecules are highly packed in symmetrical lattices and have many interactions between adjacent molecules. In nature, a symmetric arrangement of molecules does not have a functional role for most biological proteins. Compared to crystallography, the molecular interactions of the target molecules are so unconstrained in the electron microscopy samples that molecules show a high variability of configurations. To sort out the variations, single particle analysis is adopted and has achieved great success. Thus, our understanding of the functions of dyneins has also benefited considerably from the application of electron microscopy to isolated dynein molecules.

1.3.3 Brief description of the procedure of negative staining techniques

Although cryo–electron microscopy has showed great progress for the structural analysis of proteins, the negative staining technique is still used for high-resolution electron microscopy of macromolecules, owing to its simplicity and versatility to provide images of macromolecules with high contrast. Especially, three-dimensional reconstruction of the frozen-hydrated molecule almost always uses a negatively stained specimen as an important first step and obtains the initial model for the reconstruction.

The negative staining technique has been widely used because it is simple and versatile. Typically, an aqueous suspension of macromolecules is mixed with 1%–2% uranyl acetate and applied to a carbon-coated electron microscopy grid. Our group, however, has adopted the procedure described by Roberts and Burgess [77]. The carbon-filmed grid was treated with UV irradiation or a plasma cleaner, immediately prior to use, to make the carbon surface hydrophilic. A small amount of the sample solution (5 µL) was added onto the grid. To adsorb the molecules on the carbon film, the sample on the grid was incubated for 30 s. After the incubation, the grid was washed with two droplets of washing buffer. Then, the sample was stained with 1% (w/v) of filtered aqueous uranyl acetate. The excess liquid was absorbed by a piece of filter paper placed against the grid and the grid was air dried.

The increase in surface tension during drying and the interaction between the carbon-coated surface and molecules distort the shape of the molecule to some extent and such artefacts as onesidedness and variability of the configurations of the molecule may be introduced into the shape of the molecules. Nevertheless, appropriate caution and attention to these potential artifacts can eliminate the distortion and biased adsorption of the molecule and provide reliable structural information. Despite the distortion of the molecules, for example, this technique has been used with great success in numerous computer reconstructions of viruses and other large macromolecule assemblies including dynein molecules.

Resolving the details of the molecules depends on the depth of the stain. If the stain layer is narrower than a certain depth of the particles, the parts of the particle beyond this depth are often incompletely stained. As a result, the unstained parts would not contribute to the projection image. This would lead to a one-sided staining and a poor overall contrast. Alternatively, deep staining obscures the details of the molecules and a lower resolution of the fine parts of the molecule is the result. Optimal imaging conditions should be found in the intermediate areas, where molecules are delineated with a clear outline of the stain [72].

1.4 Properties of dynein molecules revealed by advanced electron microscopic techniques

1.4.1 Purification of axonemal dyneins from *Chlamydomonas* axonemes

The absence of an expression system for axonemal dyneins has forced us to purify them from axonemes isolated from the cell body or sperm heads. The availability of a biochemical amount of the proteins thus limits the organisms for axonemal dynein preparation: *Tetrahymena* [78], *Chlamydomonas* [40], *Paramecium*, and sea urchin sperm have previously been used. Our group has focused on *Chlamydomonas* axonemes [13,37,79–81]. *Chlamydomonas* is one of the most useful experimental systems for axonemal dynein studies since this organism has a large repository of dynein mutants and axonemal mutants and much knowledge has been accumulated concerning the flagellar movement and waveforms of these mutants.

For the preparation of axonemal dyneins, our group usually employs a 60 L culture of *Chlamydomonas* in TAP medium. For culturing and harvesting large quantities of the cells, the readers are recommended to consult Gorman and

Levine [82], Witman [83], or King [84]. To harvest such large cultures, a centrifuge with a high-capacity rotor that is able to process 12 L of sample in a single run (Sorvall RC12BP with H-12,000 rotor, 750 g for 9 min, Kendro Laboratory Products, Newton, Connecticut, USA) is used in our laboratory, as well as a tangential-flow filtration system that has been used by other groups.

Outer-arm dynein is usually isolated from wild-type axonemes. However, the subspecies of inner-arm dyneins from flagella of *Chlamydomonas* are purified from flagellar axonemes of an outer-armless mutant (*oda1*) [85,86]. If the readers are considering to conduct single molecule measurements on axonemal dyneins, attention should be paid to the purity of the samples since single molecule sensitivity is such that the properties of contaminating molecules will unintentionally influence the results if they are more active than the target molecule. The ability of electron microscopy to view individual molecules means that a high purity of the specimen is essential for the same reason as that for the single molecule studies. Therefore, for inner-arm dynein studies, the investigation of an outer-armless mutant (*oda1*) is highly recommended to avoid contamination by outer-arm dyneins.

Flagellar axonemes are prepared largely based on the method described by King [87]. The harvested and washed cells are deflagellated with dibucaine-HCl. Flagella are collected through a series of centrifugations, and demembranated with 25 KHME solution containing 25 mM KCl, 10 mM HEPES/KOH, 5 mM $MgSO_4$, 1 mM EGTA, pH 7.4, 0.2% (w/v) Nonidet P40 (NP40), and proteinase inhibitors (0.5 mM phenylmethylsulfonyl fluoride (PMSF), 10 μg/mL aprotinin, and 10 μg/mL leupeptin). Unfortunately, NP40 is no longer commercially available, but it has been shown that Igepal CA-630 (Sigma Chemical Co., St Louis, MO) is an acceptable substitute [87] though our group is yet to test it. The resultant axonemes are precipitated by centrifugation at 10,000 rpm, resuspended in 2 mL of HMDE (30 mM HEPES/KOH, 5 mM $MgSO_4$, 1 mM dithiothreitol, 1 mM EGTA, pH 7.4, and the same proteinase inhibitors as those used in demembranation) containing 0.6 M KCl. This mixture is left on ice for 15 min and then precipitated again by centrifugation. The supernatant containing crude dyneins is retained. The pellet is reextracted and centrifuged in the same manner once more. Two supernatant fractions containing crude dyneins are combined and diluted up to 20 mL with the HMDE solution containing the proteinase inhibitors to yield 20 mL of the extract with a final KCl concentration of 120 mM. The solution is centrifuged at 27,000 × g for 30 min at 4°C before it is loaded onto a column.

The crude dynein extract is then fractionated by high-performance liquid chromatography with a MonoQ 5/50 GL anion exchange column (GE Healthcare Japan, Tokyo, Japan) by elution with a linear gradient from 120 to 500 mM KCl in HMDE solution containing 0.5 mM PMSF, with a flow rate of 600 μL/min with a 115 mL gradient (Fig. 1.4). For the purification of the inner dynein subspecies,

Dyneins

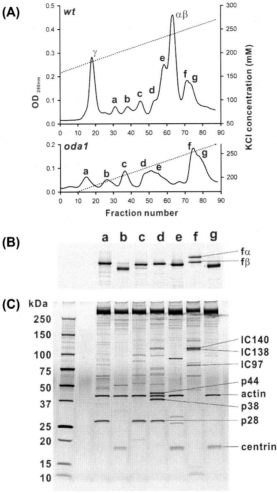

Figure 1.4 Purification of inner-arm dynein subspecies by ionic exchange chromatography. (A) Elution profiles of 0.6 M KCl extracts of wild type (wt) and *oda1* axonemes by ion exchange chromatography on a Mono-Q column. (B) High molecular region of a silver-stained 3% polyacrylamide gel. The peak fractions in (A) further purified with a Mini-Q column are loaded on the gel. (C) Chromatograms of inner-arm dyneins. Peak fractions on a Mono-Q column are further purified with the Mini-Q column and loaded on a 5%–20% polyacrylamide gel. Silver staining reveals the composition of inner-arm dyneins. Dynein-f is composed of two distinct heavy chains, containing IC140, IC138, IC97, and other small components. The other inner-arm dyneins consist of monomeric heavy chains, one actin molecule and either the centrin (dynein-b, -e, and -g) or dimer of the essential light chain, p28 (dynein-a, -c, and -d). Dynein-d has additional light chains, p44, and p38.

the pooled fractions of each subspecies are further purified by a second HPLC fractionation (Mini Q PC3.2/3 column, GE Healthcare Japan, Tokyo, Japan) with a linear gradient from 120 to 400 mM KCl in HMDE solution with a flow rate of 200 µL/min with a 2.2 mL gradient. A recent report suggests that the use of a stronger anion exchange column, such as Uno-Q (Bio-Rad, Hercules, CA), may provide improved separations of the inner-arm subspecies dynein-d [87,88].

Purity and aggregation of the subspecies can be examined using 3% sodium dodecyl sulfate polyacrylamide gel electrophoresis (SDS-PAGE) with silver staining and by sucrose density gradient centrifugation. The yields of dyneins from 60 L of the *Chlamydomonas* culture are approximately 300 µg/mL^{-1} × 50 µL of dynein-c, 250 µg/mL × 50 µL of dynein e and 600 µg/mL × 200 µL of dynein-f. Although some purified dyneins can be stored in liquid nitrogen in the presence of 10% glycerol (dynein-c and -e) or 20% sucrose (dynein-f) for up to a month without loss of their motility, freshly purified samples are recommended for the best results, especially for the single molecule measurements and single particle analysis following negative staining/cryo–electron microscopy.

1.4.2 Purification of cytoplasmic dyneins

Purification procedures depend on the source of the organism. Readers are directed to consult other published literature [89]. Cytoplasmic dynein is now able to be prepared in the expression systems. The overexpression system of cytoplasmic dynein was first established by Koonce and Samsó [8] in *D. discoideum*. They characterized the configuration of the motor domain of cytoplasmic dynein and showed the ringlike configuration with a central hole of the motor domain by using negative staining electron microscopy [11]. Nishiura et al. succeeded in the expression of active cytoplasmic dynein from *D. discoideum* [9]. Another source of cytoplasmic dynein was a budding yeast, *S. cerevisiae*. Reck-Peterson and her colleagues succeeded in overexpressing yeast cytoplasmic dynein and carried out remarkable studies on the cytoplasmic dynein regarding their mechanical properties and structures [7]. The establishment of cytoplasmic dynein expression systems including human cells [4,16] has changed the game. The combination of these dynein expression systems with the structural analysis and/or single molecule studies [4,7,16,65,90,91] has paved a new way to understand the mechanism of the force generation of dynein molecule.

1.4.3 Negative staining electron microscopy with image processing including masking, classifying, and averaging of inner-arm dyneins

In this section, we will briefly describe the procedure of image processing using electron micrographs of inner-arm dynein subspecies, as the details of the image processing have been described elsewhere [77]. Negative staining electron micrographs contain rich structural information of the molecules, which is concealed by the noise in the raw micrographs (Fig. 1.5A). To retrieve this

Dyneins

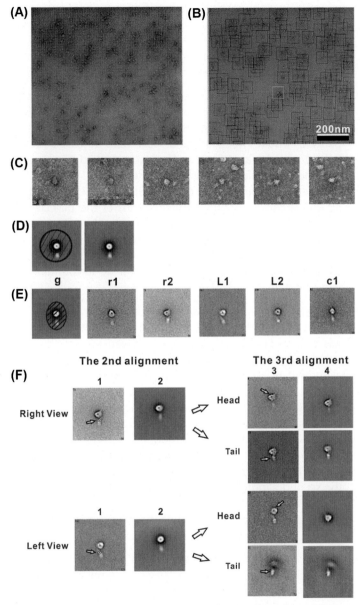

Figure 1.5 Procedures of single particle analysis on inner-arm dynein subspecies e. (A) Raw micrographs of negatively stained full-length dynein-e molecules purified from *Chlamydomonas* flagella. (B) A screen shot of the BOXER. Dynein particles in the digitized micrographs are first identified and windowed out. (C) Excised dynein subspecies e particles in individual windows. The window is sufficiently large to cover the whole molecule, which is important for successive alignment. (D) Concentric circles with inner and outer radii are overlaid on an image average of dynein. (E) The classification of views of an entire dynein molecule in the right view (r), the left view (L) and the center (c). (F) The second alignment is done in each class of view to align the tail better. For the second alignment, we marked the point uppermost part of the base in each class average. The marked point of each window is set to the center. The third alignment is performed on the images with the second alignment. Focusing on the head domain or tail domain in each view allows the detailed analysis of the head and tail domains. *Arrows* in the panels in (F) indicate marked points, which are set to the center of respective windows. To avoid misalignment in rotational alignment of the head domain, we mark rather the eccentric points of head domains than the center. In this alignment, we marked the middle of upper edge on each head image.

information, high-resolution images are needed. For single particle analysis, we prefer to use electron microscopy films and an image scanner for digitization instead of a CCD camera, owing to their wider observation area than cameras, even though electron micrographs can be acquired using a transmission electron microscope equipped with a CCD camera.

Here we describe the procedure of image analysis using negative staining images of inner-arm dynein subspecies e. The single particle images of dynein e molecules are detected semiautomatically by BOXER [92] (Fig. 1.5A and B) and partly by visual inspection, and more than at least 5000 (more than 10,000 is better) particles are windowed out and stacked (Fig. 1.5C). To obtain the best alignment and classification, it is necessary to start with a sufficient number of particles. According to Roberts and Burgess [77], the size depends on the quality of staining, the range of orientations of the particles on the grid and their conformational variability. A minimum number of particles would typically be approximately 2000. For dynein, they started with approximately 10,000 particles [77]. The size of the window should be large enough to contain the centered molecule with a sufficient region of surrounding background to allow image alignment.

Among the various software applications for single particle image processing available, such as SPIDER [93], IMAGIC [94], and EMAN [92], our group has chosen SPIDER to process windowed images. Images of a dynein molecule are brought into mutual alignment by a reference-free strategy. For the alignment of a whole molecule, the head domain was first excluded from the rotational alignment since the head domain has a strong impact on the process (Fig. 1.5D). In this alignment, the head is centralized and the tail points downward. The resultant aligned molecules are classified into numerous classes using K-means clustering. Among these classes, we segregate three characteristic views of a whole dynein molecule: left views, side views, and right views [13,77] (Fig. 1.5E). All molecules in a particular view are combined for the second alignment and subsequently classified by multireference algorithms with iterative processes. The images were subclassified according to the head domains or tail domains by applying a mask to the tail domain or to the head domain, respectively (Fig. 1.5F).

Fig. 1.6 summarizes the single particle image analysis on the inner-arm dynein subspecies of *Chlamydomonas* flagellar axonemes. These dyneins show that their head domains have an asymmetric ringlike morphology, as reported previously [13,77,79,81], and show conserved, structural landmarks. The deposit of the stain in the center of the head domain is clearly visible. The right view (C-terminal face) shows three pronounced stain-excluding globular domains around the right side of the head. The grooves are visible on the lower and left sides of the head in the right view and on the lower and right sides of the head in the left view. These dynein subspecies are shown to retain their molecular

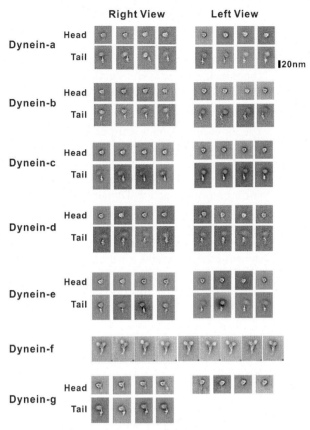

Figure 1.6 Gallery of electron micrographs of the inner-arm dynein subspecies of *Chlamydomonas* after single particle processing. Two characteristic views (the right view and the left view) of dynein molecules are shown for each subspecies. Since dynein-f has two heads, the head alignment was not performed. The head domains in each subspecies look similar to each other, while the tail domains show a variation in their morphologies. Analyses of the head or tail domains for the right and left views were done in each subspecies (a–e and g). Four typical class averages are shown in each category. The nine class averages of whole molecules are shown with dynein subspecies f since molecular configurations is different from the other subspecies. Adsorption of subspecies g to carbon film is clearly biased to the right view. In dynein subspecies g, only 182 particles out of 11,150 were classified to the left view.

configuration and their ringlike motor domains are very similar to each other as well as other dyneins [11,13,72,74].

Visualizing small and flexible domains in image averaging is challenging since the relative position of the flexible domains varies and, thus, averaging smears the structure out. To solve this problem, variance images have been used [14,73]. Image classification is used to group molecules with flexible domains, like stalks, by using a mask that surrounds all positions of flexible domains while excluding the stable domain, like the head, entirely. Class averages can reveal the flexible domain structure clearly. For example, Roberts and his colleagues showed that the positions of the green fluorescent protein (GFP)-tag introduced into specific sites of the cytoplasmic dynein motor domain or the linker could be identified with negative staining combined with single particle analysis and they succeeded in showing the swing motion of the linker coupled with the ATPase cycle [14].

The inner-arm dynein subspecies have different light and intermediate chains. Dynein subspecies a and c have a single heavy chain and three light chains (two copies of 28 kDa protein, p28, and one monomeric actin). Dynein subspecies b, e, and g have a single heavy chain and two light chains (centrin and monomeric actin). Dynein subspecies d has a single heavy chain and five light chains (monomeric actin, two copies of p28, p44, and p38). The tail domain shows the variability of the morphology in each subspecies. Chemical cross-linking experiments [95] showed that the N-terminal region of the dynein-c heavy chain binds two copies of p28 and the actin subunit binds to p28. From the gallery of the tail domains of the inner-arm dynein subspecies (Fig. 1.6), the morphology of the tail of dynein-d looks different from the others. This difference may reflect the light chain composition of dynein-d since dynein subspecies a and c with the same light chain composition show a similar morphology of their tail domains and dynein subspecies b, e, and g with the other light chain composition show a similar morphology of their tail domains.

1.4.4 The flexibility of the tail revealed by negative stain electron microscopy

Alignment of the tail domain of dynein subspecies shows its flexibility. Global averaging of the whole dynein molecule smears the tail and the relative position of the tail to the head domain varies. Image analysis of dynein subspecies e also shows the flexibility of its tail domain. The tail domain consists of three distinct subdomains: neck, shaft, and base. The origin of the flexibility has been proposed to be localized in the neck domain [72]. This flexibility of the tail may provide the geometrical freedom of the head domain configuration and the intermolecular head-to-head interaction [65].

Since an expressed cytoplasmic dynein without a native tail but with a substituted tail shows intact processive movement in vitro [7], the native tail and its associated components are not essential for dynein motility in vitro. However, they may regulate dynein motility in vivo and recent data indicate that the noncatalytic subunits link dynein to cargos and to several adaptor proteins that regulate dynein function. For example, missense mutations in the tail domain of cytoplasmic dynein in mice are known to cause neurodegenerative disease. The best-characterized model of dynein dysfunction is the Legs-at-odd-angles mouse [96]. This mutation is thought to affect homodimerization of the dynein heavy chains and/or association of the heavy and intermediate chains. Single molecule nanometry on the mutant dynein showed that dynein purified from the mutant mice has lower processivity and more frequently shows bidirectional motility along a microtubule and sidesteps to adjacent protofilaments compared with the wild-type dynein [97]. The results suggest that the mutation in the tail domain of dynein results in increased flexibility of the dynein molecule and diminished gating between the motor domains.

1.4.5 The autoinhibition of cytoplasmic dynein 1

In single molecule studies, mammalian cytoplasmic dynein often shows a diffusive bidirectional motion on a microtubule [98]. This behavior has been a topic of controversy for a long time [99]. Our group has presented evidence for the existence of an autoinhibited state in mammalian cytoplasmic dynein [65]. We observed the molecular configuration of dynein molecules using negative staining techniques under various nucleotide conditions and measured the movement of single dynein molecules on microtubules at the corresponding nucleotide conditions. In the absence of ATP or in the presence of ADP–vanadate, most of dynein molecules were observed to be in a phi-shaped configuration [67,68], where the dynein heads stacked in a back-to-back manner and the stalks were crossed. However, in the absence of nucleotides, the two heads were separated. The single molecule measurements suggest that the stacked head configuration correlates with a diffusive movement of the dynein molecules on microtubules. In contrast, the configuration of the separated heads correlates with the activation of a processive movement. By varying the nucleotide conditions and by physically separating two heads with either a rigid peptide or a DNA scaffold, we demonstrated a good correlation between the fraction of dynein observed to adopt the "stacked" conformation and the diffusion coefficient of single dimers, and that the separation of the two heads restores directional motility. The structural basis for the autoinhibition of cytoplasmic dynein has recently been revealed [100] (see also Chapter 2).

1.5 Force generating mechanism of dynein

1.5.1 Swing motion of the linker and changes in head morphology

The linker is a structure located in the portion of the tail proximal to AAA1, which serves as a connection between AAA1 and the main part of the tail (Fig. 1.1B). The existence of the linker was first described in images of negatively stained monomeric axonemal dynein [13]. While the linker is normally docked onto the head ring, it is shown to be a relatively large structure about 2 nm wide and 10 nm long when the linker is disrupted and undocked from the head ring [13–15].

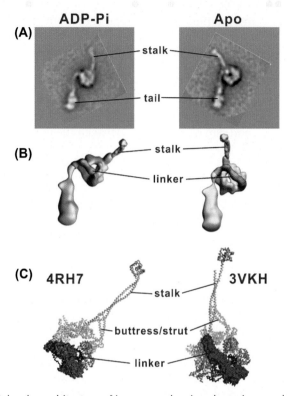

Figure 1.7 Molecular architecture of inner-arm dynein subspecies c and its structural changes. (A) The left view of dynein subspecies c in the ADP–vanadate state (left panel) and in the APO state (right panel). The particles of the negatively stained sample were aligned and averaged with single particle analysis. (B) Complete 3D models of dynein-c in the ADP–vanadate state (left panel) and APO state (right panel). Modified after [15]. (C) Crystal structure of the motor domain of human cytoplasmic dynein 2 in the ADP–vanadate state (PDB ID: 4RH7, in the left view) and *D. discoideum* cytoplasmic dynein motor domain in ADP state (PDB ID: 3AY1, in the right view). The six individual AAA modules are highlighted in same colors used in Fig. 1.1A. *The figures were prepared using PyMOL provided by DeLano Scientific LLC (*http://www.pymol.org*).*

On the basis of electron microscopy observations, it has been suggested that the linker is involved in force generation through its interaction with the head ring [13,15]. Two-dimensional analysis on negatively stained axonemal dynein described the conformations of the dynein molecule in two different nucleotide states, which mimic the post- and prepower stroke conformations of the motor (Fig. 1.7A). In the absence of nucleotide (postpower stroke conformation, state I), the tail emerges near the base of the stalk. In the presence of ATP and vanadate, which forms a dynein–ADP–Vi complex that mimics the dynein–ADP–Pi conformation (prepower stroke conformation, state II), the tail emerges further away from the stalk base. These observations were interpreted as the linker being able to swing relative to the head ring [13].

The movement of the linker has subsequently been confirmed in cytoplasmic dynein, identified as the N-terminal region of the motor domain using GFP- and blue fluorescent protein-tagged constructs by negative stain electron microscopy and FRET [51]. In the absence of nucleotide or in the presence of ADP, GFP inserted at the linker's N-terminus lies close to AAA4, at the base of the stalk, in the so-called unprimed position, whereas in the presence of ATP and vanadate, the GFP lies close to AAA2, in the primed position.

The cryo–electron microscopy [15] and crystal structures of dynein showed that the linker is composed of helical bundles and does not sit flat on the head ring but rather arches over it (Fig. 1.1D). The linker is composed of four predominantly helical subdomains (from the N-terminus, subdomains 1, 2, 3, and 4). The C-terminus subdomain 4 interacts with AAA1 and part of AAA6 and is connected to AAA1. The N-terminal subdomain 1 contacts with AAA5 in the yeast motor domain (APO state), or with AAA4 in the *Dictyostelium* head ring (ADP state). The human cytoplasmic dynein-2 motor domain has been crystallized in the presence of ADP–vanadate [4]. The overall configuration of the motor domain of this dynein is similar to the motor domains previously reported. In this crystal, the motor domain is trapped in a prepower stroke state and the ATPase site at AAA1 binds ADP–Vi [4]. The crystal structure has revealed that the linker kinks between subdomains 2 and 3 of the linker. This observation is consistent with the previous electron microscope observations [15]. Therefore, the linker can be divided into static and mobile subdomains. The mobile subdomains 1 and 2 undergo a rigid-body movement relative to the static subdomains 3 and 4. These static subdomains interact with AAA2 to stabilize the subdomains 3 and 4 in the ADP–vanadate state [4] (Fig. 1.7B and C).

1.5.2 The change in size and shape of the head domain

When the AAA ring of dynein is viewed from the linker face, the AAA modules display an uneven spacing. Large gaps are present between the large domains of AAA1 and AAA2 and also between those of AAA5 and AAA6 (Fig. 1.1 and 1.8B). In contrast, in the AAA ring of the *Dictyostelium* structure, the gap between AAA5 and AAA6 is not present, and the gap between AAA1 and AAA2 is partially filled by the linker. The negative stain electron micrographs of axonemal dynein subspecies c showed a dense deposit of the stain on the dynein head ring. The deposit may correspond to the gap between AAA1 and AAA2. Additionally, the *Dictyostelium* dynein motor domain has a more symmetrical and planar ring than that of the yeast.

Although there are many possibilities to explain these differences, one fascinating idea is that the differences are the result of the different nucleotide states of the two molecules [101]. It is noted that the head ring of axonemal dynein subspecies c in the ADP–vanadate state observed in the negative staining electron microscopy images is much more symmetric than that in the APO state. This observation is not

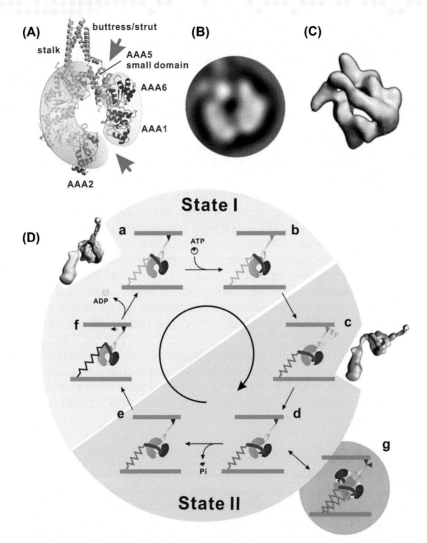

Figure 1.8 A model of the mechanochemical cycle of dynein. (A) Atomic structure of *Dictyostelium* dynein in the ADP state. The head domain is depicted as a combination of two discs because there are large gaps between AAA1 and AAA2 and between AAA5 and AAA6. (B) Electron microscopy of the negatively stained motor domain followed by single particle analysis succeeded in capturing two distinct domains of inner-arm dynein-c. The large deposit of the stain corresponds to the gap between AAA1 and AAA2. (C) Cryo–electron microscopy reconstruction of the dynein-c head in the absence of nucleotide. (D) Putative mechanochemical cycle of dynein. The cycle is divided into two major states, state I (APO state, yellow) and state II (ADP–Pi state, light green). On ATP binding to the ATPase site in AAA1, the origin of the tail emerging moves far from the stalk owing to the kink of the linker (a–b–c). As the result, the molecule is elongated. The microtubule-binding domain (MTBD) extended forward undergoes a Brownian search for the next binding site (c). Once the MTBD binds to the microtubules, the head ring starts remodeling and facilitates the product release (d–e). On product release, the linker docks on the AAA4/5 (state I, e–f–g). The docking of the linker to the motor ring leads to the contraction of the whole dynein molecule. The microtubule is thus dragged by this contraction. Mammalian cytoplasmic dyneins are thought to have the auto-inhibition state [65]. The dynein in this state is depicted as two molecules stacking in a back-to-back configuration (g). *(C) Modified after A.J. Roberts, B. Malkova, M.L. Walker, H. Sakakibara, N. Numata, T. Kon, R. Ohkura, T.A. Edwards, P.J. Knight, K. Sutoh, K. Oiwa, S.A. Burgess, ATP-driven remodeling of the linker domain in the dynein motor, Structure 20 (2012) 1670–1680.*

coincidental with the differences found in the crystal structure between yeast and *Dictyostelium*. The distortion of the head ring might represent the signal pathway between the MTBD of the stalk and the principal ATPase site in AAA1.

A recent report on the crystal structure of dynein motor domain in the ADP–vanadate state suggests how the motor domain is primed to produce force and how the head ring is remodeled coupled with ATP hydrolysis [4]. Closure of the head ring occurs on binding of ATP at the primary ATPase site, and it makes the head compact. This closure induces a steric clash between the linker at its N-terminal region and the head ring and the clash causes the linker to adopt a kinked configuration. The closure also changes the interface between the stalk and buttress/strut. This results in the helix sliding in the stalk that causes the MTBD at its tip to be released from the microtubule. This remodeling of the motor domain and the linker revealed from the crystal structure confirms the previous electron microscopy observations.

1.5.3 The flexibility of the stalk revealed by negative stain electron microscopy

The stiffness of a 15-nm length of coiled-coil peptide clamped at one end can be estimated to be approximately 0.4 pN/nm. This figure provides a good estimation of the stalk stiffness. The range of conformations for an axonemal dynein molecule observed in negatively stained samples can also provide an estimation of the stiffness of the stalk, which is 0.5 pN/nm in APO molecules and 0.14 pN/nm in ATP–vanadate molecules. These estimations imply that the stalk is too flexible to work as a rigid lever [102] and also that the variation of the stalk stiffness is dependent on the nucleotide conditions.

The crystallographic analyses of the motor domain of *Dictyostelium* dynein [6] and that of human dynein-2 [4] have provided evidence for the variation of the stalk stiffness and for the structural information pathway between AAA1 and MTBD. In the structure of ADP–Vi-bound dynein-2, closure of the AAA1 site and the rigid-body movement of AAA2–AAA4 cause the AAA4/AAA5 interface gap to narrow and the AAA6L subdomain to rotate toward the ring center. This change is accompanied by the change of the position of the buttress/strut. The helix CC2 of the stalk was then pulled by the buttress/strut and the base of the stalk deviates from the symmetrical, regular coiled coil observed in ADP-bound dynein-1. Thus, dynein coordinates AAA1 ATPase and MTBD by switching the stalk–buttress/strut structure between the straight and kinked conformations and showed the change in stiffness of the stalk.

1.5.4 Force generating mechanism proposed by electron microscopy

The observations of negatively stained electron microscopy samples, as summarized above, combined with the structural information obtained from the crystal structures of the motor domain [2–6] can provide a model of dynein force generation (Fig. 1.8): in the APO state, dynein tightly binds to a microtubule through the MTBD and the AAA1–AAA2 gap is wide. The N-terminal end of the linker is docked at the stable position near AAA4/5 (Figure 1.1A–C, state I in D). On binding of Mg-ATP to the AAA1 ATPase site, the gap is then closed and a large rearrangement among AAA modules occurs (Fig. 1.8D(a–c)), which is referred to as the "domino effect" [2]. The information of ATP binding thus propagates along the ring.

The rearrangement of the modules in the AAA ring changes the interaction between the buttress/strut and the stalk. This change elicits the relative shear of CC1 and CC2 of the stalk leading to the conformational change of the MTBD and to dissociation of the MTBD from the microtubule (Fig. 1.8D(c)). The linker shows a steric clash with the head ring owing to the closure of the head ring and the clash causes the linker to detach from the AAA4/5 docking point and change its orientation on the head ring, causing the tail to emerge far from the stalk (state II, Fig. 1.8D(c–e)).

When ATP is hydrolyzed at AAA1, the gap between AAA1–2 increases. This change induces the rotational shift in the position of the AAA5 large domain, which pushes the buttress/strut toward the AAA4 domain. Thus, the buttress/strut–stalk interaction is modified and leads to an increase in the affinity of MTBD to the microtubule. The MTBD then finds the docking site on the microtubule and subsequently binds to it. This binding leads to an increase of the AAA1–2 gap, which facilitates the product release. On release of the products, the linker orientation on the ring changes, which brings the tail emergence point closer to the stalk (state I). The linker changes its orientation by switching between two differently docked positions on the head ring, thus producing a rotation of the head ring that causes the stalk to swing or to be winched.

1.5.5 Winch and power stroke mechanism for force generation of dynein

The cryo–electron micrograph of entire dynein molecules interacting with microtubules has revealed that, even though the tail and linker shift relative to the ring and stalk, as observed in isolated dynein molecules, the stalk orientation on the microtubule remains fixed irrespective of the nucleotide state [103,104]. Cryo–electron tomograms of axonemes in the APO state were compared to those obtained in the presence of ADP–vanadate. Global changes of the dynein arm complexes were shown and several key changes in the dynein structures were determined. Although the stalks are not clearly visible in the tomograms, close examinations showed that the stalks typically tilt towards the proximal end of the axoneme (the base of the axoneme) in both nucleotide conditions. The dynein head rings were observed to move 8 nm toward the distal end of the axoneme on the release of the nucleotide [104]. Since the MTBD attaches to the adjacent microtubule, the movement results in dragging the adjacent microtubule distally, which produces the shear.

These observations provide evidence for the concept of dynein as an ATP-dependent winch. The winch model explains the result of Carter et al. [105], in which cytoplasmic dyneins, with their stalk coiled coil either lengthened or shortened by seven heptads, move toward the minus end of a microtubule, irrespective of the length of the stalk. This result is remarkable since these changes in the stalk length would be predicted to rotate the head ring by 180 degrees and reverse the direction of dynein movement according to the lever arm model. To explain the directionality of dynein, it was proposed that the head ring does not elicit a leverlike rotation of the linker domain perpendicular to the stalk; rather, contraction where the force vector of the linker domain's conformational change is directed parallel to an angled stalk.

However, the electron density map of cryo-EM data of axonemal dynein sub-species c was well fitted to an averaged tomogram of the axoneme if the head and stalk are rotated. The head rotation and winch mechanisms are controversial issues that were raised recently. Recently, Imai and his colleagues [73] reported their cryo-EM observations of tail-truncated, artificially dimerized *D. discoideum* cytoplasmic dynein bound to microtubules in the presence of physiological concentrations of ATP. This dynein dimerized through an N-terminally fused GST-tag has the ability to move processively along a microtubule and was rapidly frozen during its movement. Thus, the image obtained is thought to be of a dynamic dynein molecule in action. Using single particle image processing, they show that both dynein stalks point toward the MT minus end. The stalk angle of the rear head becomes shallower while the stalk angle of the front head becomes steeper when the distance between both MTBDs increases. These observations suggest that the dynein head rotates and conducts a Brownian search for the next binding site by extension of the stalk.

These two mechanisms, rotation or winch, may not be exclusive since the dynein molecule is so flexible with two flexible extensions that, under the tension in the axoneme, can be aligned with the head domain in a straight line even if the head rotates.

1.5.6 Perspectives

Advanced electron microscopy has provided the intuitive qualities of images of dynein molecules to promote our understanding about the structure and function of dynein molecules. These techniques have bridged the gap between optical microscopy functional analysis and atomic resolution crystal structures. In fact, recently, our knowledge of the dynein molecule has drastically increased. For example, a combination of the electron microscopy observations and single molecule measurements of cytoplasmic dynein has resolved the longstanding controversial argument on the mode of the movement and force generated by mammalian cytoplasmic dynein by presenting evidence for the existence of an autoinhibited state [65,100].

For ciliary and flagellar movements, the ensemble of various types of dynein orchestrates their distinct functions to coordinate the sliding and generate the bending motion of the axoneme. For example, in the flagellar bend formation, the active sliding of doublet microtubules is proposed to switch from one side of the axoneme to the other: the effective stroke is generated mainly by the sliding between doublet 3 and 4 and the recovery stroke is generated mainly by that between 7 and 8. To achieve this, dynein molecules on the doublets are controlled precisely in space and time. Switching of the active side seems to be achieved through antagonistic inhibition of dynein activity. On the other hand, the bend propagation along the axoneme length seems to be generated by sequential activation of dynein arms. The structural basis of these coordinations of dynein arms has been proposed [106]. Two connections between the outer and inner dynein arms were reported in the 96-nm structural repeat. One is between DRC and the "internal bulb," assigned as N-terminal tails of α- and β-heavy chains of the outer dynein arm, whereas the second connection, which is 48 nm away from the first, binds to the LC–IC complex of dynein I1(dynein-f).

Electron microscopy has now provided not only the structure of single dynein molecules but also the structural basis of the ensemble behavior of dynein molecules. The information provided by the electron microscopy should be tested by functional analysis with optical microscopy. Therefore, the analysis requires an experimental system working under the geometrically well defined and mechanically constrained conditions, which should be comparable to the axoneme structure. DNA-templated assembly techniques combined with the single molecule technique will provide a good experimental system for determining the ensemble behavior of dynein molecules [107,108].

References

[1] I.R. Gibbons, Studies on the protein components of cilia from *Tetrahymena pyriformis*, Proc. Natl. Acad. Sci. USA 50 (1963) 1002–1010.

[2] A.P. Carter, C. Cho, L. Jin, R.D. Vale, Crystal structure of the dynein motor domain, Science 331 (2011) 1159–1165.

[3] H. Schmidt, E.S. Gleave, A.P. Carter, Insights into dynein motor domain function from a 3.3-Å crystal structure, Nat. Struct. Mol. Biol. 19 (2012) 492–497.

[4] H. Schmidt, R. Zalyte, L. Urnavicius, A.P. Carter, Structure of human cytoplasmic dynein-2 primed for its power stroke, Nature 518 (7539) (2015) 435–438.

[5] T. Kon, K. Sutoh, G. Kurisu, X-ray structure of a functional full-length dynein motor domain, Nat. Struct. Mol. Biol. 18 (2011) 638–642.

[6] T. Kon, T. Oyama, R. Shimo-Kon, K. Imamula, T. Shima, K. Sutoh, G. Kurisu, The 2.8 Å crystal structure of the dynein motor domain, Nature 484 (2012) 345–350.

[7] S.L. Reck-Peterson, A. Yildiz, A.P. Carter, A. Gennerich, N. Zhang, R.D. Vale, Single-molecule analysis of dynein processivity and stepping behavior, Cell 126 (2) (2006) 335–348.

[8] M.P. Koonce, M. Samsó, Overexpression of cytoplasmic dynein's globular head causes a collapse of the interphase microtubule network in Dictyostelium, Mol. Biol. Cell 7 (1996) 935–948.

[9] M. Nishiura, T. Kon, K. Shiroguchi, R. Ohkura, T. Shima, Y.Y. Toyoshima, K. Sutoh, A single-headed recombinant fragment of Dictyostelium cytoplasmic dynein can drive the robust sliding of microtubules, J. Biol. Chem. 279 (22) (2004) 22799–22802.

[10] A.F. Neuwald, L. Aravind, J.L. Spouge, E.V. Koonin, AAA+: a class of chaperone-like ATPases associated with the assembly, operation, and disassembly of protein complexes, Genome Res. 9 (1999) 27–43.

[11] M. Samsó, M. Radermacher, J. Frank, M.P. Koonce, Structural characterization of a dynein motor domain, J. Mol. Biol. 276 (1998) 927–937.

[12] M.A. Gee, J.E. Heuser, R.B. Vallee, An extended microtubule-binding structure within the dynein motor domain, Nature 390 (1997) 636–639.

[13] S.A. Burgess, M.L. Walker, H. Sakakibara, P.J. Knight, K. Oiwa, Dynein structure and power stroke, Nature 421 (2003) 715–718.

[14] J.A. Roberts, N. Numata, M.L. Walker, Y.S. Kato, B. Malkova, T. Kon, R. Ohkura, F. Arisaka, P.J. Knight, K. Sutoh, S.A. Burgess, AAA+ ring and linker swing mechanism in the dynein motor, Cell 136 (2009) 485–495.

[15] A.J. Roberts, B. Malkova, M.L. Walker, H. Sakakibara, N. Numata, T. Kon, R. Ohkura, T.A. Edwards, P.J. Knight, K. Sutoh, K. Oiwa, S.A. Burgess, ATP-driven remodeling of the linker domain in the dynein motor, Structure 20 (2012) 1670–1680.

[16] A.A. Furuta, M. Yoshio, K. Oiwa, H. Kojima, K. Furuta, Creating biomolecular motors based on dynein and actin-binding proteins, Nat. Nanotechnol. (2016).

[17] Z.H. Zhou, Atomic resolution cryo electron microscopy of macromolecular complexes, Adv. Protein Chem. Struct. Biol. 82 (2011) 1–35.

[18] I. Gibbons, A. Rowe, Dynein: a protein with adenosine triphosphatase activity from cilia, Science 149 (1965) 424–426.

[19] M. Takahashi, Y. Tonomura, Binding of 30s dynein with the B-tubule of the outer doublet of axonemes from *Tetrahymena pyriformis* and adenosine triphosphate-induced dissociation of the complex, J. Biochem. (Tokyo) 84 (6) (1978) 1339–1355.

[20] L.T. Haimo, B.R. Telzer, J.L. Rosenbaum, Dynein binds to and crossbridges cytoplasmic microtubules, Proc. Natl. Acad. Sci. USA 76 (11) (1979) 5759–5763.

[21] D.R. Mitchell, F.D. Warner, Interactions of dynein arms with b subfibers of *Tetrahymena* cilia: quantitation of the effects of magnesium and adenosine triphosphate, J. Cell Biol. 87 (1) (1980) 84–97.

[22] M.E. Porter, K.A. Johnson, Characterization of the ATP-sensitive binding of *Tetrahymena* 30 S dynein to bovine brain microtubules, J. Biol. Chem. 258 (10) (1983) 6575–6581.

[23] M.E. Porter, K.A. Johnson, Transient state kinetic analysis of the ATP-induced dissociation of the dynein-microtubule complex, J. Biol. Chem. 258 (10) (1983) 6582–6587.

[24] M.E. Porter, K.A. Johnson, The interaction of Tetrahymena 30S dynein with bovine brain microtubules, J. Submicrosc. Cytol. 15 (1) (1983) 199–200.

[25] T. Oda, N. Hirokawa, M. Kikkawa, Three-dimensional structures of the flagellar dynein-microtubule complex by cryoelectron microscopy, J. Cell Biol. 177 (2) (2007) 243–252.

[26] S. Aoyama, R. Kamiya, Strikingly fast microtubule sliding in bundles formed by *Chlamydomonas* axonemal dynein, Cytoskelet. Hob. 67 (6) (2010) 365–372.

[27] R. Weisenberg, E.W. Taylor, Studies on ATPase activity of sea urchin eggs and the isolated mitotic apparatus, Exp. Cell Res. 53 (2) (1968) 372–384.

[28] D.J. Asai, An antiserum to the sea urchin 20 S egg dynein reacts with embryonic ciliary dynein but it does not react with the mitotic apparatus, Dev. Biol. 118 (2) (1986) 416–424.

[29] M.M. Pratt, The identification of a dynein ATPase in unfertilized sea urchin eggs, Dev. Biol. 74 (2) (1980) 364–378.

[30] B.M. Paschal, R.B. Vallee, Retrograde transport by the microtubule-associated protein MAP 1C, Nature 330 (1987) 181–183.

[31] R.B. Vallee, J.S. Wall, B.M. Paschal, H.S. Shpetner, Microtubule-associated protein 1C from brain is a two-headed cytosolic dynein, Nature 332 (6164) (1988) 561–563.

[32] A. Mikami, B.M. Paschal, M. Mazumdar, R.B. Vallee, Molecular cloning of the retrograde transport motor cytoplasmic dynein (MAP 1C), Neuron 10 (5) (1993) 787–796.

[33] B. Afzelius, Electron microscopy of the sperm tail; results obtained with a new fixative, J. Biophys. Biochem. Cytol. 5 (2) (1959) 269–278.

[34] J. Avolio, A.N. Glazzard, M.E. Holwill, P. Satir, Structures attached to doublet microtubules of cilia: computer modeling of thin-section and negative-stain stereo images, Proc. Natl. Acad. Sci. USA 83 (13) (1986) 4804–4808.

[35] J. Avolio, S. Lebduska, P. Satir, Dynein arm substructure and the orientation of arm-microtubule attachments, J. Mol. Biol. 173 (3) (1984) 389–401.

[36] D. Nicastro, C. Schwartz, J. Pierson, R. Gaudette, M.E. Porter, J.R. McIntosh, The molecular architecture of axonemes revealed by cryoelectron tomography, Science 313 (5789) (2006) 944–948.

[37] T. Ishikawa, H. Sakakibara, K. Oiwa, The architecture of outer dynein arms in situ, J. Mol. Biol. (2007).

[38] R.D. Allen, A reinvestigation of cross-sections of cilia, J. Cell Biol. 37 (3) (1968) 825–831.

[39] F.D. Warner, D.R. Mitchell, C.R. Perkins, Structural conformation of the ciliary ATPase dynein, J. Mol. Biol. 114 (3) (1977) 367–384.

[40] G.B. Witman, N. Minervini, Dynein arm conformation and mechanochemical transduction in the eukaryotic flagellum, Symp. Soc. Exp. Biol. 35 (1982) 203–223.

[41] I.R. Gibbons, Cilia and flagella of eukaryotes, J. Cell Biol. 91 (3 Pt 2) (1981) 107s–124s.

[42] B.H. Gibbons, I.R. Gibbons, Properties of flagellar "rigor waves" formed by abrupt removal of adenosine triphosphate from actively swimming sea urchin sperm, J. Cell Biol. 63 (3) (1974) 970–985.

[43] S.M. Penningroth, G.B. Witman, Effects of adenylyl imidodiphosphate, a nonhydrolyzable adenosine triphosphate analog, on reactivated and rigor wave sea urchin sperm, J. Cell Biol. 79 (3) (1978) 827–832.

[44] S.M. Penningroth, K. Olehnik, A. Cheung, ATP formation from adenyl-5'-yl imidodiphosphate, a nonhydrolyzable ATP analog, J. Biol. Chem. 255 (20) (1980) 9545–9548.

[45] I.R. Gibbons, M.P. Cosson, J.A. Evans, B.H. Gibbons, B. Houck, K.H. Martinson, W.S. Sale, W.J. Tang, Potent inhibition of dynein adenosinetriphosphatase and of the motility of cilia and sperm flagella by vanadate, Proc. Natl. Acad. Sci. USA 75 (5) (1978) 2220–2224.

[46] T. Shimizu, K.A. Johnson, Presteady state kinetic analysis of vanadate-induced inhibition of the dynein ATPase, J. Biol. Chem. 258 (22) (1983) 13833–13840.

[47] W.S. Sale, I.R. Gibbons, Study of the mechanism of vanadate inhibition of the dynein cross-bridge cycle in sea urchin sperm flagella, J. Cell Biol. 82 (1) (1979) 291–298.

[48] M. Okuno, Y. Hiramoto, Mechanical stimulation of starfish sperm flagella, J. Exp. Biol. 65 (2) (1976) 401–413.

[49] R.W. Lymn, E.W. Taylor, Mechanism of adenosine triphosphate hydrolysis by actomyosin, Biochemistry 10 (25) (1971) 4617–4624.

[50] K.A. Johnson, Pathway of the microtubule–dynein ATPase and the structure of dynein: a comparison with actomyosin, Annu. Rev. Biophys. Biophys. Chem. 14 (1985) 161–188.

[51] T. Kon, T. Mogami, R. Ohkura, M. Nishiura, K. Sutoh, ATP hydrolysis cycle-dependent tail motions in cytoplasmic dynein, Nat. Struct. Mol. Biol. 12 (2005) 513–519.

[52] T. Mogami, T. Kon, K. Ito, K. Sutoh, Kinetic characterization of tail swing steps in the ATPase cycle of *Dictyostelium* cytoplasmic dynein, J. Biol. Chem. 282 (2007) 21639–21644.

[53] U.W. Goodenough, J.E. Heuser, Substructure of the outer dynein arm, J. Cell Biol. 95 (3) (1982) 798–815.

[54] U. Goodenough, J. Heuser, Structural comparison of purified dynein proteins with in situ dynein arms, J. Mol. Biol. 180 (4) (1984) 1083–1118.

[55] U.W. Goodenough, J.E. Heuser, Outer and inner dynein arms of cilia and flagella, Cell 41 (2) (1985) 341–342.

[56] U.W. Goodenough, J.E. Heuser, Substructure of inner dynein arms, radial spokes, and the central pair/projection complex of cilia and flagella, J. Cell Biol. 100 (6) (1985) 2008–2018.

[57] W.S. Sale, U.W. Goodenough, J.E. Heuser, The substructure of isolated and in situ outer dynein arms of sea urchin sperm flagella, J. Cell Biol. 101 (4) (1985) 1400–1412.

[58] S.A. Burgess, Rigor and relaxed outer dynein arms in replicas of cryofixed motile flagella, J. Mol. Biol. 250 (1) (1995) 52–63.

[59] S. Tsukita, S. Tsukita, J. Usukura, H. Ishikawa, ATP-dependent structural changes of the outer dynein arm in *Tetrahymena* cilia: a freeze-etch replica study, J. Cell Biol. 96 (5) (1983) 1480–1485.

[60] P. Lupetti, C. Mencarelli, M. Rosetto, J.E. Heuser, R. Dallai, Structural and molecular characterization of dynein in a gall-midge insect having motile sperm with only the outer arm, Cell Motil. Cytoskelet. 39 (4) (1998) 303–317.

[61] D. Nicastro, J.R. McIntosh, W. Baumeister, 3D structure of eukaryotic flagella in a quiescent state revealed by cryo-electron tomography, Proc. Natl. Acad. Sci. USA 102 (44) (2005) 15889–15894.

[62] K.H. Bui, H. Sakakibara, T. Movassagh, K. Oiwa, T. Ishikawa, Molecular architecture of inner dynein arms in situ in *Chlamydomonas reinhardtii* flagella, J. Cell Biol. 183 (5) (2008) 923–932.

[63] K.A. Johnson, J.S. Wall, Structure and molecular weight of the dynein ATPase, J. Cell Biol. 96 (3) (1983) 669–678.

[64] G.B. Witman, K.A. Johnson, K.K. Pfister, J.S. Wall, Fine structure and molecular weight of the outer arm dyneins of Chlamydomonas., J. Submicrosc. Cytol. 15 (1983) 193–197.

[65] T. Torisawa, M. Ichikawa, A. Furuta, K. Saito, K. Oiwa, H. Kojima, Y.Y. Toyoshima, K. Furuta, Autoinhibition and cooperative activation mechanisms of cytoplasmic dynein, Nat. Cell Biol. 16 (11) (2014) 1118–1124.

[66] B.M. Paschal, H.S. Shpetner, R.B. Vallee, MAP 1C is a microtubule-activated ATPase which translocates microtubules in vitro and has dynein-like properties, J. Cell Biol. 105 (1987) 1273–1282.

[67] L.A. Amos, Brain dynein crossbridges microtubules into bundles, J. Cell Sci. 93 (Pt 1) (1989) 19–28.

[68] J. Fan, L.A. Amos, Antibodies to cytoplasmic dynein heavy chain map the surface and inhibit motility, J. Mol. Biol. 307 (5) (2001) 1317–1327.

[69] X. Yu, L. Jin, Z.H. Zhou, 3.88 A structure of cytoplasmic polyhedrosis virus by cryo-electron microscopy, Nature 453 (7193) (2008) 415–419.

[70] J. Liu, J.K. Howell, S.D. Bradley, Y. Zheng, Z.H. Zhou, S.J. Norris, Cellular architecture of *Treponema pallidum*: novel flagellum, periplasmic cone, and cell envelope as revealed by cryo electron tomography, J. Mol. Biol. 403 (4) (2010) 546–561.

[71] M. Eisenstein, The field that came in from the cold, Nat. Methods 13 (1) (2016) 19–22.

[72] S.A. Burgess, M.L. Walker, H. Sakakibara, K. Oiwa, P.J. Knight, The structure of dynein-c by negative stain electron microscopy, J. Struct. Biol. 146 (2004) 205–216.

[73] H. Imai, T. Shima, K. Sutoh, M.L. Walker, P.J. Knight, T. Kon, S.A. Burgess, Direct observation shows superposition and large scale flexibility within cytoplasmic dynein motors moving along microtubules, Nat. Commun. 6 (2015) 8179.

[74] N. Mizuno, S. Toba, M. Edamatsu, J. Watai-Nishii, N. Hirokawa, Y.Y. Toyoshima, M. Kikkawa, Dynein and kinesin share an overlapping microtubule-binding site, EMBO J. 23 (13) (2004) 2459–2467.

[75] S. Chowdhury, S.A. Ketcham, T.A. Schroer, G.C. Lander, Structural organization of the dynein-dynactin complex bound to microtubules, Nat. Struct. Mol. Biol. 22 (4) (2015) 345–347.

[76] V. Belyy, M.A. Schlager, H. Foster, A.E. Reimer, A.P. Carter, A. Yildiz, The mammalian dynein-dynactin complex is a strong opponent to kinesin in a tug-of-war competition, Nat. Cell Biol. 18 (9) (2016) 1018–1024.

[77] A.J. Roberts, S.A. Burgess, Electron microscopic imaging and analysis of isolated dynein particles, Methods Cell Biol. 91 (2009) 41–61.

[78] K.A. Johnson, The pathway of ATP hydrolysis by dynein. Kinetics of a presteady state phosphate burst, J. Biol. Chem. 258 (22) (1983) 13825–13832.

[79] N. Kotani, H. Sakakibara, S.A. Burgess, H. Kojima, K. Oiwa, Mechanical properties of inner-arm dynein-f (dynein I1) studied with in vitro motility assays, Biophys. J. 93 (3) (2007) 886–894.

[80] S. Toba, L.A. Fox, H. Sakakibara, M.E. Porter, K. Oiwa, W.S. Sale, Distinct roles of 1alpha and 1beta heavy chains of the inner arm dynein I1 of *Chlamydomonas* flagella, Mol. Biol. Cell 22 (2011) 342–353.

[81] Y. Shimizu, H. Sakakibara, H. Kojima, K. Oiwa, Slow axonemal dynein e facilitates the motility of faster dynein c, Biophys. J. 106 (10) (2014) 2157–2165.

[82] D.S. Gorman, R.P. Levine, Cytochrome f and plastocyanin: their sequence in the photosynthetic electron transport chain of *Chlamydomonas reinhardi*, Proc. Natl. Acad. Sci. USA 54 (6) (1965) 1665–1669.

[83] G.B. Witman, Isolation of Chlamydomonas flagella and flagellar axonemes, Methods Enzymol. 134 (1986) 280–290.

[84] S.M. King, Large-scale isolation of *Chlamydomonas* flagella, Methods Cell Biol. 47 (1995) 9–12.

[85] O. Kagami, R. Kamiya, Translocation and rotation of microtubules caused by multiple species of *Chlamydomonas* inner-arm dynein, J. Cell Sci. 103 (1992) 653–664.

[86] H. Sakakibara, H. Kojima, Y. Sakai, E. Katayama, K. Oiwa, Inner-arm dynein c of Chlamydomonas flagella is a single-headed processive motor, Nature 400 (6744) (1999) 586–590.

[87] S.M. King, Purification of axonemal dyneins and dynein-associated components from *Chlamydomonas*, Methods Cell Biol. 92 (2009) 31–48.

[88] R. Yamamoto, T. Yagi, R. Kamiya, Functional binding of inner-arm dyneins with demembranated flagella of *Chlamydomonas* mutants, Cell Motil. Cytoskelet. 63 (5) (2006) 258–265.

[89] B.M. Paschal, H.S. Shpetner, R.B. Vallee, Purification of brain cytoplasmic dynein and characterization of its in vitro properties, Methods Enzymol. 196 (1991) 181–191.

[90] A. Gennerich, S.L. Reck-Peterson, Probing the Force Generation and Stepping Behavior of Cytoplasmic Dynein, In: E. Peterman, G. Wuite (eds), Single Molecule Analysis. Methods in Molecular Biology (Methods and Protocols), Humana Press, 783 (2011) 63–80.

[91] A. Gennerich, A.P. Carter, S.L. Reck-Peterson, R.D. Vale, Force-induced bidirectional stepping of cytoplasmic dynein, Cell 131 (2007) 952–965.

[92] S.J. Ludtke, P.R. Baldwin, W. Chiu, EMAN: semiautomated software for high-resolution single-particle reconstructions, J. Struct. Biol. 128 (1) (1999) 82–97.

[93] J. Frank, M. Radermacher, P. Penczek, J. Zhu, Y. Li, M. Ladjadj, A. Leith, SPIDER and WEB: processing and visualization of images in 3D electron microscopy and related fields, J. Struct. Biol. 116 (1) (1996) 190–199.

[94] M. van Heel, G. Harauz, E.V. Orlova, R. Schmidt, M. Schatz, A new generation of the IMAGIC image processing system, J. Struct. Biol. 116 (1) (1996) 17–24.

[95] H.A. Yanagisawa, R. Kamiya, Association between actin and light chains in *Chlamydomonas* flagellar inner-arm dyneins, Biochem. Biophys. Res. Commun. 288 (2) (2001) 443–447.

[96] M. Hafezparast, R. Klocke, C. Ruhrberg, A. Marquardt, A. Ahmad-Annuar, S. Bowen, G. Lalli, A.S. Witherden, H. Hummerich, S. Nicholson, P.J. Morgan, R. Oozageer, J.V. Priestley, S. Averill, V.R. King, S. Ball, J. Peters, T. Toda, A. Yamamoto, Y. Hiraoka, M. Augustin,

D. Korthaus, S. Wattler, P. Wabnitz, C. Dickneite, S. Lampel, F. Boehme, G. Peraus, A. Popp, M. Rudelius, J. Schlegel, H. Fuchs, M. Hrabe de Angelis, G. Schiavo, D.T. Shima, A.P. Russ, G. Stumm, J.E. Martin, E.M. Fisher, Mutations in dynein link motor neuron degeneration to defects in retrograde transport, Science 300 (5620) (2003) 808–812.

[97] K.M. Ori-McKenney, J. Xu, S.P. Gross, R.B. Vallee, A cytoplasmic dynein tail mutation impairs motor processivity, Nat Cell Biol. 12 (2010) 1228–1234.

[98] J.L. Ross, K. Wallace, H. Shuman, Y.E. Goldman, E.L. Holzbaur, Processive bidirectional motion of dynein-dynactin complexes in vitro, Nat. Cell Biol. 8 (2006) 562–570.

[99] S. Toba, T.M. Watanabe, L. Yamaguchi-Okimoto, Y.Y. Toyoshima, H. Higuchi, Overlapping hand-over-hand mechanism of single molecular motility of cytoplasmic dynein, Proc. Natl. Acad. Sci. USA 103 (2006) 5741–5745.

[100] K. Zhang, et al., Cryo-EM reveals how human cytoplasmic dynein is auto-inhibited and activated, Cell 169 (2017) 1303–1314. DOI: http://dx.doi.org/10.1016/j.cell.2017.05.025.

[101] C. Cho, R.D. Vale, The mechanism of dynein motility: insight from crystal structures of the motor domain, Biochim. Biophys. Acta 1823 (2011) 182–191.

[102] K. Oiwa, H. Sakakibara, Recent progress in dynein structure and mechanism, Curr. Opin. Cell Biol. 17 (1) (2005) 98–103.

[103] H. Ueno, T. Yasunaga, C. Shingyoji, K. Hirose, Dynein pulls microtubules without rotating its stalk, Proc. Natl. Acad. Sci. USA 105 (2008) 19702–19707.

[104] T. Movassagh, K.H. Bui, H. Sakakibara, K. Oiwa, T. Ishikawa, Nucleotide-induced global conformational changes of flagellar dynein arms revealed by in situ analysis, Nat. Struct. Mol. Biol. 17 (2010) 761–767.

[105] A.P. Carter, Structure and functional role of dynein's microtubule-binding domain, Science 322 (2008) 1691–1695.

[106] D. Nicastro, Cryo-electron tomography reveals conserved features of doublet microtubules in flagella, Proc. Natl. Acad. Sci. USA 108 (2011) E845–E853.

[107] N.D. Derr, B.S. Goodman, R. Jungmann, A.E. Leschziner, W.M. Shih, and S.L. Reck-Peterson, Tug-of-war in motor protein ensembles revealed with a programmable DNA origami scaffold, Science 338 (2012) 662–665.

[108] K. Furuta, A. Furuta, Y.Y. Toyoshima, M. Amino, K. Oiwa, H. Kojima, Measuring collective transport by defined numbers of processive and nonprocessive kinesin motors, Proc. Natl. Acad. Sci. USA 110 (2) (2013) 501–506.

In this chapter

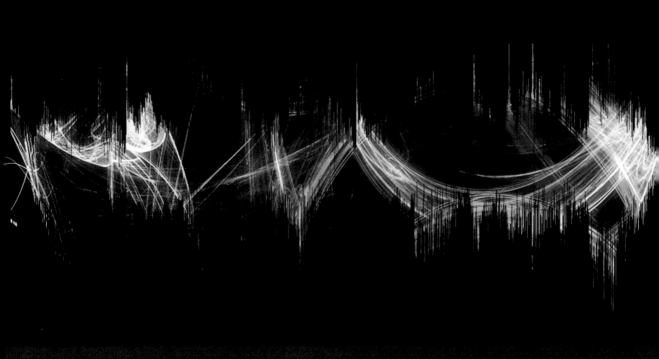

Mechanism and regulation of dynein motors

Helgo Schmidt[1], Andrew P. Carter[2]

[1]Institut de Génétique et de Biologie Moléculaire et Cellulaire (IGBMC), Illkirch, France;
[2]Medical Research Council Laboratory of Molecular Biology, Cambridge, United Kingdom

2.1 Overall architecture of the dynein motor

The dynein complex consists of heavy, intermediate, light–intermediate, and light chains. The N-terminal heavy-chain part is called the tail. It acts as an assembly platform for the other chains and is also responsible for heavy-chain multimerization. The C-terminal ~3500 amino acid residues of the heavy chain form the dynein motor, which generates the motility for the movement along microtubules. The dynein motor consists of a linker domain, a ring of six AAA+ (ATPases associated with various cellular activities) domains and a C-terminal domain [1–4] (Fig. 2.1A and B). This general architecture is conserved in all dynein isoforms.

Various crystal as well as electron microscopy (EM) structures have elucidated the molecular structure of the dynein motor. The linker domain is mainly α-helical and divided into N- and C-terminal subdomains of roughly equal size [5–8] that enclose a prominent central cleft spanned by a single hinge helix [9]. Each of the AAA+ domains of the AAA+ ring can be subdivided into α/β large (AAAL) and α-helical small domains (AAAS). The first four AAA+ modules can bind ATP. Each of the ATP-binding sites is formed by amino acid residues that are located at the interface between neighboring AAA+ domains. ATP binding triggers gap closure between these neighboring AAA+ domains (Fig. 2.1C). Whereas the active site amino acid residues of AAA1 are strictly conserved, the ability to hydrolyze ATP at sites AAA3 and AAA4 varies among dynein isoforms [4]. AAA2 seems to be in a permanent ATP bound state and unable to hydrolyze ATP [6,8]. The α/β fold of AAA1L, AAA2L, AAA3L, AAA4L, and AAA5L is extended by α-helical or β-hairpin inserts [10]. There are also modifications of the AAAS basic architecture. AAA4S and AAA5S have two coiled-coil extensions, the stalk and the buttress (or strut) [5,7]. At the tip of the ~150-Å stalk sits the microtubule-binding domain (MTBD). The architecture and size of the dynein motor C-terminal domain varies between species. In yeast, it consists of only one α-helix whereas in higher eukaryotes it is extended to include six additional α-helices and a β-barrel [10].

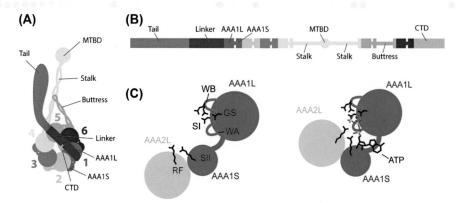

Figure 2.1 Architecture of the dynein motor. (A) Schematic cartoon representation of the dynein heavy chain. The cargo-binding dynein tail and the dynein motor are shown in transparent and solid colors, respectively. (B) Arrangement of the dynein heavy-chain domains. (C) Architecture of the AAA1 ATP-binding site. Left panel: In the absence of nucleotide, the AAA1 ATP-binding site, located at the interface between AAA1L and AAA2L, is open. Right panel: The binding of ATP induces the closure of the AAA1 site. Conserved amino acid residues of the active site are shown in stick representation. *CTD*, C-terminal domain; *GS*, glutamate-switch asparagine; *RF*, arginine finger; *SI*, sensor I asparagine; *SII*, sensor II arginine; *WA*, Walker-A motif; *WB*, Walker-B motif.

2.2 The mechanochemical cycle of the dynein motor

Among the four nucleotide-binding sites of the dynein motor AAA+ ring, AAA1 is the only site that is indispensable for motility generation [11,12]. EM [13–15], electron-tomography [16,17], crystallographic [5–9,18], kinetic [19], biochemical [11,12], and fluorescence resonance energy transfer (FRET) studies [20] have revealed how ATP turnover at AAA1 is linked to movement along the microtubule. When AAA1 is empty, the MTBD binds tightly to the microtubule [11] and the linker is straight with its N-terminal domain contacting AAA5 [8]. Upon ATP binding to AAA1, the MTBD switches into a low microtubule affinity state and releases from the microtubule [12,19]. Another consequence of ATP binding is that the linker bends at its central cleft so that the N-terminal domain lies above AAA2/AAA3 [9,18,20]. The dynein motor is able to rebind to the microtubule after the hydrolysis of ATP and the linker switches from the bent into the straight conformation. This straightening of the linker generates the force for dynein motility and has been termed the linker power stroke [13]. After the power stroke, the linker N-terminal domain lies close to AAA4 [6]. The return to its original position at AAA5 correlates with the release of ADP and resets the ATP hydrolysis cycle so that the dynein motor is able to carry out the next step [4,8].

2.3 Conformational changes in the AAA+ ring drive the mechanochemical cycle

The analysis of various crystal and EM structures suggests that the geometry of the AAA+ ring plays a key role in correlating ATP turnover at AAA1 with MTBD and linker remodeling [5–9,14,18,21] (Fig. 2.2A). These structures were obtained from different dynein isoforms of *Saccharomyces cerevisiae*, *Dictyostelium discoideum*, and *Homo sapiens* raising the question of isoform and species-dependent differences. However, the conserved architecture of the dynein motor makes it likely that the mechanistic principles presented in the following paragraphs can be generalized.

When the AAA1 site is in the APO state, there are two large gaps between AAA1/AAA2 and AAA5/AAA6 that split the AAA+ ring into two halves consisting of AAA1+AAA6 and AAA2–AAA5 [8]. ATP binding to AAA1 brings the conserved active site residues, located on AAA1 and AAA2 (Fig. 2.1C), in close proximity and thus closes the AAA1/AAA2 gap [18]. At the same time, the closure of the whole AAA+ ring is triggered, because AAA1 is part of the AAA1+AAA6 half and AAA2 is part of the AAA2–AAA5 half [18].

The closed AAA+ ring conformation has two consequences. Firstly, it causes the change in MTBD microtubule affinity from high to low via a sliding movement of the stalk coiled-coil helices [9] (Fig. 2.2A). The two α-helices that make up the stalk, CC1 and CC2, show the characteristic coiled-coil heptad repeat pattern of hydrophobic and hydrophilic residues. However, whereas CC2 has a single heptad repeat, CC1 features two possible heptad registries [22] giving rise to alternative coiled-coil arrangements, α and β, that stabilize the MTBD high and low microtubule affinity states, respectively [2,22,23]. In the closed AAA+ ring conformation, the buttress has caused CC2 to slide with respect to CC1, which leads to a stalk that supports the low microtubule affinity state of the MTBD [9]. This buttress induced CC2 sliding is caused by AAA6L and the tightly associated AAA5S, which harbors the buttress. To accommodate to the closed AAA+ ring conformation, the AAA6L/AAA5S/buttress unit is forced to undergo a rotational movement that allows the buttress to pull on CC2 to change the MTBD microtubule affinity from high to low [9] (Fig. 2.2A).

The conformational changes in the MTBD associated with this transition have been elucidated by a cryo-EM [24] and two crystal structures [25,26] of stalk-MTBD fragments. In the high-microtubule affinity state, the MTBD H1, H3, and H6 α-helices contact the tubulin dimer [24]. Mutating the amino acid residues involved in these contacts severely reduced the microtubule affinity of the MTBD suggesting critical contributions to microtubule binding [22,24,27]. In

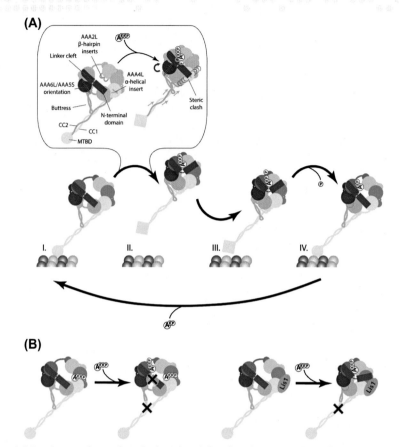

Figure 2.2 Model for the mechanochemical cycle of the dynein motor. (A) Schematic cartoon representation of the cycle. I. In the absence of ATP the AAA+ ring of the dynein motor is split into two halves. The linker is straight and the MTBD is in the high-microtubule affinity state leading to tight association with the microtubule. II. The binding of ATP to the AAA1 site closes the AAA+ ring. The linker adopts the bent pre–power stroke conformation and the MTBD switches to the low microtubule affinity state. The inset between I and II provides more details about the ATP-induced conformational changes. The closure of the AAA+ ring enforces a rotation of AAA6L/AAA5S (*red arrow*, *dotted line* indicates the orientation of AAA6L/AAA5S). This allows the buttress to pull on the stalk coiled-coil CC2 α-helix (*orange arrow*). The subsequent sliding of CC2 with respect to CC1 (*dark yellow arrows*) switches the MTBD microtubule affinity state from high (*circle*) to low (*square*), which causes the microtubule detachment of the dynein motor. The closed AAA+ ring conformation also leads to a steric clash between the AAA4L α-helical insert and the N-terminal domain of the linker (*black half circles*), which enforces the bending of the linker. The AAA2L β-hairpin inserts might transiently interact with the linker cleft to support the remodeling process. III. After ATP hydrolysis, the dynein motor is able to rebind to the microtubule. IV. Upon rebinding, the linker undergoes the power stroke, the force-producing event of the motor cycle. It straightens and switches from the pre– into the post–power stroke conformation. The linker power stroke is associated with a partial opening of the AAA+ ring. Further opening of the AAA+ ring leads to the ejection of ADP from the AAA1 site to restart the cycle. (B) The mechanochemical cycle can be influenced by the AAA3 ATP–binding site and the regulator Lis1. Left panel: When the AAA3 site is in the ATP state, ATP binding to AAA1 no longer results in linker remodeling and microtubule release (*black cross*). Right panel: When Lis1 binds to the dynein motor ATP binding to AAA1 still leads to linker remodeling, but the MTBD does not switch into the low microtubule affinity state anymore (*black cross*). For clarity the dynein tail is not shown.

the low microtubule affinity state of the MTBD [25,26], the H1 and H3 α-helices relocate to a new position where the interactions with the tubulin dimer are no longer possible [24].

A second consequence of the closed AAA+ ring conformation is a potential steric clash between the α-helical insert of AAA4L and the linker N-terminal domain (Fig. 2.2A). This clash was proposed to trigger the remodeling of the linker into the bent conformation [9].The "steric clash" model is supported by negative-stain EM studies, which demonstrated that the deletion of the AAA4L α-helical insert impairs the ability of the linker to adopt the bent conformation [9]. Also a recent cryo-EM study on the dynein motor ATP state revealed a sub-population of motors with partially bent linkers that contact a region of AAA4 consistent with the AAAL helical insert [18].

In addition to the AAA4L α-helical insert, two β-hairpins of AAA2L, the H2 (helix 2) and PS I (presensor I) inserts, were also predicted to contribute to linker bending [6,8]. The deletion of the H2 insert affected the ATP-induced bending of the linker in an FRET assay [6]. Based on this finding and a simulated ATP-induced closure of the AAA1 site, it was suggested that the AAA2L inserts would come into close contact with the central cleft between the N- and C-terminal linker subdomains to induce linker bending [6,8]. However, a recent crystal structure of a dynein motor where the linker adopts the bent conformation did not show this interaction [9]. Thus, it is currently unclear how exactly the AAA2L inserts could contribute to the remodeling of the linker. One possibility is that they support linker bending by transiently interacting with the linker cleft during the closure of the AAA+ ring. The different orientations of the AAA2L inserts with respect to the linker cleft observed in two crystal structures of the dynein motor ADP state [2] might define a potential trajectory path [28]. Linker bending could subsequently be induced in combination with an AAA4L insert–mediated steric clash [21].

2.4 The remodeling of the linker during the mechanochemical cycle

The linker bending during the mechanochemical cycle consists of a rigid-body movement of the N-terminal relative to the C-terminal domain with the central cleft as a pivot point [9,14,18]. During the remodeling, the linker hinge helix, which connects both linker domains, is forced into a distorted conformation [9]. It is tempting to speculate that the hinge helix in its distorted form acts as a store of mechanical energy. The linker power stroke, characterized by the straightening of the bent linker, might therefore be driven by the release of mechanical energy in the hinge helix [21].

The bent linker features hydrophobic interactions across the central linker cleft and contacts the AAA2 and AAA3 domains of the AAA+ ring [9]. Both interactions could potentially stabilize the bent linker conformation. However, deleting of the AAA+ ring contacts did not affect the ability of the linker to adopt the bent conformation [9] suggesting that the linker in its bent form is primarily stabilized by intramolecular hydrophobic interactions.

The bent linker conformation seems not to be static. Cryo-EM structures of the dynein motor revealed that the bent linker shows considerable flexibility [14,18]. This might reflect an equilibrium of remodeling substeps from the straight to the bent linker conformation or a dynamic wiggling of the linker N-terminal domain caused by the opposing forces of the distorted hinge helix and the hydrophobic linker cleft interactions [28].

2.5 ATP hydrolysis primes the dynein motor for microtubule rebinding and the linker power stroke

After microtubule release and linker bending induced by ATP binding to AAA1, ATP hydrolysis occurs. In its posthydrolysis state the dynein motor is capable of rebinding to the microtubule [19] and to carry out the linker power stroke. Microtubule binding also requires the MTBD to switch back to the high-affinity state, which triggers the power stroke, the straightening of the bent linker [19]. After the power stroke, the dynein motor is bound to ADP (Fig. 2.2A).

Although dynein motor crystal and cryo-EM structures are available that show the linker before and after the power stroke [6,8,9,18], the molecular events causing the linker power stroke are currently not understood. One of the most prominent features of the ADP bound post–power stroke dynein motor is the reopening of the AAA+ ring between AAA1/AAA2 and AAA4/AAA5 [6] (Fig. 2.2A). It has been speculated that a steric clash between the AAA2 and AAA3 domains of the opening AAA+ ring and the N-terminal domain of the bent linker might cause the power stroke. The potential clash might trigger the power stroke by disturbing the hydrophobic interactions at the linker cleft that stabilize the bent linker [21]. However, experimental evidence for this hypothesis is lacking.

To release ADP from the AAA1 site, the AAA+ ring opens further to widen the gap between AAA1 and AAA2 as indicated by a crystal structure of the dynein APO state [8]. In the ADP-state, there are gaps between AAA1/AAA2 and AAA4/AAA5, which split the AAA+ ring into two halves, consisting of AAA2–AAA4 and AAA1+AAA5–AAA6. In the APO structure, the position of AAA5 has changed, it has now become part of a AAA2–AAA5 block of AAA+ domains. The new position of AAA5 allows the linker to push both halves further apart by contacting AAA5 and AAA1 at the same time. This opens up the gap between AAA1 and AAA2 to promote the release of ADP so that the AAA1 site is able to start a new ATP hydrolysis cycle [8,21] (Fig. 2.2A).

2.6 Variations of the dynein mechanochemical cycle

The basic scheme of dynein motility can be modified by the AAA3 site [1] and regulators, such as Lis1 [3], to fine tune dynein motor activity for different biological environments (Fig. 2.2B). When AAA3 is in the ADP state, the dynein motor carries out the standard mechanochemical cycle as outlined previously. However, if the AAA3 site is in the ATP state, the AAA1 ATP–binding event is no longer communicated around the AAA+ ring [18] and, as a result, there is no linker remodeling [18] and only impaired microtubule release [11,29,30]. A dynein motor crystal structure with AMPPNP bound at the AAA3 site has provided insights into the molecular basis of this regulation [18]. In this structure, the linker is in contact with AAA5 and also AAA2. The linker contact with AAA2 is thought to prevent AAA2 from moving toward AAA1 upon ATP binding to AAA1, which interferes with linker bending and release from the microtubule.

The dynein motor regulator Lis1 prevents the ATP-induced release from the microtubule [31], decreases the velocity of dynein molecules in single-molecule assays [31,32], and keeps dynein in a prolonged force-producing state [32]. The molecular basis for these Lis1 effects has been investigated by EM. Lis1 binds at the interface between AAA3 and AAA4 [31,33] and prevents the linker N-terminal domain from docking onto AAA5 [33] during the dynein motor cycle. Shortening the linker so that it is no longer blocked by Lis1 renders the dynein motor Lis1 insensitive [33]. This suggests that Lis1 exerts its effects by blocking linker binding to AAA5. However, it is not clear how this would prevent the ATP-induced release from the microtubule and give rise to the increase in motor force [3]. The AAA3 and Lis1 mediated variations of the dynein motor cycle might allow the dynein motor to hold on to the microtubule against high loads [31,33,34], which is especially important when dynein has to transport big cargos like nuclei.

2.7 How dynein motors walk along the microtubule

To prevent track detachment, most motor proteins are organized as dimers that coordinate the motility cycles between the two partners so that one motor is always attached to the track while the other is taking a step.

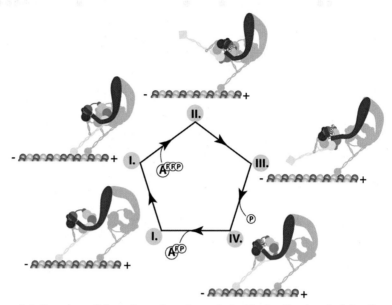

Figure 2.3 **Dynein walking along the microtubule.** The leading head of the dimeric dynein motor (colored) is taking a step. I. ATP binding to the AAA1 site leads to the bending of the linker. II–III. This allows the leading head to explore binding sites located closer to the microtubule minus end. IV. The dynein motor rebinds to the microtubule and the linker switches back from the bent to the straight conformation ("power stroke"). After the release of ADP, the cycle can start again and the leading or trailing head might take the next step. Microtubule-binding sites of the leading head are highlighted in red. The tail domain is responsible for the dimerization of the dynein motor. All domains of the dynein motor are color-coded like in Fig. 2.1.

Single-molecule studies have demonstrated such processive movements also for dimeric dynein motors, which walk along the microtubule in the minus end direction [35] (Fig. 2.3). However, in contrast to the strict hand-over-hand mechanism reported for kinesin and myosin dimers, the two dynein motors show a more complex stepping behavior. Leading and trailing heads are able to take multiple steps without passing each other, heads step frequently backward, and the distance between both heads shows considerable variability [35–38]. When the motor domains are in close proximity, both heads are equally likely to take a step forward [36,38] (Fig. 2.4A). Coordinating the motility cycles of the two heads might not be necessary because of the high dynein motor duty ratio, the time fraction of the motility cycle spent on the track. It would make the concurrent microtubule detachment of the stepping and nonstepping head very unlikely [39]. However, there is also evidence for direct communication between the two motors. A recent study has demonstrated that deleting parts of dynein motor C-terminal

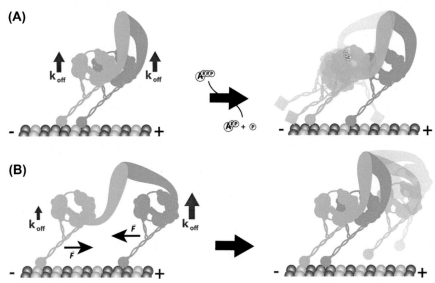

Figure 2.4 Coordination of dynein stepping and determinants of microtubule minus end directionality. (A) Leading (green) and trailing (blue) head are equally likely to take the next step at short interhead distances. As an example, the ATP-induced detachment of the leading head is shown. The search for a binding site is biased toward the minus end because of the preferred 50-degree angle between the stalk/MTBD and the microtubule. The stepping head explores different binding sites (*transparent green cartoon* representations), but only binds at a site where the preferred angle is possible (*solid green cartoon* representation). (B) When the distance between both heads increases, the trailing head becomes more likely to take the next step. The increased distance causes tension between both heads, which inhibits the ATP-dependent mechanism of microtubule release. The leading and trailing head experience a force directed toward plus and minus end (*black arrows*), respectively. Stepping of the trailing is more likely, because the MTBD shows an increased microtubule release rate when under the influence of minus end directed forces. The tension relief would bias the detached trailing head toward a binding site closer to the minus end (*transparent and solid blue cartoon* representations).

domain severely impairs the processive movement of dynein dimers [40]. Increasing or decreasing the conformational flexibility of the C-terminal domain led to shorter or longer processive movements, suggesting the C-terminal domain is involved in interhead communication [40].

When the intermotor distance between dimeric dynein motors is large, the trailing head becomes more likely to detach from the microtubule and to step forward, suggesting force between the heads aides coordination [36,38] (Fig. 2.4B). This dynein motor coordination might be explained by two effects of interhead force. Firstly, when microtubule detachment is probed in an optical trap assay where force is applied via the dynein linker, the presence of ATP does not increase the microtubule release rate [41]. Secondly, it has been demonstrated that the microtubule release rate of dynein motor depends on the direction of the applied force. When force is applied toward the microtubule minus end, the release rate is high. When it is applied in the microtubule plus end direction, the release rate is low [41,42]. The combined effect of the interhead force is the inhibition of an ATP-mediated stochastic microtubule detachment and an increase of the microtubule detachment and stepping rate of the trailing head, which experiences a minus end directed force [41] (Fig. 2.4B).

Surprisingly, processive movement could also be obtained by tethering a wild-type dynein motor to a construct consisting of a stalk and MTBD genetically fused to seryl-tRNA synthetase. This chimera showed velocities and microtubule run lengths similar to the wild-type dynein dimer, indicating that a single active dynein motor head tethered to the microtubule is sufficient for processive movement [41].

2.8 Determinants of dynein directionality

The mechanism of dynein's microtubule minus end directionality is still under debate [1]. Early suggestions referred to other motor proteins where directionality is determined by the power stroke of the motor cycle. In myosins the motile element that produces force in response to ATP hydrolysis is called the "lever arm." Reversing the direction of the lever arm power stroke by protein engineering resulted in a reversal of the direction of the myosin motor suggesting the direction of the power stroke controls the translocation of the stepping head [43]. Similar models based on the direction of the linker power stroke were suggested for the dynein motor [13]. However, reversing the power stroke of the dynein motor by extending or shrinking the stalk by one heptad repeat to introduce a 180-degree rotation of the AAA+/linker region with respect to the MTBD did not change the directionality of the dynein motor [25].

As an alternative to the linker power stroke model it has been suggested that the angle between the stalk/MTBD and the microtubule might participate in the determination of dynein directionality [25]. EM studies have revealed that this angle is around 50 degrees [25,44]. This preferred binding orientation of the stalk/MTBD would bias the diffusional search of the stepping head toward a binding site closer to the microtubule minus end (Fig. 2.4A). Such a mechanism might predominantly determine the directionality when the two dynein motors are close together and interhead tension is low so that microtubule detachment is largely driven by ATP binding [41]. In addition to the angle between the stalk/MTBD and the microtubule, the force induced microtubule detachment of dynein motors might also contribute to directionality. The interhead tension between the two motors of the dynein dimer favors the release of the trailing head and would bias its next step toward the minus end [42] (Fig. 2.4B), especially when both motor heads are far apart.

2.9 Dynein motor activation by cargo binding

The activity of dimeric kinesin motors is inhibited in the absence of cargo to prevent futile ATP hydrolysis and overcrowding of microtubules with empty kinesin motors. The reason for this inhibition is the kinesin tail that folds back onto the motor domain [45]. In the case of kinesin-1 and maybe also other kinesin family members, the tail–motor interaction interferes with the ATP hydrolysis activity by holding the motors in a conformation that inhibits the release of ADP from the active site [46]. Cargo binding to the kinesin tail overcomes this autoinhibition and activates dimeric kinesin [47,48]. Until recently, no autoinhibition mechanisms had been described for the dynein motor. Full length dynein motors purified from *S. cerevisiae* are able to walk processively along the microtubule as revealed by single-molecule studies [35]. However, similar experiments carried out on recombinant full length human dynein did not lead to processive movement [49]. Processive movements were only observed for a ternary complex between dynein, the ubiquitous dynein activator dynactin [50], and cargo adapters such as Bicaudal-D 2 (BICD2) [51,52]. This suggests that dynein motors of higher eukaryotes also predominantly exist in an autoinhibited form and that activation requires cargo binding. How might this autoinhibited dynein motor form look like? In the presence of ATP or the ATP hydrolysis transition state analog ADP.Vi, dyneins predominantly exist in a conformation where both motor domains are stacked together as revealed by recent EM studies [53]. In this stacked form, the two stalks of the motor heads are crossed, and the corresponding MTBDs most likely have opposite orientations with respect to each other. The result would be that effectively only one MTBD is able to bind to the microtubule, so that the dynein motor cannot undergo processive movements. The authors also demonstrated that pulling the two motor domains apart by introducing a DNA rod between them rescues processive movement [53]. Together with the processive movements observed in the presence of dynactin and cargo adapters this suggests that cargo binding to the dynein tail might activate the motor activity by physically pulling apart the stacked motor heads. A cryo-EM structure of the ternary dynein tail/BICD2/dynactin complex has provided insights into how this activation mechanism might work [54]. In this

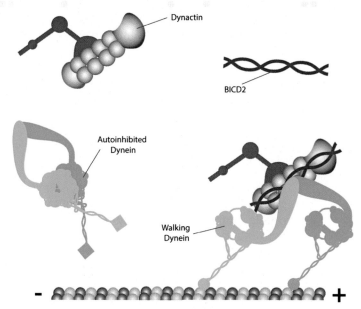

Figure 2.5 Dynein activation. Human dynein is not a processive motor on its own. In its inhibited form, the stalks are crossed with the microtubule domains facing in opposite directions. Processive dynein movements can only be observed in complex with dynactin and cargo adapters such as BICD2.

structure both tails bind asymmetrically to dynactin. This asymmetry could be subsequently imposed onto the motor heads, which extend from the tail (Fig. 2.1A), to force them apart (Fig. 2.5) [54]. A recent EM study of the full length dynein/BICD2/dynactin complex bound to microtubules revealed that both motor heads are orientated in a way compatible with walking along the microtubule [55]. This suggests that the increased conformational flexibility of the motor heads caused by dynactin/cargo adapter binding is ultimately used to reorient them for processive movement [56].

2.10 Conclusions

The last years have seen unprecedented progress in our understanding of the dynein motor architecture and mechanism. We are now at a stage where we have high-resolution information of various nucleotide states and are able to develop detailed mechanistic models for linker remodeling and the regulation of microtubule affinity. However, to get a more comprehensive mechanistic understanding, high-resolution information on the microtubule bound APO and ADP states is needed. The first cryo-EM structures of the ubiquitous dynein regulator Lis1 bound to the dynein motor have provided exciting insights into its function, but more work is needed before we fully understand its molecular effects. In terms of the dynein complex the ultimate goal must be to visualize it walking along the microtubule at high resolution. The recent breakthrough developments in the field of cryo-EM and electron tomography might make such difficult studies possible in the future. These studies will help us to get insights into so far unknown key aspects of the motor mechanism like the structural consequences of microtubule binding on the dynein mechanochemical cycle, the events triggering the linker power stroke, the molecular basis of dynein directionality ,and the coordination of dynein motors at large interhead separation. Combined with high-resolution work on the autoinhibited state of the dynein complex such studies would also finally allow us to understand how dynein is activated upon cargo binding.

References

[1] G. Bhabha, G.T. Johnson, C.M. Schroeder, R.D. Vale, Trends Biochem. Sci. 41 (2016) 94–105.

[2] A.P. Carter, J. Cell Sci. 126 (2013) 705–713.

[3] M.A. Cianfrocco, M.E. DeSantis, A.E. Leschziner, S.L. Reck-Peterson, Annu. Rev. Cell Dev. Biol. 31 (2015) 83–108.

[4] A.J. Roberts, T. Kon, P.J. Knight, K. Sutoh, S.A. Burgess, Nat. Rev. Mol. Cell Biol. 14 (2013) 713–726.

[5] A.P. Carter, C. Cho, L. Jin, R.D. Vale, Science 331 (2011) 1159–1165.

[6] T. Kon, T. Oyama, R. Shimo-Kon, K. Imamula, T. Shima, K. Sutoh, G. Kurisu, Nature 484 (2012) 345–350.

[7] T. Kon, K. Sutoh, G. Kurisu, Nat. Struct. Mol. Biol. 18 (2011) 638–642.

[8] H. Schmidt, E.S. Gleave, A.P. Carter, Nat. Struct. Mol. Biol. 19 (2012) 492–497 S491.

[9] H. Schmidt, R. Zalyte, L. Urnavicius, A.P. Carter, Nature 518 (2015) 435–438.

[10] E.S. Gleave, H. Schmidt, A.P. Carter, J. Struct. Biol. 186 (2014) 367–375.

[11] T. Kon, M. Nishiura, R. Ohkura, Y.Y. Toyoshima, K. Sutoh, Biochemistry 43 (2004) 11266–11274.

[12] S.L. Reck-Peterson, R.D. Vale, Proc. Natl. Acad. Sci. U.S.A. 101 (2004) 1491–1495.

[13] S.A. Burgess, M.L. Walker, H. Sakakibara, P.J. Knight, K. Oiwa, Nature 421 (2003) 715–718.

[14] A.J. Roberts, B. Malkova, M.L. Walker, H. Sakakibara, N. Numata, T. Kon, R. Ohkura, T.A. Edwards, P.J. Knight, K. Sutoh, S.A. Burgess, Structure 20 (2012) 1670–1680.

[15] A.J. Roberts, N. Numata, M.L. Walker, Y.S. Kato, B. Malkova, T. Kon, R. Ohkura, F. Arisaka, P.J. Knight, K. Sutoh, S.A. Burgess, Cell 136 (2009) 485–495.

[16] J. Lin, K. Okada, M. Raytchev, M.C. Smith, D. Nicastro, Nat. Cell Biol. 16 (2014) 479–485.

[17] T. Movassagh, K.H. Bui, H. Sakakibara, K. Oiwa, T. Ishikawa, Nat. Struct. Mol. Biol. 17 (2010) 761–767.

[18] G. Bhabha, H.C. Cheng, N. Zhang, A. Moeller, M. Liao, J.A. Speir, Y. Cheng, R.D. Vale, Cell 159 (2014) 857–868.

[19] K. Imamula, T. Kon, R. Ohkura, K. Sutoh, Proc. Natl. Acad. Sci. U.S.A. 104 (2007) 16134–16139.

[20] T. Kon, T. Mogami, R. Ohkura, M. Nishiura, K. Sutoh, Nat. Struct. Mol. Biol. 12 (2005) 513–519.

[21] H. Schmidt, Bioessays 37 (2015) 532–543.

[22] I.R. Gibbons, J.E. Garbarino, C.E. Tan, S.L. Reck-Peterson, R.D. Vale, A.P. Carter, J. Biol. Chem. 280 (2005) 23960–23965.

[23] T. Kon, K. Imamula, A.J. Roberts, R. Ohkura, P.J. Knight, I.R. Gibbons, S.A. Burgess, K. Sutoh, Nat. Struct. Mol. Biol. 16 (2009) 325–333.

[24] W.B. Redwine, R. Hernandez-Lopez, S. Zou, J. Huang, S.L. Reck-Peterson, A.E. Leschziner, Science 337 (2012) 1532–1536.

[25] A.P. Carter, J.E. Garbarino, E.M. Wilson-Kubalek, W.E. Shipley, C. Cho, R.A. Milligan, R.D. Vale, I.R. Gibbons, Science 322 (2008) 1691–1695.

[26] Y. Nishikawa, T. Oyama, N. Kamiya, T. Kon, Y.Y. Toyoshima, H. Nakamura, G. Kurisu, J. Mol. Biol. 426 (2014) 3232–3245.

[27] M.P. Koonce, I. Tikhonenko, Mol. Biol. Cell 11 (2000) 523–529.

[28] H. Schmidt, A.P. Carter, Biopolymers 105 (2016) 557–567.

[29] M.A. DeWitt, C.A. Cypranowska, F.B. Cleary, V. Belyy, A. Yildiz, Nat. Struct. Mol. Biol. 22 (2015) 73–80.

[30] M.P. Nicholas, F. Berger, L. Rao, S. Brenner, C. Cho, A. Gennerich, Proc. Natl. Acad. Sci. U.S.A. 112 (2015) 6371–6376.

[31] J. Huang, A.J. Roberts, A.E. Leschziner, S.L. Reck-Peterson, Cell 150 (2012) 975–986.

[32] R.J. McKenney, M. Vershinin, A. Kunwar, R.B. Vallee, S.P. Gross, Cell 141 (2010) 304–314.

[33] K. Toropova, S. Zou, A.J. Roberts, W.B. Redwine, B.S. Goodman, S.L. Reck-Peterson, A.E. Leschziner, Elife 3 (2014).

[34] M.A. DeWitt, C.A. Cypranowska, F.B. Cleary, V. Belyy, A. Yildiz, Nat. Struct. Mol. Biol. 22 (2015) 73–80.

[35] S.L. Reck-Peterson, A. Yildiz, A.P. Carter, A. Gennerich, N. Zhang, R.D. Vale, Cell 126 (2006) 335–348.

[36] M.A. DeWitt, A.Y. Chang, P.A. Combs, A. Yildiz, Science 335 (2012) 221–225.

[37] H. Imai, T. Shima, K. Sutoh, M.L. Walker, P.J. Knight, T. Kon, S.A. Burgess, Nat. Commun. 6 (2015) 8179.

[38] W. Qiu, N.D. Derr, B.S. Goodman, E. Villa, D. Wu, W. Shih, S.L. Reck-Peterson, Nat. Struct. Mol. Biol. 19 (2012) 193–200.

[39] T. Shima, K. Imamula, T. Kon, R. Ohkura, K. Sutoh, J. Struct. Biol. 156 (2006) 182–189.

[40] N. Numata, T. Shima, R. Ohkura, T. Kon, K. Sutoh, FEBS Lett. 585 (2011) 1185–1190.

[41] F.B. Cleary, M.A. Dewitt, T. Bilyard, Z.M. Htet, V. Belyy, D.D. Chan, A.Y. Chang, A. Yildiz, Nat. Commun. 5 (2014) 4587.

[42] A. Gennerich, A.P. Carter, S.L. Reck-Peterson, R.D. Vale, Cell 131 (2007) 952–965.

[43] Z. Bryant, D. Altman, J.A. Spudich, Proc. Natl. Acad. Sci. U.S.A. 104 (2007) 772–777.

[44] H. Ueno, T. Yasunaga, C. Shingyoji, K. Hirose, Proc. Natl. Acad. Sci. U.S.A. 105 (2008) 19702–19707.

[45] K.J. Verhey, J.W. Hammond, Nat. Rev. Mol. Cell Biol. 10 (2009) 765–777.

[46] H.Y. Kaan, D.D. Hackney, F. Kozielski, Science 333 (2011) 883–885.

[47] D.L. Coy, W.O. Hancock, M. Wagenbach, J. Howard, Nat. Cell Biol. 1 (1999) 288–292.

[48] M. Imanishi, N.F. Endres, A. Gennerich, R.D. Vale, J. Cell Biol. 174 (2006) 931–937.

[49] M. Trokter, N. Mucke, T. Surrey, Proc. Natl. Acad. Sci. U.S.A. 109 (2012) 20895–20900.

[50] T.A. Schroer, Annu. Rev. Cell Dev. Biol. 20 (2004) 759–779.

[51] R.J. McKenney, W. Huynh, M.E. Tanenbaum, G. Bhabha, R.D. Vale, Science 345 (2014) 337–341.

[52] M.A. Schlager, H.T. Hoang, L. Urnavicius, S.L. Bullock, A.P. Carter, EMBO J. 33 (2014) 1855–1868.

[53] T. Torisawa, M. Ichikawa, A. Furuta, K. Saito, K. Oiwa, H. Kojima, Y.Y. Toyoshima, K. Furuta, Nat. Cell Biol. 16 (2014) 1118–1124.

Dyneins

[54] L. Urnavicius, K. Zhang, A.G. Diamant, C. Motz, M.A. Schlager, M. Yu, N.A. Patel, C.V. Robinson, A.P. Carter, Science 347 (2015) 1441–1446.

[55] S. Chowdhury, S.A. Ketcham, T.A. Schroer, G.C. Lander, Nat. Struct. Mol. Biol. 22 (2015) 345–347.

[56] A.P. Carter, A.G. Diamant, L. Urnavicius, Curr. Opin. Struct. Biol. 37 (2016) 62–70.

In this chapter

Structural analysis of dynein intermediate and light chains

John C. Williams, Amanda E. Siglin, Christine M. Lightcap,
Amrita Dawn
Beckman Research Institute at the City of Hope, Duarte, CA, United States

3.1 Introduction

Cytoplasmic [1,2] and axonemal dynein [3] are multiple component complexes that share some common features but participate in fundamentally different processes (Chapter 2 (vol. 1 of this book)). In either complex, the heavy chain (HC) subunits hydrolyze ATP and interact with microtubules (MTs) to generate force [1] (Chapter 2). The N-termini of the dynein HCs act as scaffolds for the accessory proteins, which include the dynein intermediate chain (IC), the dynein light–intermediate chains (LICs), and the dynein light chains (LCs) [1,2]. These accessory subunits interact with other target proteins or macromolecules and have also been suggested to directly and indirectly regulate dynein ATPase activity [4,5]. In the case of cytoplasmic dynein, cellular targets that interact with the dynein ICs, LICs, and LCs are frequently considered as cargo for retrograde MT transport [2,6]. While the IC and LIC make specific, but limited, interactions with cargo (e.g., dynactin [7,8] and pericentrin [9]), the dynein LCs bind to a large number of proteins [2,10] including signaling molecules (TrkA [11]), transcription factors [12], and viruses (see Tables 3.1 and 3.2). With the limited number of cytoplasmic dyneins (e.g., nearly all tissues appear to express only one isoform [13,14]) and noting that cytoplasmic dynein is the dominant retrograde motor protein in the cell (Chapter 14 (vol. 1 of this book)), the diversity of LC-binding partners originally lead to the notion that they bridge cargo to dynein [15]. Structural and thermodynamic studies of the LCs and their different targets shortly followed these initial in vitro experiments. These studies of the LCs bound to fragments of the IC do not support this bridging model [16], rather, they argue for a role in the assembly of the dynein complex and/or a regulatory role of the IC subunits. Moreover, recent studies have identified additional proteins that not only work in conjunction with dynein and dynactin, but also target specific organelles such as BicD2 [17].

Dyneins. https://doi.org/10.1016/B978-0-12-809470-9.00003-5

Table 3.1 LC8 targets and their associated sequences

Target	LC8 sequence	Sequence location	Dimer[a]	NLS/Nuclear[b]	NLS location	References
53BP1	VSAATQTI	167–174	Yes	Yes/Yes	1395–1401 1626–1629 1668–1685	[35,151]
Amsacta moorei entomopox-virus AMV179 ORF	NIKSTQTC	152–159	N/A	Yes/No	194–210	[152,153]
African swine fever virus p54	QNTASQTM	256–263	Yes	Yes/Yes	55–58	[86,154]
Ana2 centriole duplication factor	TICAGTQTDP SSTTGTQCDI	159–168 237–246	Yes			[93]
Bassoon	TANYGSQTQ TAEFSTQTP MVAQGTQTP	1425–1433 1501–1509 1529–1537	Yes*	Yes/No	791–893 1012–1028 1072–1088 2600–2604 2625–2636	[155]
BimEL	CDKSTQTP	110–117	N/A	No/No	–	[101,142]
BimL	CDKSTQTP	50–57	N/A	No/No	–	[101,142]
Bmf	EDKATQTL	143–150	N/A	No/Yes	–	[101,156]
Cdk2	Unknown	–	N/A	No/Yes	–	[157,158]
Ciz1	Unknown	–	Yes*	Yes/Yes	689–714	[158,159]
Dazl	VDRSIQTV	268–275	Yes	No/Yes	–	[160,10]
Dynein HC	QDKLVQTP	1876–1883	Yes*	No/No	–	[161,162]
Dynein IC	YTKETQTP	166–173	Yes*	Yes/Yes	13–29 42–58	[16,84]
Egl	VDAESQTL	950–957	Yes	No/Yes	–	[12,163]
Estrogen receptor alpha	Unknown	AB domain	Yes	Yes/Yes	243–272	[164,165]
Gephyrin	EDKGVQCE	206–213	Yes	Yes/Yes	347–350	[161,166]
GKAP	LSIGIQVD	650–657	N/A	Yes/No	338–344 904–910	[167,78]
GRINL1A	REIGVGCDL	424–432	Yes*	Yes/Ukn	365–381	[168]
Grp3	TSQATQTE	610–617	N/A	No/No	–	[169]
Human adeno-virus protease	LIKSTQTV	107–114	Yes	No/Yes	–	[152,170]

Table 3.1 LC8 targets and their associated sequences—cont'd

Target	LC8 sequence	Sequence location	Dimer[a]	NLS/Nuclear[b]	NLS location	References
Human herpes simplex virus helicase	LAKSTQTF	760–767	Yes	No/Yes	–	[171,152]
Human herpesvirus U19 Gene	FSRHTQTD	260–267	N/A	No/Yes	–	[172,152]
IκBα	Disulfide?	–	N/A	Yes/Yes	83–128	[104,173]
Kibra	Unknown	–	Yes*	Yes/Yes	361–376	[174]
Kid-1	TTKSTQTQ	93–100	Yes	Yes/Yes	180–186	[175,10,176]
MAP-4	GSKSTQTV	627–634		No/No	–	[176,10]
Mokola virus	EDKSTQTP	140–147	Yes	No/Yes	–	[152,177]
Myosin Va	EDKNTMTD	1284–1291	Yes	Yes/Yes	858–874 1467–1473 1643–1649	[178,10]
Nek9	VGMHSKGTQTA	940–950	Yes	Yes	Yes/Yes	[95]
nNOS	KDTGIQVD	234–241	Yes	Yes/Yes	728–734 993–996 1323–1326	[40,179]
NRF-1	EHGVTQTE	3–10	Yes	Yes/Yes	89–108	[180,12]
Nup159	ADRDVQTS AESGIQTD KENEVQTD KHNSTQTV VDNGLQTR CNFSVQTP	1105–1112 1118–1125 1131–1138 1143–1150 1155–1162 1167–1174	Yes*	No/Yes	–	[181]
nudE	KRTDMAVQAT	200–211	Yes	No/No	–	[182]
Pak1	RDVATSPI	212–224	Yes	Yes/Yes	243–245	[183,33]
Rabies virus	EDKSTQTT	142–149	Yes	Yes/Yes	211–214	[151,50]
RASGRP1	RAVAHKTQTE	618–628	Yes	Yes/No	142–153	[184]
RSV attachment glycoprotein	TTTSTQTN	137–144	Yes	Yes/No	155–162 190–200	[152,185]
Swallow	SAKATQTD	280–287	Yes*	Yes/Yes	349–352 485–488	[186,187]
Trypanosoma CITFA	MPEVGTQV	1–8	Yes	Yes/Yes	324–338	

Continued

Dyneins

Table 3.1 LC8 targets and their associated sequences—cont'd

Target	LC8 sequence	Sequence location	Dimer[a]	NLS/Nuclear[b]	NLS location	References
TRP14	Unknown	–	No	No/No	–	[188]
TRPS1	VDRSTQDE	1219–1226	Yes	Yes/Yes	900–903	[82,189]
Viral protein VP35 of Ebola virus (EBOV)	*Zaire* SQTQT *Sudan* KQVQT *Reston* SSSQT	71–75 All species	Yes	No/Yes	–	[85,190]

[a] *Either reported the literature, noted in public databases, or contains a highly predicted coiled coil, (http://www.isrec.isb-sib.ch/webmarcoil/webmarcoilC1.html) and denoted by*.*
[b] *Reported nuclear localization sequences (NLS) and their sequence locations. The PSORT II NLS prediction program (http://psort.hgc.jp/form2.html) was used to determine potential NLS sequences for LC8 binding proteins that have not been experimentally verified.*

Table 3.2 Reported TcTex-1 binding proteins[a]

Target	Dimerization domain[b]	References
BMPR-II	Yes	[79]
CD5	Yes	[54]
CD155	Yes	[191]
DOC2a	N/A	[53]
Dynein IC	Yes	[16]
FIP-1	Unknown	[192]
p59 Fyn Kinase	N/A	[134]
G Protein βγ Subunit	N/A	[193]
HSV1	Yes	[194]
Lfc	Yes	[195,196]
PTH Receptor	Yes	[98]
Rhodopsin	Yes	[81]
SATB1	Yes	[84]
TrkA	Yes	[197]
VDAC	Yes	[80]
VP26	Yes	[198]

[a] *The reported TcTex-1 binding site in these binding partners is not clear based on the structural data and thus not determined.*
[b] *Either reported the literature, noted in public databases, or contains a highly predicted coiled coil (http://www.isrec.isb-sib.ch/webmarcoil/webmarcoilC1.html), and denoted by*.*

This chapter focuses on these structural and biochemical investigations of the LCs and ICs. It also addresses the promiscuity of the dynein LCs, LC8, and TcTex-1, as well as their potential role on dynein function. Since much of the structural and biochemical efforts focus on cytoplasmic dynein, and in particular on dynein of animal origin (rat and *Drosophila*), the primary emphasis will be on cytoplasmic dynein. Information from other dynein complexes will be used to illustrate potential structural differences among the different dynein family members.

3.2 Abbreviated background of light chains

The dynein LCs were identified as accessory subunits of axonemal and cytoplasmic dynein isolated and purified from sea urchins [18], *Chlamydomonas* flagella [19], and bovine brain tissue [20]. Treatment of isolated dynein with potassium iodide produced two distinct fractions, one containing the dynein HCs and LICs and the other containing the dynein ICs and LCs [21]. LCs isolated from cytoplasmic dynein include three polypeptides, TcTex-1 (DYNLT1/3) [22], LC8 (DYNLL1/2) [23], and LC7/Robl (DYNLRB1/2) [24]. In addition to these LCs, other low molecular weight polypeptides, also denoted as LCs, have been isolated from axonemal dynein complexes [25].

Among the cytoplasmic LCs, LC8 (DYNLL1) remains the best characterized dynein subunit. It is an essential gene that is expressed in all tissues but found at higher levels in brain and testes [26]. LC8 is not only highly conserved across all eukaryotes, but also found in plants, which are devoid of the dynein HC or IC [27]. Initial studies characterizing LC8 function did so by impairing or completely blocking its expression. These studies found that complete loss of LC8 function in *Drosophila* results in embryonic lethality, and partial loss-of-function produces apoptotic signatures (e.g., nuclear envelope breakdown) [28]. Additional genetic screens in *Drosophila* showed that mutations of LC8 disrupt axon trajectories [29]. In *Aspergillus nidulans*, deletion of LC8 prevented asexual reproduction and affected the localization and function of the dynein HC [30]. RNA-mediated interference of LC8 in *Caenorhabditis elegans* inhibited pronuclear migration in males as well as reduction or absence of the mitotic spindle in early embryonic cells [31]. Studies have shown that overexpression of LC8 may play a critical role in breast cancer [32,33]. LC8 has also been demonstrated to directly bind to a number of transcription factors and regulators of transcription. Its interaction with zinc finger transcription factor, ASCIZ, synergizes with Myc in Myc-dependent lymphomas to promote tumorigenicity [34]. Its interaction with 53BP1 leads to nuclear accumulation of p53 upon DNA damage [35]. LC8 binds to class I transcription factor A and is required for RNA Polymerase 1 activity in *Trypanosoma brucei* [36]. In addition, LC8 has been shown to play a critical role in cell stress [37] and autophagy [38,39]. For instance, it has been implicated in regulating neuronal nitric oxide synthase (nNOS) activity [40]; however, recent studies have suggested an alternative explanation [41].

Finally, LC8 plays important roles in a number of viruses and their infectivity. It has been implicated in the early transcription and localization of the rabies virus [32,42]. It has been demonstrated to bind to the Ebola virus protein VP35, regulating RNA synthesis [43]. It has been demonstrated to bind the integrase encoded by human immunodeficiency virus type 1 (HIV-1), affecting uncoating, and is reverse transcription of the virus [44].

TcTex-1 (DYNLT1) was originally identified as part of the t-complex and its mutation causes male sterility [45,46]. Unlike LC8, it is not as well conserved across eukaryotes and is absent in yeast and plants. Viable *Drosophila* were obtained upon the deletion of TcTex-1 suggesting it is not essential for development; however, males missing the TcTex-1 gene were completely sterile [47]. Additionally, cytoplasmic dynein failed to localize to the nuclear membrane in early spermatids [47]. In addition to binding to the dynein IC, TcTex-1 has been observed to bind to membrane-associated receptors. Most recently, TcTex-1 was demonstrated to bind to Activin receptor IIB [48], between residues 428 and 512, and reduce its activity. It has also been described to bind PTHR receptor [49], CD155 rabies receptor [50], and the voltage-dependent anion channel 1 (VDAC1) [51]. TcTex-1 has been demonstrated to bind guanine nucleotide exchange factor, Lfc, with moderate affinity [52]. Other cellular targets include the small GTPase RagA [52], Doc2α [53], and CD5 [54]. Likewise, TcTex-1 has shown to bind to NUP98, NIP153, and NUP62, and colocalize with the nuclear pore complex [55]. Finally, depletion of TcTex-1 (and LC8) was recently shown to play an important role in ER-to-Golgi transport [56].

Like LC8, TcTex-1 is also frequently observed to play critical roles in viral pathogenesis. For instance, depletion of TcTex-1 by siRNA affected the pericentriolar targeting of Mason-Pfizer monkey virus [57]. TcTex-1 was also identified in a two-hybrid screen to bind to the human papillomavirus (HPV) protein, HPV16 L2 [58]. Depletion of TcTex-1 by siRNA blocked infection of the cell by the virus [58].

LC7, also known as DYNLRB, roadblock, and km23, is the least studied of the dynein LCs. Like LC8 and TcTex-1, it was found to bind to both axonemal and cytoplasmic dynein [24]. There are two isoforms that are expressed in a tissue-dependent manner [24]. Studies aimed at identifying proteins that regulate axonal transport in *Drosophila* identified a 193-bp deletion in LC7 in EMS-treated flies. Axons isolated from these flies showed axon swelling, the accumulation of synaptotagmin and choline acetyltransferase, and nerve degeneration [24]. LC7 has also been implicated in cancer. In patients with hepatocellular carcinoma the LC7 isoform, DYNLRB1, is upregulated, while DYNLRB2 is suppressed [59]. Mutations in LC7 have also been reported at a high frequency in tissue samples isolated from patients with ovarian cancer [60]; however, an independent study did not confirm these initial findings [61].

3.3 Structure of the apo light chains

Since dynein participates in multiple processes and disruption of its function either through mutation or suppression of its subunits produces pleiotropic effects, structural investigations were initiated to more fully understand the function of the LCs in terms of target recognition as well as the regulation of LC function. The solution structure of LC8 was first published in 1998 [62] and was subsequently shown by NMR to interact with a 17 residue fragment of nNOS [63]. The crystal structure showed that LC8 forms a symmetric homodimer with a core consisting of two β-sheets, each flanked by two antiparallel α-helices (Fig. 3.1). Each β-sheet is formed by five antiparallel β-strands: β1, β4, β5, β2, and β3′. Strand β3′ originates from the other monomer and thus there are two swapped β-strands in the dimer [64]. In addition to the interactions arising from the swapped β-strands, there are a number of hydrogen bonds and hydrophobic interactions at the dimer interface. Of interest are His55 and Ser88. At low pH, His55 becomes protonated, placing two charges adjacent to each other and creating a stable monomer [65]. Similarly, Ser88 was recently proposed to be phosphorylated by Pak1 [33] placing charges adjacent to each other and affording in a stable monomer [66–68]. While the stability of the LC8 monomer provides insight to the molecular mechanism of its folding and assembly, it is not clear whether a monomeric form is physiologically relevant (see below).

Like LC8, TcTex-1 was also reported to bind to multiple proteins (Table 3.2), but these binding proteins are distinct from those identified for LC8. In addition, there is no sequence similarity between LC8 and TcTex-1. On the other hand, both LC8 and TcTex-1 showed the same order of predicted secondary elements (e.g., β1-β2-α2-α3-β3-β4-β5). NMR studies confirmed the secondary structure arrangement and suggested that TcTex-1 and LC8 shared a common fold [69]. Crystallographic [70] and NMR [71] investigations showed that TcTex-1 is structurally homologous to LC8 including the dimer swapped strand β3 (Fig. 3.1). Superposition of the LC8 and TcTex-1 structures indicates that TcTex-1 is extended along the twofold axis (e.g., the α-helices and β-strands are longer). Consistent with this observation, each TcTex-1 protomer buries ~1500 Å^2 of surface area, compared to ~750 Å^2 in LC8 [70]. In contrast to LC8, the symmetric TcTex-1 homodimer does not contain histidines at the dimer interface and its dimerization is not pH sensitive [70,72].

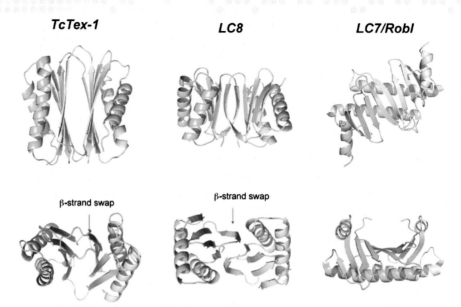

TcTex-1 **LC8** **LC7/Robl**

β-strand swap   β-strand swap

Figure 3.1 Ribbon diagrams of each member of the dynein LCs. All three LCs are homodimeric. One protomer is colored pale green and the other is colored pale cyan. TcTex-1 and LC8 are structural homologues, though they share very little in terms of sequence similarity. In both cases, TcTex-1 and LC8 form homodimers through a swapped strand β3. The fold of LC7 is clearly different than TcTex-1 and LC8. The second α-helix and third β-strand of each protomer makes extensive interactions to the same elements of the other protomer.

The third class of dynein LCs, LC7/Robl, shows no sequence similarity to either LC8 or TcTex-1 and has a limited number of known binding partners (dynein ICs, TGFβ, and Rab6). In addition, secondary structure predictions indicated that its arrangement of secondary structure elements was significantly different from LC8 and TcTex-1 (e.g., α1-β2-α2-β3-β4-β5). NMR [73] and diffraction studies [74] proved in fact that the LC7 fold is distinctly different than TcTex-1 and LC8 (Fig. 3.1). It too is a symmetric homodimer with extensive contacts across the dimer interface. Specifically, the 10-strand β-sheet of the dimer is formed by antiparallel strands β2, β1, β5, β4, and β3 of one monomer connected to the strand β3′ of the other monomer via a twofold axis. This extended β-sheet wraps around a long, antiparallel helical bundle formed by helix α2 of each monomer. Helices α1 and α1′ are located on the other side of the β-sheet and run perpendicular to the coiled-coil, producing a "saddlelike shape." LC7 also shows structural similarity to MglB from *Thermus thermophilus*, a protein proposed modulate the Ras/Rab/Rho GTPase [73,75].

3.4 Structure of liganded light chains

Both LC8 and TcTex-1 bind to numerous targets whereas the number of binding partners of LC7 is far more limited (see Tables 3.1 and 3.2). Reported LC8-binding partners include cytoplasmic partners such as dynein IC [76], myosin Va [77], Pak1 [33], GKAP [78], and 53BP1 [35]. Likewise, TcTex-1 binds to a large number of cytoplasmic proteins including membrane-associated proteins. These include dynein IC [22], Fyn and Trk kinases [11], bone morphogenetic receptor type II [79], VDAC [80], and rhodopsin [81]. Both LC8 and TcTex-1 have been shown to bind to transcription factors as well, including TRPS1 [82], Swallow [83], and SATB1 [84]. In addition, a large number of viruses have been reported to bind to LC8 and TcTex-1. The expanding list includes Ebola virus VP38 [85], lyssavirus [42], African swine fever [86], herpes VP26 caspid protein [87], and HPV type 16 L2 capsid protein [88]. While LC8 and TcTex-1 have been shown to bind to a large number of proteins, LC7 is much more restricted. Beyond the dynein IC [24], the only other proteins reported to bind to LC7 are TGFβ [89] and Rab6 [90].

There are currently nine distinct X-ray structures of LC8 bound to unique ligands: dynein IC peptide [16], NOS peptide [64], Swallow peptide [91], Pak1 peptide [67], Nup159 [92], Ana2 Centriole Duplication Factor peptide [93], two Chica peptides [94], phosphorylated and nonphosphorylated Nek9 peptide [95], and an "evolved" dimeric IC sequence [96]. In addition, the binding site of several peptides derived from LC8-binding proteins has been identified through NMR perturbation methods. Superposition of the X-ray structures shows that the LC8-binding peptide forms an edge on β-strand to the swapped β3-strand of LC8 (Fig. 3.2). In addition, a large number of the LC8-binding peptides encode a glutamine flanked by hydrophobic residues (Fig. 3.2D). The structures of the dynein IC, nNOS, Swallow, Nup159, and the Ana2 peptides show that the side chain amine group of this glutamine caps helix α2 of LC8 through a backbone hydrogen bond to Lys36. The flanking hydrophobic groups are buried in pockets formed by strands β2, β4, and β5 of LC8. These structures indicate that these two pockets can accommodate small hydrophobic side chains (e.g., Thr, Val, Ile), sequences from reported LC8-binding proteins also suggest that these hydrophobic pockets can also accommodate Ser, Cys, and Leu. We observe that reported LC8-binding sequence in TRPS1 would place an Asp at the second (C-terminal) pocket. The effect of this substitution remains to be characterized.

We also observe that a small number of reported LC8-binding sequences encode a different residue at the predominantly conserved glutamine found in dynein IC, nNOS, Swallow, etc. This includes myosin Va, Pak1, and GrinL1A. To test whether these "noncanonical" ligands could bind to distinct regions on

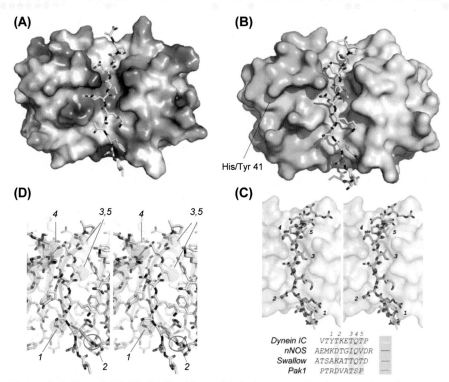

(A)

(B)

His/Tyr 41

(D)

(C)

	1 2 3 4 5	
Dynein IC	VTYTKETQTP	
nNOS	AEMKDTGIQVDR	—
Swallow	ATSAKATTQTD	—
Pak1	PTRDVATSP	

Figure 3.2 **Liganded LC8.** (A) The IC peptide is shown on the electrostatic surface of LC8 (contoured at ±5 kT [199]). (B) *A stereo view of the interaction between the IC ligand and LC8.* Important interactions are highlighted and denoted by numbers from the N-terminus to the C-terminus of the peptide. Thr67 of LC8 is highly conserved and makes a critical salt bridge to a number of LC8-binding proteins (position 1). All cytoplasmic dynein ICs from yeast to humans encode a tyrosine (position 2). The hydrophobic pockets that intercalate the hydrophobic groups surrounding the highly conserved glutamine are highlighted in green (positions 3 and 5). Finally, the highly conserved glutamine makes a hydrogen bond to Lys36 and caps the N-terminus of helix α2. (C) Superposition of liganded LC8, determined by diffraction methods, is shown in stereo. nNOS (green carbons), Swallow (cyan carbons), and Pak1 (wheat carbons) are superimposed on the dynein IC (yellow carbons). Note that the backbone of each target is nearly indistinguishable. (D) The sequence difference between each isoform is shown on the surface of each LC. Surfaces colored pale cyan or pale green are identical. Surfaces colored yellow indicate similar residues.

LC8 (i.e., the flanking α-helices) and constitute a unique class of dynein cargo, the point mutation, Lys36→Pro, was created in LC8 to block the capping interaction between the target peptide and helix α2 backbone. Binding studies confirmed that the dynein IC failed to interact with the mutant LC8 whereas the Pak1 peptide bound to the mutant at a similar level [67]. The X-ray structure of this "noncanonical" peptide bound to LC8 indicated that the Pak1 peptide binds to LC8 in the same manner as the other LC8-binding peptide. In this structure, the highly conserved glutamine is replaced by a serine that could potentially cap helix α2 at Lys36.

The Pak1-LC8 structure did draw attention to a second important interaction. Namely, an aspartate that is located i–4 residues from the predominantly conserved glutamine that is present in Pak1 as well as many other LC8-binding peptides including Myosin Va. The aspartate side chain interacts with hydroxyl group of Thr67 located in the swapped strand β3 of LC8. Mutation of Thr67 to Ala in LC8 significantly reduced the binding affinity of the Pak1 peptide but not of the dynein IC [67]. Lastly, there appears to be some preference for a lysine/arginine at the i–3 position (underlined in Table 3.1). Taken together, these data suggest a potential hierarchy of LC8-binding peptides based on affinity.

Finally, we note that titration experiments of ^{15}N labeled LC8 followed by NMR using unlabeled myosin Va or GrinL1A peptides found that both of these peptides bind to the same groove as the other peptide ligands. Myosin Va not only encodes a Met at the conserved glutamine position but also encodes an aspartate at the i–4 position. GrinL1A encodes a glycine at this position, but does not encode an aspartate. The GrinL1A peptide was modeled on to the structure of the nNOS peptide and places an Ile at the i–4 position, Gly at the i–3 position, and a Cys at the i+1 position. However, shifting the GrinL1A peptide by two residues to the C-terminus also seems to produce a plausible model. This would also place a glycine at the predominantly conserved glutamine, but would place a Thr at the i–4 position, an Arg at the i–3 position, as well as an Ile and Val at the hydrophobic positions. This shift would mimic the dynein IC, Swallow, and other structures. Additional studies including structural and biochemical investigations are required to determine the register and affinities of these targets.

In the case of TcTex-1, there are now three liganded structures. These include the dynein IC peptide bound to TcTex-1 and LC8 [16], a second complex of the dynein IC bound to LC8 and TcTex-1 using different species homologues [97], and an NMR structure Activin Receptor IIB peptide fused the C-terminus of TcTex-1 [48]. Not surprisingly, the IC fragment binds to TcTex-1 in a similar manner as the IC binds to LC8. Specifically, the IC also forms an edge on interaction with the swapped strand β3. As TcTex-1 is slightly extended compared to LC8, there are additional backbone hydrogen bonds between the IC peptide and TcTex-1. The IC peptide also buries hydrophobic residues, V110 and F112, in TcTex-1 at similar positions found in LC8. However, the helix α2 in TcTex-1 is extended compared to LC8 and self-capped. Thus, there is no obvious selection pressure for a glutamine or a different "capping" residue at this position as observed for LC8. This observation may partially account for the lack of overlap between TcTex-1 and LC8 reported targets. While an arginine-rich sequence was originally proposed to be the "TcTex-1" binding motif [69], the IC peptide used in the structural studies included these N-terminal arginines. However, none of these arginines were observed in the electron density maps with the IC peptide, suggesting that they are not required for TcTex-1 binding. These observations are further confirmed upon superposition of the structure Activin Receptor IIB peptide—TcTex-1 fusion [48]. The fused peptide lies edge on to the swapped strand β3, makes similar hydrophobic interactions, and there are no specific interactions with arginines. However, the superposition of this fused construct shows some distortion of the flanking alpha helices as well as fewer interactions between the N-terminus of the fused ligand and the swapped strand β3, suggesting the possibility of strain due to the five residue Gly–Ser linker. Consequently, it still remains difficult to identify a TcTex-1 "canonical" sequence from the reported TcTex-1 binding partners.

Without a clear signature, we have mutated a number of residues in the IC peptide and used size exclusion chromatography to characterize the effect of these point mutations (data not published, A Dawn and JC Williams). We observe Leu112, Met114, and Phe122 mutated individually to alanine effectively abrogated the interaction (numbering from rat IC2, isoform C). Mutation of the other residues to alanine, on the other hand, did not strongly affect the IC–TcTex-1 interaction. These preliminary binding studies suggest that the specificity of TcTex-1 for its peptide targets require hydrophobic residues at the N- and C-terminus spanned by eight residues, and further explain the observation that TcTex-1 and LC8 bind distinct and non-overlapping targets. Applying these observations to reported TcTex-1 targets (e.g., rhodopsin [81], PTHRa [98], CD155 [99], etc.) to determine the TcTex-1 binding site remains challenging. In one case, there is a reasonable agreement between the reported TcTex-1 binding peptide in Doc2A [53] and the structure and mutagenesis data of the IC–TcTex-1 interaction (Fig. 3.3). Clearly, more structural and binding assays are required to define a TcTex-1 binding motif.

Lastly, the structure of a dynein IC fragment bound to LC7 was solved [74]. The peptide fragment spans residues 182–219 of the IC (rat IC2, isoform C) and forms a short α-helix, followed by a two residue break and a long α-helix, which wraps along the β-sheet and helix α1. The first α-helix from IC places the Ile190 side chain into shallow pocket formed by strands β1, β2, and β5. The second α-helix from the IC–LC7 interface predominantly consists of hydrophobic residues from the IC peptide fitting into a deep groove formed by the helix α1 and strands β3, β4, and β5 of LC7. In addition, the end of the IC peptide encodes a glutamate that interacts with the completely conserved Arg located in strand β3. Unlike LC8 or TcTex-1 where the target peptides bind in a parallel fashion, the IC ligand binds to LC7 in an antiparallel manner.

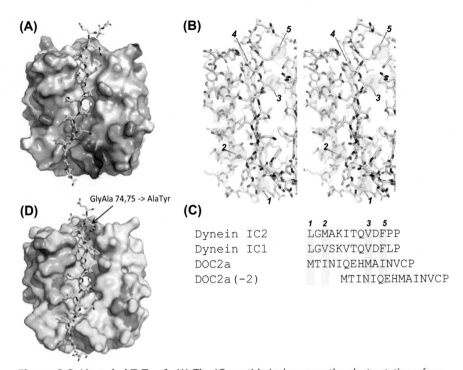

(A)

(B)

(D) GlyAla 74,75 -> AlaTyr

(C)

```
             1  2        3  5
Dynein IC2   LGMAKITQVDFPP
Dynein IC1   LGVSKVTQVDFLP
DOC2a        MTINIQEHMAINVCP
DOC2a(-2)      MTINIQEHMAINVCP
```

Figure 3.3 Liganded TcTex-1. (A) The IC peptide is shown on the electrostatic surface of TcTex-1. Note that the charge distribution differs from LC8, which may also affect ligand selection (contoured at ±5 kT [199]). (B) *Stereo view of the IC bound to TcTex-1—* Residues found to be required for binding are shadowed in green. Mutation of Leu112 (position 1), Met114 (position 2), and Phe122 (position 5) to alanine significantly reduces the IC–TcTex-1 interaction. Positions 3 and 5, respectively, occupy positions analogous to the hydrophobic positions on LC8. Position 4, Asp121 occupies a position similar to the conserved glutamine found in the LC ligands, but does not cap the helix α2 of TcTex-1. Its mutation to alanine does not strongly affect the IC binding. The requirement of hydrophobic residues, leucine and methionine, at positions 1 and 2, likely accounts for the difference in ligand selection between TcTex-1 and LC8. (C) *Potential TcTex-1 ligand—*based on the alanine mutagenesis, the reported binding region of Doc2 aligns well to both cytoplasmic IC isoforms. An alternative alignment requires a shift of two residues towards the N-terminus. (D) *Isoform differences between TcTex-1 and RP3.* Unlike the LC8 isoforms, there are number of sequences differences between TcTex-1, and its family member, RP3. Surfaces colored orange indicate no similarity. Sequence differences in the ligand binding site are marked by the *arrow*.

3.5 LC8 and TcTex-1 promiscuity

These liganded atomic structures of the LCs provide a plausible explanation for the promiscuity of LC8 and TcTex-1 and the limited number of LC7-binding proteins. The interface between the LC8 and TcTex-1 and their respective peptide ligands consists primarily of backbone hydrogen

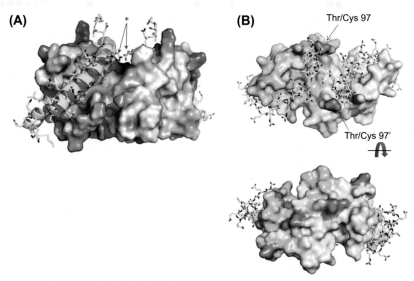

Figure 3.4 Liganded LC7. (A) The IC peptide (yellow) is shown on the electrostatic surface of LC7. Note that the IC–LC7 interface is predominantly through side chain interactions and is distinctly different than the IC and TcTex-1 or LC8 interfaces. (B) *Differences between LC7 isoforms*. There are several minor sequence differences lining the IC-binding site on LC7. Highlighted is residue 97, which is either a threonine or a cysteine. The side chain does not contact the IC peptide; however, oxidation of the sulfhydryl group could affect ligand binding. Otherwise, the remaining differences are located opposite surface of the IC-binding site.

bonds through the edge on strand interaction and the placement of several hydrophobic interactions (Fig. 3.5A and B). Neither backbone nor hydrophobic interactions are strongly sequence specific. In this sense, the interaction between LC8 and its targets and TcTex-1 and its targets resembles the mechanism by which MHC recognizes a large number of antigenic peptides [100]. This is in stark contrast to the LC7-IC interaction. In this case, the interface is dominated by side chain interactions between the IC peptide and LC7. While a number of the side chain interactions are hydrophobic (Fig. 3.5C), there are several conserved hydrogen bonds and salt bridges. Specifically, a mixed hydrogen bond/salt bridge network, formed by residues Glu252 and Asn253 in the IC and Arg71 and Gln93 in LC7, is present at the C-terminal end of the IC peptide (Fig. 3.5, D—*Drosophila* numbering). In addition, Glu247 of the IC (denoted by *) also forms a salt bridge to the backbone carbonyl of Arg10 in LC7. Taken together, these observations indicate that the LC7-binding site is far more selective than TcTex-1 or LC8. In addition, Blast searches using the LC7 peptide failed to identify additional LC7 targets. On the other hand, using dominant LC8-binding sequences from biopanning LC8 to identify new targets produced a large set of hits, some of them shown experimentally to bind to LC8 ([96] please see supplemental data for an extensive list).

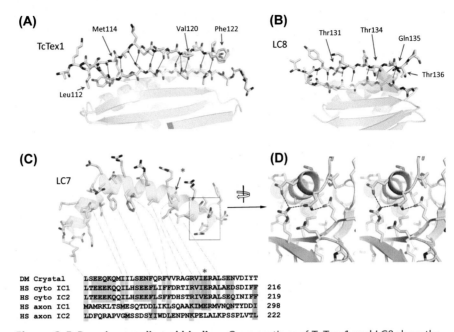

Figure 3.5 Promiscuous ligand binding. Cross-sections of TcTex-1 and LC8 show the IC (yellow) binds to the swapped strand β3 (cyan) of TcTex-1 (A) and LC8 (B) primarily through backbone hydrogen bonds (*black dashes*). Important side chain interactions are also highlighted. (C) The IC fragment that binds to LC7 encodes a large number of hydrophobic groups along one face of the α-helix. At the C-terminal end, the IC makes several side chain interactions (*boxed region*). An extended hydrogen bond/salt bridge network (*red dashes*) is shown in stereo (D). Sequence alignment of the human cytoplasmic and axonemal dynein ICs indicates that axonemal IC1 encodes similar residues that form the extended side chain network and encodes hydrophobic residues at many of the same sites as cytoplasmic IC. Thus, axonemal IC1 likely binds LC7. Axonemal IC2, however, differs significantly at these sites.

3.6 Light chain isoforms

In mammals, there are two isoforms of each LC, each located on a different chromosome [14]. For TcTex-1, LC8, and LC7, the isoforms are 54% identical and 76% similar, 93% identical and 98% similar, and 77% identical and 87% similar, respectively. Mapping these differences to the surface of each structure suggests that the sequence differences in LC8 and LC7 should not affect IC binding. While phenotypic and biochemical differences between the LC7 isoforms have yet to be identified, residue 41 located on helix α2 has been demonstrated to differentially direct the LC8 isoforms to bind to dynein or myosin Va [101]. Moreover, BimL and Bmf, which encode nearly identical LC8-binding sequences, preferentially bind to one isoform over the other [101–103]. While detailed biochemical studies demonstrated that both LC8 isoforms bound to peptides derived from Bmf with equal affinity, in vivo assays demonstrated that the exogenous expression of the His41→Tyr LC8 mutant in HEK-293 cells redirected the first LC8 isoform (DYNLL1) to myosin. The exogenous expression of the complementary mutation in the second isoform encoding, Tyr41→His mutation, redirects this mutant isoform to dynein [101]. The origin of this differential localization remains unclear and should be reconfirmed in other cell lines; however, these findings suggest that additional cellular factors are required for this differential localization [101].

In addition, the first five residues of LC8 are disordered and not visible in the electron density maps; however, the second residue in this disordered region also differs between the two isoforms. Specifically, the first isoform (DYNLL1) encodes a cysteine and second isoform (DYNLL2) encodes a serine. The cysteine isoform of LC8 was shown to bind to IκBα [104] and inhibit its phosphorylation by IKK through a redox mechanism [105,106]. Trp14, a novel disulfide reductase, was shown to reduce DYNLL1 through disulfide trapping experiments [105,106]. Furthermore, the depletion of Trp14 or LC8 by siRNA was shown to interfere with TNFα-induced NF-κB activation. Collectively, these results suggest that binding activity of the DYNLL1 isoform is regulated, in part, by the redox state of the cell [107].

The differences in TcTex-1 and RP3 appear to be more significant and may affect ligand selection. While both isoforms bind to the dynein IC, TcTex-1, but not RP3, binds to the PTH receptor [98]. In addition, RP3, but not TcTex-1, was demonstrated to bind to and modulate the activity of the nuclear matrix-associated protein SATB1 [108]. This differential binding may be explained by substitutions at residues 74 and 75, where glycine and alanine in TcTex-1 are substituted by alanine and tyrosine in RP3. In addition, the flanking helices also show a number of differences. These could affect TcTex-1/RP3 distributions in a similar manner observed in the LC8 isoforms [101]. Finally, TcTex-1 and RP3 have been demonstrated to form heterodimers, although these heterodimers were not capable of binding to the dynein IC [109]. Currently no structural data are available for RP3, which may help to elucidate the isoform specific binding properties.

Lastly, while the differential expression of LC7 has been associated with hepatocellular carcinoma, there are no significant substitutions between the LC7 isoforms that should strongly affect IC binding. The second isoform of LC7, DYNLRB2, encodes a cysteine located at the C-terminus, residue 97. Oxidation of this cysteine could affect its interaction with the IC (Fig. 3.4). Many of the remaining sequence differences in LC7 map to the opposite surface, formed in part by the antiparallel α-helix bundle (Fig. 3.4), suggesting that this surface interacts with TGFβ or Rab6. Reversion of these surface residues to the other isoform followed by in vivo assays could provide insight into both cellular and isoform specific functions of LC7.

3.7 Mammalian dynein intermediate chains

There are six dynein IC homologues identified in the human genome IC [27]. Four including DNAI1 and DNAI2 are associated with axonemal dynein. The other two, DYNC1I1 (IC1) and DYNC1I2 (IC2), are associated with cytoplasmic dynein. Of the two cytoplasmic dynein ICs, there are two splice variants for each, both occurring at essentially equivalent positions. Of these, the IC2 isoform C (the shortest isoform) is ubiquitously expressed. The remaining isoforms are primarily expressed in brain tissues [110]. All IC homologues encode a C-terminal WD domain (amino acids ~220–580; ~2/3 of molecule located at the C-terminus) that has been demonstrated to bind to the dynein HC [111]. In addition, secondary structure prediction algorithms predict short α-helices immediately before and after the WD domain of each IC [112] (Fig. 3.6).

The N-termini of the ICs, however, differ significantly between the cytoplasmic and axonemal IC isoforms. Specifically, the first 42 residues of both cytoplasmic ICs are highly charged (17 lysines and arginines and 12 aspartates and glutamates in the first 44 residues of rat IC). This region is also strongly predicted to form a coiled-coil and has been demonstrated by multiple groups to interact with dynactin, a critical multisubunit complex that is essential for many of the functions of cytoplasmic dynein (Chapter 17 (vol. 1 of this book)) [8,113–115]. Immediately following this stretch is a disordered region that is rich in proline and serine residues. Several of these serines have been shown to be phosphorylated in a cell cycle–dependent manner [116,117] (Chapters 18 and 19 (vol. 1 of this book)). In addition, the splice sites for both cytoplasmic ICs in mammals are located in this region. Immediately following this serine and proline-rich stretch is the binding site of the LCs, TcTex-1, and LC8 (residues 112 to 138—numbering based on Rat IC2 isoform C). The LC7-binding site (residues 182–218) follows a short acidic stretch. Immediately following the LC7 site is the WD domain.

The N-termini of the axonemal IC isoforms not only differ significantly from cytoplasmic dynein ICs, they also differ significantly between themselves. Using Jpred3 to predict secondary structure [112] and RONN to predict disorder [118], the first 90 residues of axonemal IC2 (e.g., DNAI2) encode a significant number of β-strands and show little disorder, suggesting it is a folded domain. Similar predictions of the N-terminus of axonemal IC1 (e.g., DNAI1) show that the N-terminus is largely disordered, with the exception of a small region, residues 55–130, that contains a mix of predicted α-helices and β-strands. As noted earlier, the extreme N-terminus of cytoplasmic dynein IC specifically binds to dynactin and is critical to many of cytoplasmic dynein functions. By analogy, differences in secondary structure of the N-termini of each axonemal IC isoform suggest that each interacts with a different macromolecule and likely act in different processes.

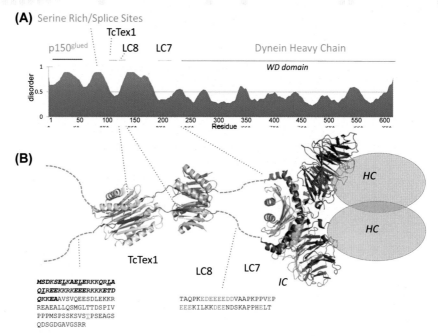

(A) Serine Rich/Splice Sites

Figure 3.6 Model of cytoplasmic IC–LC complex. (A) The binding sites of the LCs, p150^{Glued}, and the HC domain are shown above the predicted disorder of cytoplasmic dynein IC2. The first 180 residues of the IC are predicted to be disordered (e.g., values above 0.5). Of interest, this region encodes the binding sites of the LCs, LC8 and TcTex-1, as well as p150^{Glued}. This region also contains the serine rich region, which has been shown to be phosphorylated in a cell dependent manner. (B) The structures of the IC (purple) bound to TcTex-1 (cyan) and LC8 (light blue) (2PG1.pdb) and the IC bound to LC7 (green) (3L9K.pdb) were combined with the WD domain of Rbbp4 (purple) (3GFC. pdb) to produce a spatial model of the LC–IC complex. The IC α-helix was manually superimposed on the N-terminal helix preceding the WD domain of Rbb4. The position of the TcTex-1 and LC8 with respect to LC7 and the WD domains is not known, but likely constrained through the dimer of dimers. The position of the HC with respect to the WD domains is based on peptide–WD domain complexes, but is also tentative. Also shown are the amino acid sequences of IC N-terminus and the region between the LC8- and LC7-binding sites. The predicted coiled-coil region and the p150^{Glued} binding site at the N-terminus are highlighted in *bold italics*. The phosphorylation site, T89, is highlighted in green. AUC, CD, and NMR studies indicate that the first 112 residues are disordered and do not dimerize. Consequently, these are depicted as random coils.

The precise position of the LC-binding sites within mammalian axonemal ICs remains undetermined. There are several sites within the axonemal ICs that could potentially bind to LC8. For instance, searching the human DNAI1 coding sequence for all glutamines immediately surrounded by threonine, valine, or isoleucine and within a disordered region produces several hits including 207-RDRECQT-213. Using similar criteria for human DNAI2, several potential sites were identified including 47-VDTGIQCS-54. On the other hand, without a clear canonical sequence it is difficult to identify a TcTex-1 binding site. Finally, using the secondary structure prediction and the cytoplasmic IC–LC7 structure (Fig. 3.5), it appears that DNAI1, but not DNAI2, encodes an LC7-binding site. Specifically, many of the hydrophobic residues in DNAI1 closely match the hydrophobic residues in the cytoplasmic IC sequences. The same region in DNAI2, however, encodes a number of proline residues that should destabilize the formation of an α-helix (Fig. 3.5). While there have been in roads made in understanding cytoplasmic dynein, there is a significant lack of structural data for axonemal dynein, especially concerning mammalian axonemal dynein.

3.8 Molecular model of the light–intermediate chain structure

With the structures of TcTex-1, LC8, and LC7 bound to their respective fragments of the dynein IC, it is possible to generate a plausible model of the cytoplasmic IC–LC complex. At the C-terminus, a number of the WD domains deposited in the protein database (PDB) encode an N-terminal and C-terminal helix, similar to what is predicted from the primary sequences of the dynein IC. A BLAST search [119] using the rat cytoplasmic IC2C sequence and the PDB identified the WD domain of histone-binding protein, Rbbp4 (3GFC.pdb), as a close structural homologue. It is 43% similar to the IC WD domain over residues 307–534. Moreover, the "sequence register" of the N-terminal helix of Rbbp4 with respect to the WD domain closely matches the sequence register of the IC α-helix that binds to LC7. Superposition of the N-terminal α-helix of Rbbp4 domain to each α-helix of the ICs that is bound to LC7 affords a model with minimal steric clashes and no overlap between the WD domains. The relative orientation of the two β-propeller domains also appears to be well positioned to bind to the dimeric dynein HC (Fig. 3.6), and is consistent with recent EM studies [120,121].

At the N-terminus of the cytoplasmic ICs, the crystal structure of the co-complex of TcTex-1, LC8, and the IC shows that ICs bind as extended strands along each LC-binding site in a parallel manner [16]. The remaining N-terminal region of the ICs, residues 1–111 and 138–214, are predicted to be disordered, and are highly, but oppositely charged. The isoelectric point of the IC fragment that binds to dynactin, residues 1–44, is 10.5, whereas the isoelectric point of the region between the LC8-binding site and the LC7-binding site, residues 138–182, is 3.9. This observation suggests that their potential interaction with other cellular components will be dominated by electrostatic interactions. While the IC–LC8–TcTex-1 complex also indicates that there is no contact between LC8 and TcTex-1 LCs, their position in the middle of a highly disordered region suggests that they restrict the conformation and/or radius of gyration of the IC, potentially enhancing the apparent affinity of potential targets binding to these disordered regions (e.g., p150^{glued}—see below).

3.9 Light chains and cargo

The observation that LC8 and TcTex-1 bind to a large number of proteins led to the hypothesis that they act as adapters to bind cargo to the dynein motor complex [15,122]. This hypothesis was partially bolstered by the observation that in mammals, there is essentially one cytoplasmic dynein [123]. Thus, it was hypothesized that different combinations of the promiscuous LCs could potentially expand the repertoire of cargo that required retrograde transport. As structural data became available, the validity of this model became suspect. Specifically, each characterized cargo peptide binds in precisely the same groove as the IC peptide (demonstrated for LC8). However, since the homodimeric LCs encode two binding surfaces, it is possible to envision a model where one interface binds to the IC and the other binds to the cargo peptide.

There are, however, compelling thermodynamic and kinetic arguments against this model. First, the cytoplasmic dynein motor complex is composed of homodimeric subunits. The dynein HC is homodimeric through its N-terminal domain [111], and this region acts as a scaffold for the dynein IC and LICs [124]. In turn, these chains are either dimeric in of themselves or dimeric through their association of the dynein HC. As a consequence, the dimeric ICs can act as a bivalent ligand for the bivalent LCs. Bivalent–bivalent interactions afford energy additivity and higher apparent affinities. Specifically, the free energy at each site can be added modulated by a "linkage" term [125,126]. Mathematically, $\Delta G_{Total} = \Delta G_1 + \Delta G_2 + \Delta G_{linker}$. While the contribution of the linker invariably counteracts energy additivity, in the case where it is negligible ($\Delta G_{linker} = 0$), the binding constants of each interaction are multiplicative (e.g., $-RT \ln (k_{total}) = -RT \ln (k_{total}) - RT \ln (k_{total})$, which simplifies to $k_{total} = k_1 k_2$). A powerful example of this principle, demonstrated by the Whitesides' group, shows that coupling vancomycin and D-Ala-D-Ala to trimeric scaffolds dramatically reduced the dissociation constant of the monomeric interaction from $\sim 10^{-6}$ M to $\sim 10^{-17}$ M [127], which far exceeds the dissociation constant of biotin and avidin. While geometric and entropic effects of the linker interfere with simple multiplication of the individual affinities, modest but significant gains in the apparent affinity are observed in most cases [125].

Enhanced affinity due to multivalency has been observed for LC8 and TcTex-1 [16,128]. First, SUPREX studies, a mass spectrometry method that follows hydrogen–deuterium exchange and extracts thermodynamic information, showed substantial gains in the stability of LC8 in the presence of an IC peptide encoding both LC8 and TcTex-1, prebound to TcTex-1. These original experiments have been independently verified [97]. The affinity for the LC8-binding peptide derived from the IC is $\sim 10\,\mu M$ for a single site on LC8. The affinity of an "optimal" dimer of this peptide for LC8 would be $100\,pM$ (e.g., 10^{-5} M $\times 10^{-5}$ M). A peptide that encodes tandem LC8-binding sites afforded 1000-fold increase in the apparent affinity for LC8 compared to a peptide with a single LC8-binding site [97]. Likewise, the artificial dimerized IC peptide was shown by ITC to significantly enhance the binding affinity to LC8 compared to the monomeric analogue. An upper limit for the dissociation constant, K_D, was placed at $0.003\,\mu M$ for a dimeric GSGRGTQTE molecule compared to $K_D = 1.64\,\mu M$ for Ac-GRGTQTE peptide. Of note, the difference for this pair is predominantly reflected in a reduced entropic value for the dimeric moiety, consistent the additive effects of a dimeric ligand binding to a dimeric substrate.

Second, molecular traps developed to determine the role of LC8 and TcTex-1 on dynein in vivo were developed based on this additivity principle [128]. Specifically, peptide fragments based on the LC8-binding site or TcTex-1 binding site of the dynein IC were fused to FKBP. Dimerization of FKBP using the small molecule AP20187 [129] dramatically enhanced the weak affinity of the monomeric LC-binding peptides, producing high affinity, bivalent "receptors" or traps for the LCs [128]. The expression of these traps in vivo did not produce significant defects associated with acute dynein inhibition (which would be expected for the bridging model). The addition of AP20187 to cells expressing these traps, however, led to rapid endosome and lysosome dispersion presumably by sequestering the LCs from endogenous dynein. These observations reinforce the proposal that bivalent–bivalent interactions out compete an uncoupled, trivalent interaction.

Lastly, nearly all characterized proteins that bind to the LCs, LC8 or TcTex-1, are dimeric (see Tables 3.1 and 3.2) [16]. Thus, the interaction of these LC-binding proteins to their respective LCs (LC8 or TcTex-1) is also likely to be exclusive, or dynein independent. In fact, replacing the LC8-binding region in Pak1 with FKBP and using AP20187 to dimerize Pak1 permitted rapid nuclear localization of Pak1 [130]. In other words, dimerization of Pak1, mediated by LC8, is required for its nuclear import. In this sense, LC8 has recently been described as a "dimerization hub" [131]. However, this term must be used carefully. The LC8-binding partner must be dimeric or encode a region that will dimerize after sufficiently high concentrations are achieved to saturate both LC8-binding sites [132], otherwise energy additivity is lost. Thus, it is more accurate to consider the LCs as allosteric effectors. Regardless, these data strongly argue against the original cargo or bridging models.

3.10 Posttranslational modifications

Posttranslational modification of the LCs and/or ICs by phosphorylation, ubiquitination, etc. is a potential means to regulate their interaction with targeted proteins. While sumoylation, acetylation, and other modifications have not been reported for the LCs or ICs, phosphorylation has been reported for each LCs and the ICs.

Interest in the phosphorylation of LC8 stems from a recent report indicating that Pak1 binds to and phosphorylates Ser88 and plays an important role in breast cancer [33]. As noted earlier, LC8 bearing the phosphomimic, Ser88→Glu or Ser88→Asp, is monomeric and fails to bind IC peptides encoding the LC8-binding site. However, careful examination of the LC8 structure indicates that Ser88, part of strand β5, is buried at the dimer interface. As such, it is sterically occluded from the kinase. Moreover, analytical ultracentrifugation [67] and NMR [133] studies indicate that LC8 is a relatively tight dimer ($K_D < 50\,nM$). While wild-type LC8 could not be phosphorylated by Pak1 in in vitro assays using purified reagents, phosphorylation was observed using His-tagged LC8 encoding a thrombin cleavage site [67]. The thrombin cleavage site, coincidentally, closely matches the canonical Pak1 phosphorylation consensus sequence. Thus, the inclusion of tags for immunochemistry and purification can lead to false positives and may account for the original observation that LC8 can be phosphorylated.

TcTex-1 and LC7 have also been proposed to be phosphorylated in vivo [89,108,134]. In each case these sites are remote from the peptide-binding interface. Moreover, a number of these reported sites are located in ordered secondary structure elements, analogous to Ser88 in LC8. Recently, it has been shown that the majority of well-characterized phosphorylation sites are located within intrinsically disordered regions [135]. In fact, no phosphorylation sites were identified when applying a disorder-enhanced phosphorylation site prediction algorithm to each LC sequence [135]. While these sites may very well be *bona fide* kinase targets, they remain tenuous without further investigation.

Phosphorylation of cytoplasmic dynein ICs in a cell cycle–dependent manner has also been reported. In these studies, phosphorylation of Ser84 was found to block p150[Glued] binding using blot overlay assays [117]. However, a separate study demonstrated that phosphorylation of Ser84 does not affect the dynein–dynactin interaction [136]. This latter study is also consistent with our biophysical studies indicating that a fragment containing residues 1–44 of the IC is sufficient to bind to p150[Glued] [137]. Rather than physically blocking dynactin binding it is possible that phosphorylation of this site creates a binding site for a separate protein and this interaction effectively decouples the IC from p150[Glued]. Recent studies support this notion [116,138,139].

Of note, a number of sites at the N-terminus of both cytoplasmic ICs were identified using the same disorder-enhanced phosphorylation site prediction algorithm, including Ser84 and Thr89 [135].

3.11 The role of LC8 and TcTex-1 on dynein

In the absence of a "bridging model," the role of the LCs on dynein function remains unresolved; however, there are several alternative hypotheses [109]. First, it is possible that the LCs are required for the assembly of the dynein complex. As shown in Fig. 3.6, the use of different LCs would ensure the register of the IC, enforce a parallel architecture, and lead to the efficient binding of the dimeric HC through the WD domains. This is supported in part by the observation that treatment of purified cytoplasmic dynein with potassium iodide produces two distinct fractions, one containing the dynein HC and LICs and the other containing the ICs and LCs [21]. Moreover, in *Chlamydomonas*, the LCs have been observed to bind to axonemal ICs [140,141]. Collectively, these data support the notion that the LCs are necessary for the assembly of the dynein complex.

A second potential role for the LCs is to modulate the phosphorylation of the sequences encoded by LC-binding proteins. For instance, Thr56 located within the LC8-binding site on BimL/EL, *DKSTQTSP*, has been shown to be phosphorylated by JNK kinase [142]. Phosphorylation of this site was demonstrated to abrogate LC8 binding and increases apoptotic activity [142]. Likewise, Cdk5, Erk2, and Cdc2 have been shown to phosphorylate Pak1 at Thr212, the residue that immediately precedes the LC8-binding site on Pak1 (212-TP*TRDVTSPI*). While the phosphomimic of the Pak1-LC8–binding site, T212E, still bound LC8 in vitro [67], it is possible that LC8 occupancy of these targets could sterically occlude the site to kinase activity. In fact, an LC8-binding site within Nek9 was recently identified and within this sequence is an autophosphorylation site, Ser944. It was demonstrated through structural and biophysical methods that phosphorylation of Ser944 reduced the monomeric affinity by sixfold [95]. In the context of a bivalent–bivalent interaction, phosphorylation of Ser944 in both protomers would lead to a much greater reduction in affinity of the complex. In this sense, LC8 could regulate the activity of these molecules; however, experimental evidence for this mechanism is still required. By analogy, TcTex-1 could operate through a similar mechanism on its target proteins.

Lastly, LC occupancy could regulate the function of the dynein ICs. This hypothesis stems from several observations. First, the N-terminal region of the dynein IC is intrinsically disordered [137,143]. While the first ~42 residues are predicted to form a coiled-coil, CD, Analytical UltraCentrifugation studies, and NMR studies clearly indicate that it is unfolded [137]. Disordered regions in other systems have been shown recently to "fold" on their target [144]. Such interactions can be highly specific as the disordered region makes multiple contacts over an extended region of the target. A disordered to coiled-coil folding transition has been observed for the IC in the presence of dimeric p150[Glued], the principle dynactin [137]. In addition, these disordered/ordered interactions are frequently of low affinity due to an entropic penalty of structuring an unfolded polypeptide reference. However, both the dynein IC and dynactin p150[Glued] are dimeric under physiological conditions and the loss in binding affinity due to entropy should be compensated in part by bivalency. More importantly, the binding site for the LCs, TcTex-1 and LC8, are located in the middle of the disordered IC N-terminus, residues 112–138. By binding the IC at this point, the

LCs can restrict the "radius of gyration" of the N-terminal 44 residues, which in turn, can enhance the apparent affinity of the interaction [137]. The dynein LC traps were initially designed to test this hypothesis [128]. Specifically, if LC8 or TcTex-1 act as allosteric regulators of dynein function, sequestering the LCs from *preassembled* dynein (e.g., not possible with RNAi) should produce defects commonly associated with dynein interference (e.g., microinjection techniques). Such effects were observed. As noted, chemically induced dimerization of the LC8 or the TcTex-1 traps expressed in Cos1 or Cos7 cells produced rapid endosome and lysosome dispersion [128], consistent with the reduction of the IC–p150^{Glued} interaction. Furthermore, using a photocleavable form chemical dimerizer, AP20187, we demonstrated that the dispersed organelles returned to their steady state distribution upon photocleavage of the trap [145].

The regulatory role of the LCs was further demonstrated using yeast as a model system [146]. First, a second LC8-binding site on the dynein IC was identified adjacent to the canonical LC8 site, offering an explanation for the absence of a TcTex-1 homologue in yeast. Second, mutation that abrogated LC8 binding at the first or second LC8-binding site within the IC demonstrated affected spindle positioning. Mutation of both sites in the IC was effectively equivalent to the deletion of the entire IC. Moreover, imaging studies of these mutants indicated that LC8 interaction plays a role in the recruitment of the dynactin complex, again consistent with a regulatory role [146].

Finally, it was observed that the affinity for LC7 was not significantly enhanced using an IC preassembled with TcTex-1 and LC8 compared to the same IC construct without the LCs [74]. The threefold difference suggests the slight increase in affinity is due to local concentration effects as opposed to energy additivity. In other words, TcTex-1 and LC8 appear to act independently of LC7. This observation also points to a regulatory role for LC8 and TcTex-1 and an assembly role for LC7.

Collectively, these examples are consistent with the hypothesis that LC occupancy of the cytoplasmic IC affects p150^{Glued} and potentially nudE binding. These investigations also suggest a different model for retrograde transport of critical macromolecules that addresses the lack of multiple members in the cytoplasmic dynein family. Specifically, it is possible to locally sort macromolecules into vesicles, load these vesicles onto dynactin or other adapter proteins like Bicd2, and use dynein to carry these vesicles along MTs. This mitigates conceptual issues of using one cytoplasmic dynein to transport one molecule (e.g., TrkA bridges to the IC through TcTex-1). In this sense, this model is very similar to conventional cargo transport where dynein is the tractor and dynactin is the trailer. Evidence supporting such a model is beginning to appear [147–149]. In fact, a "hitch-hiking" model, where macromolecules and organelles bind to targets on the surface of endosomes and subsequently transported by dynein, has been put forth to explain cargo diversity [150].

3.12 Summary

It has been an extremely exciting time over the past few years in terms of the structural biology of dynein. These studies have cleared up many questions and significantly aided in the design of biophysical and cell-based assays. Structural investigations of the LCs, initially performed to understand their role in the context of the cytoplasmic dynein, have met their mark. These investigations provide compelling evidence that the LCs do not bridge cargo to cytoplasmic dynein, but rather play important roles in assembly in the dynein complex (e.g., LC7) and potentially in the regulation of the N-terminus of cytoplasmic dynein IC (e.g., LC8 and TcTex-1).

References

[1] A.P. Carter, A.G. Diamant, L. Urnavicius, How dynein and dynactin transport cargos: a structural perspective, Curr. Opin. Struct. Biol. 37 (2016) 62–70.

[2] R.B. Vallee, J.C. Williams, D. Varma, L.E. Barnhart, Dynein: an ancient motor protein involved in multiple modes of transport, J. Neurobiol. 58 (2004) 189–200.

[3] S.M. King, Axonemal dynein arms, Cold Spring Harb. Perspect. Biol. (August 2016). http://dx.doi.org/10.1101/cshperspect.a028100.

[4] A.R. Kini, C.A. Collins, Modulation of cytoplasmic dynein ATPase activity by the accessory subunits, Cell Motil. Cytoskeleton 48 (2001) 52–60.

[5] R.J. McKenney, M. Vershinin, A. Kunwar, R.B. Vallee, S.P. Gross, LIS1 and NudE induce a persistent dynein force-producing state, Cell 141 (2010) 304–314.

[6] S.M. King, The dynein microtubule motor, Biochim. Biophys. Acta 1496 (2000) 60–75.

[7] K.T. Vaughan, E.L. Holzbaur, R.B. Vallee, Subcellular targeting of the retrograde motor cytoplasmic dynein, Biochem. Soc. Trans. 23 (1995) 50–54.

[8] S. Karki, E.L. Holzbaur, Affinity chromatography demonstrates a direct binding between cytoplasmic dynein and the dynactin complex, J. Biol. Chem. 270 (1995) 28806–28811.

[9] A. Purohit, S.H. Tynan, R. Vallee, S.J. Doxsey, Direct interaction of pericentrin with cytoplasmic dynein light intermediate chain contributes to mitotic spindle organization, J. Cell Biol. 147 (1999) 481–492.

[10] I. Rodriguez-Crespo, B. Yelamos, F. Roncal, J.P. Albar, P.R. Ortiz de Montellano, F. Gavilanes, Identification of novel cellular proteins that bind to the LC8 dynein light chain using a pepscan technique, FEBS Lett. 503 (2001) 135–141.

[11] H. Yano, F.S. Lee, H. Kong, J. Chuang, J. Arevalo, P. Perez, et al., Association of Trk neurotrophin receptors with components of the cytoplasmic dynein motor, J. Neurosci. 21 (2001) RC125.

[12] R.P. Herzig, U. Andersson, R.C. Scarpulla, Dynein light chain interacts with NRF-1 and EWG, structurally and functionally related transcription factors from humans and drosophila, J. Cell Sci. 113 (Pt 23) (2000) 4263–4273.

[13] R.D. Vale, The molecular motor toolbox for intracellular transport, Cell 112 (2003) 467–480.

[14] K.K. Pfister, P.R. Shah, H. Hummerich, A. Russ, J. Cotton, A.A. Annuar, et al., Genetic analysis of the cytoplasmic dynein subunit families, PLoS Genet. 2 (2006) e1.

[15] S.J. King, M. Bonilla, M.E. Rodgers, T.A. Schroer, Subunit organization in cytoplasmic dynein subcomplexes, Protein Sci. 11 (2002) 1239–1250.

[16] J.C. Williams, P.L. Roulhac, A.G. Roy, R.B. Vallee, M.C. Fitzgerald, W.A. Hendrickson, Structural and thermodynamic characterization of a cytoplasmic dynein light chain-intermediate chain complex, Proc. Natl. Acad. Sci. U.S.A. 104 (2007) 10028–10033.

[17] R.J. McKenney, W. Huynh, M.E. Tanenbaum, G. Bhabha, R.D. Vale, Activation of cytoplasmic dynein motility by dynactin-cargo adapter complexes, Science 345 (2014) 337–341.

[18] C.W. Bell, E. Fronk, I.R. Gibbons, Polypeptide subunits of dynein 1 from sea urchin sperm flagella, J. Supramol. Struct. 11 (1979) 311–317.

[19] K.K. Pfister, R.B. Fay, G.B. Witman, Purification and polypeptide composition of dynein ATPases from *Chlamydomonas* flagella, Cell Motil. 2 (1982) 525–547.

[20] R.B. Vallee, S.E. Davis, Low molecular weight microtubule associated proteins are light chains of MAP 1, Proc. Natl. Acad. Sci. U.S.A. 80 (1983) 1342–1346.

[21] S.R. Gill, D.W. Cleveland, T.A. Schroer, Characterization of DLC-A and DLC-B, two families of cytoplasmic dynein light chain subunits, Mol. Biol. Cell 5 (1994) 645–654.

[22] S.M. King, J.F. Dillman 3rd, S.E. Benashski, R.J. Lye, R.S. Patel-King, K.K. Pfister, The mouse t-complex-encoded protein Tctex-1 is a light chain of brain cytoplasmic dynein, J. Biol. Chem. 271 (1996) 32281–32287.

[23] S.M. King, R.S. Patel-King, The M_r=8,000 and 11,000 outer arm dynein light chains from *Chlamydomonas* flagella have cytoplasmic homologues, J. Biol. Chem. 270 (1995) 11445–11452.

[24] A.B. Bowman, R.S. Patel-King, S.E. Benashski, J.M. McCaffery, L.S. Goldstein, S.M. King, Drosophila roadblock and *Chlamydomonas* LC7: a conserved family of dynein-associated proteins involved in axonal transport, flagellar motility, and mitosis, J. Cell Biol. 146 (1999) 165–180.

[25] K.K. Pfister, G.B. Witman, Subfractionation of *Chlamydomonas* 18 S dynein into two unique subunits containing ATPase activity, J. Biol. Chem. 259 (1984) 12072–12080.

[26] S.M. King, E. Barbarese, J.F. Dillman 3rd, R.S. Patel-King, J.H. Carson, K.K. Pfister, Brain cytoplasmic and flagellar outer arm dyneins share a highly conserved Mr 8,000 light chain, J. Biol. Chem. 271 (1996) 19358–19366.

[27] B. Wickstead, K. Gull, Dyneins across eukaryotes: a comparative genomic analysis, Traffic 8 (2007) 1708–1721.

[28] T. Dick, K. Ray, H.K. Salz, W. Chia, Cytoplasmic dynein (ddlc1) mutations cause morphogenetic defects and apoptotic cell death in *Drosophila melanogaster*, Mol. Cell Biol. 16 (1996) 1966–1977.

[29] R. Phillis, D. Statton, P. Caruccio, R.K. Murphey, Mutations in the 8 kDa dynein light chain gene disrupt sensory axon projections in the Drosophila imaginal CNS, Development 122 (1996) 2955–2963.

[30] B. Liu, X. Xiang, Y.R. Lee, The requirement of the LC8 dynein light chain for nuclear migration and septum positioning is temperature dependent in *Aspergillus nidulans*, Mol. Microbiol. 47 (2003) 291–301.

[31] P. Gonczy, C. Echeverri, K. Oegema, A. Coulson, S.J. Jones, R.R. Copley, et al., Functional genomic analysis of cell division in C. elegans using RNAi of genes on chromosome III, Nature 408 (2000) 331–336.

[32] G.S. Tan, M.A. Preuss, J.C. Williams, M.J. Schnell, The dynein light chain 8 binding motif of rabies virus phosphoprotein promotes efficient viral transcription, Proc. Natl. Acad. Sci. U.S.A. 104 (2007) 7229–7234.

[33] R.K. Vadlamudi, R. Bagheri-Yarmand, Z. Yang, S. Balasenthil, D. Nguyen, A.A. Sahin, et al., Dynein light chain 1, a p21-activated kinase 1-interacting substrate, promotes cancerous phenotypes, Cancer Cell 5 (2004) 575–585.

[34] D.M. Wong, L. Li, S. Jurado, A. King, R. Bamford, M. Wall, et al., The transcription factor ASCIZ and its target DYNLL1 are essential for the development and expansion of MYC-driven B cell lymphoma, Cell Reports 14 (2016) 1488–1499.

[35] K.W. Lo, H.M. Kan, L.N. Chan, W.G. Xu, K.P. Wang, Z. Wu, et al., The 8-kDa dynein light chain binds to p53-binding protein 1 and mediates DNA damage-induced p53 nuclear accumulation, J. Biol. Chem. 280 (2005) 8172–8179.

[36] J.K. Kirkham, S.H. Park, T.N. Nguyen, J.H. Lee, A. Gunzl, Dynein light chain LC8 is required for RNA Polymerase I-Mediated transcription in *Trypanosoma brucei*, facilitating assembly and promoter binding of class I transcription factor A, Mol. Cell. Biol. 36 (2016) 95–107.

Dyneins

[37] Y.W. Chang, R. Jakobi, A. McGinty, M. Foschi, M.J. Dunn, A. Sorokin, Cyclooxygenase 2 promotes cell survival by stimulation of dynein light chain expression and inhibition of neuronal nitric oxide synthase activity, Mol. Cell Biol. 20 (2000) 8571–8579.

[38] S. Luo, D.C. Rubinsztein, BCL2L11/BIM: a novel molecular link between autophagy and apoptosis, Autophagy 9 (2013) 104–105.

[39] Y. Batlevi, D.N. Martin, U.B. Pandey, C.R. Simon, C.M. Powers, J.P. Taylor, et al., Dynein light chain 1 is required for autophagy, protein clearance, and cell death in Drosophila, Proc. Natl. Acad. Sci. U.S.A. 107 (2010) 742–747.

[40] S.R. Jaffrey, S.H. Snyder, PIN: an associated protein inhibitor of neuronal nitric oxide synthase, Science 274 (1996) 774–777.

[41] S.S. Parhad, D. Jaiswal, K. Ray, S. Mazumdar, The protein inhibitor of nNOS (PIN/DLC1/LC8) binding does not inhibit the NADPH-dependent heme reduction in nNOS, a key step in NO synthesis, Biochem. Biophys. Res. Commun. 472 (2016) 189–193.

[42] Y. Jacob, H. Badrane, P.E. Ceccaldi, N. Tordo, Cytoplasmic dynein LC8 interacts with lyssavirus phosphoprotein, J. Virol. 74 (2000) 10217–10222.

[43] P. Luthra, D.S. Jordan, D.W. Leung, G.K. Amarasinghe, C.F. Basler, Ebola virus VP35 interaction with dynein LC8 regulates viral RNA synthesis, J. Virol. 89 (2015) 5148–5153.

[44] K.D. Jayappa, Z. Ao, X. Wang, A.J. Mouland, S. Shekhar, X. Yang, et al., Human immunodeficiency virus type 1 employs the cellular dynein light chain 1 protein for reverse transcription through interaction with its integrase protein, J. Virol. 89 (2015) 3497–3511.

[45] E. Lader, H.S. Ha, M. O'Neill, K. Artzt, D. Bennett, Tctex-1: a candidate gene family for a mouse t complex sterility locus, Cell 58 (1989) 969–979.

[46] S. Indu, S.C. Sekhar, J. Sengottaiyan, A. Kumar, S.M. Pillai, M. Laloraya, et al., Aberrant expression of dynein light chain 1 (DYNLT1) is associated with human male factor infertility, Mol. Cell. Proteomics 14 (2015) 3185–3195.

[47] M.G. Li, M. Serr, E.A. Newman, T.S. Hays, The Drosophila tctex-1 light chain is dispensable for essential cytoplasmic dynein functions but is required during spermatid differentiation, Mol. Biol. Cell 15 (2004) 3005–3014.

[48] J. Merino-Gracia, H. Zamora-Carreras, M. Bruix, I. Rodriguez-Crespo, Molecular basis for the protein recognition specificity of the dynein light chain DYNLT1/Tctex1: characterization of the interaction with Activin receptor IIB, J. Biol. Chem. 291 (2016) 20962–20975.

[49] D.M. Roth, G.W. Moseley, D. Glover, C.W. Pouton, D.A. Jans, A microtubule-facilitated nuclear import pathway for cancer regulatory proteins, Traffic 8 (2007) 673–686.

[50] H. Raux, A. Flamand, D. Blondel, Interaction of the rabies virus P protein with the LC8 dynein light chain, J. Virol. 74 (2000) 10212–10216.

[51] Y.D. Fang, X. Xu, Y.M. Dang, Y.M. Zhang, J.P. Zhang, J.Y. Hu, et al., MAP4 mechanism that stabilizes mitochondrial permeability transition in hypoxia: microtubule enhancement and DYNLT1 interaction with VDAC1, PLoS One 6 (2011) e28052.

[52] J. Merino-Gracia, M.F. Garcia-Mayoral, P. Rapali, R.A. Valero, M. Bruix, I. Rodriguez-Crespo, DYNLT (Tctex-1) forms a tripartite complex with dynein intermediate chain and RagA, hence linking this small GTPase to the dynein motor, FEBS J. 282 (2015) 3945–3958.

[53] F. Nagano, S. Orita, T. Sasaki, A. Naito, G. Sakaguchi, M. Maeda, et al., Interaction of Doc2 with tctex-1, a light chain of cytoplasmic dynein. Implication in dynein-dependent vesicle transport, J. Biol. Chem. 273 (1998) 30065–30068.

[54] A. Bauch, K.S. Campbell, M. Reth, Interaction of the CD5 cytoplasmic domain with the Ca2+/calmodulin-dependent kinase IIdelta, Eur. J. Immunol. 28 (1998) 2167–2177.

[55] N.J. Sarma, N.R. Yaseen, Dynein light chain 1 (DYNLT1) interacts with normal and oncogenic nucleoporins, PLoS One 8 (2013) e67032.

[56] K.J. Palmer, H. Hughes, D.J. Stephens, Specificity of cytoplasmic dynein subunits in discrete membrane trafficking steps, Mol. Biol. Cell 20 (12) (2009) 2885–2899.

[57] J. Vlach, J. Lipov, M. Rumlova, V. Veverka, J. Lang, P. Srb, et al., D-retrovirus morphogenetic switch driven by the targeting signal accessibility to Tctex-1 of dynein, Proc. Natl. Acad. Sci. U.S.A. 105 (2008) 10565–10570.

[58] M.A. Schneider, G.A. Spoden, L. Florin, C. Lambert, Identification of the dynein light chains required for human papillomavirus infection, Cell Microbiol. 13 (2011) 32–46.

[59] J. Jiang, L. Yu, X. Huang, X. Chen, D. Li, Y. Zhang, et al., Identification of two novel human dynein light chain genes, DNLC2A and DNLC2B, and their expression changes in hepatocellular carcinoma tissues from 68 Chinese patients, Gene 281 (2001) 103–113.

[60] W. Ding, Q. Tang, V. Espina, L.A. Liotta, D.T. Mauger, K.M. Mulder, A transforming growth factor-beta receptor-interacting protein frequently mutated in human ovarian cancer, Cancer Res. 65 (2005) 6526–6533.

[61] I.G. Campbell, W.A. Phillips, D.Y. Choong, Genetic and epigenetic analysis of the putative tumor suppressor km23 in primary ovarian, breast, and colorectal cancers, Clin. Cancer Res. 12 (2006) 3713–3715.

[62] H. Tochio, S. Ohki, Q. Zhang, M. Li, M. Zhang, Solution structure of a protein inhibitor of neuronal nitric oxide synthase, Nat. Struct. Biol. 5 (1998) 965–969.

[63] J.S. Fan, Q. Zhang, M. Li, H. Tochio, T. Yamazaki, M. Shimizu, et al., Protein inhibitor of neuronal nitric-oxide synthase, PIN, binds to a 17-amino acid residue fragment of the enzyme, J. Biol. Chem. 273 (1998) 33472–33481.

[64] J. Liang, S.R. Jaffrey, W. Guo, S.H. Snyder, J. Clardy, Structure of the PIN/LC8 dimer with a bound peptide, Nat. Struct. Biol. 6 (1999) 735–740.

[65] A. Nyarko, L. Cochrun, S. Norwood, N. Pursifull, A. Voth, E. Barbar, Ionization of His 55 at the dimer interface of dynein light-chain LC8 is coupled to dimer dissociation, Biochemistry 44 (2005) 14248–14255.

[66] Y. Song, G. Benison, A. Nyarko, T.S. Hays, E. Barbar, Potential role for phosphorylation in differential regulation of the assembly of dynein light chains, J. Biol. Chem. 282 (2007) 17272–17279.

[67] C.M. Lightcap, S. Sun, J.D. Lear, U. Rodeck, T. Polenova, J.C. Williams, Biochemical and structural characterization of the Pak1- LC8 interaction, J. Biol. Chem. 283 (40) (2008) 27314–27324.

[68] C. Song, W. Wen, S.K. Rayala, M. Chen, J. Ma, M. Zhang, et al., Serine 88 phosphorylation of the 8-kDa dynein light chain 1 is a molecular switch for its dimerization status and functions, J. Biol. Chem. 283 (2008) 4004–4013.

[69] Y.K. Mok, K.W. Lo, M. Zhang, Structure of Tctex-1 and its interaction with cytoplasmic dynein intermediate chain, J. Biol. Chem. 276 (2001) 14067–14074.

[70] J.C. Williams, H. Xie, W.A. Hendrickson, Crystal structure of dynein light chain TcTex-1, J. Biol. Chem. 280 (2005) 21981–21986.

[71] H. Wu, M.W. Maciejewski, S. Takebe, S.M. King, Solution structure of the Tctex1 dimer reveals a mechanism for dynein-cargo interactions, Structure (Camb) 13 (2005) 213–223.

[72] M. Talbott, M. Hare, A. Nyarko, T.S. Hays, E. Barbar, Folding is coupled to dimerization of Tctex-1 dynein light chain, Biochemistry 45 (2006) 6793–6800.

[73] J. Song, R.C. Tyler, M.S. Lee, E.M. Tyler, J.L. Markley, Solution structure of isoform 1 of Roadblock/LC7, a light chain in the dynein complex, J. Mol. Biol. 354 (2005) 1043–1051.

[74] J. Hall, Y. Song, P.A. Karplus, E. Barbar, The crystal structure of dynein intermediate chain-light chain roadblock complex gives new insights into dynein assembly, J. Biol. Chem. 285 (29) (2010) 22566–22575.

[75] E.V. Koonin, L. Aravind, Dynein light chains of the Roadblock/LC7 group belong to an ancient protein superfamily implicated in NTPase regulation, Curr. Biol. 10 (2000) R774–R776.

[76] J. Fan, Q. Zhang, H. Tochio, M. Li, M. Zhang, Structural basis of diverse sequence-dependent target recognition by the 8 kDa dynein light chain, J. Mol. Biol. 306 (2001) 97–108.

[77] F.S. Espindola, D.M. Suter, L.B. Partata, T. Cao, J.S. Wolenski, R.E. Cheney, et al., The light chain composition of chicken brain myosin-Va: calmodulin, myosin-II essential light chains, and 8-kDa dynein light chain/PIN, Cell Motil. Cytoskeleton 47 (2000) 269–281.

[78] S. Naisbitt, J. Valtschanoff, D.W. Allison, C. Sala, E. Kim, A.M. Craig, et al., Interaction of the postsynaptic density-95/guanylate kinase domain-associated protein complex with a light chain of myosin-V and dynein, J. Neurosci. 20 (2000) 4524–4534.

[79] R.D. Machado, N. Rudarakanchana, C. Atkinson, J.A. Flanagan, R. Harrison, N.W. Morrell, et al., Functional interaction between BMPR-II and Tctex-1, a light chain of Dynein, is isoform-specific and disrupted by mutations underlying primary pulmonary hypertension, Hum. Mol. Genet. 12 (2003) 3277–3286.

[80] C. Schwarzer, S. Barnikol-Watanabe, F.P. Thinnes, N. Hilschmann, Voltage-dependent anion-selective channel (VDAC) interacts with the dynein light chain Tctex1 and the heat-shock protein PBP74, Int. J. Biochem. Cell Biol. 34 (2002) 1059–1070.

[81] A.W. Tai, J.Z. Chuang, C. Bode, U. Wolfrum, C.H. Sung, Rhodopsin's carboxy-terminal cytoplasmic tail acts as a membrane receptor for cytoplasmic dynein by binding to the dynein light chain Tctex-1, Cell 97 (1999) 877–887.

[82] F.J. Kaiser, K. Tavassoli, G.J. Van den Bemd, G.T.G. Chang, B. Horsthemke, T. Moroy, et al., Nuclear interaction of the dynein light chain LC8a with the TRPS1 transcription factor suppresses the transcriptional repression activity of TRPS1, Hum. Mol. Genet. 12 (2003) 1349–1358.

[83] F. Schnorrer, K. Bohmann, C. Nusslein-Volhard, The molecular motor dynein is involved in targeting swallow and bicoid RNA to the anterior pole of Drosophila oocytes, Nat. Cell Biol. 2 (2000) 185–190.

[84] T.Y. Yeh, J.Z. Chuang, C.H. Sung, Dynein light chain rp3 acts as a nuclear matrix-associated transcriptional modulator in a dynein-independent pathway, J. Cell Sci. 118 (2005) 3431–3443.

[85] T. Kubota, M. Matsuoka, T.H. Chang, M. Bray, S. Jones, M. Tashiro, et al., Ebolavirus VP35 interacts with the cytoplasmic dynein light chain 8, J. Virol. 83 (2009) 6952–6956.

[86] C. Alonso, J. Miskin, B. Hernaez, P. Fernandez-Zapatero, L. Soto, C. Canto, et al., African swine fever virus protein p54 interacts with the microtubular motor complex through direct binding to light-chain dynein, J. Virol. 75 (2001) 9819–9827.

[87] M.W. Douglas, R.J. Diefenbach, F.L. Homa, M. Miranda-Saksena, F.J. Rixon, V. Vittone, et al., Herpes simplex virus type 1 capsid protein VP26 interacts with dynein light chains RP3 and Tctex1 and plays a role in retrograde cellular transport, J. Biol. Chem. 279 (2004) 28522–28530.

[88] M.A. Schneider, G.A. Spoden, L. Florin, C. Lambert, Identification of the dynein light chains required for human papillomavirus infection, Cell Microbiol. 13 (1) (2010) 32–46.

[89] Q. Tang, C.M. Staub, G. Gao, Q. Jin, Z. Wang, W. Ding, et al., A novel transforming growth factor-beta receptor-interacting protein that is also a light chain of the motor protein dynein, Mol. Biol. Cell 13 (2002) 4484–4496.

[90] B. Wanschers, R. van de Vorstenbosch, M. Wijers, B. Wieringa, S.M. King, J. Fransen, Rab6 family proteins interact with the dynein light chain protein DYNLRB1, Cell Motil. Cytoskeleton 65 (2008) 183–196.

[91] G. Benison, P.A. Karplus, E. Barbar, Structure and dynamics of LC8 complexes with KXTQT-motif peptides: swallow and dynein intermediate chain compete for a common site, J. Mol. Biol. 371 (2007) 457–468.

[92] E.M. Romes, A. Tripathy, K.C. Slep, Structure of a yeast Dyn2-Nup159 complex and molecular basis for dynein light chain-nuclear pore interaction, J. Biol. Chem. 287 (2012) 15862–15873.

[93] L.K. Slevin, E.M. Romes, M.G. Dandulakis, K.C. Slep, The mechanism of dynein light chain LC8-mediated oligomerization of the Ana2 centriole duplication factor, J. Biol. Chem. 289 (2014) 20727–20739.

[94] S. Clark, A. Nyarko, F. Lohr, P.A. Karplus, E. Barbar, The anchored flexibility model in LC8 motif recognition: insights from the Chica complex, Biochemistry 55 (2016) 199–209.

[95] P. Gallego, A. Velazquez-Campoy, L. Regue, J. Roig, D. Reverter, Structural analysis of the regulation of the DYNLL/LC8 binding to Nek9 by phosphorylation, J. Biol. Chem. 288 (2013) 12283–12294.

[96] P. Rapali, L. Radnai, D. Suveges, V. Harmat, F. Tolgyesi, W.Y. Wahlgren, et al., Directed evolution reveals the binding motif preference of the LC8/DYNLL hub protein and predicts large numbers of novel binders in the human proteome, PLoS One 6 (2011) e18818.

[97] J. Hall, P.A. Karplus, E. Barbar, Multivalency in the assembly of intrinsically disordered dynein intermediate chain, J. Biol. Chem. 284 (2009) 33115–33121.

[98] M. Sugai, M. Saito, I. Sukegawa, Y. Katsushima, Y. Kinouchi, N. Nakahata, et al., PTH/PTH-related protein receptor interacts directly with Tctex-1 through its COOH terminus, Biochem. Biophys. Res. Commun. 311 (2003) 24–31.

[99] S. Mueller, X. Cao, R. Welker, E. Wimmer, Interaction of the poliovirus receptor CD155 with the dynein light chain Tctex-1 and its implication for poliovirus pathogenesis, J. Biol. Chem. 277 (2002) 7897–7904.

[100] D.H. Fremont, M. Matsumura, E.A. Stura, P.A. Peterson, I.A. Wilson, Crystal structures of two viral peptides in complex with murine MHC class I H-2Kb, Science 257 (1992) 919–927.

[101] C.L. Day, H. Puthalakath, G. Skea, A. Strasser, I. Barsukov, L.Y. Lian, et al., Localization of dynein light chains 1 and 2 and their pro-apoptotic ligands, Biochem. J. 377 (2004) 597–605.

[102] H. Puthalakath, A. Villunger, L.A. O'Reilly, J.G. Beaumont, L. Coultas, R.E. Cheney, et al., Bmf: a proapoptotic BH3-only protein regulated by interaction with the myosin V actin motor complex, activated by anoikis, Science 293 (2001) 1829–1832.

[103] H. Puthalakath, D.C. Huang, L.A. O'Reilly, S.M. King, A. Strasser, The proapoptotic activity of the Bcl-2 family member Bim is regulated by interaction with the dynein motor complex, Mol. Cell 3 (1999) 287–296.

[104] P. Crepieux, H. Kwon, N. Leclerc, W. Spencer, S. Richard, R. Lin, et al., I kappaB alpha physically interacts with a cytoskeleton-associated protein through its signal response domain, Mol. Cell Biol. 17 (1997) 7375–7385.

[105] W. Jeong, T.S. Chang, E.S. Boja, H.M. Fales, S.G. Rhee, Roles of TRP14, a thioredoxin-related protein in tumor necrosis factor-alpha signaling pathways, J. Biol. Chem. 279 (2004) 3151–3159.

[106] Y. Jung, H. Kim, S.H. Min, S.G. Rhee, W. Jeong, Dynein light chain LC8 negatively regulates NF-kappaB through the redox-dependent interaction with IkappaBalpha, J. Biol. Chem. 283 (2008) 23863–23871.

[107] S.M. King, Dynein-independent functions of DYNLL1/LC8: redox state sensing and transcriptional control, Sci. Signal 1 (2008) pe51.

[108] T.Y. Yeh, D. Peretti, J.Z. Chuang, E. Rodriguez-Boulan, C.H. Sung, Regulatory dissociation of Tctex-1 light chain from dynein complex is essential for the apical delivery of rhodopsin, Traffic 7 (2006) 1495–1502.

[109] K.W. Lo, J.M. Kogoy, B.A. Rasoul, S.M. King, K.K. Pfister, Interaction of the DYNLT (TCTEX1/RP3) light chains and the intermediate chains reveals novel intersubunit regulation during assembly of the dynein complex, J. Biol. Chem. 282 (2007) 36871–36878.

[110] S.J. Susalka, K.K. Pfister, Cytoplasmic dynein subunit heterogeneity: implications for axonal transport, J. Neurocytol. 29 (2000) 819–829.

[111] S.H. Tynan, M.A. Gee, R.B. Vallee, Distinct but overlapping sites within the cytoplasmic dynein heavy chain for dimerization and for intermediate chain and light intermediate chain binding, J. Biol. Chem. 275 (2000) 32769–32774.

[112] C. Cole, J.D. Barber, G.J. Barton, The Jpred 3 secondary structure prediction server, Nucleic Acids Res. 36 (2008) W197–W201.

[113] K.T. Vaughan, R.B. Vallee, Cytoplasmic dynein binds dynactin through a direct interaction between the intermediate chains and p150Glued, J. Cell Biol. 131 (1995) 1507–1516.

[114] K.T. Vaughan, S.H. Tynan, N.E. Faulkner, C.J. Echeverri, R.B. Vallee, Analysis of p150[Glued] as the cytoplasmic dynein-binding component of dynactin: evidence for colocalization with CLIP-170 at microtubule distal-ends, Mol. Biol. Cell 7 (1996) 403a.

[115] C.M. Waterman-Storer, S. Karki, E.L.F. Holzbaur, The p150[Glued] component of the dynactin complex binds to both microtubules and the actin-related protein centractin (Arp-1), Proc. Natl. Acad. Sci. U.S.A. 92 (1995) 1634–1638.

[116] J. Whyte, J.R. Bader, S.B. Tauhata, M. Raycroft, J. Hornick, K.K. Pfister, et al., Phosphorylation regulates targeting of cytoplasmic dynein to kinetochores during mitosis, J. Cell Biol. 183 (2008) 819–834.

[117] P.S. Vaughan, J.D. Leszyk, K.T. Vaughan, Cytoplasmic dynein intermediate chain phosphorylation regulates binding to dynactin, J. Biol. Chem. 276 (2001) 26171–26179.

[118] Z.R. Yang, R. Thomson, P. McNeil, R.M. Esnouf, RONN: the bio-basis function neural network technique applied to the detection of natively disordered regions in proteins, Bioinformatics 21 (2005) 3369–3376.

[119] S.F. Altschul, T.L. Madden, A.A. Schaffer, J. Zhang, Z. Zhang, W. Miller, et al., Gapped BLAST and PSI-BLAST: a new generation of protein database search programs, Nucleic Acids Res. 25 (1997) 3389–3402.

[120] L. Urnavicius, K. Zhang, A.G. Diamant, C. Motz, M.A. Schlager, M. Yu, et al., The structure of the dynactin complex and its interaction with dynein, Science 347 (2015) 1441–1446.

[121] S. Chowdhury, S.A. Ketcham, T.A. Schroer, G.C. Lander, Structural organization of the dynein-dynactin complex bound to microtubules, Nat. Struct. Mol. Biol. 22 (2015) 345–347.

[122] K.K. Pfister, Dynein cargo gets its groove back, Structure 13 (2005) 172–173.

[123] R.D. Vale, R.A. Milligan, The way things move: looking under the hood of molecular motor proteins, Science 288 (2000) 88–95.

[124] S.H. Tynan, A. Purohit, S.J. Doxsey, R.B. Vallee, Light intermediate chain 1 defines a functional subfraction of cytoplasmic dynein which binds to pericentrin, J. Biol. Chem. 275 (2000) 32763–32768.

[125] M. Mammen, S.K. Choi, W. GM, Polyvalent interactions in biological systems: implications for design and use of multivalent ligands and inhibitors, Angew. Chem. Int. Ed. Eng. 37 (1998) 2755–2794.

[126] W.P. Jencks, On the attribution and additivity of binding energies, Proc. Natl. Acad. Sci. U.S.A. 78 (1981) 4046–4050.

[127] J. Rao, J. Lahiri, L. Isaacs, R.M. Weis, G.M. Whitesides, A trivalent system from vancomycin.D-ala-D-Ala with higher affinity than avidin.biotin, Science 280 (1998) 708–711.

[128] D. Varma, A. Dawn, A. Ghosh-Roy, S.J. Weil, K.M. Ori-McKenney, Y. Zhao, et al., Development and application of in vivo molecular traps reveals that dynein light chain occupancy differentially affects dynein-mediated processes, Proc. Natl. Acad. Sci. U.S.A. 107 (2010) 3493–3498.

[129] T. Clackson, W. Yang, L.W. Rozamus, M. Hatada, J.F. Amara, C.T. Rollins, et al., Redesigning an FKBP-ligand interface to generate chemical dimerizers with novel specificity, Proc. Natl. Acad. Sci. U.S.A. 95 (1998) 10437–10442.

[130] C.M. Lightcap, G. Kari, L.E. Arias-Romero, J. Chernoff, U. Rodeck, J.C. Williams, Interaction with LC8 is required for Pak1 nuclear import and is indispensable for zebrafish development, PLoS One 4 (2009) e6025.

[131] E. Barbar, Dynein light chain LC8 is a dimerization hub essential in diverse protein networks, Biochemistry 47 (2008) 503–508.

[132] E.T. Mack, R. Perez-Castillejos, Z. Suo, G.M. Whitesides, Exact analysis of ligand-induced dimerization of monomeric receptors, Anal. Chem. 80 (2008) 5550–5555.

[133] G. Benison, M. Chiodo, P.A. Karplus, E. Barbar, Structural, thermodynamic, and kinetic effects of a phosphomimetic mutation in dynein light chain LC8, Biochemistry 48 (2009) 11381–11389.

[134] K.S. Campbell, S. Cooper, M. Dessing, S. Yates, A. Buder, Interaction of p59fyn kinase with the dynein light chain, Tctex-1, and colocalization during cytokinesis, J. Immunol. 161 (1998) 1728–1737.

[135] L.M. Iakoucheva, P. Radivojac, C.J. Brown, T.R. O'Connor, J.G. Sikes, Z. Obradovic, et al., The importance of intrinsic disorder for protein phosphorylation, Nucleic Acids Res. 32 (2004) 1037–1049.

[136] S.J. King, C.L. Brown, K.C. Maier, N.J. Quintyne, T.A. Schroer, Analysis of the dynein-dynactin interaction in vitro and in vivo, Mol. Biol. Cell 14 (2003) 5089–5097.

[137] A.E. Siglin, S. Sun, J.K. Moore, S. Tan, M. Poenie, J.D. Lear, et al., Dynein and dynactin leverage their bivalent character to form a high-affinity interaction, PLoS One 8 (2013) e59453.

[138] F.J. Gao, S. Hebbar, X.A. Gao, M. Alexander, J.P. Pandey, M.D. Walla, et al., GSK-3beta phosphorylation of cytoplasmic dynein reduces Ndel1 binding to intermediate chains and alters dynein motility, Traffic 16 (2015) 941–961.

[139] K. Ikeda, O. Zhapparova, I. Brodsky, I. Semenova, J.S. Tirnauer, I. Zaliapin, et al., CK1 activates minus-end-directed transport of membrane organelles along microtubules, Mol. Biol. Cell 22 (2011) 1321–1329.

[140] L.M. DiBella, M. Sakato, R.S. Patel-King, G.J. Pazour, S.M. King, The LC7 light chains of *Chlamydomonas* flagellar dyneins interact with components required for both motor assembly and regulation, Mol. Biol. Cell 15 (2004) 4633–4646.

[141] C.A. Tanner, P. Rompolas, R.S. Patel-King, O. Gorbatyuk, K. Wakabayashi, G.J. Pazour, et al., Three members of the LC8/DYNLL family are required for outer arm dynein motor function, Mol. Biol. Cell 19 (2008) 3724–3734.

[142] K. Lei, R.J. Davis, JNK phosphorylation of Bim-related members of the Bcl2 family induces Bax-dependent apoptosis, Proc. Natl. Acad. Sci. U.S.A. 100 (2003) 2432–2437.

[143] E.J. Barbar, S. Norwood, M. Hare, Increase in flexibility upon dissociation of dynein light chain LC8 dimer as probed by NMR and mass spectrometry, Biophys. J. 82 (2002) 1701.

[144] H.J. Dyson, P.E. Wright, Coupling of folding and binding for unstructured proteins, Curr. Opin. Struct. Biol. 12 (2002) 54–60.

[145] S. Ahmed, J. Xie, D. Horne, J.C. Williams, Photocleavable dimerizer for the rapid reversal of molecular trap antagonists, J. Biol. Chem. 289 (2014) 4546–4552.

[146] M.D. Stuchell-Brereton, A. Siglin, J. Li, J.K. Moore, S. Ahmed, J.C. Williams, et al., Functional interaction between dynein light chain and intermediate chain is required for mitotic spindle positioning, Mol. Biol. Cell 22 (2011) 2690–2701.

[147] N. Sharma, C.D. Deppmann, A.W. Harrington, C. St Hillaire, Z.Y. Chen, F.S. Lee, et al., Long-distance control of synapse assembly by target-derived NGF, Neuron 67 (2010) 422–434.

[148] M.P. McShane, M. Zerial, Survival of the weakest: signaling aided by endosomes, J. Cell Biol. 182 (2008) 823–825.

[149] S. Kermorgant, P.J. Parker, Receptor trafficking controls weak signal delivery: a strategy used by c-Met for STAT3 nuclear accumulation, J. Cell Biol. 182 (2008) 855–863.

[150] J. Salogiannis, S.L. Reck-Peterson, Hitchhiking: a non-canonical mode of microtubule-based transport, Trends Cell Biol. 27 (2) (2016) 141–150.

[151] G.W. Moseley, D.M. Roth, M.A. Dejesus, D.L. Leyton, R.P. Filmer, C.W. Pouton, et al., Dynein light chain association sequences can facilitate nuclear protein import, Mol. Biol. Cell 18 (8) (2007) 3204–3213.

[152] M. Martinez-Moreno, I. Navarro-Leida, F. Roncal, J.P. Albar, C. Alonso, F. Gavilanes, et al., Recognition of novel viral sequences that associate with the dynein light chain LC8 identified through a pepscan technique, FEBS Lett. 544 (2003) 262–267.

[153] T.G. Senkevich, L.S. Wyatt, A.S. Weisberg, E.V. Koonin, B. Moss, A conserved poxvirus NlpC/P60 superfamily protein contributes to vaccinia virus virulence in mice but not to replication in cell culture, Virology 374 (2008) 506–514.

[154] B. Hernaez, G. Diaz-Gil, M. Garcia-Gallo, J. Ignacio Quetglas, I. Rodriguez-Crespo, L. Dixon, et al., The African swine fever virus dynein-binding protein p54 induces infected cell apoptosis, FEBS Lett. 569 (2004) 224–228.

[155] A. Fejtova, D. Davydova, F. Bischof, V. Lazarevic, W.D. Altrock, S. Romorini, et al., Dynein light chain regulates axonal trafficking and synaptic levels of Bassoon, J. Cell Biol. 185 (2009) 341–355.

[156] M.D. Show, J.S. Folmer, M.D. Anway, B.R. Zirkin, Testicular expression and distribution of the rat bcl2 modifying factor in response to reduced intratesticular testosterone, Biol. Reprod. 70 (2004) 1153–1161.

[157] D.A. Blanchard, S. Mouhamad, M.T. Auffredou, A. Pesty, J. Bertoglio, G. Leca, et al., Cdk2 associates with MAP kinase in vivo and its nuclear translocation is dependent on MAP kinase activation in IL-2-dependent Kit 225 T lymphocytes, Oncogene 19 (2000) 4184–4189.

[158] P. den Hollander, R. Kumar, Dynein light chain 1 contributes to cell cycle progression by increasing cyclin-dependent kinase 2 activity in estrogen-stimulated cells, Cancer Res. 66 (2006) 5941–5949.

[159] A. Lukasik, K.A. Uniewicz, M. Kulis, P. Kozlowski, Ciz1, a p21 cip1/Waf1-interacting zinc finger protein and DNA replication factor, is a novel molecular partner for human enhancer of rudimentary homolog, FEBS J. 275 (2008) 332–340.

[160] R.A. Anderson, N. Fulton, G. Cowan, S. Coutts, P.T. Saunders, Conserved and divergent patterns of expression of DAZL, VASA and OCT4 in the germ cells of the human fetal ovary and testis, BMC Dev. Biol. 7 (2007) 136.

[161] A.D. Lajoix, R. Gross, C. Aknin, S. Dietz, C. Granier, D. Laune, Cellulose membrane supported peptide arrays for deciphering protein-protein interaction sites: the case of PIN, a protein with multiple natural partners, Mol. Divers 8 (2004) 281–290.

[162] X. Xiang, C. Roghi, N.R. Morris, Characterization and localization of the cytoplasmic dynein heavy chain in *Aspergillus nidulans*, Proc. Natl. Acad. Sci. U.S.A. 92 (1995) 9890–9894.

[163] W.M. Saxton, Microtubules, motors, and mRNA localization mechanisms: watching fluorescent messages move, Cell 107 (2001) 707–710.

[164] S.K. Rayala, P. den Hollander, S. Balasenthil, Z. Yang, R.R. Broaddus, R. Kumar, Functional regulation of oestrogen receptor pathway by the dynein light chain 1, EMBO Rep. 6 (6) (2005) 538–544.

[165] Y. Xu, R.J. Traystman, P.D. Hurn, M.M. Wang, Membrane restraint of estrogen receptor alpha enhances estrogen-dependent nuclear localization and genomic function, Mol. Endocrinol. 18 (2004) 86–96.

[166] R. Studer, L. von Boehmer, T. Haenggi, C. Schweizer, D. Benke, U. Rudolph, et al., Alteration of GABAergic synapses and gephyrin clusters in the thalamic reticular nucleus of GABAA receptor alpha3 subunit-null mice, Eur. J. Neurosci. 24 (2006) 1307–1315.

[167] E. Kim, S. Naisbitt, Y.P. Hsueh, A. Rao, A. Rothschild, A.M. Craig, et al., GKAP, a novel synaptic protein that interacts with the guanylate kinase-like domain of the PSD-95/SAP90 family of channel clustering molecules, J. Cell Biol. 136 (1997) 669–678.

[168] M.F. Garcia-Mayoral, M. Martinez-Moreno, J.P. Albar, I. Rodriguez-Crespo, M. Bruix, Structural basis for the interaction between dynein light chain 1 and the glutamate channel homolog GRINL1A, FEBS J. 277 (2010) 2340–2350.

[169] S.M. Okamura, C.E. Oki-Idouchi, P.S. Lorenzo, The exchange factor and diacylglycerol receptor RasGRP3 interacts with dynein light chain 1 through its C-terminal domain, J. Biol. Chem. 281 (2006) 36132–36139.

[170] A. Webster, I.R. Leith, R.T. Hay, Activation of adenovirus-coded protease and processing of preterminal protein, J. Virol. 68 (1994) 7292–7300.

[171] A.D. Kwong, B.G. Rao, K.T. Jeang, Viral and cellular RNA helicases as antiviral targets, Nat. Rev. Drug Discov. 4 (2005) 845–853.

[172] E. Kofod-Olsen, K. Ross-Hansen, J.G. Mikkelsen, P. Hollsberg, Human herpesvirus 6B U19 protein is a PML-regulated transcriptional activator that localizes to nuclear foci in a PML-independent manner, J. Gen. Virol. 89 (2008) 106–116.

[173] S. Sachdev, A. Hoffmann, M. Hannink, Nuclear localization of IkappaB alpha is mediated by the second ankyrin repeat: the IkappaB alpha ankyrin repeats define a novel class of cis-acting nuclear import sequences, Mol. Cell Biol. 18 (1998) 2524–2534.

[174] S.K. Rayala, P. den Hollander, B. Manavathi, A.H. Talukder, C. Song, S. Peng, et al., Essential role of KIBRA in co-activator function of dynein light chain 1 in mammalian cells, J. Biol. Chem. 281 (2006) 19092–19099.

[175] J.D. Huang, S.T. Brady, B.W. Richards, D. Stenolen, J.H. Resau, N.G. Copeland, et al., Direct interaction of microtubule- and actin-based transport motors, Nature 397 (1999) 267–270.

[176] N. Tekki-Kessaris, J.V. Bonventre, C.A. Boulter, Characterization of the mouse Kid1 gene and identification of a highly related gene, Kid2, Gene 240 (1999) 13–22.

[177] D. Pasdeloup, N. Poisson, H. Raux, Y. Gaudin, R.W. Ruigrok, D. Blondel, Nucleocytoplasmic shuttling of the rabies virus P protein requires a nuclear localization signal and a CRM1-dependent nuclear export signal, Virology 334 (2005) 284–293.

[178] M.C. Pranchevicius, M.M. Baqui, H.C. Ishikawa-Ankerhold, E.V. Lourenco, R.M. Leao, S.R. Banzi, et al., Myosin Va phosphorylated on Ser1650 is found in nuclear speckles and redistributes to nucleoli upon inhibition of transcription, Cell Motil. Cytoskeleton 65 (2008) 441–456.

[179] Z. Yuan, B. Liu, L. Yuan, Y. Zhang, X. Dong, J. Lu, Evidence of nuclear localization of neuronal nitric oxide synthase in cultured astrocytes of rats, Life Sci. 74 (2004) 3199–3209.

[180] M. Hertel, S. Braun, S. Durka, C. Alzheimer, S. Werner, Upregulation and activation of the Nrf-1 transcription factor in the lesioned hippocampus, Eur. J. Neurosci. 15 (2002) 1707–1711.

[181] P. Stelter, R. Kunze, D. Flemming, D. Hopfner, M. Diepholz, P. Philippsen, et al., Molecular basis for the functional interaction of dynein light chain with the nuclear-pore complex, Nat. Cell Biol. 9 (2007) 788–796.

[182] R.J. McKenney, S.J. Weil, J. Scherer, R.B. Vallee, Mutually exclusive cytoplasmic dynein regulation by NudE-Lis1 and dynactin, J. Biol. Chem. 286 (2011) 39615–39622.

[183] R.R. Singh, C. Song, Z. Yang, R. Kumar, Nuclear localization and chromatin targets of p21-activated kinase 1, J. Biol. Chem. 280 (2005) 18130–18137.

[184] E. Salzer, D. Cagdas, M. Hons, E.M. Mace, W. Garncarz, O.Y. Petronczki, et al., RASGRP1 deficiency causes immunodeficiency with impaired cytoskeletal dynamics, Nat. Immunol. 17 (2016) 1352–1360.

[185] F.P. Polack, P.M. Irusta, S.J. Hoffman, M.P. Schiatti, G.A. Melendi, M.F. Delgado, et al., The cysteine-rich region of respiratory syncytial virus attachment protein inhibits innate immunity elicited by the virus and endotoxin, Proc. Natl. Acad. Sci. U.S.A. 102 (2005) 8996–9001.

[186] J. Hegde, E.C. Stephenson, Distribution of swallow protein in egg chambers and embryos of *Drosophila melanogaster*, Development 119 (1993) 457–470.

[187] L. Wang, M. Hare, T.S. Hays, E. Barbar, Dynein light chain LC8 promotes assembly of the coiled-coil domain of swallow protein, Biochemistry 43 (2004) 4611–4620.

[188] W. Jeong, Y. Jung, H. Kim, S.J. Park, S.G. Rhee, Thioredoxin-related protein 14, a new member of the thioredoxin family with disulfide reductase activity: implication in the redox regulation of TNF-alpha signaling, Free Radic. Biol. Med. 47 (2009) 1294–1303.

[189] F.J. Kaiser, P. Brega, M.L. Raff, P.H. Byers, S. Gallati, T.T. Kay, et al., Novel missense mutations in the TRPS1 transcription factor define the nuclear localization signal, Eur. J. Hum. Genet. 12 (2004) 121–126.

[190] W.B. Cardenas, Y.M. Loo, M. Gale Jr., A.L. Hartman, C.R. Kimberlin, L. Martinez-Sobrido, et al., Ebola virus VP35 protein binds double-stranded RNA and inhibits alpha/beta interferon production induced by RIG-I signaling, J. Virol. 80 (2006) 5168–5178.

[191] S. Ohka, N. Matsuda, K. Tohyama, T. Oda, M. Morikawa, S. Kuge, et al., Receptor (CD155)-dependent endocytosis of poliovirus and retrograde axonal transport of the endosome, J. Virol. 78 (2004) 7186–7198.

[192] S.A. Lukashok, L. Tarassishin, Y. Li, M.S. Horwitz, An adenovirus inhibitor of tumor necrosis factor alpha-induced apoptosis complexes with dynein and a small GTPase, J. Virol. 74 (2000) 4705–4709.

[193] P. Sachdev, S. Menon, D.B. Kastner, J.Z. Chuang, T.Y. Yeh, C. Conde, et al., G protein beta gamma subunit interaction with the dynein light-chain component Tctex-1 regulates neurite outgrowth, EMBO J. 26 (2007) 2621–2632.

[194] K. Dohner, K. Radtke, S. Schmidt, B. Sodeik, Eclipse phase of herpes simplex virus type 1 infection: efficient dynein-mediated capsid transport without the small capsid protein VP26, J. Virol. 80 (2006) 8211–8224.

[195] A. Gauthier-Fisher, D.C. Lin, M. Greeve, D.R. Kaplan, R. Rottapel, F.D. Miller, Lfc and Tctex-1 regulate the genesis of neurons from cortical precursor cells, Nat. Neurosci. 12 (2009) 735–744.

[196] D. Meiri, M.A. Greeve, A. Brunet, D. Finan, C.D. Wells, J. LaRose, et al., Modulation of Rho guanine exchange factor Lfc activity by protein kinase A-mediated phosphorylation, Mol. Cell Biol. 29 (2009) 5963–5973.

[197] H. Yano, M.V. Chao, Biochemical characterization of intracellular membranes bearing Trk neurotrophin receptors, Neurochem. Res. 30 (2005) 767–777.

[198] J.H. Carter, L.E. Douglass, J.A. Deddens, B.M. Colligan, T.R. Bhatt, J.O. Pemberton, et al., Pak-1 expression increases with progression of colorectal carcinomas to metastasis, Clin. Cancer Res. 10 (2004) 3448–3456.

[199] N.A. Baker, D. Sept, S. Joseph, M.J. Holst, J.A. McCammon, Electrostatics of nanosystems: application to microtubules and the ribosome, Proc. Natl. Acad. Sci. U.S.A. 98 (2001) 10037–10041.

In this chapter

Biochemical purification of axonemal and cytoplasmic dyneins

Kazuo Inaba
University of Tsukuba, Shimoda, Japan

4.1 Introduction

The initial observation of cilia and flagella revealed two projections extending from the A-tubule of the doublet microtubules, the "arms" [1]. Although several attempts were made to characterize the ATPase activity for motility of cilia and flagella in the 1960s, the first biochemical characterization of the ATPase was carried out with *Tetrahymena* cilia by Gibbons and Rowe [2] and the motor was given the name "dynein." They extracted ATPases from ciliary axonemes by dialysis against a low ionic strength solution containing EDTA, followed by density gradient centrifugation, and obtained two fractions at 30S and 14S. The extraction of ATPase paralleled the disappearance of the arms [3]. Subsequently, axonemal dyneins were isolated from the flagella of sea urchin sperm, and other ciliates, such as *Tetrahymena* and *Paramecium*, and the unicellular alga *Chlamydomonas*. Isolation and purification of the outer arm dyneins is relatively easy because they apparently represent a single molecular species, whereas multiple species are present in inner arm dyneins. The application of ion exchange column chromatography provided a method to efficiently separate these multiple inner arm dynein species into distinct fractions.

Electron microscopy, including negative staining, the quick-freeze deep-etching method, and more recently cryo–electron tomography, has revealed the organized arrangement of outer and inner arm dyneins in the axonemes of cilia and flagella. Outer arm dyneins are positioned at 24-nm intervals on the A-tubule of doublet microtubule in situ [4–8], whereas different species of inner arm dyneins are arranged in a 96-nm repeat [9–12]. Each species of axonemal dynein is arranged in a specific association and order with respect to each other, often via some structural components that are not extracted during dynein preparation or that dissociate during purification.

Initial efforts to detect and purify cytoplasmic dynein was first attempted from sea urchin eggs using sucrose density gradient centrifugation or column chromatography [13,14]. However, discrimination of isolated cytoplasmic dynein

Dyneins. https://doi.org/10.1016/B978-0-12-809470-9.00004-7

89

ATPase from the cytoplasmic precursors of ciliary dyneins was often argued. In mammals, cytoplasmic dynein was first described as one of the microtubule-associated proteins (MAPs), which associated with microtubules during polymerization or sedimented with paclitaxel (taxol)-stabilized microtubules. One major MAP, termed MAP1C, became dissociated from microtubules following the addition of ATP. MAP1C exhibited microtubule-activated ATPase and microtubule-gliding properties, and hence was concluded to be the long-sought cytoplasmic dynein [15,16].

Despite the substantial genomic information that is now available from quite a large number of organisms, direct isolation of active molecules from cells provides the strongest evidence for the molecular shape, protein complex substructure, biological activity, and posttranslational modification of the candidate dyneins. Furthermore, novel proteins that are associated with dyneins can be identified through direct biochemical isolation of the complexes.

This chapter broadly focuses on the biochemical isolation of axonemal and cytoplasmic dyneins. Although dyneins have been isolated from many organisms, detailed subunit composition, the mechanism of force generation, and its regulation have been studied only in a very limited number of species. Here, I overview the procedures that have been employed over the last 50 years for the fractionation of cells and organelles, and the extraction and purification of dynein motors.

4.2 Axonemal dyneins

4.2.1 Isolation of cilia and flagella

Cilia and flagella are filamentous organelles that extrude from cells; therefore they can be dissociated from the cell body by mechanical shear. In addition, induction of deflagellation/deciliation by physical (cold shock and osmolality shock) and chemical (divalent cation, ethanol, dibucaine) stimuli can be applied to flagellar or ciliary isolation for those organisms or tissues that deflagellate or deciliate during the cell cycle or spontaneously. Exceptions include several stimuli that terminate the cell cycle and highly differentiated cells such as spermatozoa (Table 4.1).

Although sperms are morphologically differentiated cells with a reduced volume of cytoplasm, they have acrosomes containing cytosolic proteases that can rapidly degrade dyneins. Outer arm dynein directly isolated from ejaculated sperm were observed to have varying levels of degradation [82]. To avoid contamination from other cytoplasmic components and unexpected proteolysis, cilia/flagella are best separated from cell bodies before isolation of axonemal dynein is initiated. A simple method to separate cilia and flagella from other parts of the cell is mechanical shear. Sperm flagella are most commonly obtained by homogenization in a glass/Teflon Potter–Elvehjem-type homogenizer, followed by differential centrifugation to pellet the sperm heads and

Dyneins

Table 4.1 Strategies for dynein isolation

Organism	Source	Category	Dynein species	Cell fractionation	Extraction	Purification	References
Chlamydomonas	Flagella	Axonemal	Outer arm dynein	DF/DM	HE	SDGC, GF, IECC, HAP	[17–22]
	Flagella	Axonemal	Inner arm dynein (I1/f)	DF/DM	HE	SDGC, IECC	[23–29]
	Flagella	Axonemal	Inner arm dynein (others)	DF/DM	HE	SDGC, IECC	[23,30,31]
	Flagella	Cytoplasmic	Dynein 2 (IFT dynein)	DF/DM	freeze/thaw	SDGC, GF, IECC	[32]
Tetrahymena	Cilia	Axonemal	Outer arm dynein	DC/DM	HE, LE	SDGC	[2,3,33–36]
Paramecium	Cilia	Axonemal	Outer arm dynein	DC/DM	HE	SDGC	[37–43]
Giant amoebae		Cytoplasmic	Dynein 1	HM		MT, SDGC	[44]
Slime mold		Cytoplasmic	Dynein 1	HM		MT, SDGC	[45]
Filamentous fungi		Cytoplasmic	Dynein 1	freeze/thaw, HM		MT, SDGC, GE	[46]
Sea anemone	Sperm flagella	Axonemal	Outer arm dynein	HM/DM	HE	SDGC	[47]
Nematode		Cytoplasmic	Dynein 1	HM		MT, SDGC	[48]
Mussel	Sperm flagella	Axonemal	Outer arm dynein	HM/DM	HE	SDGC	[49]
	Gill cilia	Axonemal	Outer arm dynein	DC/DM	HE	SDGC	[49]
Oyster	Sperm flagella	Axonemal	Outer arm dynein	HM/DM	HE	SDGC	[50]
Scallop	Gill cilia	Axonemal	Outer arm dynein	DC/DM	HE	SDGC	[51]
Squid	Optic lobe	Cytoplasmic	Dynein 1	HM		MT, SDGC	[52]
Drosophila	Embryo	Cytoplasmic	Dynein 1	HM		MT, SDGC	[53]
Sea urchin	Sperm flagella	Axonemal	Outer arm dynein	HM/DM	HE	SDGC	[54–61]
	Sperm flagella	Axonemal	Inner arm dynein	HM/DM	HE, LE	SDGC, IEC, HAP	[62–64]
	Embryonic cilia	Axonemal	Outer arm dynein	DC	HE	SDGC	[65,66]
	Eggs	Cytoplasmic	Dynein 1	HM		MT, SDGC, IEC, HAP	[13,14,67–70]

Continued

Dyneins

Table 4.1 Strategies for dynein isolation—cont'd

Organism	Source	Category	Dynein species	Cell fractionation	Extraction	Purification	References
Starfish	Sperm flagella	Axonemal	Outer arm dynein	HM/DM	HE/LE	SDGC	[71]
Sea squirt	Sperm flagella	Axonemal	Outer arm dynein	HM/DM	HE	SDGC, GE, IECC	[72–75]
Trout, Salmon	Sperm flagella	Axonemal	Outer arm dynein	HM/DM	HE	SDGC	[76–78]
Black rockcod	Sperm flagella	Axonemal	Outer arm dynein	HM/DM	HE	SDGC	[79]
			Inner arm dynein	HM/DM		SDGC	[79]
Chick	Embryo brain	Cytoplasmic	Dynein 1	HM		MT, SDGC, IECC	[80,81]
Bull	Sperm flagella	Axonemal	Outer arm dynein	DM	LE	SDGC	[82–84]
Calf	Brain	Cytoplasmic	Dynein 1	HM		MT, SDGC, IECC	[85,86]
Pig	Trachea cilia	Axonemal	Outer arm dynein	DC/DM	HE	SDGC	[87]
	Brain	Cytoplasmic	Dynein 1	HM		MT, GF	[88]
Human	Sperm	Axonemal	Outer arm dynein	DM	HE	SDGC	[89]

Note: all the reports on the isolation of dynein are not included in the list. *DC*, deciliation; *DF*, deflagellation; *DM*, demembranation; *GF*, gel filtration; *HAP*, hydroxylapatite chromatography; *HE*, high-salt extraction; *HM*, homogenization; *IECC*, ion exchange column chromatography; *LE*, low-salt extraction; *MT*, cosedimentation with microtubules; *SDGC*, sucrose density gradient centrifugation.

recover the flagella into the supernatant. Mammalian sperm have accessory structures, such as the fibrous sheath and outer dense fibers, so that it is hard to detach flagella using a homogenizer. Instead, sonication is an effective way to detach flagella from the head and subsequent centrifugation yields a relatively pure flagella fraction [83,84].

For isolation of *Chlamydomonas* flagella for biochemical purification of dynein, flagella excision induced by dibucaine is most often used [17]. Flagellar release is induced by an influx of Ca^{2+} and excess influx is usually terminated by adding EGTA. Flagellar excision can also be caused by pH shock using acetic acid [90]. Contamination by cell bodies in the flagellar fraction is best removed by centrifuge through a 25% sucrose cushion. Because cells do not survive well after dibucaine treatment, deflagellation by pH shock is preferentially employed in experiments examining flagellar regeneration. Dibucaine-induced deciliation is also applicable to ciliates, such as *Paramecium* [37,38] and *Tetrahymena* [33,34].

Deciliation can also be induced by a high concentration of Ca^{2+} in *Paramecium* [39,40]. Replacement of Ca^{2+} with Ba^{2+} or Mn^{2+} minimizes cell damage and contamination by trichocysts [41,42].

For isolation of cilia from the epithelia of multicellular organisms, dibucaine is less effective. Instead, several alternative methods have been devised. In trachea, cilia are detached by vigorous shaking or vortexing of the tissue in a low ionic strength solution with a high concentration of Ca^{2+}, in the presence of nonionic detergents such as Triton X-100 or CHAPS [87,91]. Although the ciliary membrane is damaged or extracted by this method, vortexing in a high concentration of Ca^{2+} and the presence of detergent has been effective in isolating ciliary or flagellar axonemes from other organisms or tissues, including *Euglena* [92] and rabbit oviduct [93].

Osmolality shock is generally used to isolate cilia from marine invertebrate tissues and embryos/larvae. Deciliation of sea urchin embryos is induced by exposure to a hypertonic solution, as first described by Auclair and Siegel [94]. Hypertonic solutions, usually made by adding extra amounts of NaCl or 2X artificial seawater, have been employed in sea urchin embryos [65,66], scallop gill [51], mussel gill [49], and ascidian branchial sacs [95]. Intriguingly, brief exposure of mussel gill to a hypertonic solution followed by transfer to a serotonin-containing normal seawater results in the isolation of only stimulated lateral cilia [96].

4.2.2 Demembranation

To remove the plasma membrane of cilia and flagella, a nonionic detergent digitonin was first used in *Tetrahymena* cilia [3]. However, Triton X-100 demembranated cell models of sea urchin sperm were found to be efficiently reactivated by Mg-ATP [97], since then this detergent became used to demembranate cilia and flagella for the isolation of dyneins [54]. Because the outer arm dynein shows "latency" that can be activated by Triton X-100, a method to avoid Triton X-100 was devised using sucrose-based osmotic shock [55]. Overall, demembranation by Triton X-100 is most commonly used in the isolation of dynein from sperm flagella.

On the other hand, *Chlamydomonas* flagella are generally demembranated by NP-40, a detergent that differs slightly different from Triton X-100 in the length of the polyethylene oxide chain [18]. NP-40 is not commercially available anymore, but a substitute Igepal CA-630 can be used [17]. Either Triton X-100 or NP-40 is used to demembranate cilia or flagella for isolation of dynein from protists, e.g., NP-40 for *Tetrahymena* [35] and Triton X-100 for *Paramecium* [39].

4.2.3 Isolation of outer arm dynein

Since the first isolation and definition of axonemal dynein from *Tetrahymena* cilia [3], outer arm dyneins have been biochemically isolated and characterized in many organisms. Single molecular species of outer arm dynein are arranged at 24-nm intervals on the A-tubule of doublet microtubules, although differences in the dynein heavy chain are reported between the proximal and distal parts of tracheal cilia [98]. Thus, it is relatively easy to purify outer arm dynein and consequently, its molecular components are the best understood. The outer arm dynein is characterized well in *Chlamydomonas* flagella [19,20], *Tetrahymena* cilia [35], *Paramecium* cilia [39,43], and in sperm flagella of sea urchin [54–61], the tunicate *Ciona* [72,73], trout and salmon [76–78], and bull [82] (Table 4.1). Although the subunits have been less characterized, outer arm dyneins were also isolated from sperm or their flagella of starfish [71], Antarctic black rockcod [79], oyster [50], mussels [49], annelids [99], sea anemone [47], and human [89], and porcine tracheal epithelial cilia [87].

The outer arm dynein is composed of two or three motor subunits (heavy chains, HCs) and associated subunits (intermediate chains, ICs, and light chains, LCs) [19,100–102] (Fig. 4.1A,B). The heavy chains constitute head and stem domains. The number of the heavy chains is phylogenetically distinct [108,109]. There are three in *Chlamydomonas, Tetrahymena,* and *Paramecium* outer arm dynein, whereas only two are present in sea urchins and tunicates. Because the major part of the head domain is constructed from the heavy chains, the former are three-headed (α, β, and γ heavy chains) and the latter two-headed (α and β heavy chains) macromolecular complexes. Electron microscopic comparison of the numbers of dynein heads among multiple organisms demonstrates that the three- and two-headed dyneins are present in two major supergroups of eukaryotes—bikonts and unikonts, respectively [109]. Bikonts and unikonts were originally defined as organisms whose ancestors had either two cilia or a single cilium, respectively [110]. The latter group includes the Metazoa, Choanozoa, Fungi, and Amoebozoa, whereas the former contains Plantae, Chromalveolata, Excavata, and Rhizaria.

The heavy chains of outer arm dynein were originally named based on their electrophoretic mobility; therefore the name of various heavy chains in one species does not always correspond to the orthologue of another species [101,102]. The composition of the intermediate and light chains has also been conserved among organisms during evolution. The presence of opisthokont-specific intermediate chain (TNDK-IC) [72,111], a phosphorylated LC (TcTex-2) [78,112], a Ca^{2+}-sensor LC (calaxin) [73,113], and bikont-specific LC (Ca^{2+}-binding LC4) [114] demonstrates that the structure and regulation of the outer arm dynein are phylogenetically related [109].

The outer arm dynein is selectively extracted from the axonemes by a high-salt solution containing 0.6 M KCl or NaCl and Mg^{2+} [97]. This selective extraction is applicable to sperm from several organisms [55,63,72,74,76,82], *Tetrahymena* [33], and *Paramecium* [39,42]. Subsequent sucrose density gradient centrifugation sediments the outer arm dynein at around 19-21S, the

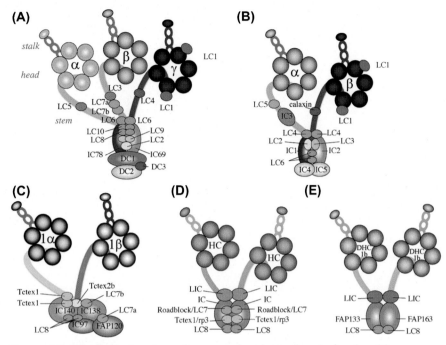

Figure 4.1 Schematic drawings of axonemal and cytoplasmic dyneins. (A) *Chlamydomonas* outer arm dynein. (B) Outer arm dynein from metazoan sperm flagella. (C) *Chlamydomonas* inner arm dynein f/I1. (D) Cytoplasmic dynein 1. (E) Cytoplasmic dynein 2 (DHC1b, IFT dynein). *These models are based on the references (A) S.M. King, R.S. Patel-King, C.G. Wilkerson, G.B. Witman, The 78,000-Mr intermediate chain of* Chlamydomonas *outer arm dynein is a microtubule binding protein, J. Cell Biol. 131 (1995) 399–409; M. Sakato, S.M. King, Design and regulation of the AAA+ microtubule motor dynein, J Struct Biol. 146 (2004) 58–71; M. Sakato, H. Sakakibara, S.M. King,* Chlamydomonas *outer arm dynein alters conformation in response to Ca²⁺, Mol. Biol. Cell 18 (2007) 3620–3634; (B) A. Hozumi, Y. Satouh, Y. Makino, T. Toda, H. Ide, K. Ogawa, S.M. King, K. Inaba, Molecular characterization of Ciona sperm outer arm dynein reveals multiple components related to outer arm docking complex protein 2, Cell Motil. Cytoskeleton 63 (2006) 591–603; K. Mizuno, P. Padma, A. Konno, Y. Satouh, K. Ogawa, K. Inaba, A novel neuronal calcium sensor family protein, calaxin, is a potential Ca²⁺-dependent regulator for the outer arm dynein of metazoan cilia and flagella, Biol. Cell 101 (2009) 91–103; S.M. King, J.L. Gatti, A.G. Moss, G.B. Witman, Outer arm dynein from trout spermatozoa: substructural organization, Cell Motil. Cytoskeleton 16 (1990) 266–278; (C) M. Wirschell, C. Yang, P. Yang, L. Fox, H.A. Yanagisawa, R. Kamiya, G.B. Witman, M.E. Porter, W.S. Sale, IC97 is a novel intermediate chain of I1 dynein that interacts with tubulin and regulates interdoublet sliding, Mol. Biol. Cell 20 (2009) 3044–3054; K. Inaba, Regulatory subunits of axonemal dynein, in: K. Hirose, L.A. Amos (Eds.), Handbook of Dynein, Pan Stanford Publishing Pte. Ltd, 2012, pp. 303–324; (D) V.J. Allan, Cytoplasmic dynein, Biochem. Soc. Trans. 39 (2011) 1169–1178; (E) P. Rompolas, L.B. Pedersen, R.S. Patel-King, S.M. King,* Chlamydomonas *FAP133 is a dynein intermediate chain associated with the retrograde intraflagellar transport motor, J. Cell Sci. 120 (2007) 3653–3665.*

S-values depending on an organism (Fig. 4.2A). The outer arm dynein from sucrose density gradient contains a trace amount of tubulin, which can be removed by subsequent Mono Q anion-exchange column chromatography (Fig. 4.2B) [74]. An exception is that discrete two-headed outer arm dynein is

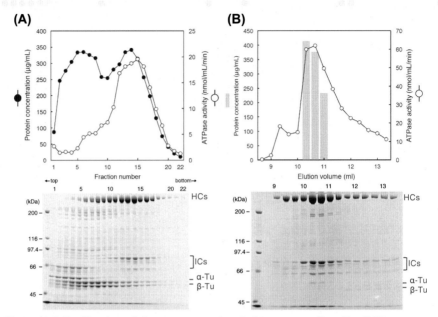

Figure 4.2 Purification of the outer arm dynein from sperm flagella of *Ciona intestinalis*. (A) Sucrose density gradient centrifugation of a high-salt extract of the axonemes. The upper profile shows protein concentration and ATPase activity; the lower shows SDS-PAGE with 6% polyacrylamide gel. The positions of the dynein HC, ICs, and tubulins are indicated. (B) Mono Q anion-exchange column chromatography of pooled fraction from the sucrose density gradient. The upper panel shows the protein elution profile with ATPase activity; the lower shows SDS-PAGE with 6% polyacrylamide gel. Note that components such as tubulins are separated from outer arm dynein. *The data are kindly provided by Dr. Osamu Kutomi.*

difficult to isolate from sperm flagella of Antarctic rockcod [79]. A similar pattern is observed in the flatfish (*Pleuronectes yokohama*) (Inaba, unpublished), implying that dyneins from marine teleost sperm might have a salt-dependent regulatory system.

During the early stages of dynein research, sucrose density gradients were preferentially used to isolate axonemal dyneins because of their large molecular mass [2]. Hence, the S-values were conventionally used to compliment the nomenclature of dynein species. However, the substructure of outer arm dynein can be partially dissociated during preparation. Sucrose density gradient centrifugation of high-salt extracts of axonemes usually yield intact and/or dissociated outer arm dyneins: *Tetrahymena* dynein, 30S and 14S [33], or 22S and 14S [36]; *Paramecium*, 22S and 14S [39,40]; sea urchin dynein, 21S and 12–14S [56]; trout and salmon, 19S [77,78]; *Chlamydomonas*, 18S and 12S [17,21,22]. The nomenclature using "S" value has been substituted with that expressing the dynein species with "outer" or "inner" arm dynein but is still useful to compare molecular sizes between different organisms and to describe the dissociated components of dyneins.

Outer arm dyneins from sea urchin sperm can be dissociated into subcomponents in a low-ionic strength solution [115]. Subcomponents with different heavy chains show distinct ATPase, microtubule sliding, and molecular structure properties [58,115–117]. This method for dissociation of outer arm dynein into subcomponents is applicable to *Chlamydomonas* [17,21] and trout outer arm dyneins [77].

High-salt extraction removes both outer and inner arm dyneins from the axonemes of *Chlamydomonas* [17]. In the absence of Mg^{2+}, sucrose density gradient centrifugation accelerates the dissociation of three-headed outer arm dynein into two subcomponents containing α–β heavy chains (18S) and γ heavy chain (12S) due to hydrostatic pressure imparted by centrifugation. The addition of Mg^{2+} in the gradient and the use of smaller rotors reducing hydrostatic pressure help avoid dissociation [118]. Use of mutants lacking inner arm dynein (e.g., *ida1*) improves the purity of outer arm dynein in sucrose density gradient fractions [118].

Outer arm dyneins have several associated proteins that are necessary for regulation and anchoring to the axonemal microtubules. In *Chlamydomonas*, an ~7S component is released by high-salt extraction from axonemes [22]. It was found to be a docking complex (DC), which is composed of DC1 (83 kDa), DC2 (62 kDa), and DC3 (25 kDa) and is directly associated with the outer arm dynein in situ. The orthologues of DC1 and DC3 are not found in animals, but multiple DC2 species are present in the axonemes of the ascidian *Ciona* and are isolated in association with the outer arm dynein [75]. It should be kept in mind that other proteins associated with outer arm dynein may also dissociate during extraction and purification, as seen in the case of TPR protein Ap58 [57].

4.2.4 Isolation of inner arm dynein

Inner arm dyneins are more diverse than outer arm dynein in their molecular species and specialized arrangement in the axoneme. A discrete two-headed species and several other forms of inner arms were found by electron microscopy with the quick-freeze deep-etch method [119,120], thin-sections [23], and image analysis [9]. Comparison between the electron microscopic images and dynein heavy chains in several *Chlamydomonas* mutants identified three species of inner dynein arm, I1, I2, and I3, in every 96-nm repeat [23]. Separation of KCl extracts of axonemes by Mono Q anion-exchange column chromatography results in several discrete ATPase peaks [25,26] (see Fig. 4.3). The extract from the axonemes of outer armless mutant (*oda1*) is separated into seven ATPase peaks, designated dyneins a–g [24]. Dynein-f is a two-headed species that corresponds to I1, comprising two heavy chains (1α and 1β), three intermediate chains (IC140, IC138, and IC97), and five LCs (TcTex-1, TcTex-2b, LC7a and LC7b, LC8) and an associated protein FAP120 [26]. Others

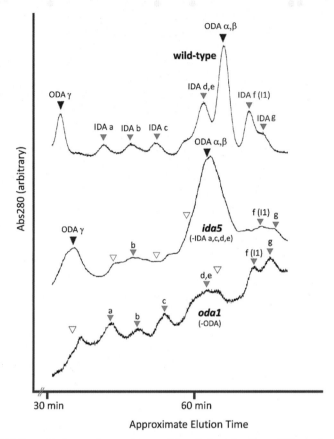

Figure 4.3 Mono Q chromatographic profiles of dyneins from *Chlamydomonas* **wild type and mutants.** Flagella are prepared from wild-type *ida5* (lacking inner arm dynein a, c, d, and e) and *oda1* (lacking outer arm dynein) strains, demembranated and extracted by a high-salt solution. The extract is separated by Mono Q column by a linear gradient of KCl [121]. Closed and open triangles represent the presence and absence of each dynein species, confirmed by SDS-PAGE; red, outer arm dyneins; blue, inner arm dyneins. *The data are kindly provided by Dr. Ryosuke Yamamoto.*

(a–e and g) are single-headed dyneins, which additionally contain actin and p28 (a, c, d), or actin and centrin (b, e, g) [122] (Fig. 4.2C). Cryo–electron tomography has specified the topological arrangement of eight heavy chains on the A-tubule [12].

Because f/I1 is the only two-headed inner arm dynein, it is purified as a 21S particle from high-salt extracts of mutant axonemes, which lack both outer arm dynein and radial spokes, by sucrose density gradient centrifugation [27]. Mono Q–based separation of the axonemal extract from inner armless mutants has been useful to define missing inner arm species [17,24,28] (Fig. 4.3). Analysis of the protein components in each dynein fraction resulted in identification of novel dynein subunits, for example, both 38 kDa and 44 kDa proteins were identified in the dynein-d fraction [30,31]. However, a purified dynein does not always contain the whole complement of components that are arranged on the microtubule in situ. For example, an ankyrin repeat protein (FAP120) is associated with f/I1 but dissociates during Mono Q column chromatography [29].

Column chromatography with several different resins are also effective tools for isolation of dyneins. However, careful resin selection is a key to obtain high recovery of dyneins. For example, cross-linked dextran (e.g., Sephadex) or cellulose (e.g., DE52) resins strongly adsorb dynein and the recovery of dynein is, therefore, extremely low (unpublished observation). Gel filtration with a large size exclusion may be applicable for dynein isolation but this property of the resin affects the recovery of the outer arm dynein. The resins that should be considered for purification step are Sepharose CL-6B [77,123], Superdex 200 [72], and Superose 6 [29,32,74] for gel filtration; DEAE-Sephacel [35], Mono Q [23,24,74], and Uno Q [30] for anion-exchange column chromatography; and hydroxyapatite [18,62].

As compared to *Chlamydomonas* flagella, isolation and characterization of inner arm dyneins in other cells or organisms have been done only in a few cases. Potentially single-headed inner arm dynein was isolated from sea urchin sperm by sucrose density gradient centrifugation and column chromatography [64]. KCl-treated outer armless axonemes were extracted by a low-ionic strength solution [124] or higher-salt solution, followed by hydroxyapatite and sucrose density gradient centrifugation [62]. These purification steps resulted in isolating a fraction of sea urchin sperm axonemes rich in the heavy chain "D" and "B." On the other hand, potentially two-headed inner arm dynein, possibly equivalent to f/I1 dynein, was isolated from sea urchin sperm flagella [63]. In this case, KCl-treated outer armless axonemes were extracted with a higher-salt solution containing 5 mM ATP, followed by sucrose density gradient centrifugation. The ATPase showed a peak at 22–23S, a little larger than that of outer arm dynein. In this fraction, the heavy chains "C" and "A" as well as two major intermediate chains are included [63]. Application of a similar method for isolation of inner arm dynein from the ascidian *Ciona* results in the purification of two-headed f/I1 dynein with orthologues of *Chlamydomonas* IC140, IC138, and IC97 (Kutomi, Mizuno, Inaba, unpublished).

4.3 Cytoplasmic dynein

Cytoplasmic dynein is responsible for microtubule-dependent minus-ended transport. In addition to sucrose density gradient centrifugation and gel filtration, methods that take advantages of the binding to microtubules and ATP-dependent release are used to isolate cytoplasmic dynein. Immunoprecipitation or tag-based pull down methods are also powerful strategy for isolation.

4.3.1 Cytoplasmic dynein 1

Cytoplasmic dynein is responsible for the microtubule-dependent minus-ended transport in cytoplasm. It was first isolated from sea urchin eggs [13,14], but there was much argument over whether it was a precursor of ciliary axonemal dyneins. Then one of the MAPs in brain, MAP1C, was shown to be released by ATP from microtubules and its ATPase was activated by microtubules. Isolated molecules had a two-headed structure, similar to axonemal two-headed dyneins, demonstrating that MAP1C is the cytoplasmic dynein [85,86]. Cytoplasmic dynein contains the heavy chain (motor domains; DYNC1H) and five sub-units, including the intermediate chain (DYNC1I), the light–intermediate chain (DYNC1LI), and three LCs: LC8 (DYNLL), Tctex1/rp3 (DYNLT), and roadblock (DYNLRB) [107]. All the subunits form a homodimer, making up the two-headed motor. Cytoplasmic dynein functions in association with a large protein complex, called dynactin, which is composed of 23 subunits including p150/Glued, p50/dynamitin, Arp1 filament, Bicoidal D, RZZ complex, Lis-1, NDE1, and other proteins [80,81,125,126] (Fig. 4.1D).

Biochemical isolation of cytoplasmic dynein was started first using sea urchin eggs in 1980s [14,67]. Egg cytosol was precipitated by ammonium sulfate and sequentially separated by gel filtration (Sepharose 4B), calmodulin-affinity, hydroxyapatite chromatography. and sucrose density gradient centrifugation [68]. The purified dynein showed a sedimentation coefficient of 12.3S, suggesting that it was a dissociate form of two-headed cytoplasmic dynein. Sucrose density gradient centrifugation of the high-salt extract of the proteins associated with taxol-stabilized microtubules also sediments an ATPase at around 12S in the sucrose gradient [69]. Later, a 20S dynein ATPase, different from ciliary precursor in the immunoreactivity, was isolated from the ATP extract of the proteins associated with taxol-stabilized microtubules [70].

In parallel, attempts to isolate cytoplasmic dynein were carried out from brain. MAPs from calf white matter were separated by SDS-PAGE into five bands, MAP1A, 1B, 1C, 2A, and 2B [15], among which MAP1C is extracted from the microtubules by GTP and shows a microtubule-activated ATPase activity [85]. The ATPase in the ATP extract of MAPs sediments through sucrose density gradient centrifugation at 20S. This ATPase has a heavy chain with the same electrophoretic mobility as that of dynein heavy chain of ciliary dynein, and has a two-headed dynein structure [85,86].

By using a similar principle, cytoplasmic dyneins have been isolated from several sources, including pig brain [88], chick embryo brain [80,81], *Drosophila* embryo homogenates [53], squid optic lobe [52], the cellular slime mold *Dictyostelium discoideum* [45], the nematode *Caenorhabditis elegans* [48], the giant amoeba *Reticulomyxa* [44], and the filamentous fungus *Neurospora crassa* [46]. Incubation of the cytosol at 37°C in the presence of ATP, GTP, and taxol sediments microtubules with MAPs, thus the supernatant is rich in cytoplasmic dynein. The supernatant is incubated with microtubules purified from bovine brain in the presence of AMP-PNP. AMP-PNP, a nonhydrolyzable analogue of ATP, stabilizes the binding of kinesin to the microtubule but does not affect the binding of cytoplasmic dynein to the microtubules in the absence of ATP. In addition, depletion of ATP by the addition of apyrase or glucose/hexokinase accelerates the binding of dyneins to the microtubules, resulting in higher yields. The ATP extract of microtubules is subjected to sucrose density gradient centrifugation. Cytoplasmic dynein sediments at 20S, whereas small amounts of kinesin are separated from dynein and sediment at ~9S. The vesicle transport activity in the fraction released by ATP is partly decreased by sucrose density centrifugation. Following purification by Mono Q anion-exchange column chromatography, this activity dramatically decreased due to dissociation of the dynactin complex [80].

4.3.2 Cytoplasmic dynein 2 (intraflagellar transport dynein)

Intraflagellar transport (IFT) is a bidirectional movement in cilia and flagella, composed of the anterograde transport driven by kinesin-2 and retrograde transport driven by cytoplasmic dynein 1b (also called cytoplasmic dynein 2 or the IFT dynein) [127–130]. Cytoplasmic dynein 2 is isolated from the flagellar matrix by a method similar to that used in the isolation of cytoplasmic dynein 1 [32]. To obtain flagellar matrix proteins, isolated flagella were subjected to repeated freeze/thaw to facilitate disruption of the membrane, and resulting supernatants are used for further purification steps. Flagellar membrane-matrix proteins are also extracted by a solution containing 1% Tergitol NP-40 or 0.1% Igepal CA-630. Most of the cytoplasmic dynein 2 is extracted in the flagellar matrix fraction by freeze–thaw extraction, along with kinesin-2, IFT-A, and IFT-B complexes. However, some fraction of it still remains with the resultant axonemes, and can be further extracted by 0.6 M KCl solution along with the axonemal dyneins.

Sucrose density gradient centrifugation of the flagellar matrix proteins separates the extract into two peaks around ~18S and ~10S. The former contains two-headed cytoplasmic dynein, composed of the heavy chain DHC1b (DYNC2H1); intermediate chains FAP133 (WDR34), and FAP163 (WDR60); light–intermediate chain D1bLIC (DYNC2LI1); and the light chain LC8 (Fig. 4.2E). The latter peak contains a dissociated subcomplex of cytoplasmic dynein 1b/2. Mono Q anion-exchange chromatography of the flagellar matrix proteins causes the dissociation of cytoplasmic dynein 1b/2 into two subcomplexes: DHC1b-D1bLIC and FAP133-LC8 [32]. A larger complex associated with cytoplasmic dynein 1b/2 is isolated by gel filtration on a Superose 6 gel filtration column of the freeze–thaw extract prepared in a low-ionic strength solution. This fraction contains cytoplasmic dynein 2, associated with FLA10 kinesin-2 and IFT-complex A (but not B complex) [32].

4.3.3 Recombinant dynein

Progress in the isolation of recombinant cytoplasmic dynein or native dynein heavy chain with tagged-dynein components has brought about a powerful new system for studying dynein structure and molecular function, as represented by recent single-molecule analysis [131] and X-ray crystal structures [132,133]. Because dynein has a very large molecular size, several attempts with different constructs and host cells for transfection have been tried to obtain soluble recombinant protein. Studies have been reported on the isolation of cytoplasmic dyneins with several hosts for transformation, including mammalian cells, *Dictyostelium, Escherichia coli,* and yeast (Table 4.1). In contrast, there are only a few reports on the isolation of recombinant axonemal dyneins [134], in which the DYH4 gene tagged with His- and FLAG-tagged GFP at the N-terminus was subcloned into pGBKT7 vector and transformed into a wild-type *Tetrahymena thermophila* strains. The recombinant dynein could be expressed and isolated from cilia.

Mazumdar et al. [135] constructed a full-length cDNA of rat cytoplasmic dynein from clones of dynein fragments, and tagged it with FLAG-tag or His-tag at the C-terminus. Plasmid was transfected into insect cells (Sf9) using baculovirus. The tagged cDNA was also subcloned into the pCMV vector and transfected into mammalian COS-7 cells (African green monkey kidney cells). The expressed proteins were recovered from the cells by homogenization in a buffer containing protease inhibitors. The supernatant after centrifugation was used for further purification by Ni^{2+}-affinity column or by sucrose density gradient centrifugation [135]. Recombinant human cytoplasmic dynein 1 and 2 were also reported to be isolated from HEK (human embryonic kidney)-293 cells [136]. The clones were inserted into pcDNA5/FRT/TO vector with streptavidin-binding peptide (SBP)-tag (for purification by avidin-biotin binding) and SNAP-tag (for fluorescent dye labeling by enzyme-mediated covalent binding). Cells were homogenized and centrifuged, then the supernatant was applied to an affinity resin with streptavidin ligand [136]. Recombinant intermediate chain of cytoplasmic dynein with a purification tag is reported to be effective in isolating cytoplasmic dynein from Hela cells [137].

Dictyostelium AX-2 cells were transfected by a plasmid containing the full-length or C-terminal motor region plus a part of the stem region of the cytoplasmic dynein heavy chain [138]. Cells were sonicated and centrifuged to obtain a clear supernatant, which was subsequently mixed with taxol-stabilized bovine microtubules in the presence of apyrase. Recombinant dynein was extracted by Mg-ATP and purified by sucrose density gradient centrifugation [138]. GFP could be inserted at the N-terminus of a truncated dynein heavy chain with a Gly–Gly–Gly linker [139,140]. Using anti-GFP antibody, the recombinant dynein could be attached to the glass slide through biotinyl protein G and streptavidin for motility assays.

For isolation of recombinant cytoplasmic dynein from the budding yeast *Saccharomyces cerevisiae*, a truncated yeast cytoplasmic dynein fragment and various mutants were subcloned into the pGEX vector and expressed in *E. coli*. To increase the solubility, GST and BCCP (biotin carboxyl carrier protein) were fused with the dynein at the N- and C-terminus, respectively. His-tag was located at the C-terminus of BCCP. Proteins were affinity purified by glutathione agarose and Ni^{2+}-affinity column [141].

Recombinant cytoplasmic dynein was also produced in yeast by homologous recombination [126,142]. Tags for a purification strategy were added to the 5'-end of full-length or truncated dynein constructs fused with GFP, including a ZZ tag (two copies of the IgG-binding domain of protein A), TEV protease cleavage site, and a short sequence to enhance the protease cleavage [142]. GFP was used for tethering dynein onto the glass slide for motility analysis. HA tags were introduced for immunoblotting, and HaloTag (a modified halo-alkane dehalogenase tag) for covalent linkage of fluorescent dyes or biotin. Sequences of GST, FRB (FKBP- and rapamycin-binding domain), and FKBP12 (a 12 kDa FK506-binding protein) were added to the N-terminus of truncated dynein to artificially dimerize the truncated dynein heavy chains [142]. After cell lysis, recombinant dyneins were recovered in the clear cytosol, then bound to IgG Sepharose, released by TEV protease and purified by microtubule-affinity method.

4.4 Storage of dynein

Compared to other molecular motors, dyneins have highly complex structures with very large motor subunits. Therefore, they are basically not stable and cannot be stored for long periods after isolation. Experiments concerning dynein activity or molecular structure should therefore be done within several days. Although a thiophosphorylated form of *Tetrahymena* 22S dynein was reported to be able to be stored at −70°C [39], storage in a freezer may cause a partial or drastic changes of dynein structure, resulting in denaturation. However, several reports demonstrate procedures for dynein storage. The method most often used is quick freezing in liquid N_2 for both cytoplasmic dynein [142,143] and axonemal dynein [144]. Additives for stabilization of dynein might improve the storage of dynein activity, e.g., addition of 10% sucrose and 1 mM DTT [74,144] or 10% (v/v) glycerol [145].

4.5 Conclusion and perspective

Biochemical purification of native dynein provides information of the subunit composition, associated proteins, and motor activity. Sources for biochemical purification of dynein are limited to organisms that provide a large quantity of starting material. Recent genetic approaches enable us to use high-affinity tags for isolation of dynein complexes from lesser quantities of material. If the genome of the source organism has been sequenced, the subunits and the associated proteins of dyneins can be identified by mass spectrometry–based analysis of small amounts of protein. However, it is always possible that introduction of tags to dynein components might change the structure and activity of dynein. In addition, biochemical isolation using several steps of separation might cause the dissociation of protein complexes or changes in motor activity. Both single-particle analysis and X-ray crystal analysis have revealed the fine molecular structure of dyneins. Progress in cryo–electron tomography clearly visualizes the dynein structure in situ in axonemes or within cells. Combining these techniques would allow researchers to explore the difference between the biochemically isolated dyneins and the dyneins in situ and specify the missing components. Several genetic or biochemical tools for protein–protein interactions are expected to fill the missing gaps and thus we anticipate knowing how these interactions dynamically change in dynein-mediated cellular activities.

Acknowledgments

I thank Osamu Kutomi (Yamanashi University) and Ryosuke Yamamoto (Osaka University) for their help in making figures. I am grateful to Steve King (University of Connecticut Health Center) for critical reading of the manuscript. A part of this chapter includes work done under the support of Grant-in-Aid 15H01201 for Scientific Research on Innovative Areas and 22370023 for Scientific Research (B) from the Ministry of Education, Culture, Sports, Science, and Technology of Japan (MEXT).

References

[1] B.B.J. Afzelius, Electron microscopy of the sperm tail; results obtained with a new fixative, J. Biophys. Biochem. Cytol. 5 (1959) 269–278.

[2] I.R. Gibbons, A.J. Rowe, Dynein: a protein with adenosine triphosphatase activity from cilia, Science 149 (1965) 424–426.

[3] I.R. Gibbons, Studies on the protein components of cilia from *Tetrahymena* performs, Proc. Natl. Acad. Sci. U.S.A. 50 (1963) 1002–1010.

[4] F.D. Warner, D.R. Mitchell, Structural conformation of ciliary dynein arms and the generation of sliding forces in *Tetrahymena* cilia, J. Cell Biol. 76 (1978) 261–277.

[5] U.W. Goodenough, J.E. Heuser, Substructure of the outer dynein arm, J. Cell Biol. 95 (1982) 798–815.

[6] P. Satir, J. Wais-Steider, S. Lebduska, A. Nasr, J. Avolio, The mechanochemical cycle of the dynein arm, Cell Motil. 1 (1981) 303–327.

[7] D. Nicastro, J.R. McIntosh, W. Baumeister, 3D structure of eukaryotic flagella in a quiescent state revealed by cryo-electron tomography, Proc. Natl. Acad. Sci. U.S.A. 102 (2005) 15889–15894.

[8] T. IshikawA, H. Sakakibara, K. Oiwa, The architecture of outer dynein arms in situ, J. Mol. Biol. 368 (2007) 1249–1258.

[9] D.N. Mastronarde, E.T. O'Toole, K.L. McDonald, J.R. McIntosh, M.E. Porter, Arrangement of inner dynein arms in wild-type and mutant flagella of *Chlamydomonas*, J. Cell Biol. 118 (1992) 1145–1162.

[10] M.E. Porter, Axonemal dyneins: assembly, organization, and regulation, Curr. Opin. Cell Biol. 8 (1996) 10–17.

[11] D. Nicastro, C. Schwartz, J. Pierson, R. Gaudette, M.E. Porter, J.R. McIntosh, The molecular architecture of axonemes revealed by cryoelectron tomography, Science 313 (2006) 944–948.

[12] K.H. Bui, H. Sakakibara, T. Movassagh, K. Oiwa, T. Ishikawa, Molecular architecture of inner dynein arms in situ in *Chlamydomonas reinhardtii* flagella, J. Cell Biol. 183 (2008) 923–932.

[13] R.C. Weisenberg, E.W. Taylor, Studies on the ATPase activity of sea urchin eggs and the isolated mitotic apparatus, Exp. Cell Res. 53 (1968) 372–384.

[14] M.M. Pratt, The identification of a dynein ATPase in unfertilized sea urchin eggs, Dev. Biol. 74 (1980) 364–378.

[15] G.S. Bloom, T.A. Schoenfeld, R.B. Vallee, Widespread distribution of the major polypeptide component of MAP 1 (microtubule-associated protein 1) in the nervous system, J. Cell Biol. 98 (1984) 320–330.

[16] B.M. Paschal, R.B. Vallee, Retrograde transport by the microtubule-associated protein MAP 1C, Nature 330 (1987) 181–183.

[17] S.M. King, Purification of axonemal dyneins and dynein-associated components from *Chlamydomonas*, Methods Cell Biol. 92 (2009) 31–48.

[18] K.K. Pfister, R.B. Fay, G.B. Witman, Purification and polypeptide composition of dynein ATPases from *Chlamydomonas* flagella, Cell Motil. 2 (1982) 525–547.

[19] S.M. King, The dynein microtubule motor, Biochim. Biophys. Acta 1496 (2000) 60–75.

[20] K. Wakabayashi, Regulation of axonemal outer-arm dyneins in cilia, in: S.M. King (Ed.), Dyneins: Structure, Biology and Disease, 2011, pp. 296–311.

[21] K.K. Pfister, G.B. Witman, Subfractionation of *Chlamydomonas* 18 S dynein into two unique subunits containing ATPase activity, J. Biol. Chem. 259 (1984) 12072–12080.

[22] S. Takada, R. Kamiya, Functional reconstitution of *Chlamydomonas* outer dynein arms from alpha-beta and gamma subunits: requirement of a third factor, J. Cell Biol. 126 (1994) 737–745.

[23] G. Piperno, Z. Ramanis, E.F. Smith, W.S. Sale, Three distinct inner dynein arms in *Chlamydomonas* flagella: molecular composition and location in the axoneme, J. Cell Biol. 110 (1990) 379–389.

[24] U.W. Goodenough, B. Gebhart, V. Mermall, D.R. Mitchell, J.E. Heuser, High-pressure liquid chromatography fractionation of *Chlamydomonas* dynein extracts and characterization of inner-arm dynein subunits, J. Mol. Biol. 194 (1987) 481–494.

[25] O. Kagami, S. Takada, R. Kamiya, Microtubule translocation caused by three subspecies of inner-arm dynein from *Chlamydomonas* flagella, FEBS Lett. 264 (1990) 179–182.

[26] M. Wirschell, C. Yang, P. Yang, L. Fox, H.A. Yanagisawa, R. Kamiya, G.B. Witman, M.E. Porter, W.S. Sale, IC97 is a novel intermediate chain of I1 dynein that interacts with tubulin and regulates interdoublet sliding, Mol. Biol. Cell. 20 (2009) 3044–3054.

[27] G. Habermacher, W.S. Sale, Regulation of flagellar dynein by phosphorylation of a 138-kD inner arm dynein intermediate chain, J. Cell Biol. 136 (1997) 167–176.

[28] R. Yamamoto, K. Song, H.A. Yanagisawa, L. Fox, T. Yagi, M. Wirschell, M. Hirono, R. Kamiya, D. Nicastro, W.S. Sale, The MIA complex is a conserved and novel dynein regulator essential for normal ciliary motility, J. Cell Biol. 201 (2013) 263–278.

[29] K. Ikeda, R. Yamamoto, M. Wirschell, T. Yagi, R. Bower, M.E. Porter, W.S. Sale, R. Kamiya, A novel ankyrin-repeat protein interacts with the regulatory proteins of inner arm dynein f (I1) of *Chlamydomonas* reinhardtii, Cell Motil. Cytoskeleton 66 (2009) 448–456.

[30] R. Yamamoto, H.A. Yanagisawa, T. Yagi, R. Kamiya, A novel subunit of axonemal dynein conserved among lower and higher eukaryotes, FEBS Lett. 580 (2006) 6357–6360.

[31] R. Yamamoto, H.A. Yanagisawa, T. Yagi, R. Kamiya, Novel 44-kilodalton subunit of axonemal Dynein conserved from *Chlamydomonas* to mammals, Eukaryot. Cell 7 (2008) 154–161.

[32] P. Rompolas, L.B. Pedersen, R.S. Patel-King, S.M. King, *Chlamydomonas* FAP133 is a dynein intermediate chain associated with the retrograde intraflagellar transport motor, J. Cell Sci. 120 (2007) 3653–3665.

[33] M.E. Porter, K.A. Johnson, Characterization of the ATP-sensitive binding of *Tetrahymena* 30 S dynein to bovine brain microtubules, J. Biol. Chem. 258 (1983) 6575–6581.

[34] T.M. Gibson, D.J. Asai, Isolation and characterization of 22S outer arm dynein from *Tetrahymena* cilia, Methods Cell Biol. 62 (2000) 433–440.

[35] K.A. Johnson, J.S. Wall, Structure and molecular weight of the dynein ATPase, J. Cell Biol. 96 (1983) 669–678.

[36] Y.Y. Toyoshima, Chymotryptic digestion of *Tetrahymena* 22S dynein. I. Decomposition of three-headed 22S dynein to one- and two-headed particles, J. Cell Biol. 105 (1987) 887–895.

[37] Y. Mogami, K. Takahashi, Calcium and microtubule sliding in ciliary axonemes isolated from *Paramecium caudatum*, J. Cell Sci. 61 (1983) 107–121.

[38] O. Kutomi, M. Hori, M. Ishida, T. Tominaga, H. Kamachi, F. Koll, J. Cohen, N. Yamada, M. Noguchi, Outer dynein arm light chain 1 is essential for controlling the ciliary response to cyclic AMP in *Paramecium tetraurelia*, Eukaryot. Cell 11 (2012) 645–653.

[39] T. Hamasaki, K. Barkalow, J. Richmond, P. Satir, cAMP-stimulated phosphorylation of an axonemal polypeptide that copurifies with the 22S dynein arm regulates microtubule translocation velocity and swimming speed in *Paramecium*, Proc. Natl. Acad. Sci. U.S.A. 88 (1991) 7918–7922.

[40] K. Barkalow, T. Hamasaki, P. Satir, Regulation of 22S dynein by a 29-kD light chain, J. Cell Biol. 126 (1994) 727–735.

[41] T. Fukushi, K. Hiwatashi, Preparation of mating reactive cilia from *Paramecium* caudate by MnCl2, J. Protozool. 17 (1970) s21.

[42] D.L. Nelson, Preparation of cilia and subciliary fractions from *Paramecium*, Methods Cell Biol. 47 (1995) 17–24.

[43] S.M. Travis, D.L. Nelson, Purification and properties of dyneins from *Paramecium* cilia, Biochim. Biophys. Acta 966 (1988) 73–83.

[44] U. Euteneuer, M.P. Koonce, K.K. Pfister, M. Schliwa, An ATPase with properties expected for the organelle motor of the giant amoeba, *Reticulomyxa*, Nature 332 (1988) 176–178.

[45] M.P. Koonce, J.R. McIntosh, Identification and immunolocalization of cytoplasmic dynein in *Dictyostelium*, Cell Motil. Cytoskeleton 15 (1990) 51–62.

[46] S. Kumar, I.H. Lee, M. Plamann, Two approaches to isolate cytoplasmic dynein ATPase from *Neurospora crassa*, Biochimie 82 (2000) 229–236.

[47] H. Mohri, K. Inaba, M. Kubo-Irie, H. Takai, Y. Yano-Toyoshima, Characterization of outer arm dynein in sea anemone, *Anthopleura midori*, Cell Motil. Cytoskeleton 44 (1999) 202–208.

[48] R.J. Lye, M.E. Porter, J.M. Scholey, J.R. McIntosh, Identification of a microtubule-based cytoplasmic motor in the nematode C. elegans, Cell 51 (1987) 309–318.

[49] R.E. Stephens, G. Prior, Dynein inner arm heavy chain identification in cAMP-activated flagella using class-specific polyclonal antibodies, Cell Motil. Cytoskeleton 30 (1995) 261–271.

[50] S. Wada, M. Okuno, K. Nakamura, H. Mohri, Dynein of sperm flagella of oyster belonging to protostomia also has a two-headed structure, Biol. Cell. 76 (1992) 311–317.

[51] R.W. Linck, Comparative isolation of cilia and flagella from the lamellibranch mollusc, *Aequipecten irradians*, J. Cell Sci. 12 (1973) 345–367.

[52] B.J. Schnapp, T.S. Reese, Dynein is the motor for retrograde axonal transport of organelles, Proc. Natl. Acad. Sci. U.S.A. 86 (1989) 1548–1552.

[53] T.S. Hays, M.E. Porter, M. McGrail, P. Grissom, P. Gosch, M.T. Fuller, J.R. McIntosh, A cytoplasmic dynein motor in *Drosophila*: identification and localization during embryogenesis, J. Cell Sci. 107 (1994) 1557–1569.

[54] I.R. Gibbons, E. Fronk, Some properties of bound and soluble dynein from sea urchin sperm flagella, J. Cell Biol. 54 (1972) 365–381.

[55] C.W. Bell, C.L. Fraser, W.S. Sale, W.J. Tang, I.R. Gibbons, Preparation and purification of dynein, Methods Enzymol. 85 (1982) 450–474.

[56] C.W. Bell, E. Fronk, I.R. Gibbons, Polypeptide subunits of dynein 1 from sea urchin sperm flagella, J. Supramol. Struct. 11 (1979) 311–317.

[57] K. Ogawa, K. Inaba, Ap58: a novel in situ outer dynein arm-binding protein, Biochem. Biophys. Res. Commun. 343 (2006) 385–390.

[58] W.S. Sale, U.W. Goodenough, J.E. Heuser, The substructure of isolated and in situ outer dynein arms of sea urchin sperm flagella, J. Cell Biol. 101 (1985) 1400–1412.

[59] K. Inaba, H. Mohri, Dynamic conformational changes of 21 S dynein ATPase coupled with ATP hydrolysis revealed by proteolytic digestion, J. Biol. Chem. 264 (1989) 8384–8388.

[60] I.R. Gibbons, Dynein ATPases as microtubule motors, J. Biol. Chem. 263 (1988) 15837–15840.

[61] K. Ogawa, H. Mohri, A dynein motor superfamily, Cell Struct. Funct. 21 (1996) 343–349.

[62] K. Inaba, T. Mohri, H. Mohri, B-band protein in sea urchin sperm flagella, Cell Motil. Cytoskeleton 10 (1988) 506–517.

[63] E. Yokota, I. Mabuchi, Isolation and characterization of a novel dynein that contains C and A heavy chains from sea urchin sperm flagellar axonemes, J. Cell Sci. 107 (1994) 345–351.

[64] S. Wada, M. Okuno, H. Mohri, Inner arm dynein ATPase fraction of sea urchin sperm flagella causes active sliding of axonemal outer doublet microtubule, Biochem. Biophys. Res. Commun. 175 (1991) 173–178.

[65] R.E. Stephens, Tubulin and tektin in sea urchin embryonic cilia: pathways of protein incorporation during turnover and regeneration, J. Cell Sci. 107 (1994) 683–692.

[66] Y. Jin, S. Yaguchi, K. Shiba, L. Yamada, J. Yaguchi, D. Shibata, H. Sawada, K. Inaba, Glutathione transferase theta in apical ciliary tuft regulates mechanical reception and swimming behavior of sea urchin embryos, Cytoskeleton (Hoboken) 70 (2013) 453–470.

[67] D.J. Asai, L. Wilson, A latent activity dynein-like cytoplasmic magnesium adenosine triphosphatase, J. Biol. Chem. 260 (1985) 699–702.

[68] S. Hisanaga, H. Sakai, Cytoplasmic dynein of the sea urchin egg. II. Purification, characterization and interactions with microtubules and Ca-calmodulin, J. Biochem. 93 (1983) 87–98.

[69] J.M. Scholey, B. Neighbors, J.R. McIntosh, E.D. Salmon, Isolation of microtubules and a dynein-like MgATPase from unfertilized sea urchin eggs, J. Biol. Chem. 259 (1984) 6516–6525.

[70] M.E. Porter, P.M. Grissom, J.M. Scholey, E.D. Salmon, J.R. McIntosh, Dynein isoforms in sea urchin eggs, J. Biol. Chem. 263 (1988) 6759–6771.

[71] I. Mabuchi, T. Shimizu, Y. Mabuchi, A biochemical study of flagellar dynein from starfish spermatozoa: protein components of the arm structure, Arch. Biochem. Biophys. 176 (1976) 564–576.

[72] P. Padma, A. Hozumi, K. Ogawa, K. Inaba, Molecular cloning and characterization of a thioredoxin/nucleoside diphosphate kinase related dynein intermediate chain from the ascidian, *Ciona intestinalis*, Gene 275 (2001) 177–183.

[73] K. Mizuno, P. Padma, A. Konno, Y. Satouh, K. Ogawa, K. Inaba, A novel neuronal calcium sensor family protein, calaxin, is a potential Ca^{2+}-dependent regulator for the outer arm dynein of metazoan cilia and flagella, Biol. Cell. 101 (2009) 91–103.

[74] K. Inaba, K. Mizuno, Purification of dyneins from sperm flagella, Methods Cell Biol. 92 (2009) 49–63.

[75] A. Hozumi, Y. Satouh, Y. Makino, T. Toda, H. Ide, K. Ogawa, S.M. King, K. Inaba, Molecular characterization of Ciona sperm outer arm dynein reveals multiple components related to outer arm docking complex protein 2, Cell Motil. Cytoskeleton 63 (2006) 591–603.

[76] J.L. Gatti, S.M. King, A.G. Moss, G.B. Witman, Outer arm dynein from trout spermatozoa. Purification, polypeptide composition, and enzymatic properties, J. Biol. Chem. 264 (1989) 11450–11457.

[77] S.M. King, J.L. Gatti, A.G. Moss, G.B. Witman, Outer-arm dynein from trout spermatozoa: substructural organization, Cell Motil. Cytoskeleton 16 (1990) 266–278.

[78] K. Inaba, S. Morisawa, M. Morisawa, Proteasomes regulate the motility of salmonid fish sperm through modulation of cAMP-dependent phosphorylation of an outer arm dynein light chain, J. Cell Sci. 111 (1998) 1105–1115.

[79] S.M. King, S.P. Marchese-Ragona, S.K. Parker, H.W. Detrich 3rd, Inner and outer arm axonemal dyneins from the Antarctic rockcod *Notothenia coriiceps*, Biochemistry 36 (1997) 1306–1314.

[80] T.A. Schroer, M.P. Sheetz, Two activators of microtubule-based vesicle transport, J. Cell Biol. 115 (1991) 1309–1318.

[81] S.R. Gill, T.A. Schroer, I. Szilak, E.R. Steuer, M.P. Sheetz, D.W. Cleveland, Dynactin, a conserved, ubiquitously expressed component of an activator of vesicle motility mediated by cytoplasmic dynein, J. Cell Biol. 115 (1991) 1639–1650.

[82] M. Belles-Isles, C. Chapeau, D. White, C. Gagnon, Isolation and characterization of dynein ATPase from bull spermatozoa, Biochem. J. 240 (1986) 863–869.

[83] H.I. Calvin, Isolation of subfractionation of mammalian sperm heads and tails, Methods Cell Biol. 13 (1976) 85–104.

[84] J.T. San Agustin, G.B. Witman, Isolation of ram sperm flagella, Methods Cell Biol. 47 (1995) 31–36.

[85] B.M. Paschal, H.S. Shpetner, R.B. Vallee, MAP 1C is a microtubule-activated ATPase which translocates microtubules in vitro and has dynein-like properties, J. Cell Biol. 105 (1987) 1273–1282.

[86] R.B. Vallee, J.S. Wall, B.M. Paschal, H.S. Shpetner, Microtubule-associated protein 1C from brain is a two-headed cytosolic dynein, Nature 332 (1988) 561–563.

[87] A.T. Hastie, D.T. Dicker, S.T. Hingley, F. Kueppers, M.L. Higgins, G. Weinbaum, Isolation of cilia from porcine tracheal epithelium and extraction of dynein arms, Cell Motil. Cytoskeleton 6 (1986) 25–34.

[88] L.A. Amos, Brain dynein crossbridges microtubules into bundles, J. Cell Sci. 93 (1989) 19–28.

[89] C. Cibert, Identification of an ATPase activity associated with the high molecular weight proteins of the human spermatozoon axonemes, Biol. Cell. 55 (1985) 63–69.

[90] L.M. Quarmby, H.C. Hartzell, Two distinct, calcium-mediated, signal transduction pathways can trigger deflagellation in *Chlamydomonas* reinhardtii, J. Cell Biol. 124 (1994) 807–815.

[91] A.T. Hastie, Isolation of respiratory cilia, Methods Cell Biol. 47 (1995) 93–98.

[92] Z.P. Kabututu, M. Thayer, J.H. Melehani, K.L. Hill, CMF70 is a subunit of the dynein regulatory complex, J. Cell Sci. 123 (2010) 3587–3595.

[93] R.G. Anderson, Isolation of ciliated or unciliated basal bodies from the rabbit oviduct, J. Cell Biol. 60 (1974) 393–404.

[94] W. Auclair, B.W. Siegel, Cilia regeneration in the sea urchin embryo: evidence for a pool of ciliary proteins, Science 154 (1966) 913–915.

[95] A. Konno, K. Shiba, C. Cai, K. Inaba, Branchial cilia and sperm flagella recruit distinct axonemal components, PLoS One 10 (2015) e0126005.

[96] E.W. Stommel, R.E. Stephens, Cyclic AMP and calcium in the differential control of Mytilus gill cilia, J. Comp. Physiol. A 157 (1985) 451–459.

[97] B.H. Gibbons, I.R. Gibbons, The effect of partial extraction of dynein arms on the movement of reactivated sea-urchin sperm, J. Cell Sci. 13 (1973) 337–357.

[98] A. Fliegauf, H. Olbrich, J. Horvath, J.H. Wildhaber, M.A. Zariwala, M. Kennedy, M.R. Knowles, H. Omran, Mislocalization of DNAH5 and DNAH9 in respiratory cells from patients with primary ciliary dyskinesia, Am. J. Respir. Crit. Care Med. 171 (2005) 1343–1349.

[99] K. Nakamura, E. Masuyama, T. Suzaki, Y. Shigenaka, Flagellar adenosine triphosphatases from annelid spermatozoa: electrophoretic identification of dyneins, Arch. Biochem. Biophys. 214 (1982) 172–179.

[100] S.M. King, Sensing the mechanical state of the axoneme and integration of Ca^{2+} signaling by outer arm dynein, Cytoskeleton (Hoboken) 67 (2010) 207–213.

[101] K. Inaba, Molecular basis of sperm flagellar axonemes: structural and evolutionary aspects, Ann. N. Y. Acad. Sci. 1101 (2007) 506–526.

[102] K. Inaba, Sperm flagella: comparative and phylogenetic perspectives of protein components, Mol. Hum. Reprod. 17 (2011) 524–538.

[103] S.M. King, R.S. Patel-King, C.G. Wilkerson, G.B. Witman, The 78,000-Mr intermediate chain of *Chlamydomonas* outer arm dynein is a microtubule binding protein, J. Cell Biol. 131 (1995) 399–409.

[104] M. Sakato, S.M. King, Design and regulation of the AAA+ microtubule motor dynein, J. Struct. Biol. 146 (2004) 58–71.

[105] M. Sakato, H. Sakakibara, S.M. King, *Chlamydomonas* outer arm dynein alters conformation in response to Ca^{2+}, Mol. Biol. Cell 18 (2007) 3620–3634.

[106] K. Inaba, Regulatory subunits of axonemal dynein, in: K. Hirose, L.A. Amos (Eds.), Handbook of Dynein, Pan Stanford Publishing Pte. Ltd, 2012, pp. 303–324.

[107] V.J. Allan, Cytoplasmic dynein, Biochem. Soc. Trans. 39 (2011) 1169–1178.

[108] H. Mohri, K. Inaba, S. Ishijima, B.A. Baba, Tubulin-dynein system in flagellar and ciliary movement, Proc. Jpn. Acad. Ser. B Phys. Biol. Sci. 88 (2012) 397–415.

[109] K. Inaba, Calcium sensors of ciliary outer arm dynein: functions and phylogenetic considerations for eukaryotic evolution, Cilia 4 (2015) 6.

[110] T. Cavalier-Smith, The phagotrophic origin of eukaryotes and phylogenetic classification of Protozoa, Int. J. Syst. Evol. Microbiol. 52 (2002) 297–354.

[111] K. Ogawa, H. Takai, A. Ogiwara, E. Yokota, T. Shimizu, K. Inaba, H. Mohri, Is outer arm dynein intermediate chain 1 multifunctional? Mol. Biol. Cell. 7 (1996) 1895–1907.

[112] K. Inaba, O. Kagami, K. Ogawa, Tctex2-related outer arm dynein light chain is phosphorylated at activation of sperm motility, Biochem. Biophys. Res. Commun. 256 (1999) 177–183 Erratum in: Biochem Biophys Res Commun 268 (2000) 952.

[113] K. Mizuno, K. Shiba, M. Okai, Y. Takahashi, Y. Shitaka, K. Oiwa, M. Tanokura, K. Inaba, Calaxin drives sperm chemotaxis by Ca^{2+}±mediated direct modulation of a dynein motor, Proc. Natl. Acad. Sci. U.S.A. 109 (2012) 20497–20502.

[114] S.M. King, R.S. Patel-King, Identification of a Ca^{2+}-binding light chain within *Chlamydomonas* outer arm dynein, J. Cell Sci. 108 (1995) 3757–3764.

[115] W.J. Tang, C.W. Bell, W.S. Sale, I.R. Gibbons, Structure of the dynein-1 outer arm in sea urchin sperm flagella. I. Analysis by separation of subunits, J. Biol. Chem. 257 (1982) 508–515.

[116] A.G. Moss, J.L. Gatti, G.B. Witman, The motile beta/IC1 subunit of sea urchin sperm outer arm dynein does not form a rigor bond, J. Cell Biol. 118 (1992) 1177–1188.

[117] A.G. Moss, W.S. Sale, L.A. Fox, G.B. Witman, The alpha subunit of sea urchin sperm outer arm dynein mediates structural and rigor binding to microtubules, J. Cell Biol. 118 (1992) 1189–1200.

[118] K. Nakamura, C.G. Wilkerson, G.B. Witman, Functional interaction between *Chlamydomonas* outer arm dynein subunits: the gamma subunit suppresses the ATPase activity of the alpha beta dimer, Cell Motil. Cytoskeleton 37 (1997) 338–345.

[119] U.W. Goodenough, J.E. Heuser, Substructure of inner dynein arms, radial spokes, and the central pair/projection complex of cilia and flagella, J. Cell Biol. 100 (1985) 2008–2018.

[120] S.A. Burgess, D.A. Carter, S.D. Dover, D.M. Woolley, The inner dynein arm complex: compatible images from freeze-etch and thin section methods of microscopy, J. Cell Sci. 100 (1991) 319–328.

[121] O. Kagami, R. Kamiya, Translocation and rotation of microtubules caused by multiple species of *Chlamydomonas* inner-arm dynein, J. Cell Sci. 103 (1992) 653–664.

[122] R. Kamiya, T. Yagi, Functional diversity of axonemal dyneins as assessed by in vitro and in vivo motility assays of *Chlamydomonas* mutants, Zoolog Sci. 31 (2014) 633–644.

[123] A.G. Moss, J.L. Gatti, S.M. King, G.B. Witman, Purification and characterization of Salmo gairdneri outer arm dynein, Methods Enzymol. 196 (1991) 201–222.

[124] K. Ogawa, I.R. Gibbons, Dynein 2. A new adenosine triphosphatase from sea urchin sperm flagella, J. Biol. Chem. 251 (1976) 5793–5801.

[125] T.A. Schroer, Dynactin, Annu. Rev. Cell Dev. Biol. 20 (2004) 759–779.

[126] J.R. Kardon, S.L. Reck-Peterson, R.D. Vale, Regulation of the processivity and intracellular localization of *Saccharomyces cerevisiae* dynein by dynactin, Proc. Natl. Acad. Sci. U.S.A. 106 (2009) 5669–5674.

[127] K.G. Kozminski, K.A. Johnson, P. Forscher, J.L. Rosenbaum, A motility in the eukaryotic flagellum unrelated to flagellar beating, Proc. Natl. Acad. Sci. U.S.A. 90 (1993) 5519–5523.

[128] D.G. Cole, D.R. Diener, A.L. Himelblau, P.L. Beech, J.C. Fuster, J.M. Rosenbaum, *Chlamydomonas* kinesin-II-dependent intraflagellar transport (IFT): IFT particles contain proteins required for ciliary assembly in *Caenorhabditis elegans* sensory neurons, J. Cell Biol. 141 (1998) 993–1008.

[129] G.J. Pazour, B.L. Dickert, G.B. Witman, The DHC1b (DHC2) isoform of cytoplasmic dynein is required for flagellar assembly, J. Cell Biol. 144 (1999) 473–481.

[130] J.L. Rosenbaum, G.B. Witman, Intraflagellar transport, Nat. Rev. Mol. Cell Biol. 3 (2002) 813–825.

[131] A.J. Roberts, B. Malkova, M.L. Walker, H. Sakakibara, N. Numata, T. Kon, R. Ohkura, T.A. Edwards, P.J. Knight, K. Sutoh, K. Oiwa, S.A. Burgess, Structure 20 (2012) 1670–1680.

[132] T. Kon, T. Oyama, R. Shimo-Kon, K. Imamula, T. Shima, K. Sutoh, G. Kurisu, The 2.8 Å crystal structure of the dynein motor domain, Nature 484 (2012) 345–350.

[133] A.P. Carter, C. Cho, L. Jin, R.D. Vale, Crystal structure of the dynein motor domain, Science 331 (2011) 1159–1165.

[134] M. Edamatsu, The functional expression and motile properties of recombinant outer arm dynein from *Tetrahymena*, Biochem. Biophys. Res. Commun. 447 (2014) 596–601.

[135] M. Mazumdar, A. Mikami, M.A. Gee, R.B. Vallee, In vitro motility from recombinant dynein heavy chain, Proc. Natl. Acad. Sci. U.S.A. 93 (1996) 6552–6556.

[136] M. Ichikawa, Y. Watanabe, T. Murayama, Y.Y. Toyoshima, Recombinant human cytoplasmic dynein heavy chain 1 and 2: observation of dynein-2 motor activity in vitro, FEBS Lett. 585 (2011) 2419–2423.

[137] T. Kobayashi, T. Murayama, Cell cycle-dependent microtubule-based dynamic transport of cytoplasmic dynein in mammalian cells, PLoS One 4 (2009) e7827.

[138] M.P. Koonce, M. Samsó, Overexpression of cytoplasmic dynein's globular head causes a collapse of the interphase microtubule network in *Dictyostelium*, Mol. Biol. Cell. 7 (1996) 935–948.

[139] M. Nishiura, T. Kon, K. Shiroguchi, R. Ohkura, T. Shima, Y.Y. Toyoshima, K. Sutoh, A single-headed recombinant fragment of *Dictyostelium* cytoplasmic dynein can drive the robust sliding of microtubules, J. Biol. Chem. 279 (2004) 22799–22802.

[140] T. Kon, M. Nishiura, R. Ohkura, Y.Y. Toyoshima, K. Sutoh, Distinct functions of nucleotide-binding/hydrolysis sites in the four AAA modules of cytoplasmic dynein, Biochemistry 43 (2004) 11266–11274.

[141] Y. Takahashi, M. Edamatsu, Y.Y. Toyoshima, Multiple ATP-hydrolyzing sites that potentially function in cytoplasmic dynein, Proc. Natl. Acad. Sci. U.S.A. 101 (2004) 12865–12869.

[142] S.L. Reck-Peterson, A. Yildiz, A.P. Carter, A. Gennerich, N. Zhang, R.D. Vale, Single-molecule analysis of dynein processivity and stepping behavior, Cell 126 (2006) 335–348.

[143] T. Kon, T. Shima, K. Sutoh, Protein engineering approaches to study the dynein mechanism using a *Dictyostelium* expression system, Methods Cell Biol. 92 (2009) 65–82.

[144] K. Shiroguchi, Y.Y. Toyoshima, Regulation of monomeric dynein activity by ATP and ADP concentrations, Cell Motil. Cytoskeleton 49 (2001) 189–199.

[145] M.A. Schlager, H.T. Hoang, L. Urnavicius, S.L. Bullock, A.P. Carter, In vitro reconstitution of a highly processive recombinant human dynein complex, EMBO J. 33 (2014) 1855–1868.

In this chapter

Single-molecule dynein motor mechanics in vitro

Ahmet Yildiz
University of California, Berkeley, CA, United States

5.1 Introduction

Landmark discoveries in the study of cytoskeletal motors have been made through advances in single-molecule biophysics. Measurements on individual isolated proteins in vitro have contributed significantly to our understanding of how motors take steps in a unidirectional manner, and generate force and motility along their track. Detailed mechanistic models now exist for kinesin-1 [1], myosin V [2], and myosin VI [3], dimeric motors that take 8–36 nm successive steps in a hand-over-hand manner [1,4,5]. Kinesin and myosin share many similarities because they are both members of the G-protein family [6]. Dynein, however, is a divergent branch of the AAA family of ATPases [7], which mostly function as chaperons or unfoldases [8,9]. As a result, very little can be inferred from the study of other motors, and dynein's unusual structure suggests that it has unique mechanistic features. Studies of cytoskeletal motors have been aided by the ability to produce recombinant motors that can be mutated and assayed at the single-molecule level. The production of a recombinant dynein capable of processive motion was considerably more challenging due to its large size and complexity. Recombinant expression of dynein has been achieved recently in yeast [10], *Dictyostelium* [11], and insect cells [12]. The in vitro analysis of these motors played a major role in dissecting the mechanism of dynein motility.

5.2 The mechanochemical cycle of dynein

Molecular motors utilize common mechanical principles to generate force and motility along linear tracks. Coordinated with their cycle of ATP hydrolysis, an individual head of a typical cytoskeletal motor (1) binds strongly to its track, (2) produces a force-generating conformational change (power stroke), (3) releases from its track, (4) performs a recovery stroke, and (5) rebinds its track in a forward direction, resulting in unidirectional movement.

Structural [13–17], biochemical [18,19], and bulk FRET studies [20] of *S. cerevisiae* and *Dictyostelium* dynein revealed that nucleotide binding and hydrolysis

113

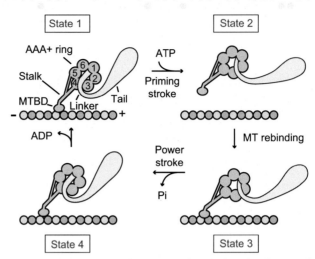

Figure 5.1 Domain organization and mechanochemical cycle of DHC. The AAA+ ring is composed of six AAA domains and binds to microtubules through a coiled-coil stalk with a small MTBD. The linker undergoes ATP-dependent conformational changes. Synchronization of these conformational changes with the microtubule binding/unbinding cycle generates force and motility.

at AAA1, coupled with the rearrangement of the linker domain at the surface of the AAA+ ring are the principal drivers that govern the mechanochemical cycle of dynein [21,22]. Briefly, in the absence of a nucleotide in the AAA1 site (APO state, State 1 in Fig. 5.1), dynein tightly attaches to a microtubule [13] and the linker is in a straight (postpower stroke) conformation, which runs at the surface of the ring and exits the ring near AAA5 [23,24]. In State 2, ATP binding to AAA1 closes the gap between AAA1 and AAA2, which triggers rigid body motions of the AAA subunits within the ring [23–25]. Remarkably, these conformational arrangements cause the buttress, a short coiled-coil emerging from AAA5, to alter the registry of the stalk and trigger the release of the motor from microtubule (see below) [14–17,26]. ATP binding to AAA1 also forces the linker to move from its straight conformation into a bent conformation, referred to as the prepower stroke state (also called as the priming stroke) [16,17,25,27]. In this conformation, the linker exits the ring at AAA2. Although we do not yet know the exact sequence of microtubule release, priming stroke, and ATP hydrolysis after ATP binding to AAA1, to produce net movement, the motor must release from microtubule first before it undergoes the priming stroke [22]. In State 3, the head hydrolyzes ATP before rebinding to a microtubule, which triggers the release of the inorganic phosphate [19]. In the ADP-bound state (State 4), dynein tightly attaches to microtubule and the linker undergoes isomerization from pre- to postpower stroke conformation, returning to its original straight conformation (Fig. 5.1) [16,17,24,25]. This transition is believed to generate the force required to drive dynein motility (power stroke) [25,28]. The release of ADP from AAA1 is the rate limiting step of the mechanochemical cycle [19,29], and after ADP release, dynein is primed for ATP binding (State 1) to initiate the next cycle.

5.3 Processivity of a dynein dimer

Since the discovery of the motors that walk along linear tracks, several assays were developed to test their activity in vitro. In filament gliding assays, motors are fixed to a microscope slide and fluorescently labeled microtubules are then added to the flow chamber. These microtubules bind to the motors and are translated relative to the surface in the presence of ATP. A large number of motors can bind simultaneously to the same filament and work cooperatively. In motility assays, microtubule tracks are immobilized to the glass surface, and movement of fluorescently labeled motors on these tracks is examined in a flow chamber. These assays are better suited to determine whether a single motor can take many consecutive steps before dissociating from microtubules, termed processivity. Similar to kinesin motors, dynein monomers attached to a surface through their linker glide microtubule filaments [10,11], attesting to the fact that single-motor domains are capable of producing force and unidirectional motility independent of their partner. Motility assays of beads that are driven by cytoplasmic dynein suggested that single dynein dimers are processive motors [30,31]. Previously, motility assays on inner-arm dynein c from *Chlamydomonas* suggested that a single dynein head could be sufficient for processive motility [32]. However, more recent studies on recombinant *S. cerevisiae* dynein showed that dynein monomers are not processive [10]. This observation is consistent with the model shown in Fig. 5.1 that a monomer binds to a microtubule, performs a power stroke, and diffuses away from the microtubule after binding to ATP.

Engineering approaches to *S. cerevisiae* dynein revealed the minimal requirements of processive motility. Several observations demonstrated that *the linker swing mechanism is essential for motility*. Dynein monomers are not able to glide microtubules when they are immobilized to the glass surface through their C-terminus [10]. Consistently, linking of the two motor heads that lack the dimerization domain through the C-terminus of the ring failed to produce any motility [10]. Mutations that disrupt ATPase activity at the AAA1 site [10,26] or connections between the linker and the surface of the ring also abolish dynein motility [16,17]. When two *S. cerevisiae* dynein monomers are paired through natural or forced dimerization through the tail region, they walk processively for ~1.5 microns on an average (referred to as run length) before dissociating from microtubules [10]. *Processivity does not necessarily require two catalytically active motor domains*. Instead, dynein is able to walk at near wild-type speeds when one of the heads is inactivated by ATPase mutations to AAA1 and AAA3 [33,34]. The entire AAA+ ring and linker domains of one head can be replaced with an inert protein retaining a microtubule tether [34]. Therefore, dynein processivity can be facilitated by one active motor domain, provided that its linker domain is connected to a microtubule tether. The tether prevents microtubule dissociation as the active monomer steps forward and provides solid support for the swinging motion of the linker domain to facilitate unidirectional motility.

Processive motility of dynein is regulated by several factors, including motor copy number and salt concentration. Increasing the motor copy number leads to dramatic increase in motor processivity [35,36], presumably because the probability of both dimers to release from a microtubule is rather small. Indeed, in vitro studies have shown that linking two or more dynein motors through a DNA scaffold [37–39] or increasing the motor density on polystyrene beads result in a dramatic increase in the motor run length even at high-salt concentrations [35,36]. Similarly, multiple nonprocessive monomers of kinesin, myosin, or dynein facilitate long range transport of a cargo bead [38,40–42]. Most dyneins do not walk processively at physiologically relevant salt concentrations in vitro (~150 mM K^+), raising the possibility that multiple dyneins are needed for long range transport of a cargo inside cells. The other possibility is that diffusion of the cargos is greatly restricted by the crowdedness and dense meshwork of the cytoplasm, enabling the motors on a cargo to quickly rebind microtubule after transient detachment. These ideas can be tested by mimicking the cellular environment in motility assays.

Dynein processivity is also regulated by dynein-associated proteins (see below) and posttranslational modifications of tubulin. In vitro motility assays on recombinant tubulin species with engineered posttranslational modifications showed that microtubule motors recognize distinct tubulin modifications [43]. For example, yeast dynein motility is not significantly affected by tyrosination of tubulin [43]. However, landing of mammalian dynein/dynactin complex is decreased multiple fold on detyrosinated microtubules. This preference is mediated by dynactin's microtubule-binding p150 subunit and α-tubulin tyrosine that facilitate initial motor-microtubule encounters, but is not needed for subsequent motility [44]. The results support the idea that the "tubulin code" may play a major role in cellular localization and activity of microtubule motors.

5.4 Velocity of dynein motors

Velocity of dynein varies greatly among species. To date, cytoplasmic dynein from yeast is the slowest dynein motor, moving at 120 nm/s [10]. While most dyneins from other species move at 1–3 μm/s along microtubules [30,45–48], axonemal dyneins were reported to slide microtubule bundles as fast as 30 μm/s [49]. While we do not yet know why dynein speed varies greatly among species, this may be related to differences in microtubule affinity and ADP release rates of these motors. Measured speed of dynein motors in vitro is either slightly lower than [50–52] or similar [30,45–48] to the speed of dynein-driven cargos in cells. Measured in vitro speed of mammalian cytoplasmic dynein also agrees with the speed of fast axonal transport (~1 μm/s) [53]. Differences between in vitro and in vivo speed measurements can be attributed to engagement of multiple motors with the transport of an intracellular cargo, as well as dynein regulatory factors that are not present in the in vitro motility assays. In addition, dynein speed is sensitive to ambient temperature, ionic strength, and viscosity of the motility buffer. Similar to kinesin, dynein speed shows Arrhenius-like dependence on temperature and the motor walks about twice as fast at 37°C than at room temperature [54].

5.5 The stepping mechanism

High-resolution tracking studies have revealed that dynein moves by taking 8–16 nm steps [10,55], corresponding to the distance between 1 and 2 tubulin heterodimers on a protofilament. Dynein step size is independent of ATP concentration [31,33,55] and each step is generated by a single ATP hydrolysis event at the AAA1 site [22]. The coupling efficiency of ATP hydrolysis to a mechanical step remains unknown, and dynein may undergo futile cycles of ATP hydrolysis. Furthermore, a fraction of dynein steps are believed to occur independent of ATP hydrolysis, powered by elastic strain energy stored in the linker domains (see below) [34].

Unlike kinesin, which almost exclusively takes 8 nm steps toward its direction of motion [56], stepping behavior of dynein is highly variable with frequent large forward steps, as well as backward and sideways steps [10,57,58]. Although dynein is able to walk along a single protofilament of a microtubule [59], it can switch from one microtubule to another [60] and even slide one microtubule relative to another [61], similar to tetrameric kinesin Eg5 [62]. This could be related to the flexible structure of the dynein motor domain [63], which gives rise to a large diffusional component in step size [34]. In comparison, kinesin-1 is forced to move to the next binding site on a protofilament because it has a short neck-linker between the heads [64–67]. Consistent with this idea, releasing kinesin from this constraint resulted in dynein-like stepping behavior [68,69].

Ability of dynein to step sideways raised the possibility that dynein explores the microtubule surface more freely than kinesin. Typically, in vitro motility assays are performed on surface-immobilized microtubules, which constrain the motors to move in two dimensions. To explore the full trajectory of dynein motility in three dimensions, these assays were performed on microtubule bridges suspended above the surface [70,71] (Fig. 5.2). Cargo beads carried by yeast dynein motors move in a helical trajectory around the microtubule bridge, with an average pitch of ~500 nm [42]. Unlike several members of kinesin and myosin motor family that rotate around their respective track in a specific direction [70,71], yeast dynein performs bidirectional helical motility without a net preference to move along a right- or left-handed helical path [42]. Similarly, mammalian cytoplasmic dynein [72] and axonemal dynein from *Tetrahymena* [73,74] were observed to move in a particular direction on a microtubule, suggesting that dynein generates torque in addition to unidirectional motility.

(A)

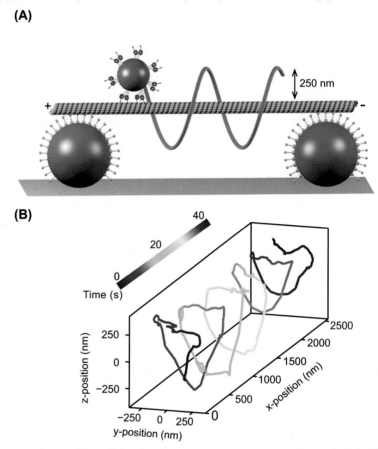

(B)

Figure 5.2 **3D tracking of dynein-driven cargo transport on microtubule bridges.**
(A) The suspended microtubule bridge is formed by attaching microtubule to surface-immobilized beads (gray) coated with dynein MTBD. A smaller "cargo" bead is coated with multiple yeast dynein motors. (B) 3D tracking of bead movement reveals helical movement of dynein along microtubules. *Figure is modified from S. Can, M.A. Dewitt, A. Yildiz, Bidirectional helical motility of cytoplasmic dynein around microtubules, eLife 3 (2014) e03205.*

Frequently stepping sideways and bidirectional helical motility may allow dynein to avoid roadblocks in dense cytoplasmic environments during cargo transport [75], while kinesin remains stuck on obstacles because of its inability to take sideways steps [65]. This may lead to differential regulation of motor activity inside cells through the expression and spatial distribution of microtubule-associated proteins (MAPs). It remains to be studied whether MAPs simply serve as obstacles on a microtubule or directly inhibit the activity of specific motors [36]. So far, in vitro studies have revealed inhibitory role of Tau isoforms on kinesin motility [76,77] and specific inhibition of yeast dynein motility by She1 [78]. She1 does not have an isoform in mammalian cells; it remains unknown whether mammalian cells have specific MAPs that inhibit dynein motility.

5.6 Stepping pattern of the two motor domains

Hand-over-hand stepping of kinesin and myosin motors requires the two heads to be kept out of phase, preventing both heads from detaching simultaneously from the cytoskeletal track. As a result, the motor can take many consecutive steps and transport cargos over long distances. This processive motility is facilitated through coordination between the two motor domains to prevent premature release from the microtubule. In kinesin and myosin, a chemical or structural transition in one head is inhibited until the partner head proceeds through a critical step in its cycle (referred to as a "gating" mechanism) [68,79,80]. Depending on the motor, such gating can be chemical in nature, e.g., ATP binding is allosterically inhibited until a head dissociates from the microtubule [81–83], or mechanical in nature, e.g., a conformational change in one head pulls or pushes the other head off from the microtubule [2,84,85]. It is possible that the actual mechanism of interhead communication utilizes both types of gating, as they are not mutually exclusive [1].

Initial studies of dynein stepping mechanics revealed that a head domain takes larger steps than the dimerization domain on average [10], similar to that of kinesin [86] and myosin [87,88]. However, high variability in the size and direction of individual steps of dynein raised the possibility that the orientation of the two heads is not fixed and dynein may utilize a more complex stepping pattern than other motors. To directly address how the heads move relative to each other, the two heads of *S. cerevisiae* dynein were labeled with different colored fluorophores and simultaneously tracked at nanometer resolution [33,89]. Remarkably, dynein moves processively without tight coordination between its heads (Fig. 5.3), a mechanism fundamentally distinct from the hand-over-hand movement of kinesin [86] and myosin [87,88]. Either head can take multiple steps before its partner moves. The leading head can take several steps before the lagging head moves forward. These observations excluded the idea that the movement of the two motor domains is tightly coordinated [55,90,91], presumably through stacking interactions between the AAA+ rings [92]. Instead, one head can step forward on its own while the other head serves as a tether to prevent dissociation of the motor from microtubule, referred to as the "tethered excursion" mechanism [33]. This model was also verified by processive motility of a heterodimeric dynein with one active head and one catalytically inactive head [33,34].

Independent movement of dynein heads has demonstrated that processive movement of a dimeric motor does not necessarily require coordination

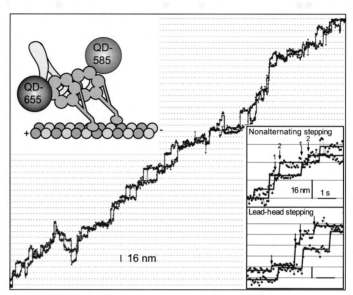

Figure 5.3 Stepping trace of dynein labeled with different colors of quantum dots (QD-585 and QD-655) shows that the heads move independently of each other during processive runs. *Figure is modified from M.A. DeWitt, A.Y. Chang, P.A. Combs, A. Yildiz, Cytoplasmic dynein moves through uncoordinated stepping of the AAA+ ring domains, Science 335 (6065) (2012) 221–225.*

between the stepping cycles of the two heads (referred to as gating). In the absence of interhead communication, the motor may walk processively if the duty ratio (percent time spent bound to a microtubule during a mechano-chemical cycle) of a head is sufficiently high (0.9) to allow the motor to take 100 steps (the average run length is 1 μm and the average step size is ~8 nm) before dissociation. The duty ratio of dynein has not been directly measured, but is estimated to be a minimum of 0.6 by the filament gliding assays using *Dictyostelium* dynein [93].

The stepping characteristic of a head in the leading and trailing positions differs significantly, and leads to high variability in dynein's step size [33]. While the overall motion is biased toward the microtubule minus end, the heads also prefer to step toward each other to minimize the elastic strain energy stored at large interhead separations. As a result, the lagging head usually takes large forward steps, while the leading head takes frequent backward steps or short forward steps. Backward stepping is not observed in kinesin and myosin V, because in these motors, the leading head remains bound to the track and pulls the lagging head in the forward direction.

Several observations suggested that stepping of the two heads is not completely independent from each other. *S. cerevisiae* dynein motility is partially gated by tension on the linker domain at high interhead separations [33,34,89]. When two heads are positioned close to each other, they are equally likely to take a step. When the heads are separated, tension on the linker triggers the release of the lagging head from a microtubule independent of ATP hydrolysis. Therefore, the lagging head becomes more likely to take a step, increasing the likelihood of stepping in a manner similar to kinesin. Optical trapping assays revealed that exerting tension on the linker domain abolishes ATP-induced release from the microtubule [34,94]. Tension may disrupt the weak contacts between the linker and the ring and indirectly inhibit ATP hydrolysis. Indeed, removing these contact sites through mutagenesis significantly reduces the ATP hydrolysis rate in the absence of external tension [16,17]. Alternatively, tension on the linker may directly affect the stalk sliding mechanism and reduce the ability of the ATP hydrolysis cycle to regulate microtubule affinity [15].

Studies on *Dictyostelium* dynein indicated different mechanisms of interhead communication. Duty ratio of monomeric constructs is surprisingly low (0.24), yet the dimer walks processively with run lengths comparable to that of *S. cerevisiae* dynein [90,95]. Dimerization also slows down the ATP hydrolysis rate [95]. It is possible that the motor domains of *Dictyostelium* dynein may allosterically affect each other through the stacking of the AAA+ rings [21,92], as in the case of other AAA+ enzymes. Interhead communication in mammalian dynein is even less clear and the effect of dynactin and cargo adapter proteins on the stepping mechanism remains to be elucidated.

5.7 The role of the AAA sites in dynein motility

Understanding the roles of each AAA+ site in dynein motility is complicated by the unique sequence and structure of these sites. Only AAA1–AAA4 have the conserved Walker A motif that facilitates nucleotide binding, whereas AAA5 and AAA6 do not bind nucleotides and they have a structural role within the AAA+ ring. The kinetics of dynein stepping are coupled to ATP hydrolysis at AAA1, and this hydrolysis is strictly required for dynein motility [10,20,26,96]. AAA2 does not have the conserved Walker B motif that facilitates nucleotide hydrolysis and this domain remains bound to ATP during the entire nucleotide cycle [17]. AAA3 and AAA4 hydrolyze ATP in cytoplasmic dynein and mutations that abrogate ATP binding and hydrolysis at the AAA4 site has moderate affect in motility [26,96], suggesting that ATPase activity at this site is not critical for the stepping cycle.

Biochemical and single-molecule studies on yeast and *Dictyostelium* dynein have revealed that ATPase activity at the AAA3 site is essential for robust motility [97]. Mutations to AAA3 significantly reduce dynein ATPase rate and speed [26,96]. AAA3 ATPase mutants interact more tightly with the microtubule [26,96,98] and inhibit prepower stroke conformation of the linker [23]. Unlike WT dynein, reducing the microtubule affinity of these mutants by increasing the salt concentration results in faster motility and increased microtubule-stimulated ATPase rates, suggesting that AAA3 allows AAA1 to control microtubule-binding affinity [98]. In the absence of AAA3 activity, AAA1 activity slows down, because it cannot "open" the tubulin gate to go to the next nucleotide cycle at its normal speed.

Using a nonhydrolyzable ATP analogue (ATP-γS) in motility assays [99] resulted in transient pausing behavior in yeast dynein motility. However, pausing was not observed in AAA3 ATPase mutants, suggesting that ATP-γS specifically inhibits the AAA3 site, but not AAA1. The analysis of the pausing behavior has shown that ATPase hydrolysis at AAA3 is decoupled from the stepping cycle and instead regulates microtubule attachment and force generation cycles [94,98]. AAA3 remains bound to ADP over the course of ~20 steps and transiently undergoes the hydrolysis cycle as dynein walks along the microtubule [98]. Because both ATP binding and hydrolysis mutants of AAA3 show similar defects in dynein motility and the linker conformation [23], this site mostly remains in a post-hydrolysis state during processive motility [16].

AAA3 may function as a regulatory switch between anchoring (slow) and transporting (fast) functions of dynein in cells (Fig. 5.4) [98]. Lis1 binds dynein near AAA3 and AAA4 [100,101], raising the possibility that Lis1 may regulate ATP binding or hydrolysis at these sites. Similar to the AAA3 ATPase mutants, Lis1 binding increases dynein's affinity for microtubules and slows down motility [100,102]. Unlike AAA3 mutants, Lis1 binding does not affect the ATP hydrolysis rate [22,100] suggesting that Lis1 uncouples the nucleotide hydrolysis cycle from the stalk sliding mechanism. Lis1 also blocks the linker from docking at the AAA5 site [101]. Further work is required to understand the molecular cues

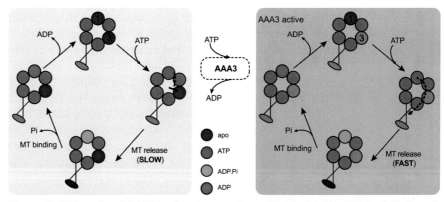

Figure 5.4 The role of AAA3 in dynein's mechanochemical cycle. ATP hydrolysis cycle at AAA1 drives dynein motility. (Left) When AAA3 is in the APO state, communication between AAA1 and the MTBD is blocked, resulting in slow progression through the AAA1 cycle. (Right) When ATP binds to AAA3 and is hydrolyzed, the allosteric circuit connecting AAA1 and the MTBD is completed, resulting in fast progression of the AAA1 cycle. *Figure is modified from M.A. DeWitt, C.A. Cypranowska, F.B. Cleary, V. Belyy, A. Yildiz, The AAA3 domain of cytoplasmic dynein acts as a switch to facilitate microtubule release, Nat. struct. mol. biol. 22 (1) (2015) 73–80.*

that govern the possible interactions between Lis1, AAA3, and the linker, and whether such regulation biophysically repurposes dynein for a diverse set of functions inside cells.

5.8 Force generation

Dynein is involved in a series of cellular functions, such as carrying large intracellular cargos and positioning of the mitotic spindle, that require the motor to produce substantial force and maintain its function under large tension. Dynein was named by the unit of force, dyne; yet force generation of dynein has been a subject of considerable debate. Optical trapping studies revealed that single *S. cerevisiae* cytoplasmic dynein stalls at 3.4 pN backward force [28,55,103]. Outer arm dynein (OAD) purified from *Tetrahymena* has been shown to produce up to 4.7 pN of force [104]. In comparison to kinesin, dynein shows less-frequent premature releases from a microtubule under load and stalls for extended periods of time. Dynein stall force is independent of ATP concentration [28,103]. The motor can also be forced to step toward the plus end under large backward forces in the presence and absence of ATP [103].

How does the two motor domains contribute to force generation of a dynein dimer? In case of kinesin, only the front head produces force and pulls the rear head forward [105]. Unlike kinesin, the two heads of a dynein dimer are not strictly coordinated during processive motility [33,89]. As a result, tail-tethered optical trapping experiments yield insufficient information about the force production of individual heads. DNA-tethered optical trap measurements were performed to measure the force generation of a single head of a yeast dynein dimer (Fig. 5.5) [105]. These measurements revealed that the stall forces of the individual heads are additive, with both heads contributing equally to the force production of the dimer [28]. Therefore, dynein heads utilize a "load-sharing" mechanism, which allows the motor to work against hindering forces larger than the maximal force produced by a single head. The bulk of the force is produced during the power stroke of a head in the microtubule-bound state. The priming stroke in the microtubule unbound state does not generate significant amount of mechanical work and only weakly biases the head to step toward the microtubule minus end [28].

Optical trapping assays measured the force-dependent microtubule release rate of dynein from a microtubule (Fig. 5.6A) [34,94]. There is a clear difference in the average time dynein remained bound to the microtubule depending on the direction of the applied force (Fig. 5.6B). When force was applied in the forward direction (minus end), dynein rapidly releases from microtubules. In contrast, release toward the plus end is significantly slower and independent of force, suggesting that dynein forms an ideal bond with a microtubule under backward tension [94]. Rapid release of a dynein head under forward tension

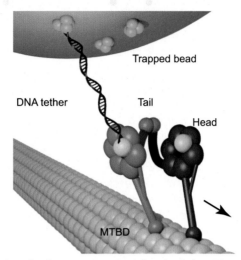

Figure 5.5 Measuring the force generation of a single head by a DNA-tethered trap. *Figure is modified from V. Belyy, N.L. Hendel, A. Chien, A. Yildiz, Cytoplasmic dynein transports cargos via load-sharing between the heads, Nat. commun. 5 (2014) 5544.*

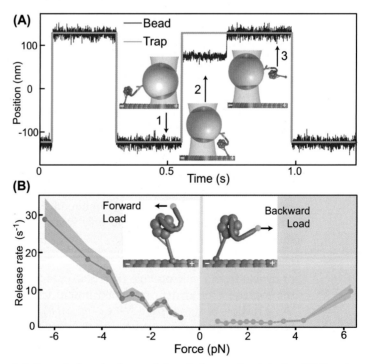

Figure 5.6 Force-induced release of dynein from microtubules. (A) Dynein-coated beads are oscillated ±125 nm along the microtubule long axis. When a monomer binds to the microtubule (1) the movement of the bead is hindered. (2) The trap exerts a constant force on the monomer until it releases the microtubule (3). (B) Dynein strongly prefers to release toward the minus end, whereas release toward the plus end is slow and independent of force. *Figure is modified from F.B. Cleary, M.A. Dewitt, T. Bilyard, Z.M. Htet, V. Belyy, D.D. Chan, et al., Tension on the linker gates the ATP-dependent release of dynein from microtubules, Nat. commun. 5 (2014) 4587.*

may allow the motor to break the symmetry of diffusion and contribute to unidirectional movement of the motor [103]. Slow and force-independent release under backward tension may enable tight anchoring of dynein to microtubule and allow these motors to efficiently function together in teams under high tension [52].

5.9 The mechanism of minus end directionality

The ability of molecular motors to move in a unidirectional manner along their track is central to their cellular function. The directionality of kinesin and myosin motors is determined by ATP-dependent conformational changes in the linker regions that connect the two motor domains. Reversing the direction of these conformational changes in kinesin [106,107] and myosin [80,108] results in reversal of motor directionality. Dynein on the other hand has more complex structural organization and the underlying mechanism of its directionality is unclear. The current model for dynein directionality [25] proposes that the priming stroke of the unbound head (Fig. 5.1) moves it toward the minus end by pushing against the microtubule-bound head. This model predicted that reversal of the priming stroke relative to the microtubule would result in plus end–directed motility. To test this model, the orientation of the AAA+ ring relative to the microtubule-binding domain (MTBD) was reversed by inserting or deleting seven heptads of the coiled-coil stalk [14]. Surprisingly, dynein maintained its minus end–directed motility when the motor domain rotated 180 degrees relative to the stalk. This result suggested that dynein directionality is not solely determined by the ATPase domain. In a departure from other motors, dynein's MTBD is separated from the catalytic head by the coiled-coil stalk. One striking feature of the stalk is the sharp kink with respect to the microtubule [14,27] and this kink region may be responsible for pointing the priming stroke of the linker toward the minus end [109] (Fig. 5.7).

This model was challenged by recent engineering approaches to dynein motor [110]. Dynein's unique domain architecture has made it possible to isolate the microtubule-binding interface from its catalytic core. Replacement of the MTBD with nonmotor actin binding proteins resulted in processive motility of this chimeric motor along actin filaments. Circular permutation of the actin binding domains resulted in reversal of dynein motility on actin, leading to the proposal that dynein directionality is not determined by the priming stroke of the linker [110]. Understanding the mechanism of dynein motility requires future engineering approaches with minimal perturbations to the dynein motor and more detailed structural and biophysical studies of reverse directionality mutants.

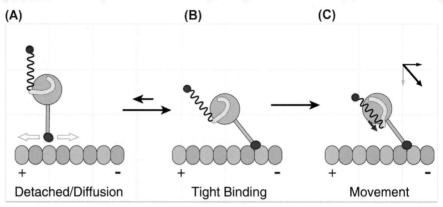

Figure 5.7 A model of dynein directionality. (A) Dynein is in a power stroke conformation when it is detached from a microtubule and diffuses to a new binding site. (B) When dynein binds to a microtubule, the stalk is tilted toward the plus end of a microtubule. (C) An ATP-driven power stroke of the linker produces force and movement whose horizontal component is directed toward the minus end (*red arrow*). The linker domain is shown in yellow and the rest of the tail is represented as a loose spring. *Figure is from A.P. Carter, J.E. Garbarino, E.M. Wilson-Kubalek, W.E. Shipley, C. Cho, R.A. Milligan, et al., Structure and functional role of dynein's microtubule-binding domain, Science 322 (5908) (2008) 1691–1695.*

5.10 Mammalian dynein/dynactin complex

Previous studies have revealed striking differences between mammalian and *S. cerevisiae* dynein. The *S. cerevisiae* dynein heavy chain is constitutively active and walks robustly in motility assays without dynactin [10,111]. Despite high retrograde transport velocities observed for mammalian dynein in vivo [51,112,113], tissue-purified mammalian dyneins showed variable results. While some studies reported that mammalian dynein is processive (reviewed in [114]), especially when it is bound to beads [30,31,115], other studies reported diffusive motility or short processive runs in vitro [46,57,116]. As a result, the mechanochemical cycle of mammalian dynein could not be studied in detail. Recent studies revealed that recombinantly expressed human dynein remains mostly in a phi conformation, in which the motor domains are stacked together [46,117], and is autoinhibited or not fully active in the absence of its adapter proteins (see Chapter 16 (vol. 1 of this book)) [47,48,116,118]. When dynein forms a complex with dynactin and a cargo adapter protein [118], the motor is released from the inhibited conformation [119,120], and the processivity and speed of its motility are dramatically increased [47,48]. Formation of the dynein/dynactin complex is proposed to rescue the heavy chain from its inactive conformation and allow the motor heads to bind simultaneously to a microtubule. Electron micrographs of microtubule-bound dynein/dynactin complexes revealed that the AAA+ rings lie parallel to each other [120]. This raises the possibility that dynactin and cargo adapter proteins may also play a regulatory role in mechanics of dynein motility.

Several studies reported that mammalian dynein has the ability to move toward the plus end of a microtubule and switches the direction of its movement depending on the concentration of ATP [60] or inorganic phosphate [121], or when the motor encounters an obstacle on a microtubule [75]. The structural and mechanistic basis of these observations remains unknown. Because mammalian dynein is not a fully active motor on its own and only displays short processive or diffusive motility [46,57,116], these switches of its directionality may be related to changes in the diffusional motion of the motor, rather than reversal of the directionality of active dynein complexes. More detailed work on dynein/dynactin complexes is required to test these observations.

Surprisingly, mammalian dyneins were found to stall at forces of ~1 pN [31,46,52,58,102], significantly weaker than other dyneins and the 6 pN force production of plus end–directed kinesin-1 motors [122]. It has been proposed that intracellular transport of cargos along microtubules requires 4–7 dynein motors per kinesin-1 to balance forces during "tug-of-war" [123]. However, this finding raised concerns about how many dyneins can simultaneously engage with a microtubule when transporting a small cargo. In addition, measured dynein to kinesin ratios on certain intracellular cargoes are inconsistent with ~6:1 coupling predicted by force-based models [124]. Optical trap recordings on dynein-driven cargos in vivo failed to observe clear stalling events, but observed peak forces with periodicity of 1–2 pN, with total forces as high as 12 pN [52]. These recordings have been interpreted as pairing-up and microclustering of dynein motors on a cargo [125], that may facilitate more effective competition against kinesin motors in a tug-of-war (see Chapter 7).

Because mammalian dynein is not fully active on its own [46–48,116,118], force measurements on purified mammalian dyneins in vitro does not reflect the force production of active dynein complexes. Optical trapping of fully active dynein/dynactin/BicD2 (DDB) complexes revealed that mammalian dynein produces 4.5 pN force, comparable to that of human kinesin-1 (Fig. 5.8) [126]. Coupling of single dynein with single human kinesin-1 in a DNA scaffold has shown that dynein is not able to compete against kinesin-1 on its own. However, the DDB complex is capable of competing against human kinesin-1 in a one-to-one mechanical tug-of-war [126]. A large increase in mammalian dynein stall force in the absence of dynactin and cargo adapter protein is also observed upon deletion of the C-terminal domain [127], but it remains unclear whether this is related to the rescue of the motor from its autoinhibited conformation.

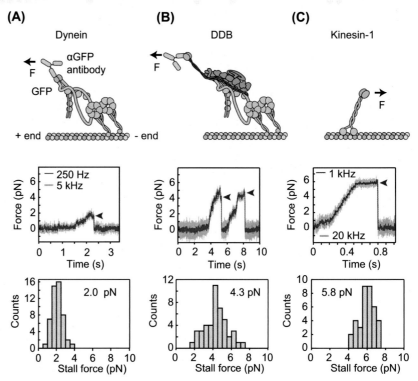

Figure 5.8 Force generation of human cytoplasmic dynein. (A) (Top) Single dynein motors are pulled with an optically trapped bead (represented with a force *arrow*). (Middle) A typical stall of a bead driven by dynein in a fixed-trap assay. After the stall, the bead snaps back to the trap center upon the detachment of the motor from a microtubule (*red arrowhead*). (Bottom) The histogram of observed stalls. Human dynein stalls at 2.0 pN backward force, on average. (B) The DDB complex stalls at 4.3 pN. (C) In comparison, human kinesin-1 stalls at 5.8 pN. *Figure is modified from V. Belyy, M.A. Schlager, H. Foster, A.E. Reimer, A.P. Carter, A. Yildiz, The mammalian dynein–dynactin complex is a strong opponent to kinesin in a tug-of-war competition, Nat. cell biol. 18 (9) (2016) 1018–1024.*

5.11 Future directions

Single-molecule studies of cytoplasmic dynein provided a detailed view of dynein's mechanochemical cycle, processivity, stepping, force generation, and directionality. These studies have revealed that the mechanism of dynein motility is distinct from kinesin and myosin motors. Because mammalian dynein is autoinhibited in the absence of a cargo adapter protein, previous in vitro studies on mammalian dynein alone should be viewed as the output of an inactive or partially inactive motor. Future studies are required to understand how dynactin and a cargo adapter proteins rescues mammalian dynein from its inhibited conformation and regulates the mechanism of its motility. Recombinant expression of dynein and dynactin complexes are required to understand the molecular basis of how single-point mutations on these complexes cause human disease [128]. Further efforts to reconstitute motor-cargo complexes are highly anticipated, as these experimental platforms would enable to how kinesin and dynein motors are recruited and their activities are regulated by cargo scaffolding proteins (such as JIP1, TRAK, Milton, egalitarian) [53,129].

In comparison to cytoplasmic dynein, the mechanisms of other dyneins remain poorly understood. Motility of dynein-2, which is specialized for intraflagellar transport (IFT) in eukaryotic cilia and flagella, has only been recently tested in vitro, and it remains unknown how dynein is adapted for the transport of IFT complexes. Intraflagellar transport dynein is autoinhibited by trapping of its mechanical and track-binding elements [129a]. Unlike cytoplasmic dynein, which maintains its processive motility over a wide range of ATP concentrations, outer arm dynein from *Tetrahymena* cilia moves processively at low ATP [130] and becomes nonprocessive at concentrations higher than $5\,\mu M$ [104]. Since axonemal dyneins bind to microtubules by their tail region and decorate microtubules in axonemes, similar to that found for myosin II in the muscle sarcomere, their collective action to slide neighboring doublet microtubules does not require processive motility under physiological conditions. It is possible that ATP-induced transition from processive to nonprocessive motility of OAD occurs because this motor has low microtubule affinity at the rate limiting transition of its mechanochemical cycle. Compared to cytoplasmic dynein, the mechanism of axonemal dyneins and their collective action to generate ciliary beating and waveforms are not well understood [131,132], and detailed single-molecule and biochemical studies are required to reveal their mechanochemical cycle [133]. Combination of structural, mechanistic, and cellular studies of dynein is poised to reveal how dyneins perform a wide variety of cellular functions and their dysfunction lead to disease.

Acknowledgments

Work in my laboratory is supported by the National Institute of Health (RO1GM094522).

References

[1] S.M. Block, Kinesin motor mechanics: binding, stepping, tracking, gating, and limping, Biophys. J. 92 (9) (2007) 2986–2995.

[2] J.A. Spudich, Molecular motors take tension in stride, Cell 126 (2) (2006) 242–244.

[3] D. Altman, H.L. Sweeney, J.A. Spudich, The mechanism of myosin VI translocation and its load-induced anchoring, Cell 116 (5) (2004) 737–749.

[4] A. Mehta, Myosin learns to walk, J. Cell Sc. 114 (Pt 11) (2001) 1981–1998.

[5] A. Yildiz, GE Prize winner. How molecular motors move, Science 311 (5762) (2006) 792–793.

[6] R.D. Vale, R.A. Milligan, The way things move: looking under the hood of molecular motor proteins, Science 288 (5463) (2000) 88–95.

[7] A.F. Neuwald, L. Aravind, J.L. Spouge, E.V. Koonin, AAA+: a class of chaperone-like ATPases associated with the assembly, operation, and disassembly of protein complexes, Genome Res. 9 (1) (1999) 27–43.

[8] S.E. Glynn, A. Martin, A.R. Nager, T.A. Baker, R.T. Sauer, Structures of asymmetric ClpX hexamers reveal nucleotide-dependent motions in a AAA+ protein-unfolding machine, Cell 139 (4) (2009) 744–756.

[9] S.Y. Lee, A. De La Torre, D. Yan, S. Kustu, B.T. Nixon, D.E. Wemmer, Regulation of the transcriptional activator NtrC1: structural studies of the regulatory and AAA+ ATPase domains, Genes Dev. 17 (20) (2003) 2552–2563.

[10] S.L. Reck-Peterson, A. Yildiz, A.P. Carter, A. Gennerich, N. Zhang, R.D. Vale, Single-molecule analysis of dynein processivity and stepping behavior, Cell 126 (2) (2006) 335–348.

[11] M. Nishiura, T. Kon, K. Shiroguchi, R. Ohkura, T. Shima, Y.Y. Toyoshima, et al., A single-headed recombinant fragment of *Dictyostelium* cytoplasmic dynein can drive the robust sliding of microtubules, J. Biol. Chem. 279 (22) (2004) 22799–22802.

[12] P. Hook, A. Mikami, B. Shafer, B.T. Chait, S.S. Rosenfeld, R.B. Vallee, Long range allosteric control of cytoplasmic dynein ATPase activity by the stalk and C-terminal domains, J. Biol. Chem. 280 (38) (2005) 33045–33054.

[13] S.A. Burgess, M.L. Walker, H. Sakakibara, P.J. Knight, K. Oiwa, Dynein structure and power stroke, Nature 421 (6924) (2003) 715–718.

[14] A.P. Carter, J.E. Garbarino, E.M. Wilson-Kubalek, W.E. Shipley, C. Cho, R.A. Milligan, et al., Structure and functional role of dynein's microtubule-binding domain, Science 322 (5908) (2008) 1691–1695.

[15] T. Kon, K. Imamula, A.J. Roberts, R. Ohkura, P.J. Knight, I.R. Gibbons, et al., Helix sliding in the stalk coiled coil of dynein couples ATPase and microtubule binding, Nat. Struct. Mol. Biol. 16 (3) (2009) 325–333.

[16] T. Kon, T. Oyama, R. Shimo-Kon, K. Imamula, T. Shima, K. Sutoh, et al., The 2.8 Å crystal structure of the dynein motor domain, Nature 484 (7394) (2012) 345–350.

[17] H. Schmidt, E.S. Gleave, A.P. Carter, Insights into dynein motor domain function from a 3.3-A crystal structure, Nat. Struct. Mol. Biol. 19 (5) (2012) 492–497 S1.

[18] K. Imamula, T. Kon, R. Ohkura, K. Sutoh, The coordination of cyclic microtubule association/dissociation and tail swing of cytoplasmic dynein, Proc. Natl. Acad. Sci. U.S.A. 104 (41) (2007) 16134–16139.

[19] T. Mogami, T. Kon, K. Ito, K. Sutoh, Kinetic characterization of tail swing steps in the ATPase cycle of *Dictyostelium* cytoplasmic dynein, J. Biol. Chem. 282 (30) (2007) 21639–21644.

[20] T. Kon, T. Mogami, R. Ohkura, M. Nishiura, K. Sutoh, ATP hydrolysis cycle-dependent tail motions in cytoplasmic dynein, Nat. Struct. Mol. Biol. 12 (6) (2005) 513–519.

[21] A.J. Roberts, T. Kon, P.J. Knight, K. Sutoh, S.A. Burgess, Functions and mechanics of dynein motor proteins, Nat. Rev. Mol. Cell Biol. 14 (11) (2013) 713–726.

[22] M.A. Cianfrocco, M.E. DeSantis, A.E. Leschziner, S.L. Reck-Peterson, Mechanism and regulation of cytoplasmic dynein, Annu. Rev. Cell Dev. Biol. 31 (2015) 83–108.

[23] G. Bhabha, H.C. Cheng, N. Zhang, A. Moeller, M. Liao, J.A. Speir, et al., Allosteric communication in the dynein motor domain, Cell 159 (4) (2014) 857–868.

[24] H. Schmidt, R. Zalyte, L. Urnavicius, A.P. Carter, Structure of human cytoplasmic dynein-2 primed for its power stroke, Nature 518 (7539) (2015) 435–438.

[25] A.J. Roberts, N. Numata, M.L. Walker, Y.S. Kato, B. Malkova, T. Kon, et al., AAA+ Ring and linker swing mechanism in the dynein motor, Cell 136 (3) (2009) 485–495.

[26] T. Kon, M. Nishiura, R. Ohkura, Y.Y. Toyoshima, K. Sutoh, Distinct functions of nucleotide-binding/hydrolysis sites in the four AAA modules of cytoplasmic dynein, Biochemistry 43 (35) (2004) 11266–11274.

[27] W.B. Redwine, R. Hernandez-Lopez, S. Zou, J. Huang, S.L. Reck-Peterson, A.E. Leschziner, Structural basis for microtubule binding and release by dynein, Science 337 (6101) (2012) 1532–1536.

[28] V. Belyy, N.L. Hendel, A. Chien, A. Yildiz, Cytoplasmic dynein transports cargos via load-sharing between the heads, Nat. Commun. 5 (2014) 5544.

[29] E.L. Holzbaur, K.A. Johnson, ADP release is rate limiting in steady-state turnover by the dynein adenosinetriphosphatase, Biochemistry 28 (13) (1989) 5577–5585.

[30] S.J. King, T.A. Schroer, Dynactin increases the processivity of the cytoplasmic dynein motor, Nat. Cell Biol. 2 (1) (2000) 20–24.

[31] R. Mallik, B.C. Carter, S.A. Lex, S.J. King, S.P. Gross, Cytoplasmic dynein functions as a gear in response to load, Nature 427 (6975) (2004) 649–652.

[32] H. Sakakibara, H. Kojima, Y. Sakai, E. Katayama, K. Oiwa, Inner-arm dynein c of *Chlamydomonas* flagella is a single-headed processive motor, Nature 400 (6744) (1999) 586–590.

[33] M.A. DeWitt, A.Y. Chang, P.A. Combs, A. Yildiz, Cytoplasmic dynein moves through uncoordinated stepping of the AAA+ ring domains, Science 335 (6065) (2012) 221–225.

[34] F.B. Cleary, M.A. Dewitt, T. Bilyard, Z.M. Htet, V. Belyy, D.D. Chan, et al., Tension on the linker gates the ATP-dependent release of dynein from microtubules, Nat. Commun. 5 (2014) 4587.

[35] R. Mallik, D. Petrov, S.A. Lex, S.J. King, S.P. Gross, Building complexity: an in vitro study of cytoplasmic dynein with in vivo implications, Curr. Biol. 15 (23) (2005) 2075–2085.

[36] M. Vershinin, B.C. Carter, D.S. Razafsky, S.J. King, S.P. Gross, Multiple-motor based transport and its regulation by Tau, Proc. Natl. Acad. Sci. U.S.A. 104 (1) (2007) 87–92.

[37] N.D. Derr, B.S. Goodman, R. Jungmann, A.E. Leschziner, W.M. Shih, S.L. Reck-Peterson, Tug-of-war in motor protein ensembles revealed with a programmable DNA origami scaffold, Science 338 (6107) (2012) 662–665.

[38] K. Furuta, A. Furuta, Y.Y. Toyoshima, M. Amino, K. Oiwa, H. Kojima, Measuring collective transport by defined numbers of processive and nonprocessive kinesin motors, Proc. Natl. Acad. Sci. U.S.A. 110 (2) (2013) 501–506.

[39] A.R. Driller-Colangelo, K.W. Chau, J.M. Morgan, N.D. Derr, Cargo rigidity affects the sensitivity of dynein ensembles to individual motor pausing, Cytoskeleton 73 (12) (2016) 693–702.

[40] W.J. Walter, I. Machens, F. Rafieian, S. Diez, The non-processive rice kinesin-14 OsKCH1 transports actin filaments along microtubules with two distinct velocities, Nat. Plants 1 (2015) 15111.

[41] E. Jonsson, M. Yamada, R.D. Vale, G. Goshima, Clustering of a kinesin-14 motor enables processive retrograde microtubule-based transport in plants, Nat. Plants 1 (7) (2015).

[42] S. Can, M.A. Dewitt, A. Yildiz, Bidirectional helical motility of cytoplasmic dynein around microtubules, eLife 3 (2014) e03205.

[43] M. Sirajuddin, L.M. Rice, R.D. Vale, Regulation of microtubule motors by tubulin isotypes and post-translational modifications, Nat. Cell Biol. 16 (4) (2014) 335–344.

[44] R.J. McKenney, W. Huynh, R.D. Vale, M. Sirajuddin, Tyrosination of alpha-tubulin controls the initiation of processive dynein-dynactin motility, EMBO J. 35 (11) (2016) 1175–1185.

[45] M. Mazumdar, A. Mikami, M.A. Gee, R.B. Vallee, In vitro motility from recombinant dynein heavy chain, Proc. Natl. Acad. Sci. U.S.A. 93 (13) (1996) 6552–6556.

[46] T. Torisawa, M. Ichikawa, A. Furuta, K. Saito, K. Oiwa, H. Kojima, et al., Autoinhibition and cooperative activation mechanisms of cytoplasmic dynein, Nat. Cell Biol. 16 (11) (2014) 1118–1124.

[47] R.J. McKenney, W. Huynh, M.E. Tanenbaum, G. Bhabha, R.D. Vale, Activation of cytoplasmic dynein motility by dynactin-cargo adapter complexes, Science 345 (6194) (2014) 337–341.

[48] M.A. Schlager, H.T. Hoang, L. Urnavicius, S.L. Bullock, A.P. Carter, In vitro reconstitution of a highly processive recombinant human dynein complex, EMBO J. 33 (17) (2014) 1855–1868.

[49] S. Aoyama, R. Kamiya, Strikingly fast microtubule sliding in bundles formed by *Chlamydomonas* axonemal dynein, Cytoskeleton 67 (6) (2010) 365–372.

[50] C. Kural, H. Kim, S. Syed, G. Goshima, V.I. Gelfand, P.R. Selvin, Kinesin and dynein move a peroxisome in vivo: a tug-of-war or coordinated movement? Science 308 (5727) (2005) 1469–1472.

[51] K.M. Ori-McKenney, J. Xu, S.P. Gross, R.B. Vallee, A cytoplasmic dynein tail mutation impairs motor processivity, Nat. Cell Biol. 12 (12) (2010) 1228–1234.

[52] A.K. Rai, A. Rai, A.J. Ramaiya, R. Jha, R. Mallik, Molecular adaptations allow dynein to generate large collective forces inside cells, Cell 152 (1–2) (2013) 172–182.

[53] S. Maday, A.E. Twelvetrees, A.J. Moughamian, E.L. Holzbaur, Axonal transport: cargo-specific mechanisms of motility and regulation, Neuron 84 (2) (2014) 292–309.

[54] W. Hong, A. Takshak, O. Osunbayo, A. Kunwar, M. Vershinin, The effect of temperature on microtubule-based transport by cytoplasmic dynein and Kinesin-1 motors, Biophy. J. 111 (8) (2016) 1816.

[55] S. Toba, T.M. Watanabe, L. Yamaguchi-Okimoto, Y.Y. Toyoshima, H. Higuchi, Overlapping hand-over-hand mechanism of single molecular motility of cytoplasmic dynein, Proc. Natl. Acad. Sci. U.S.A. 103 (15) (2006) 5741–5745.

[56] K. Svoboda, C.F. Schmidt, B.J. Schnapp, S.M. Block, Direct observation of kinesin stepping by optical trapping interferometry, Nature 365 (6448) (1993) 721–727.

[57] S. Ayloo, J.E. Lazarus, A. Dodda, M. Tokito, E.M. Ostap, E.L. Holzbaur, Dynactin functions as both a dynamic tether and brake during dynein-driven motility, Nat. Commun. 5 (2014) 4807.

[58] S.K. Tripathy, S.J. Weil, C. Chen, P. Anand, R.B. Vallee, S.P. Gross, Autoregulatory mechanism for dynactin control of processive and diffusive dynein transport, Nat. Cell Biol. 16 (12) (2014) 1192–1201.

[59] K. Shibata, M. Miura, Y. Watanabe, K. Saito, A. Nishimura, K. Furuta, et al., A single protofilament is sufficient to support unidirectional walking of dynein and kinesin, PLoS One 7 (8) (2012) e42990.

[60] J.L. Ross, H. Shuman, E.L. Holzbaur, Y.E. Goldman, Kinesin and dynein-dynactin at intersecting microtubules: motor density affects dynein function, Biophys. J. 94 (8) (2008) 3115–3125.

[61] M.E. Tanenbaum, R.D. Vale, R.J. McKenney, Cytoplasmic dynein crosslinks and slides antiparallel microtubules using its two motor domains, eLife 2 (2013) e00943.

[62] L.C. Kapitein, E.J. Peterman, B.H. Kwok, J.H. Kim, T.M. Kapoor, C.F. Schmidt, The bipolar mitotic kinesin Eg5 moves on both microtubules that it crosslinks, Nature 435 (7038) (2005) 114–118.

[63] H. Imai, T. Shima, K. Sutoh, M.L. Walker, P.J. Knight, T. Kon, et al., Direct observation shows superposition and large scale flexibility within cytoplasmic dynein motors moving along microtubules, Nat. Commun. 6 (2015) 8179.

[64] S. Ray, E. Meyhofer, R.A. Milligan, J. Howard, Kinesin follows the microtubule's protofilament axis, J. Cell Biol. 121 (5) (1993) 1083–1093.

[65] L. Scharrel, R. Ma, R. Schneider, F. Julicher, S. Diez, Multimotor transport in a system of active and inactive kinesin-1 motors, Biophys. J. 107 (2) (2014) 365–372.

[66] R. Schneider, T. Korten, W.J. Walter, S. Diez, Kinesin-1 motors can circumvent permanent roadblocks by side-shifting to neighboring protofilaments, Biophys. J. 108 (9) (2015) 2249–2257.

[67] K.J. Mickolajczyk, N.C. Deffenbaugh, J.O. Arroyo, J. Andrecka, P. Kukura, W.O. Hancock, Kinetics of nucleotide-dependent structural transitions in the kinesin-1 hydrolysis cycle, Proc. Natl. Acad. Sci. U.S.A. 112 (52) (2015) E7186–E7193.

[68] A. Yildiz, M. Tomishige, A. Gennerich, R.D. Vale, Intramolecular strain coordinates kinesin stepping behavior along microtubules, Cell 134 (6) (2008) 1030–1041.

[69] H. Isojima, R. Iino, Y. Niitani, H. Noji, M. Tomishige, Direct observation of intermediate states during the stepping motion of kinesin-1, Nat. Chem. Biol. 12 (4) (2016) 290–297.

[70] M.Y. Ali, S. Uemura, K. Adachi, H. Itoh, K. Kinosita Jr., S. Ishiwata, Myosin V is a left-handed spiral motor on the right-handed actin helix, Nat. Struct. Biol. 9 (6) (2002) 464–467.

[71] M. Brunnbauer, R. Dombi, T.H. Ho, M. Schliwa, M. Rief, Z. Okten, Torque generation of kinesin motors is governed by the stability of the neck domain, Mol. Cell 46 (2) (2012) 147–158.

[72] A. Mitra, F. Ruhnow, B. Nitzsche, S. Diez, Impact-free measurement of microtubule rotations on kinesin and cytoplasmic-dynein coated surfaces, PLoS One 10 (9) (2015) e0136920.

[73] R.D. Vale, Y.Y. Toyoshima, Rotation and translocation of microtubules in vitro induced by dyneins from *Tetrahymena* cilia, Cell 52 (3) (1988) 459–469.

[74] S. Yamaguchi, K. Saito, M. Sutoh, T. Nishizaka, Y.Y. Toyoshima, J. Yajima, Torque generation by axonemal outer-arm dynein, Biophys. J. 108 (4) (2015) 872–879.

[75] R. Dixit, J.L. Ross, Y.E. Goldman, E.L. Holzbaur, Differential regulation of dynein and kinesin motor proteins by tau, Science 319 (5866) (2008) 1086–1089.

[76] D.P. McVicker, L.R. Chrin, C.L. Berger, The nucleotide-binding state of microtubules modulates kinesin processivity and the ability of tau to inhibit kinesin-mediated transport, J. Biol. Chem. 286 (50) (2011) 42873–42880.

[77] J.L. Stern, D.V. Lessard, G.J. Hoeprich, G.A. Morfini, C.L. Berger, Phospho-regulation of tau modulates inhibition of kinesin-1 motility, Mol. Biol. Cell 28 (8) (2017) 1079–1087.

[78] S.M. Markus, K.A. Kalutkiewicz, W.L. Lee, She1-mediated inhibition of dynein motility along astral microtubules promotes polarized spindle movements, Curr. Biol. 22 (23) (2012) 2221–2230.

[79] N.R. Guydosh, S.M. Block, Backsteps induced by nucleotide analogs suggest the front head of kinesin is gated by strain, Proc. Natl. Acad. Sci. U.S.A. 103 (21) (2006) 8054–8059.

[80] Z. Bryant, D. Altman, J.A. Spudich, The power stroke of myosin VI and the basis of reverse directionality, Proc. Natl. Acad. Sci. U.S.A. 104 (3) (2007) 772–777.

[81] Z. Shang, K. Zhou, C. Xu, R. Csencsits, J.C. Cochran, C.V. Sindelar, High-resolution structures of kinesin on microtubules provide a basis for nucleotide-gated force-generation, eLife 3 (2014) e04686.

[82] M.Y. Dogan, S. Can, F.B. Cleary, V. Purde, A. Yildiz, Kinesin's front head is gated by the backward orientation of its neck linker, Cell Rep. 10 (12) (2015) 1967–1973.

[83] J.O. Andreasson, B. Milic, G.Y. Chen, N.R. Guydosh, W.O. Hancock, S.M. Block, Examining kinesin processivity within a general gating framework, eLife 4 (2015).

[84] W.O. Hancock, J. Howard, Kinesin's processivity results from mechanical and chemical coordination between the ATP hydrolysis cycles of the two motor domains, Proc. Natl. Acad. Sci. U.S.A. 96 (23) (1999) 13147–13152.

[85] L.M. Klumpp, A. Hoenger, S.P. Gilbert, Kinesin's second step, Proc. Natl. Acad. Sci. U.S.A. 101 (10) (2004) 3444–3449.

[86] A. Yildiz, M. Tomishige, R.D. Vale, P.R. Selvin, Kinesin walks hand-over-hand, Science 303 (5658) (2004) 676–678.

[87] A. Yildiz, J.N. Forkey, S.A. McKinney, T. Ha, Y.E. Goldman, P.R. Selvin, Myosin V walks hand-over-hand: single fluorophore imaging with 1.5-nm localization, Science 300 (5628) (2003) 2061–2065.

[88] A. Yildiz, H. Park, D. Safer, Z. Yang, L.Q. Chen, P.R. Selvin, et al., Myosin VI steps via a hand-over-hand mechanism with its lever arm undergoing fluctuations when attached to actin, J. Biol. Chem. 279 (36) (2004) 37223–37226.

[89] W. Qiu, N.D. Derr, B.S. Goodman, E. Villa, D. Wu, W. Shih, et al., Dynein achieves processive motion using both stochastic and coordinated stepping, Nat. Struct. Mol. Biol. 19 (2) (2012) 193–200.

[90] T. Shima, K. Imamula, T. Kon, R. Ohkura, K. Sutoh, Head-head coordination is required for the processive motion of cytoplasmic dynein, an AAA+ molecular motor, J. Struct. Biol. 156 (1) (2006) 182–189.

[91] D. Tsygankov, A.W. Serohijos, N.V. Dokholyan, T.C. Elston, Kinetic models for the coordinated stepping of cytoplasmic dynein, J. Chem. Phys. 130 (2) (2009) 025101.

[92] D. Nicastro, C. Schwartz, J. Pierson, R. Gaudette, M.E. Porter, J.R. McIntosh, The molecular architecture of axonemes revealed by cryoelectron tomography, Science 313 (5789) (2006) 944–948.

[93] T. Shima, T. Kon, K. Imamula, R. Ohkura, K. Sutoh, Two modes of microtubule sliding driven by cytoplasmic dynein, Proc. Natl. Acad. Sci. U.S.A. 103 (47) (2006) 17736–17740.

[94] M.P. Nicholas, F. Berger, L. Rao, S. Brenner, C. Cho, A. Gennerich, Cytoplasmic dynein regulates its attachment to microtubules via nucleotide state-switched mechanosensing at multiple AAA domains, Proc. Natl. Acad. Sci. U.S.A. 112 (20) (2015) 6371–6376.

[95] N. Numata, T. Shima, R. Ohkura, T. Kon, K. Sutoh, C-sequence of the *Dictyostelium* cytoplasmic dynein participates in processivity modulation, FEBS Lett. 585 (8) (2011) 1185–1190.

[96] C. Cho, S.L. Reck-Peterson, R.D. Vale, Regulatory ATPase sites of cytoplasmic dynein affect processivity and force generation, J. Biol. Chem. 283 (38) (2008) 25839–25845.

[97] A. Silvanovich, M.G. Li, M. Serr, S. Mische, T.S. Hays, The third P-loop domain in cytoplasmic dynein heavy chain is essential for dynein motor function and ATP-sensitive microtubule binding, Mol. Biol. Cell 14 (4) (2003) 1355–1365.

[98] M.A. DeWitt, C.A. Cypranowska, F.B. Cleary, V. Belyy, A. Yildiz, The AAA3 domain of cytoplasmic dynein acts as a switch to facilitate microtubule release, Nat. Struct. Mol. Biol. 22 (1) (2015) 73–80.

[99] J.R. Moffitt, Y.R. Chemla, K. Aathavan, S. Grimes, P.J. Jardine, D.L. Anderson, et al., Intersubunit coordination in a homomeric ring ATPase, Nature 457 (7228) (2009) 446–450.

[100] J. Huang, A.J. Roberts, A.E. Leschziner, S.L. Reck-Peterson, Lis1 acts as a "clutch" between the ATPase and microtubule-binding domains of the dynein motor, Cell 150 (5) (2012) 975–986.

[101] K. Toropova, S. Zou, A.J. Roberts, W.B. Redwine, B.S. Goodman, S.L. Reck-Peterson, et al., Lis1 regulates dynein by sterically blocking its mechanochemical cycle, eLife 3 (2014).

[102] R.J. McKenney, M. Vershinin, A. Kunwar, R.B. Vallee, S.P. Gross, LIS1 and NudE induce a persistent dynein force-producing state, Cell 141 (2) (2010) 304–314.

[103] A. Gennerich, A.P. Carter, S.L. Reck-Peterson, R.D. Vale, Force-induced bidirectional stepping of cytoplasmic dynein, Cell 131 (5) (2007) 952–965.

[104] E. Hirakawa, H. Higuchi, Y.Y. Toyoshima, Processive movement of single 22S dynein molecules occurs only at low ATP concentrations, Proc. Natl. Acad. Sci. U.S.A. 97 (6) (2000) 2533–2537.

[105] N.R. Guydosh, S.M. Block, Direct observation of the binding state of the kinesin head to the microtubule, Nature 461 (7260) (2009) 125–128.

[106] E.P. Sablin, R.B. Case, S.C. Dai, C.L. Hart, A. Ruby, R.D. Vale, et al., Direction determination in the minus-end-directed kinesin motor ncd, Nature 395 (6704) (1998) 813–816.

[107] S.A. Endow, K.W. Waligora, Determinants of kinesin motor polarity, Science 281 (5380) (1998) 1200–1202.

[108] L. Chen, M. Nakamura, T.D. Schindler, D. Parker, Z. Bryant, Engineering controllable bidirectional molecular motors based on myosin, Nat. Nanotech. 7 (4) (2012) 252–256.

[109] G. Bhabha, G.T. Johnson, C.M. Schroeder, R.D. Vale, How dynein moves along microtubules, Trends Biochem. Sci. 41 (1) (2016) 94–105.

[110] A. Furuta, M. Amino, M. Yoshio, K. Oiwa, H. Kojima, K. Furuta, Creating biomolecular motors based on dynein and actin-binding proteins, Nat. Nanotech. 12 (3) (2016) 233–237.

[111] J.R. Kardon, S.L. Reck-Peterson, R.D. Vale, Regulation of the processivity and intracellular localization of *Saccharomyces cerevisiae* dynein by dynactin, Proc. Natl. Acad. Sci. U.S.A. 106 (14) (2009) 5669–5674.

[112] A.J. Moughamian, G.E. Osborn, J.E. Lazarus, S. Maday, E.L. Holzbaur, Ordered recruitment of dynactin to the microtubule plus-end is required for efficient initiation of retrograde axonal transport, J. Neurosci. 33 (32) (2013) 13190–13203.

[113] M. van Spronsen, M. Mikhaylova, J. Lipka, M.A. Schlager, D.J. van den Heuvel, M. Kuijpers, et al., TRAK/Milton motor-adaptor proteins steer mitochondrial trafficking to axons and dendrites, Neuron 77 (3) (2013) 485–502.

[114] R. Jha, T. Surrey, Regulation of processive motion and microtubule localization of cytoplasmic dynein, Biochem. Soc. Trans. 43 (1) (2015) 48–57.

[115] Z. Wang, S. Khan, M.P. Sheetz, Single cytoplasmic dynein molecule movements: characterization and comparison with kinesin, Biophys. J. 69 (5) (1995) 2011–2023.

[116] M. Trokter, N. Mucke, T. Surrey, Reconstitution of the human cytoplasmic dynein complex, Proc. Natl. Acad. Sci. U.S.A. 109 (51) (2012) 20895–20900.

[117] L.A. Amos, Brain dynein crossbridges microtubules into bundles, J. Cell Sci. 93 (Pt 1) (1989) 19–28.

[118] D. Splinter, D.S. Razafsky, M.A. Schlager, A. Serra-Marques, I. Grigoriev, J. Demmers, et al., BICD2, dynactin, and LIS1 cooperate in regulating dynein recruitment to cellular structures, Mol. Biol. Cell 23 (21) (2012) 4226–4241.

[119] L. Urnavicius, K. Zhang, A.G. Diamant, C. Motz, M.A. Schlager, M. Yu, et al., The structure of the dynactin complex and its interaction with dynein, Science 347 (6229) (2015) 1441–1446.

[120] S. Chowdhury, S.A. Ketcham, T.A. Schroer, G.C. Lander, Structural organization of the dynein-dynactin complex bound to microtubules, Nat. Struct. Mol. Biol. 22 (4) (2015) 345–347.

[121] W.J. Walter, M.P. Koonce, B. Brenner, W. Steffen, Two independent switches regulate cytoplasmic dynein's processivity and directionality, Proc. Natl. Acad. Sci. U.S.A. 109 (14) (2012) 5289–5293.

[122] K. Visscher, M.J. Schnitzer, S.M. Block, Single kinesin molecules studied with a molecular force clamp, Nature 400 (6740) (1999) 184–189.

[123] A. Kunwar, M. Vershinin, J. Xu, S.P. Gross, Stepping, strain gating, and an unexpected force-velocity curve for multiple-motor-based transport, Curr. Biol. 18 (16) (2008) 1173–1183.

[124] A.G. Hendricks, E. Perlson, J.L. Ross, H.W. Schroeder 3rd, M. Tokito, E.L. Holzbaur, Motor coordination via a tug-of-war mechanism drives bidirectional vesicle transport, Curr. Biol. 20 (8) (2010) 697–702.

[125] A. Rai, D. Pathak, S. Thakur, S. Singh, A.K. Dubey, R. Mallik, Dynein clusters into lipid microdomains on phagosomes to drive rapid transport toward lysosomes, Cell 164 (4) (2016) 722–734.

[126] V. Belyy, M.A. Schlager, H. Foster, A.E. Reimer, A.P. Carter, A. Yildiz, The mammalian dynein-dynactin complex is a strong opponent to kinesin in a tug-of-war competition, Nat. Cell Biol. 18 (9) (2016) 1018–1024.

[127] M.P. Nicholas, P. Hook, S. Brenner, C.L. Wynne, R.B. Vallee, A. Gennerich, Control of cytoplasmic dynein force production and processivity by its C-terminal domain, Nat. Commun. 6 (2015) 6206.

[128] H.T. Hoang, M.A. Schlager, A.P. Carter, S.L. Bullock, DYNC1H1 mutations associated with neurological diseases compromise processivity of dynein-dynactin-cargo adaptor complexes, Proc. Natl. Acad. Sci. U.S.A. 114 (9) (2017) E1597–E606.

[129] M.M. Fu, E.L. Holzbaur, JIP1 regulates the directionality of APP axonal transport by coordinating kinesin and dynein motors, J. Cell Biol. 202 (3) (2013) 495–508..

[129a] K. Toropova, M. Mladenov, AJ. Roberts, Nat Struct Mol Biol. 24 (5) (2017 May) 461–468. http://dx.doi.org/10.1038/nsmb.3391. PMID: 28394326.

[130] M. Edamatsu, Identification of biotin carboxyl carrier protein in *Tetrahymena* and its application in in vitro motility systems of outer arm dynein, J. Microbiol. Methods 105 (2014) 150–154.

[131] J. Lin, K. Okada, M. Raytchev, M.C. Smith, D. Nicastro, Structural mechanism of the dynein power stroke, Nat. Cell Biol. 16 (5) (2014) 479–485.

[132] P. Sartori, V.F. Geyer, A. Scholich, F. Julicher, J. Howard, Dynamic curvature regulation accounts for the symmetric and asymmetric beats of *Chlamydomonas* flagella, eLife 5 (2016).

[133] M. Edamatsu, The functional expression and motile properties of recombinant outer arm dynein from *Tetrahymena*, Biochem. Biophys. Res. Commun. 447 (4) (2014) 596–601.

In this chapter

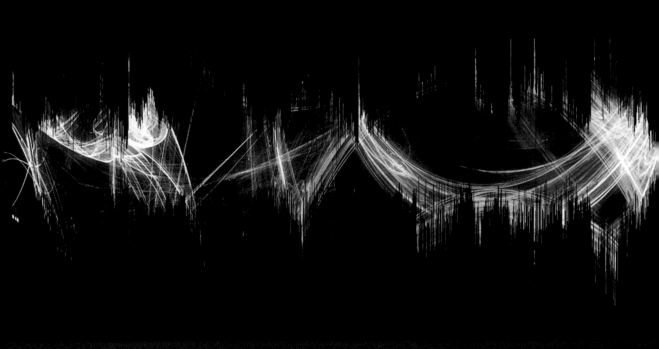

Biophysical properties of dynein in vivo

George T. Shubeita[1], Babu J.N. Reddy[2], Steven P. Gross[2]
[1]New York University Abu Dhabi, Abu Dhabi, United Arab Emirates; [2]University of California, Irvine, CA, United States

Dynein (cytoplasmic dynein-1) is a minus end–directed microtubule motor, and plays critical roles in numerous cellular functions. The motor typically functions as a large complex, composed of two heavy chains (HCs), two intermediate chains (ICs), two light–intermediate chains (LICs), and three classes of light chains (LCs; LC7, LC8, and TcTex). A rather complete discussion of the proteins in the dynein complex can be found in Ref. [1–3]. In addition, as discussed below, the motor complex frequently has its function altered by working in conjunction with two additional regulatory complexes, the BicD/dynactin complex, discussed in Ref. [4–8], and the NudE/Lis1 complex [9–19].

This chapter will discuss what is known about dynein function in vivo, especially with regards to biophysical properties such as force production, velocity, and travel distances, and also how such functions are regulated. A key factor to keep in mind is that frequently dynein works in ensembles, e.g., with multiple motors and cofactors working together to transport a cargo. This means that although regulation directly targets some single-molecule property, the ramification of the single-molecule change may not be immediately obvious, and the magnitude of the resultant ensemble effect may be a function of the number of motors present (see below, e.g., for NudE/Lis1).

6.1 Motility and regulation of dynein in vitro

Multiple copies of dyneins likely work together to move cargos, and the ensemble forces and travel distances will reflect the combined activities of all of these motors; in vivo it is often difficult to know exactly how many motors are functioning together, and also which cofactors are importantly contributing to function. Thus, interpretation of in vivo measurements can to some extent be guided by the more controlled measurements on single-molecule function made under controlled in vitro conditions. As such, we start by briefly summarizing what is known about dynein in vitro both alone and with cofactors.

Dyneins. https://doi.org/10.1016/B978-0-12-809470-9.00006-0
137

Without cofactors, single mammalian dynein molecules purified from bovine brain have a mean travel (processivity) of approximately 800 nm [20,21], take a variety of step sizes [20,22], and have a maximal force production of approximately 1.1 pN [20,23–27].

Dynein's processivity has been a bit controversial [27], with some suggesting that it was in fact unprocessive without dynactin [28], and others finding a single dynein motor as diffusing [29,29a,30]; however, this appears to be somewhat incorrect in bead assays [16,20,26,31]. When bound to beads, the motor purified from bovine brain walks approximately 800 nm, as reported above [16]. However, in the case of in vitro reconstituted human cytoplasmic dynein, when fluorescently labeled but with no bead as cargo, its travel is less [26], likely due to autoinhibition, which is relieved upon bead binding [26]. Further, its processivity is likely strongly dependent on the details of how the dimerized motor domains are held together: in murine dynein, mutations that affect the region involved in dimerization dramatically decrease the motors processivity [32], but artificial dimerization of truncated motor domains allows the motors to walk relatively well [26,28,33,34]. Finally, although quite conserved relative to bovine dynein, single murine dynein molecules exhibit much shorter mean travels (~200 nm) in vitro [32].

In vitro, cofactors are observed to significantly alter dynein's function. First, combining dynein with the largest (P150) subunit of the dynactin complex results in an approximate doubling of dynein's processivity [21,31] to approximately 2 μm. However, this dynein–dynactin complex is relatively unstable, and new work has reported that it can be stabilized by adding a fragment of BicD2; this three-member BicD2–dynactin–dynein complex is seen to have a dramatically increased mean travel distance of roughly 8 μm [28,33]. Forming a dynactin–dynein complex does not alter dynein's maximal force production [31], but the three-way BicD2–dynactin–dynein complex is reported to have significantly higher force, at ~4.3 pN [26], comparable to that of kinesin [28]. Although this seems likely to reflect how dynein might function with dynactin and BicD2 in vivo, because this force effect has only been observed for a BicD2 fragment and has not been possible to reproduce for a full-length protein, the extent to which these exciting in vitro findings reflect what actually occurs in vivo remains unknown. Finally, a domain of dynactin has also been shown to be able to inactivate dynein force production [31], inducing dynein to transition into a diffusive state, where it remains bound to microtubules but can easily diffuse along them in both directions. This ability to inactivate dynein is exciting, because it could in principle be used to control outcomes of tug-of-wars with opposite-directed kinesin motors (see below), but again, this effect involves a dynactin fragment, and the extent to which this mechanism is actually used in vivo remains unknown. The interaction of dynein and dynactin is likely regulated in multiple ways; in addition to BicD, dynein IC phosphorylation contributes to controlling the interaction [35], and its phosphorylation may be targeted by multiple enzymes, including inositol hexakisphosphate kinase-1 [36]. Further, there are multiple different ICs that may be subject to distinct such regulation (see a recent review by Kevin Pfister [37]).

The second dynein regulatory complex is the NudE/Lis1 complex. By itself, NudE inactivates dynein [15,16]; it decreases dynein's ability to bind microtubules, and also dramatically decreases dynein's processivity, so that when it does bind, under load it falls off of microtubules within a few steps. In contrast, Lis1 by itself increases the affinity of dynein for the microtubule when dynein is in its weak-binding state (the ATP-bound conformation) [16]. This has two effects. First, it allows dynein to remain bound to the microtubule while under load, allowing better combination of forces from multiple motors [16]. Second, it causes dynein to become stuck in that portion of its cycle, essentially leading to stalls [14–16]. However, when dynein functions in a three-way complex with NudE and Lis1, these issues are resolved; it is better able to remain bound to microtubules under load, allowing improved ensemble function [16], but critically the three-way complex does not get stuck, and can move normally [15,16]. Loss of Lis1 results in impaired nuclear migration [38,39]—a high-load process—so it was hypothesized that the Lis1/NudE complexes' effect to improve dynein performance under load occurred in vivo as well as in vitro, and that has now been directly demonstrated (see Ref. [17] below).

With this brief overview of in vitro experiments, we now turn to the main topic of our review: dynein biophysical function and regulation in vivo. Reflecting the historical order, we briefly touch on measurements in *Reticulomyxa*, and then discuss lipid droplet (LD) transport in *Drosophila* embryos. Finally, we turn to studies in cultured cells, and after examining force production and travel distances, will turn to regulation.

6.2 Biophysical function of dynein in vivo

The first direct measurement of in vivo cargo force production was done by Art Ashkin [40] at Bell Labs, in 1990, characterizing the force required to stop mitochondria moving in *Reticulomyxa* at about 6.24 pN, or about 2.6 pN/motor, based on an average of 2.4 of observed cross-bridges between cargo and microtubule observed via EM (each cross-bridge assumed to reflect a motor), with an error factor of 2–3. The motor was likely to be dynein, suggesting an in vivo per-motor force of 2.6 pN for cytoplasmic dynein.

The next reported in vivo force measurements were done 8 years later in *Drosophila* embryos, characterizing the forces required to stall individual moving LDs [41]. These studies found that the force to stall a LD was on the order of 3–5 pN, in both the plus end and minus end direction, and changed in increments of 1.1 pN in a developmentally regulated fashion. Since the droplets frequently reverse travel between plus end and minus end motion, and forces changed developmentally in each direction, it was suggested that either a single class of bidirectionally moving motors moved the droplets, or alternatively, that if there were two classes of opposite-polarity motors, their functions were tightly coupled. This latter interpretation turned out to be true (see below). In addition to suggesting that the unitary motor force was 1.1 pN, two key results of these initial *Drosophila* studies were that (1) forces driving cargos were reproducibly modulated during the course of development, and were kept balanced in each direction in a wild-type genetic background and (2) regulation of this force production required the protein klar (the first "kash domain" protein studied), which was hypothesized to contribute to coordination between opposite-polarity motors.

The subsequent report [42] showed that minus end droplet transport was indeed driven by cytoplasmic dynein, thus supporting the hypothesis that single dynein motors might be producing 1.1 pN of force in vivo. It also suggested that the mean travel distance in vivo was short relative to what would be expected for 3–6 processive motors moving a cargo. This is consistent with either of the following scenarios: (1) the possibility of unknown cytosolic impediments to motion, (2) the possibility that there was a tug-of-war between opposite-polarity motors, so that they interfered with each other, (3) that there was a motor coordination complex controlling motor activity, and that the engagement or disengagement of the complex determined motor activity and thus travel distances, rather than the stochastic attachment/detachment of the motors themselves. Consistent with the possibility of a coordination complex, it was found [43] that a mutation in the P150/Glued subunit of the dynactin complex impaired travel distances and force production in a way consistent with the dynactin complex usually functioning to prevent tug-of-wars, and having lost some of this ability due to the mutation. While this notion of dynactin playing a role in coordinating opposite motors is not yet proven, work in other systems has supported this hypothesis [44], and the recent identification of a dynactin domain that can turn off dynein force production [31] now points to a plausible mechanism for how such coordination might be achieved.

For the LD system in *Drosophila* embryos multiple lines of evidence confirm these early suggestions of coordination between the opposite-polarity motors, dynein and kinesin-1, present on these cargos. First, the force-induced detachment rates of the kinesins and dyneins from the microtubules suggest that a tug-of-war would not be consistent with the observed vigorous long-range transport of the LDs in both directions. Dynein's catch-bond interaction with the microtubule measured both in vivo [45] and in vitro [46] would result in cargo dynamics dominated by short excursions, biased toward minus end transport, had directionality switching been solely determined by stochastic tug-of-wars [47]. Indeed, stochastic simulations of unregulated LD transport by dynein and kinesin with single-motor properties constrained by experiment demonstrated that cargo dynamics would diverge significantly from what is observed in vivo [46].

Second, cargo directionality exhibits memory. Using the optical trap to detach LDs from the microtubules while moving either in the plus end or minus end direction, the Shubeita lab looked at the correlation between the direction of motion prior to detachment and the ensuing direction once the droplet started moving again [45]. Consistent with motor coordination, they found that it is much more likely for the cargo to continue moving in its original direction than switch directions. This finding suggests that coordination is achieved by the motors on a cargo being differentially activated. Below, we discuss some of the proteins that are likely involved in coordinating the opposite-polarity motors, both as far as directly turning the motors on or off (dynactin, klar, BicD), and involved upstream in providing the inputs to tell the direct coordinators what to do (LSD2, Huntingtin, JIP).

Finally, when transport of purified *Drosophila* LDs was reconstituted in vitro [48], LDs typically moved bidirectionally indicating the presence of both polarity endogenous motors. However, the typical run length of these bidirectional excursions is a few hundred nanometers as opposed to the micrometer run lengths observed in the embryos. This suggests that factors present in vivo and responsible for the differential activation of the opposite-polarity motors are missing in the in vitro assay resulting in extended tug-of-war instances. Similar short range in vitro transport was reported by the Holzbaur group for purified vesicles and analyzed in the context of the tug-of-war scenario [49]. For other systems, engaging in a tug-of-war may enable the opposite-polarity motors to affect biological function other than transport. An example is the motor-assisted fission of endosomes in *Dictyostelium* cells where the vesicles are stretched by the motors before fission [50].

In vitro, increasing the number of motors on an artificial cargo increases the cargo's mean travel distance and force production [23,51–53], so it is appealing to hypothesize that in vivo, regulation might occur via recruitment or release of motors to the cargo. So far, the extent to which motor recruitment/release is indeed used to control bidirectional transport remains somewhat controversial. On the one hand, there does appear to be some recruitment of motors to cargos to initiate transport. For example, bouts of retrograde transport correlate with dynein binding to early endosomes in the fungus *Ustilago maydis*, while dynein falls off during anterograde transport [54]. On the other hand, although genetically changing the number of motors per cargo did change the forces the cargos were able to produce, this change was found to have little effect on travel distances or velocities [55]. Further, other biochemical studies [56] examining changes to motor bound cargos as a function of different regulatory environments have not yielded a convincing case for the hypothesis that motor recruitment or release plays an important role in controlling the mean direction of transport for bidirectionally moving cargos. Nonetheless, some cargo initiation may involve motor recruitment, and final delivery of cargos to specific subcellular locations likely does involve release of motors. For example, signaling endosomes are stimulated to recruit dynein due to active ERK 1/2 pathway [57]. Similarly, it is reported that cessation of mito-chondrial motion during mitosis involves phosphorylation-controlled release of the dynein motors [58]. In axons, mitochondrial arrest is regulated by Ca^{2+} signaling via the Milton–Miro complex. Yet Ca^{2+} binding to this motor adapter complex has been reported to either sequester the motor heads [59] or detach the motors from the mitochondria [60]. Finally, the conformation of the small G protein, Rab3, found on synaptic vesicles appears to play a role in motor release as GTP-bound Rab3 binds the motor adapter protein DENN/MADD while GDP-bound Rab3 does not [61].

Before discussing regulation of dynein function in vivo in more depth, we first summarize the other in vivo studies of dynein biophysical properties in cells. In vitro, mammalian dynein is reported to have variable step sizes [20], and a study by the Xie group examining motion of quantum dots in endosomes in human lung cancer cells [62] detected step sizes of 8, 12, 16, 20, and 24nm in the minus end direction, but only 8nm steps in the plus end kinesin direc-tion. These findings were confirmed by the Mallik lab [27] studying dynein- and kinesin-driven phagosomes in macrophage cells. As was observed in vitro, the typical dynein step size decreased with increasing load. A later study [63] by the Xie group on LDs rather than endosomes, moving in the same lung cancer cells, found 8nm steps for plus end moving (kinesin-driven) LDs, and again, a variety of step sizes for the minus end dynein-driven LDs. Further, this lat-ter study measured cargo stall forces in both directions. Consistent with the *Drosophila* LD studies, they found opposite-polarity forces approximately bal-anced, with a magnitude of between 4 and 20pN. Interestingly, in the minus end (dynein) direction, the histogram's largest peak is at approximately 4pN, with a possible additional peak at 9pN. Thus, the simplest interpretation from these studies would be that a unitary dynein in vivo produces approximately 4pN (see Table 6.1), and takes a variety of step sizes. Consistent with this gen-eral range of forces, a study on phagosomes in macrophage cells [27] found most stall forces in the dynein direction of 6–12pN. However, there the differ-ence between peaks suggests a unitary dynein force of 1–2pN, and thus the 6–12pN forces are ascribed to multiple dynein motors functioning together. Further, a recent Gross lab study [17] (discussed more fully below) found peak plus end forces exerted by LDs to be broadly distributed between ~3 and 15pN, with roughly equal probability between 4 and 12pN, and minus end forces again between ~3 and 15pN, but roughly equal probability across the entire range (see Ref. [17], Sup. Fig. 1). From these studies a relatively consistent picture emerges for overall forces required to stall cargos moved by dyneins, although the magnitude of the per-motor force is still somewhat controversial.

Table 6.1 Measured in vivo forces for dynein

Study	System	Suggested single-motor in vivo dynein force
Welte et al. [41]	*Drosophila* embryonic lipid droplets	1.1 pN
Shubeita et al. [55]	*Drosophila* embryonic lipid droplets	2.4 pN
Ashkin et al. [40]	*Reticulomyxa* mitochondria	2.4 pN
Sims et al. [63]	Lipid droplets in A549 cells	4 pN (main peak location)
Rai et al. [27]	Phagosomes in macrophage cells	1.7–2 pN (distance between successive histogram peaks)
Blehm et al. [64]	Lipid droplets in A549 cells phagosomeS in *Dictyostelium*	~3 pN (main peak location)
Hendricks et al. [25]	Phagosomes in murine macrophages	~1.6–2.3 pN (distance between successive histogram peaks)

However, it should be noted that a somewhat different picture emerges from two other studies, by the Selvin [64] and Goldman [25] groups. These studies looked at forces required to stall LDs moving in either human (A549) lung cancer cells or in *Dictyostelium* [64], or phagocytosed latex beads in murine macrophage cells [25]. Both of these studies concluded that the single minus end dynein force was on the order of 1–3 pN, but that mostly forces were low. These low forces could be biased by the fact that frequent events of extremely small magnitude were included in the histograms [25]. These events could be dominated by noise inherent in vivo, since when the authors selected for events longer than 1 s, they observed significant counts up to about 8 pN. Further, it remains to be seen whether the choice of calibration procedure contributed to this discrepancy as both these works used similar procedures to account for the viscoelasticity of the cytosol, unlike the more straightforward calibration approaches, including a simple momentum-transfer approach [65] that is expected to be insensitive to unknown cytoplasmic details such as the local index of refraction and local viscosity.

6.3 Regulation of dynein motility in vivo

With this summary of overall dynein cargo forces and step sizes, we now turn to regulation of biophysical properties of in vivo function. Though not precisely focused on the biophysical aspects of dynein function and regulation, please also consult an excellent recent review on axonal transport [66]. Again, we start by looking at what was found by studying LD motion in *Drosophila* embryos, and then turn to other systems. In the *Drosophila* system, LD transport changes in a developmentally regulated way; early in development (~cycles 12, 13) there is bidirectional motion but with no net transport, but this changes to net plus end transport in early cycle 14, and net minus end transport in late cycle 14. Thus, by quantifying transport at these different times, one can determine which aspects of motor function are developmentally regulated. After summarizing the sequences of observed changes, we then discuss which proteins have been identified that contributes to making such changes possible. Since the droplets are continually moving bidirectionally, and frequently (every few seconds) switch direction of travel, to achieve a change in droplet distribution—e.g., to increase overall plus end transport on average—one can either increase mean plus end travel, decrease mean minus end travel, or alter both plus end and minus end transport simultaneously. It was found [41] that the minus end travel distances were predominantly unchanged, and that to tune the droplet distribution, the plus end travel was altered. Nonetheless, although minus end travel distances during these developmental phases were roughly constant, forces driving motion were not, changing from 3.3 to 5.5 to 4.4 pN. Exactly how these forces and travel distances were regulated remains unclear; however, correct function required the KASH protein klarsicht ("klar") [41]. In klar-null embryos, droplet forces were dramatically reduced to 1.1 pN, and travel distances in both directions were reduced by more than half. Whether klar coordinates opposite-polarity motors to avoid tug-of-wars, or gathers motors into groups facilitating the simultaneous engagement of multiple motors, is unknown.

In addition to klar, a second protein, Halo, also contributes to regulation of LD travel distances and force production [67]. The presence of Halo increases plus end travel distances and forces, and suppresses minus end travel distances and forces [67]. Halo appears to work in part by forming a complex with the plus end motor [68] and upregulating its function; whether its effects on dynein-driven transport are indirect, or reflect a direct interaction with the dynein motor complex remains unexplored. The mammalian homologue of Halo, if any, remains unidentified.

A number of additional proteins also contribute to regulation of droplet motion. First, the LD–specific protein LSD2 (a perilipin homologue) is required [69]; LSD2 interacts with klar and is phosphorylated in a Halo-dependent manner. Since without LSD2, the response of LD motion to Halo is impaired, it seems likely that at least some of the effect of Halo is transduced via these changes in LSD2 phosphorylation. However, while loss of LSD2 affects overall regulation of LD motion, the effects on the dynein-driven component of this motion appear quite modest, so LSD2's effects likely center on plus end transport.

Two other proteins have more significant roles in regulating dynein-driven LD motion in embryos: P150/Glued and BicD. P150 is the largest component of the dynactin complex, and interacts directly with dynein. As mentioned above, in vitro, the presence of dynactin increases dynein's travel distance twofold, or eightfold when combined with BicD. Because complete loss of P150 is lethal, the effects of a dominant-negative truncation mutant were examined in a heterozygous (Gl[1]/+) background instead. It was found [43] that the presence of the Gl[1] allele did not affect minus end force production, but did decrease minus end travel distances by roughly a factor of 2. Perhaps more intriguingly, this alteration of dynactin function had significant effects on plus end transport, consistent with its alteration having resulted in a tug-of-war between opposite-polarity motors that usually (in the wild type) did not occur; plus end forces were decreased, as were plus end travel distances, and in particular, the proportion of short-slow plus end runs (that might reflect tug-of-wars) more than doubled [43]. This apparent inducement of tug-of-wars due to loss of P150/glued function appears consistent with the recent in vitro report [31] describing a domain of P150 able to turn off dynein force production. There are also reports of Lis1 playing a role in dynein-mediated transport induced by membrane tethering of the BicD2-N fragment [70], and BicD-related proteins contributing to velocity of retrograde moving vesicles [71].

The effect of almost complete loss of BicD function on LD motion was quite significant [72], with very short plus end and minus end runs in all developmental phases, and an overall loss of the ability to regulate the average direction of droplet transport. Interestingly, the BicD[Pa66] mutant had a strong effect on minus end transport as well; during phase 2 of development (corresponding to average net plus end droplet transport) minus end travel was dramatically shorter, and statistically the same as in the BicD-null background; however, during the subsequent phase 3 of development (corresponding to average net minus end transport), minus end run lengths are substantially higher, and surprisingly, both plus end and minus end run lengths are longer in phase 3 in this mutant background than in the wild type. This is intriguing because work subsequent to this study [71] identified the PA66 mutation as likely eliminating the BicD–dynein–dynactin interaction. If this is indeed the case, keeping in mind the recent in vitro findings [28,73] that the BicD–dynactin combination increases dynein's processivity dramatically, the observation of long minus end transport in phase 3 without the BicD–dynein–dynactin contribution might suggest that in vivo there exist multiple ways to tune dynein travel distances.

Above, we have alluded to the hypothesis of a coordination complex, where multiple same-direction motors function together, and runs (cargo travel in a given direction) end not because the motors fall off the microtubule stochastically, but rather, because the coordination complex terminates the run [42]. For both LDs [43] and other cargos [44], it seems likely that dynactin plays a central role. For droplets, the central coordination/control machinery also appears to include BicD, Klar, and LSD2.

There is additional evidence that in neurons, such a coordination complex likely also involves the Huntingtin protein, which can bind to P150/Dynactin [66,74–76]. Although the full biophysical details remain to be explored, it has been shown [74] that phosphorylation of the Huntingtin protein at serine 421 can switch the average direction of bidirectional transport of multiple classes of axonal vesicles, and that this mechanism partly involves recruitment or release of kinesin-1. Huntingtin protein has also been established to play a role in positioning endosomes and lysosomes in nonneuronal cells [76], and mutant huntingtin protein impairs autophagosome dynamics [77]. Just as huntingtin can control the direction of bidirectional cargo transport, so too can Jip [78].

Since many cargos move bidirectionally (i.e., have both plus end and minus end moving motors attached, and frequently reverse travel direction), a key question is how the relative contribution of one set of motors relative to the other is controlled. So far, there is not a single unified answer to this question, though the huntingtin and Jip studies discussed above are clearly a key portion of the answer. Similarly, for the *Drosophila* LD case, the overall level of the Halo protein can be used to tune net transport [67,68], and there are other Halo family members that perhaps either target other bidirectionally moving cargos, or other kinesin family members. In addition, microtubule-associated proteins such as Tau can affect motor function, and do so differentially, with Tau inhibiting plus end kinesin transport much more than minus end dynein transport [79,80]. Conversely, Map4 appears to predominantly inhibit dynein-based transport, potentially even promoting plus end transport [81]. Additionally, signaling cascades clearly play an important role in such regulation. In addition to the AKT phosphorylation of Huntingtin [74], both dynein-driven minus end run lengths and kinesin-driven plus end run lengths of pigment granules in melanophores are regulated by PKA [82], though the exact targets of the PKA activity are unknown. Similarly, GSK3β was shown to negatively regulate both kinesin-1 and dynein in bidirectionally moving APP cargos in *Drosophila* neurons and also in *Drosophila* LDs in embryos [83]. As for PKA, the mechanistic details of how GSK3β exerts such effects remain to be fully explored, but it is clear that for these cargos GSK3β is involved in controlling the number of active motors present on the cargo and not motor–cargo attachment. Finally, CDK5 affects dynein-mediated transport [12,18,84,85], likely by affecting the dynein–NudE interaction. Though not focused in particular on the biophysical aspects of transport, a more complete general discussion of the current understanding of regulation of transport by kinases can be found in a recent review [86].

Structurally, BicD makes possible such regulated changes in transport in the *Drosophila* LD case (discussed earlier). Also, by changing utilization of BicD variants such as BicDR1, in a mammalian cell context it is possible to regulate dynein velocity in vivo [71]. Based on a variety of recent work, including the previously mentioned in vitro studies showing that the dynein–dynactin–BicD-fragment complex has extraordinary processivity and high-force production, it seems clear that BicD will emerge as a central player in regulating dynein function. For a review focused on BicD, see Ref. [4]. Further, it seems likely that BicD is a key example of a number of cargo-bound scaffolding molecules that also contribute to regulation of motion; the abovementioned excellent review discusses this possibility in significant detail [66].

Another general approach to controlling relative transport likely involves controlling motor organization. From a theoretical point of view, if motors are arranged randomly on a cargo, only a few can engage simultaneously, whereas if they are clustered at a single location, more can function together, increasing forces and run lengths [87]. Recent work [88] shows that by confining dynein motors to lipid microdomains, and changing the size of such domains, the cell indeed takes advantage of such as scheme, and pathogens can thus target such clustering to impair lysosome transport.

Another dynein regulatory strategy involves tuning overall levels of Lis1; up to a point, increased Lis1 appears to lead to increased dynein-mediated transport [89,90]. This correlation between increasing Lis1 and increasing dynein-mediated transport is intriguing, but poorly understood; as mentioned above, Lis1 by itself appears to inhibit dynein transport [15,16], as the dynein enzymatic cycle becomes incapable of driving productive motion [14], though this is not the case when dynein interacts with both NudE and Lis1 simultaneously. It seems likely that the increase in Lis1 levels contribute to dynein-mediated transport through multiple pathways. First, Lis1 can itself promote dynein recruitment to cargos such as mRNA particles [89]. Second, it is involved in initiation of some cargo motion [90]. For a more in-depth discussion of these effects, please see an excellent recent review [91]. Finally, working with NudE, Lis1 does indeed help cargos produce more robust forces in vivo [17], allowing a remarkable form of force adaptation, where stalled cargos upregulate the duration and magnitude of force production by increasingly utilizing the NudE/Lis1 system. This potentially prevents them from remaining stalled, helping them to overcome the local obstacles to their motion. However, such adaptation does not require additional recruitment of Lis1 to the cargo; at least in the LD case, the required Lis1 appears to already be present on the droplet [17], and simply engaged when needed. Another level of regulation likely occurs by using Rab6a to modulate the interaction between dynein and Lis1 [92], though whether and how this works in conjunction with NudE remains to be explored.

In conclusion, we are at an exciting time in the dynein field, with our understanding not only of its function, but also its regulation increasing rapidly. In only a few years, it seems likely that we will much better understand how activity of kinesin and dynein motors is dynamically coordinated. Also, it is probable that the pros and cons of regulation of dynein by NudE/Lis1 versus dynactin will be more fully explored, and also how these two key cofactor complexes work together to tune dynein function.

References

[1] R.D. Vale, The molecular motor toolbox for intracellular transport, Cell 112 (4) (2003) 467–480.
[2] R.B. Vallee, J.C. Williams, D. Varma, L.E. Barnhart, Dynein: an ancient motor protein involved in multiple modes of transport, J. Neurobiol. 58 (2) (2004) 189–200.

[3] K. Kevin Pfister, et al., Cytoplasmic dynein nomenclature, J. Cell Biol. 171 (3) (2005) 411–413.

[4] C.C. Hoogenraad, A. Akhmanova, Bicaudal D family of motor adaptors: linking dynein motility to cargo binding, Trends Cell Biol. 26 (5) (2016) 327–340.

[5] M. Baens, P. Marynen, A human homologue (BICD1) of the *Drosophila* bicaudal-D gene, Genomics 45 (3) (1997) 601–606.

[6] J.L. Hodgkinson, C. Peters, S.A. Kuznetsov, W. Steffen, Three-dimensional reconstruction of the dynactin complex by single-particle image analysis, Proc. Natl. Acad. Sci. U.S.A. 102 (10) (2005) 3667–3672.

[7] L. Urnavicius, et al., The structure of the dynactin complex and its interaction with dynein, Science 347 (6229) (2015) 1441–1446.

[8] S. Chowdhury, S.A. Ketcham, T.A. Schroer, G.C. Lander, Structural organization of the dynein–dynactin complex bound to microtubules, Nat. Struct. Mol. Biol. 22 (4) (2015) 345–347 Advance on (March):1–6.

[9] C. Lam, M.A. Vergnolle, L. Thorpe, P.G. Woodman, V.J. Allan, Functional interplay between LIS1, NDE1 and NDEL1 in dynein-dependent organelle positioning, J. Cell Sci. 123 (2010) 202–212.

[10] T. Torisawa, et al., Functional dissection of LIS1 and NDEL1 towards understanding the molecular mechanism of cytoplasmic dynein regulation, J. Biol. Chem. 286 (3) (2011) 1959–1965.

[11] R.J. McKenney, S.J. Weil, J. Scherer, R.B. Vallee, Mutually exclusive cytoplasmic dynein regulation by NudE-Lis1 and dynactin, J. Biol. Chem. 286 (45) (2011) 39615–39622.

[12] J.P. Pandey, D.S. Smith, A Cdk5-dependent switch regulates Lis1/Ndel1/dynein-driven organelle transport in adult axons, J. Neurosci. 31 (47) (2011) 17207–17219.

[13] P. Rompolas, R.S. Patel-King, S.M. King, Association of Lis1 with outer arm dynein is modulated in response to alterations in flagellar motility, Mol. Biol. Cell 23 (18) (2012) 3554–3565.

[14] J. Huang, A.J. Roberts, A.E. Leschziner, S.L. Reck-Peterson, Lis1 acts as a "clutch" between the ATPase and microtubule-binding domains of the dynein motor, Cell 150 (5) (2012) 975–986.

[15] M. Yamada, et al., LIS1 and NDEL1 coordinate the plus-end-directed transport of cytoplasmic dynein, EMBO J. 27 (19) (2008) 2471–2483.

[16] R.J. McKenney, M. Vershinin, A. Kunwar, R.B. Vallee, S.P. Gross, LIS1 and NudE induce a persistent dynein force-producing state, Cell 141 (2) (2010) 304–314.

[17] B.J.N. Reddy, et al., Load-induced enhancement of dynein force production by LIS1-NudE in vivo and in vitro, Nat. Commun. 7 (2016) 12259.

[18] E. Klinman, E.L. Holzbaur, Stress-induced CDK5 activation disrupts axonal transport via Lis1/Ndel1/dynein, Cell Rep. 12 (3) (2015) 462–473.

[19] T. Shu, et al., Ndel1 operates in a common pathway with LIS1 and cytoplasmic dynein to regulate cortical neuronal positioning, Neuron 44 (2) (2004) 263–277.

[20] R. Mallik, B.C. Carter, S.A. Lex, S.J. King, S.P. Gross, Cytoplasmic dynein functions as a gear in response to load, Nature 427 (6975) (2004) 649–652.

[21] S.J. King, T.A. Schroer, Dynactin increases the processivity of the cytoplasmic dynein motor, Nat. Cell Biol. 2 (1) (2000) 20–24.

[22] M.P. Singh, R. Mallik, S.P. Gross, C.C. Yu, Monte Carlo modeling of single-molecule cytoplasmic dynein, Proc. Natl. Acad. Sci. U.S.A. 102 (34) (2005) 12059–12064.

[23] R. Mallik, D. Petrov, S.A. Lex, S.J. King, S.P. Gross, Building complexity: an in vitro study of cytoplasmic dynein with in vivo implications, Curr. Biol. 15 (23) (2005) 2075–2085.

[24] M.P. Nicholas, et al., Control of cytoplasmic dynein force production and processivity by its C-terminal domain, Nat. Commun. 6 (2015) 6206.

[25] A.G. Hendricks, E.L.F. Holzbaur, Y.E. Goldman, Force measurements on cargoes in living cells reveal collective dynamics of microtubule motors, Proc. Natl. Acad. Sci. U.S.A. 109 (45) (2012) 18447–18452, http://dx.doi.org/10.1073/pnas.1215462109.

[26] V. Belyy, et al., The mammalian dynein–dynactin complex is a strong opponent to kinesin in a tug-of-war competition, Nat. Cell Biol. (2016)http://dx.doi.org/10.1038/ncb3393.

[27] A.K. Rai, A. Rai, A.J. Ramaiya, R. Jha, R. Mallik, Molecular adaptations allow dynein to generate large collective forces inside cells, Cell 152 (1–2) (2013) 172–182.

[28] R.J. McKenney, W. Huynh, M.E. Tanenbaum, G. Bhabha, R.D. Vale, Activation of cytoplasmic dynein motility by dynactin-cargo adapter complexes, Science 345 (6194) (2014) 337–341.

[29] V. Ananthanarayanan, et al., XDynein motion switches from diffusive to directed upon cortical anchoring, Cell 153 (7) (2013)http://dx.doi.org/10.1016/j.cell.2013.05.020.

[29a] Z. Wang and M.P. Sheetz, One-dimensional diffusion on microtubules of particles coated with cytoplasmic dynein and immunoglobulins, Cell Struct. Funct. 24 (5), (1999), 373–383.

[30] T. Torisawa, et al., Autoinhibition and cooperative activation mechanisms of cytoplasmic dynein, Nat. Cell Biol. 16 (11) (2014)http://dx.doi.org/10.1038/ncb3048.

[31] S.K. Tripathy, et al., Autoregulatory mechanism for dynactin control of processive and diffusive dynein transport, Nat. Cell Biol. 16 (12) (2014) 1192–1201.

[32] K.M. Ori-McKenney, J. Xu, S.P. Gross, R.B. Vallee, A cytoplasmic dynein tail mutation impairs motor processivity, Nat. Cell Biol. 12 (12) (2010) 1228–1234.

[33] M.A. Schlager, H.T. Hoang, L. Urnavicius, S.L. Bullock, A.P. Carter, In vitro reconstitution of a highly processive recombinant human dynein complex, EMBO J. 33 (17) (2014) 1–14.

[34] M. Trokter, N. Mücke, T. Surrey, Reconstitution of the human cytoplasmic dynein complex, Proc. Natl. Acad. Sci. U.S.A. 109 (51) (2012) 20895–20900.

[35] P.S. Vaughan, J.D. Leszyk, K.T. Vaughan, Cytoplasmic dynein intermediate chain phosphorylation regulates binding to dynactin, J. Biol. Chem. 276 (28) (2001) 26171–26179.

[36] M. Chanduri, et al., Inositol hexakisphosphate kinase 1 (IP6K1) activity is required for cytoplasmic dynein-driven transport, Biochem. J. (2016)http://dx.doi.org/10.1042/BCJ20160610.

[37] K.K. Pfister, Distinct functional roles of cytoplasmic dynein defined by the intermediate chain isoforms, Exp. Cell Res. 334 (1) (2015) 54–60.

[38] S. Sasaki, et al., A LIS1/NUDEL/cytoplasmic dynein heavy chain complex in the developing and adult nervous system, Neuron 28 (3) (2000) 681–696.

[39] R.B. Vallee, R.J. McKenney, K.M. Ori-McKenney, Multiple modes of cytoplasmic dynein regulation, Nat. Cell Biol. 14 (3) (2012) 224–230.

[40] A. Ashkin, K. Schütze, J.M. Dziedzic, U. Euteneuer, M. Schliwa, Force generation of organelle transport measured in vivo by an infrared laser trap, Nature 348 (6299) (1990) 346–348.

[41] M.A. Welte, S.P. Gross, M. Postner, S.M. Block, E.F. Wieschaus, Developmental regulation of vesicle transport in *Drosophila* embryos: forces and kinetics, Cell 92 (4) (1998) 547–557.

[42] S.P. Gross, M.A. Welte, S.M. Block, E.F. Wieschaus, Dynein-mediated cargo transport in vivo. A switch controls travel distance, J. Cell Biol. 148 (5) (2000) 945–956.

[43] S.P. Gross, M.A. Welte, S.M. Block, E.F. Wieschaus, Coordination of opposite-polarity microtubule motors, J. Cell Biol. 156 (4) (2002) 715–724.

[44] M. Haghnia, et al., Dynactin is required for coordinated bidirectional motility, but not for dynein membrane attachment, Mol. Biol. Cell 18 (6) (2007) 2081–2089.

[45] C. Leidel, R.A. Longoria, F.M. Gutierrez, G.T. Shubeita, Measuring molecular motor forces in vivo: implications for tug-of-war models of bidirectional transport, Biophys. J. 103 (3) (2012) 492–500.

[46] A. Kunwar, et al., Mechanical stochastic tug-of-war models cannot explain bidirectional lipid-droplet transport, Proc. Natl. Acad. Sci. U.S.A. 108 (47) (2011) 18960–18965.

[47] M.J.I. Müller, S. Klumpp, R. Lipowsky, Tug-of-war as a cooperative mechanism for bidirectional cargo transport by molecular motors, Proc. Natl. Acad. Sci. U.S.A. 105 (12) (2008) 4609–4614.

[48] T.F. Bartsch, R.A. Longoria, E.-L. Florin, G.T. Shubeita, Lipid droplets purified from *Drosophila* embryos as an endogenous handle for precise motor transport measurements, Biophys. J. 105 (5) (2013) 1182–1191.

[49] A.G. Hendricks, et al., Motor coordination via a tug-of-war mechanism drives bidirectional vesicle transport, Curr. Biol. 20 (8) (2010) 697–702.

[50] V. Soppina, A.K. Rai, A.J. Ramaiya, P. Barak, R. Mallik, Tug-of-war between dissimilar teams of microtubule motors regulates transport and fission of endosomes, Proc. Natl. Acad. Sci. U.S.A. 106 (46) (2009) 19381–19386.

[51] N.D. Derr, et al., Tug-of-war in motor protein ensembles, Science 338 (November 2, 2012) 662–666.

[52] D.K. Jamison, J.W. Driver, M.R. Diehl, Cooperative responses of multiple kinesins to variable and constant loads, J. Biol. Chem. 287 (5) (2012) 3357–3365.

[53] M. Vershinin, B.C. Carter, D.S. Razafsky, S.J. King, S.P. Gross, Multiple-motor based transport and its regulation by tau, Proc. Natl. Acad. Sci. U.S.A. 104 (1) (2007) 87–92.

[54] M. Schuster, R. Lipowsky, M.-A. Assmann, P. Lenz, G. Steinberg, Transient binding of dynein controls bidirectional long-range motility of early endosomes, Proc. Natl. Acad. Sci. U.S.A. 108 (9) (2011) 3618–3623.

[55] G.T. Shubeita, et al., Consequences of motor copy number on the intracellular transport of kinesin-1-driven lipid droplets, Cell 135 (6) (2008) 1098–1107.

[56] S.P. Gross, et al., Interactions and regulation of molecular motors in *Xenopus melanophores*, J. Cell Biol. 156 (5) (2002) 855–865.

[57] D.J. Mitchell, et al., Trk activation of the ERK1/2 kinase pathway stimulates intermediate chain phosphorylation and recruits cytoplasmic dynein to signaling endosomes for retrograde axonal transport, J. Neurosci. 32 (44) (2012) 15495–15510.

[58] J.Y.-M. Chung, J.A. Steen, T.L. Schwarz, Phosphorylation-induced motor shedding is required at mitosis for proper distribution and passive inheritance of mitochondria, Cell Rep. 16 (8) (2016) 2142–2155.

[59] X. Wang, T.L. Schwarz, The mechanism of Ca^{2+}-dependent regulation of kinesin-mediated mitochondrial motility, Cell 136 (1) (2009) 163–174.

[60] A.F. MacAskill, K. Brickley, F.A. Stephenson, J.T. Kittler, GTPase dependent recruitment of Grif-1 by miro1 regulates mitochondrial trafficking in hippocampal neurons, Mol. Cell Neurosci. 40 (3) (2009) 301–312.

[61] S. Niwa, Y. Tanaka, N. Hirokawa, KIF1Bbeta- and KIF1A-mediated axonal transport of presynaptic regulator Rab3 occurs in a GTP-dependent manner through DENN/MADD, Nat. Cell Biol. 10 (11) (2008) 1269–1279.

[62] X. Nan, P.A. Sims, X.S. Xie, Organelle tracking in a living cell with microsecond time resolution and nanometer spatial precision, ChemPhysChem 9 (5) (2008) 707–712.

[63] P.A. Sims, X.S. Xie, Probing dynein and kinesin stepping with mechanical manipulation in a living cell, ChemPhysChem 10 (9–10) (2009) 1511–1516.

[64] B.H. Blehm, T.A. Schroer, K.M. Trybus, Y.R. Chemla, P.R. Selvin, In vivo optical trapping indicates kinesin's stall force is reduced by dynein during intracellular transport, Proc. Natl. Acad. Sci. U.S.A. 110 (9) (2013) 3381–3386.

[65] Y. Jun, S.K. Tripathy, B.R.J. Narayanareddy, M.K. Mattson-Hoss, S.P. Gross, Calibration of optical tweezers for in vivo force measurements: how do different approaches compare? Biophys. J. 107 (6) (2014) 1474–1484.

[66] S. Maday, A.E. Twelvetrees, A.J. Moughamian, E.L.F. Holzbaur, Axonal transport: cargo-specific mechanisms of motility and regulation, Neuron 84 (2) (2014) 292–309.

[67] S.P. Gross, Y. Guo, J.E. Martinez, M.A. Welte, A determinant for directionality of organelle transport in *Drosophila* embryos, Curr. Biol. 13 (19) (2003) 1660–1668.

[68] G.K. Arora, S.L. Tran, N. Rizzo, A. Jain, M.A. Welte, Temporal control of bidirectional lipid-droplet motion in *Drosophila* depends on the ratio of kinesin-1 and its co-factor Halo, J. Cell Sci. 129 (7) (2016) 1416–1428.

[69] M.A. Welte, et al., Regulation of lipid-droplet transport by the perilipin homolog LSD2, Curr. Biol. 15 (14) (2005) 1266–1275.

[70] D. Splinter, et al., BICD2, dynactin, and LIS1 cooperate in regulating dynein recruitment to cellular structures, Mol. Biol. Cell 23 (21) (2012) 4226–4241.

[71] M.A. Schlager, et al., Bicaudal D family adaptor proteins control the velocity of dynein-based movements, Cell Rep. 8 (5) (2014) 1248–1256, http://dx.doi.org/10.1016/j.celrep.2014.07.052.

[72] K.S. Larsen, J. Xu, S. Cermelli, Z. Shu, S.P. Gross, BicaudalD actively regulates microtubule motor activity in lipid droplet transport, PLoS One 3 (11) (2008) e3763.

[73] M.A. Schlager, H.T. Hoang, L. Urnavicius, S.L. Bullock, A.P. Carter, In vitro reconstitution of a highly processive recombinant human dynein complex, EMBO J. 33 (17) (2014) 1855–1868.

[74] E. Colin, et al., Huntingtin phosphorylation acts as a molecular switch for anterograde/retrograde transport in neurons, EMBO J. 27 (15) (2008) 2124–2134.

[75] S. Ayloo, et al., Dynactin functions as both a dynamic tether and brake during dynein-driven motility, Nat. Commun. 5 (2014) 4807.

[76] J.P. Caviston, A.L. Zajac, M. Tokito, E.L.F. Holzbaur, Huntingtin coordinates the dynein-mediated dynamic positioning of endosomes and lysosomes, Mol. Biol. Cell 22 (4) (2011) 478–492.

[77] Y.C. Wong, E.L.F. Holzbaur, The regulation of autophagosome dynamics by huntingtin and HAP1 is disrupted by expression of mutant huntingtin, leading to defective cargo degradation, J. Neurosci. 34 (4) (2014) 1293–1305.

[78] M.M. Fu, E.L.F. Holzbaur, JIP1 regulates the directionality of APP axonal transport by coordinating kinesin and dynein motors, J. Cell Biol. 202 (3) (2013) 495–508.

[79] M. Vershinin, J. Xu, D.S. Razafsky, S.J. King, S.P. Gross, Tuning microtubule-based transport through filamentous MAPs: the problem of dynein, Traffic 9 (6) (2008) 882–892.

[80] R. Dixit, J.L. Ross, Y.E. Goldman, E.L.F. Holzbaur, Differential regulation of dynein and kinesin motor proteins by tau, Science 319 (5866) (2008) 1086–1089.

[81] I. Semenova, et al., Regulation of microtubule-based transport by MAP4, Mol. Biol. Cell 25 (20) (2014) 3119–3132.

[82] V. Rodionov, J. Yi, A. Kashina, A. Oladipo, S.P. Gross, Switching between microtubule- and actin-based transport systems in melanophores is controlled by cAMP levels, Curr. Biol. 13 (21) (2003) 1837–1847.

[83] C. Weaver, et al., Endogenous GSK-3/Shaggy regulates bidirectional axonal transport of the amyloid precursor protein, Traffic 14 (3) (2013) 295–308.

[84] G. Morfini, et al., A novel CDK5-dependent pathway for regulating GSK3 activity and kinesin-driven motility in neurons, EMBO J. 23 (11) (2004) 2235–2245.

[85] M. Niethammer, et al., NUDEL is a novel Cdk5 substrate that associates with LIS1 and cytoplasmic dynein, Neuron 28 (3) (2000) 697–711.

[86] K.L. Gibbs, L. Greensmith, G. Schiavo, Regulation of axonal transport by protein kinases, Trends Biochem. Sci. 40 (10) (2015) 597–610.

[87] R.P. Erickson, Z. Jia, S.P. Gross, C.C. Yu, How molecular motors are arranged on a cargo is important for vesicular transport, PLoS Comput. Biol. 7 (5) (2011) e1002032.

[88] A. Rai, et al., Dynein clusters into lipid microdomains on phagosomes to drive rapid transport toward lysosomes, Cell 164 (4) (2016) 722–734.

[89] C.I. Dix, et al., Lissencephaly-1 promotes the recruitment of dynein and dynactin to transported mRNAs, J. Cell Biol. 202 (3) (2013) 479–494.

[90] M.J. Egan, K. Tan, S.L. Reck-Peterson, Lis1 is an initiation factor for dynein-driven organelle transport, J. Cell Biol. 197 (7) (2012) 971–982.

[91] M.A. Cianfrocco, M.E. DeSantis, A.E. Leschziner, S.L. Reck-Peterson, Mechanism and regulation of cytoplasmic dynein, Annu. Rev. Cell Dev. Biol. 31 (2015) 83–108.

[92] M. Yamada, et al., Rab6a releases LIS1 from a dynein idling complex and activates dynein for retrograde movement, Nat. Commun. 4 (2013) 2033.

In this chapter

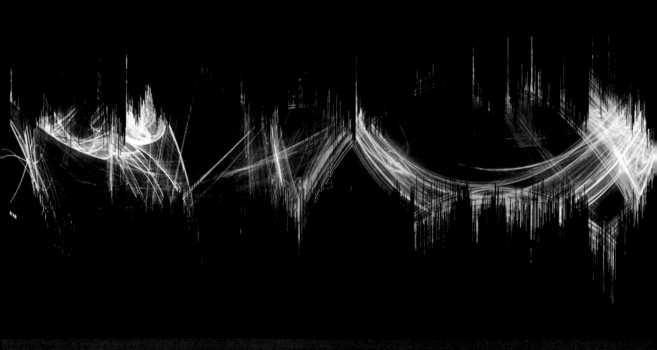

Mechanics of bidirectional cargo transport

William O. Hancock
Pennsylvania State University, University Park, PA, United States

7.1 Introduction

Microtubule-based transport of intracellular cargo is vital for cellular function and is particularly important in axons and dendrites of neurons and in cilia and flagella. This bidirectional transport is carried out by the dynein motors cytoplasmic dynein and dynein-2, and by kinesin motors, primarily from the kinesin-1, kinesin-2, and kinesin-3 families [1–3]. At the molecular level, in vitro single-molecule investigations and transient kinetics assays have generated a solid framework for understanding the fundamental principles underlying kinesin and dynein mechanochemistry. At the cell and organismal level, studies of intracellular transport in cultured cells and in vivo live imaging studies have characterized vesicle and organelle transport dynamics in both wild-type and mutant backgrounds, and identified many proteins responsible for carrying out and regulating these processes. However, there exists a gap between our mechanistic understanding of motors, adapters, and regulatory proteins at the molecular level and our ability to predict how specific perturbations at the cellular level alter bidirectional transport dynamics in disease states such as neurodegenerative diseases and ciliopathies.

Consistent with both its fundamental and clinical importance, the field of bidirectional microtubule-based transport has a rich history of experiments in cells [4–22], in vitro reconstitutions of motor and motor–cargo complexes [12,23–27], and theoretical modeling [4,24,25,28–40]. It is easy to appreciate the importance of understanding multimotor transport when considering transport deficiencies underlying specific diseases. For instance, Alzheimer's disease involves tangles of hyperphosphorylated tau protein that is thought to disrupt axonal transport, but the details of how motors, cargos, and microtubule-associated proteins (MAPs) interact during bidirectional transport is not sufficiently understood to explain what specific function is being inhibited [41–46]. Amyotrophic lateral sclerosis (ALS) and Huntington's disease also involve defects in anterograde and retrograde axonal transport [18,47–51]. Even in healthy neurons, there are a number of outstanding questions, such as: How do opposing motors interact to achieve bidirectional transport? How do regulatory proteins alter the balance of plus end- and minus end-directed motility to achieve proper vectorial transport? How do MAPs and posttranslational modifications (PTMs) of tubulin alter the speed, persistence, and directionality of transport? How does the fluidity of the phospholipid bilayer surrounding vesicles and organelles affect motor-driven transport?

This chapter focuses on the factors that control the net rate and directionality of bidirectional cargo transport along microtubules. It begins with a historical overview of this multidisciplinary field and a description of some of the model systems employed. Next a generalized model of bidirectional transport is introduced that includes specific quantitative parameters that determine transport behavior. The mechanochemical properties of kinesins and dyneins are then introduced, with reference to the motor characteristics that underlie bidirectional behavior. Finally, mechanisms that regulate bidirectional transport will be explored, focusing on the underexplored role of membrane fluidity in multimotor bidirectional transport.

7.2 Experimental and computational work to date on bidirectional transport

In 1998, Gross and Welte, working with lipid droplets in fly embryos, coined the term "tug-of-war" to describe the behavior of antagonistic kinesin and dynein motors transporting the droplets [11], and they created a foundation of both modeling and experimental work describing competition and coordination between antagonistic motors [5,8,9,52]. Working in pigment cells that offer excellent contrast under the microscope, Gelfand, Rodionov, and coworkers defined the signals that trigger dispersion and aggregation of melanosomes, and characterized cross talk between the microtubule and the actomyosin transport systems [7,10,12–14,53–56]. The system that has attracted the most attention is the transport of vesicles and organelles in neurons, where the long transport distances, the presence of both anterograde and retrograde transport, and the requirement that different cargo be transported into specific axonal and dendritic compartments provide an abundance of important mechanistic questions to address [1,2,43,48,57]. Some common themes emerge from these investigations. First, the direction of transport switches stochastically across a range of timescales. Second, transport involves pauses ranging from less than a second to longer than the experimental time frame, and the source of the pauses is not clear [3]. Third, it appears that transport dynamics do not result from motors attaching and detaching from the cargo; instead a cargo generally has both anterograde (kinesin) and retrograde (dynein) motors attached even during pauses and directional switches [16,20,23,56,58].

In vitro reconstitutions have helped define important aspects of motor–motor coordination and competition. Hendricks et al. demonstrated that neuronal vesicles could be purified from brain tissue and that their bidirectional motility reconstituted in vitro was similar to that observed in cultured neurons [23]. Moreover, they found that vesicles purified with kinesin-1, kinesin-2, and dynein motors all attached. This finding emphasizes two principles: (1) antagonistic motors generally remain bound to the cargo during directional switching, and (2) kinesins from different families are often bound to the same vesicle. Reck-Peterson and colleagues used DNA origami to reconstitute complexes containing defined numbers of human kinesin-1 and yeast dynein motors and characterized their velocity and directionality when different numbers of motors were attached [27]. This programmable system and analogous ones using protein scaffolds [26,59,60] provide an in vitro (and potentially in vivo) system for characterizing the motile behaviors of well-defined assemblies of motors, which is crucial for bridging from single-molecule in vitro studies to cellular studies. Finally, a number of theoretical models and computational approaches have been developed to uncover the molecular details of motor coordination during bidirectional transport. The Muller–Klumpp–Lipowsky mean-field model [31], which assumed equal load-sharing of all motors attached to a cargo, was able to recapitulate the stochastic directional switching observed experimentally, and was applied to in vitro experiments [23]. Stochastic approaches that do not assume equal load-sharing and incorporate richer motor dynamics have subsequently been developed and compared to experiments [4,30,32,33,37,61,62]. Thus, there are well-developed tools to study aspects of bidirectional transport at multiple length- and timescales.

7.3 Models of bidirectional transport

Kinesins and dyneins walk in opposite directions along microtubules, and so any model of this system must integrate this inherent mechanical competition. The baseline model of bidirectional transport behavior is the tug-of-war, which holds that the emergent bidirectional behavior results from mechanical competition between these antagonistic motors [3]. One clear prediction from a tug-of-war between opposing motors is that inhibiting one motor should enhance the motility in the opposite direction, and behavior consistent with a tug-of-war between kinesin and dynein has been observed in a number of studies [17,20,23,63–65]. However, one important finding from stochastic modeling of lipid droplet transport was that a tug-of-war mechanism could not quantitatively account for the observed cargo dynamics [4]. What is even more paradoxical is that a number of cell- and organism-level investigations over many years have found that inhibiting one motor diminishes transport in *both* directions, precisely the opposite of what is predicted from a tug-of-war model

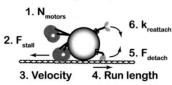

1. N_{motors}

2. F_{stall}

3. Velocity

4. Run length

5. F_{detach}

6. $k_{reattach}$

Figure 7.1 Key motor parameters that underlie bidirectional transport behavior. During transport of cargo by kinesin and dynein motors, different parameters that are derived from single-molecule experiments have different contributions to the emergent bidirectional transport behavior.

(reviewed in Ref. [3]) [9,66–80]. These are diverse systems, and in some cases simpler explanations such as transcriptional downregulation cannot be ruled out. However, the ubiquity of this phenomenon highlights the gulf between the detail at which kinesin and dynein mechanochemistry are understood at the single-molecule level and the many unknown factors that are regulating bidirectional transport in vivo. Potential factors that alter directionality include: (1) the mechanochemical properties of the specific motors carrying out the transport; (2) cargo size and the mechanical properties of the cargo and motor–cargo linkages; (3) MAPs and PTMs that alter motor–microtubule interactions in motor-specific ways; and (4) regulatory proteins that alter motor–cargo binding, motor–microtubule affinity, or motor activation.

Understanding bidirectional transport behavior in cells requires tight integration between modeling and experiments. Thus, before reviewing our current understanding of the motors involved, it is helpful to highlight six key parameters, shown in Fig. 7.1, that underlie the emergent bidirectional transport behavior of cargo carried by kinesin and dynein motors. Any quantitative model must incorporate some or all of these parameters, and it follows that the goal of experiments aimed at understanding the ensemble behavior should focus on measuring or at least constraining these parameters.

1. **Motor number (N_{motors}).** In any scenario the number of opposed motors is important for determining the overall transport behavior. For dynein, optical tweezer experiments that identify quantal peaks in stall force for beads or vesicles with many motors provide evidence that dyneins are able to coordinate their activities effectively [16,20,21]. In contrast, experiments and modeling suggest that kinesin motors cannot efficiently coordinate their activities during multimotor transport [5,22,24,25,32–34]. Thus, while motor numbers are important, activities do not necessarily linearly scale with motor number.

   The number of active motors on a cargo is also a point of regulation by adapter proteins that link motors to cargo, by kinases that may regulate motor activity through phosphorylation, and even by cargo size. Experimental methods have been developed, such as quantitative fluorescence and Western blots [23,81] to measure the number of different motors attached to a cargo. However, experimentally quantifying the number of active motors or even the number of motors that interact with the microtubule at any given time is generally quite difficult.

2. **Stall force (F_{stall}).** The stall force is defined as the force that reduces the net stepping rate to zero. The kinesin-1 stall force has been consistently found to be in the 6–8 pN range from optical tweezer experiments [82,83]. Less data are available, but kinesin-2 and kinesin-3 appear to have similar stall forces [12,84–86]. As discussed in more detail below, there are disagreements in the literature regarding the stall force of mammalian dynein, but a recent study pegs the stall force in the range of 4 pN [87], which is comparable to kinesin. During bidirectional transport, antagonistic motors attached to the same cargo will step in opposite directions, rapidly stretching their elastic tethers, and continue to step until they reach their stall force or detach. Thus, cargo directionality results from interplay between motor numbers, motor stall forces, and motor detachment kinetics, and the stall force is a key metric in any model of transport.

3. **Velocity.** For single-motor velocities, in vitro experiments consistently find kinesin-1 velocities in the range of 0.5–1 μm/s; kinesin-2 values are roughly half of that; and kinesin-3 values are two- to threefold higher [86,88]. Mammalian dynein speeds are in the range of 0.5 μm/s in the minus end direction [89,90]. The specific velocity values are obviously relevant to the in vivo transport rate, and the velocity determines the rate at which a motor pulls taut any motor–cargo compliance and begins to exert substantial forces. However, in thinking about the regulation of bidirectional transport, unloaded motor velocity can be thought of more as a scaling factor rather than a variable that regulates directional switching or other aspects of bidirectionality.

4. **Run length.** The processivity, or number of steps an individual motor takes before detachment, has been extensively studied in vitro for both kinesin and dynein [88,90]. Processivity is an important determinant of a motor's effectiveness in driving bidirectional transport because if it frequently lets go, the opposing motors will dominate transport. However, while the unloaded run length is an important metric, an even more important determinant is the motors' ability to sustain its run length against an opposing load.

5. **Detachment force, F_{detach}.** The ability of a motor to walk processively against an external load can be measured by using an optical tweezer operating in force-feedback mode [91]. This experiment yields the run length under different constant loads, but it is often easier to quantify the first-order motor detachment rate versus load, where $k_{detach} = \text{Velocity} \div \text{Run Length}$. This formulation allows the application of the "Bell Model," which is widely used to analyze the kinetics of bond rupture under mechanical loads [92–96]. Here, the detachment rate in the presence of force, $k_{detach}(F) = k_{detach}^0 e^{\frac{F}{F_{detach}}}$, where k_{detach}^0 is the unloaded detachment rate

and the detachment force, F_{detach}, is a parameter describing the propensity of the motor to detach under load. A large F_{detach} corresponds to a motor whose detachment rate is minimally affected by load, whereas a large F_{detach} indicates a motor that detaches readily under load. In this formulation $F_{det.} = \dfrac{k_B T}{\delta}$, where k_B is Boltzman's constant, T is absolute temperature ($k_B T = 4.1$ pN-nm at room temperature), and δ is a "distance parameter" (ranging from a few Angstroms to a few nanometer). The product $F^* \delta$ can be described as the work performed in breaking the bond. As described further below, there is a large range of F_{detach} across the kinesin superfamily and dynein displays complex load-dependent detachment dynamics. These features are expected to play an important role in motor competition during bidirectional transport.

6. **Reattachment rate, $k_{reattach}$.** Another important but understudied parameter is the reattachment rate of a motor following detachment from the microtubule. When a cargo becomes engaged with a microtubule, the cargo-bound motors will stochastically bind and unbind from the filament during transport, and at a given time, only a fraction of the motors will be engaged with the microtubule and actively contributing to transport. Thus, the interaction of a cargo-bound motor with its microtubule track can be treated as a reversible binding interaction with first-order transition rates k_{detach} and $k_{reattach}$. For any cargo, the fraction of motors that are attached to the microtubule at any given time is equal to $k_{reattach}/(k_{reattach} + k_{detach})$. From this relationship, we can see that the load-dependent detachment rate previously discussed is important, but the rate of reattachment is equally important. Unfortunately, this $k_{rettach}$ parameter is very difficult to measure experimentally. Its value is a function of the cargo size, the motor tether length, the scaffold rigidity, the second-order on-rate for motor–microtubule binding, the microtubule density surrounding the cargo, and potentially other parameters [40]. The influential modeling paper by Muller et al. [31] used a value of $5\,s^{-1}$ for $k_{reattach}$, and this value has propagated through the literature despite the fact that it is not well grounded in experiments. Hence, an important experimental goal (which can be assisted by theory) is to better define $k_{reattach}$ under relevant conditions for intracellular transport, and to better define the variables that regulate it.

7.4 Kinesins involved in bidirectional transport

The primary motors responsible for anterograde cargo transport in cells are kinesins in the kinesin-1, kinesin-2, and kinesin-3 families, which possess a range of mechanochemical properties. Kinesin-1 or conventional kinesin, which carries out axonal transport among other functions [97–99], is the best studied microtubule motor to date and thus provides many experimental constraints for experiments and modeling. The kinesin-2 motor KIF3A/B is heterodimeric (possessing two different motor domains) and is involved in intraflagellar transport (IFT) [58,100–105], movement of late endosomes [106,107], bidirectional transport of melanosomes [10,53], and other transport processes in cells [108–111]. Dysfunction in kinesin-2-driven transport processes is involved in polycystic kidney disease, abnormal cardiac development, retinitis pigmentosa, and other diseases [46,101,104,105,112–114]. Kinesin-3 motors are fast transport motors that were shown to move processively as monomers [115], but are now understood to act as dimers in cells with high processivity and speeds two- to threefold faster than kinesin-1 [86,116–119]. In mammals, kinesin-3 motors (including KIF1, KIF13, KIF14, KIF16, and KIF28) are involved in vesicle transport in neurons [120,121], mitosis [122–124], and other cellular functions [125–127]. Mutations in kinesin-3 motors have been linked to Charcot–Marie–Tooth disease, hereditary spastic paraplegia, and hereditary sensory and autonomic neuropathy type II [120,128–130].

Recent work showed that, in contrast to kinesin-1 that steps processively against hindering loads, kinesin-2 motors detach under load, meaning that they will be less effective in competing with the minus end–directed forces exerted by dynein attached to the same cargo [12,39,84]. Biochemical investigations explained this force-dependent behavior by demonstrating that kinesin-2 spends most of its hydrolysis cycle in a weakly bound ADP-Pi state [131]; because this state has a relatively slow microtubule detachment rate, the unloaded processivity is substantial, but because this state is highly load dependent, run lengths are strongly force sensitive. Hendricks et al. found that a vesicle population purified from axons contained kinesin-1 and kinesin-2 in a 1:4 ratio, which is consistent with kinesin-2 having a reduced mechanical performance compared to kinesin-1 [23]. Interestingly, although recent work has shown that in the absence of load, kinesin-3 motors are superprocessive (run length of 10 μm) [117,118], evidence from multimotor gliding assays [39], optical trapping [132], and the observed lack of coordination in engineered multimotor assemblies in cells [60] suggest that under load kinesin-3 motors detach from microtubules even more readily than kinesin-2. This characteristic may not be shared by all members of the kinesin-3 family [85]; but in any case, measuring the load-dependent detachment rate of different kinesin-3 motors is a high experimental priority in the field.

It has been established that some intracellular cargos are carried by combinations of motors from the kinesin-1, kinesin-2, and kinesin-3 families [23,36,71,102]. Furthermore, inhibition of the slower in the pair has been shown to lead to faster plus end transport for synaptophysin positive vesicles in neurons [71] and IFT particles in *Caenorhabditis elegans* [102], suggesting that the fast and slow motors are sharing the load during normal transport. The interplay between strong and weak motors of the same directionality, and its contribution to emergent bidirectional behavior, remains a hot topic in experimental research. To summarize, different transport kinesins have different responses to mechanical loads in both hindering and assisting directions, and these properties are expected to strongly impact their ability to work in teams and compete against dynein in bidirectional transport.

7.5 Dynein properties relevant to bidirectional transport

The study of dynein mechanochemistry has lagged behind kinesin, but due to enabling progress in heterologous expression and purification techniques [133,134] as well as the recent solution of the crystal structure [135–137], the single-molecule study of the dynein mechanism is in the midst of a golden age. Extensive work on yeast cytoplasmic dynein showed that it steps processively with little coordination between the two motor domains, and that it can generate substantial forces [133,138–145]. This yeast dynein has proven to be an excellent model system, but there are differences between yeast and mammalian dynein that must be understood before motor characteristics can be integrated into an experimentally constrained model of bidirectional transport in vertebrate cells. Yeast dynein moves more slowly than mammalian dynein and is more processive [89,133,138,142,146]. Finally, whereas the stall force of yeast dynein is in the range of 6 pN [89,133], most studies measured a stall force around 1 pN for mammalian cytoplasmic dynein [21,52,89,146–148].

Recent work has helped to resolve these discrepancies in yeast and mammalian cytoplasmic dynein properties. Gennerich and colleagues investigated the function of a C-terminal domain in mammalian dynein that lies on top of the hexameric ring and is not present in yeast dynein [89]. Removal of this "CT-cap" was found to increase the stall force of rat dynein from 0.9 to 5.5 pN, matching that of yeast. It also increased the run length, but had no effect on velocity. More importantly, the Vale and Carter labs demonstrated that a class of coiled-coil proteins (BicD2, Rab11-FIP3, hSpindly, and Hook3) acts as activators of mammalian dynein, increasing dynein run lengths to 5–10 microns in vitro [90,134]. The Yildiz lab showed that when dynein was complexed with dynactin and BicD2, the stall force was in the range of 4 pN [87], close to that of yeast cytoplasmic dynein as well as kinesin. Thus, the short run lengths and low stall forces of mammalian dynein measured in vitro single-molecule assays do not necessarily represent the performance of these motors in cells where it is presumed that BicD2 and other regulators stimulate dynein's activity.

Based on this discovery of dynein activators, for the purposes of understanding bidirectional transport in cells it is reasonable to broadly characterize dynein as a motor that moves at similar speeds as kinesins (ballpark of 0.5–1 μm/s at room temperature), and has a stall force in the same range as transport kinesins (ballpark of 5 pN). This situation contrasts with trying to model the competition of a kinesin having a 5-pN stall force with dyneins that have a 1-pN stall force [4]. However, there is one key mechanochemical property that is very important for bidirectional transport dynamics but is not well specified for mammalian cytoplasmic dynein, which is the dependence of motor detachment on load (F_{detach} in Fig. 7.1). No matter the stall force of a motor, if it tends to detach when pulled against its preferred direction of motion, then it will fare poorly in any tug-of-war scenario. This appears to be the case for kinesin-2, based on optical tweezer experiments discussed earlier [12,84]. Interestingly, there are a number of studies that demonstrate clear catch-bond behavior for dynein, meaning that the detachment rate decreases with force instead of increasing with force [4,21,149]. This catch-bond behavior is a particularly important adaptation for bidirectional transport because it means that the harder kinesin tries to pull the cargo toward the plus end, the more the dynein stays bound to resist that movement. This property could easily lead to stalled or "stuck" cargo, which is observed in cells (reviewed in Ref. [3]), and one could even imagine this property leading to oscillatory behavior.

Other studies consistent with dynein catch-bond behavior include the long pauses observed at stall forces by both yeast dynein [138] and rat dynein activated by removing the CT-cap [89]. Furthermore, activated dynein:dynactin:BicD2 complexes linked to a single kinesin-1 motor showed long duration, slow runs, consistent with catch-bond behavior of dynein (and also of kinesin, interestingly) [87]. Important goals for future single-molecule mechanical experiments of BicD-activated dynein are to characterize the force-dependent detachment kinetics of this activated dynein and to determine how this behavior contributes to the ensemble bidirectional dynamics of vesicles driven by both kinesin and dynein.

7.6 Roles of MAPs and tubulin PTMs in bidirectional transport

During bidirectional transport in cells, MAPs can alter kinesin and dynein interactions with the microtubule and can act as roadblocks to block motor stepping [50,150,151]. Tubulin PTMs such as detyrosination and glutamylation can also affect motor–microtubule interactions and binding of MAPs [150,152–158]. Despite extensive experimental investigation, the mechanisms by which MAPs and PTMs affect kinesin and dynein motility are not well defined [158–166]. Furthermore, how these activities extrapolate to multimotor bidirectional motility is still very much an open question. In a pair of notable studies, Hendricks et al. found that when intracellular vesicles were purified and their motility reconstituted on taxol-stabilized MAP-free microtubules in vitro, the essential bidirectional character of their motility was recapitulated [16,23]. Hence, it is generally appropriate to consider MAPs and PTMs as regulators that tune features of bidirectionality. One interesting hypothesis that has been put forward is that the stochastic switching that occurs during bidirectional transport may provide a "proofreading" activity (analogous to DNA polymerase), such that stalls at roadblocks such as MAPs and stationary cargo are resolved by the antagonistic motors reversing the cargo direction and allowing repeated tries to move past the blockage [74].

7.7 Potential effects of membrane fluidity on bidirectional transport

A feature of bidirectional transport that is relatively underexplored is the role played by the fluid lipid bilayer that surrounds vesicles and organelles. It has been shown that kinesins are able to pull membrane tethers from giant unilamellar vesicles (GUVs) in vitro, and that due to diffusion in the plane of the membrane, motors accumulate at the tether tips and sum their forces at these focal points [167–174]. This GUV work emphasizes important mechanical and diffusive behaviors of motors bound to intracellular cargos that are not recapitulated by attaching motors to stiff beads or other noncompliant structures in vitro. Membrane properties are expected to be particularly critical for kinesin-3 family motors such as KIF1A and Unc104. The Vale group showed that the high local concentration of Unc104 motors in membranes led to dimerization and significantly enhanced motility compared to isolated monomers [86,116,175]. In a different study Oriola and colleagues showed that engineered KIF1A motors attached to a GUV pulled out membrane tethers that spiraled around a microtubule to make a "barber pole" like structure, demonstrating the high level of intermotor coordination that can occur when motors are allowed to diffuse in the 2D bilayer and the surprising (and poorly understood) behaviors that emerge [176]. Phase separation phospholipids in the bilayer and the resulting segregation of motors into specific domains may also play potential roles. For instance, Rai et al. showed that as phagosomes mature, dynein motors cluster into cholesterol-rich microdomains, which results in a switch from bidirectional transport of early phagosomes to unidirectional minus end transport and higher stall forces in late phagosomes [177].

There is also evidence that the degree of membrane fluidity plays a role in intermotor coordination. Nelson and colleagues showed that MyosinV-functionalized vesicles made from a fluid-phase phospholipid moved faster than single-motor speeds, while gel-phase vesicles showed no such enhanced speed [178]. The enhanced speed was hypothesized to result from preferential detachment of trailing motors followed by their diffusion in the bilayer and rebinding at a position closer to the center of the vesicle. The Ostap and Tuzel groups showed that Myosin 1c ensembles can transiently generate forces in lipid bilayers, and that substantial forces can be achieved when motors are

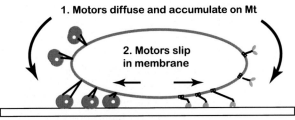

Figure 7.2 Potential effects of membrane fluidity on motor transport properties. In contrast to experiments in which motors are rigidly bound to a particle, most intracellular cargos are surrounded by a fluid lipid bilayer, in which motors can freely diffuse. It is hypothesized that this ability to diffuse in the bilayer will enhance motor function because more motors will be able to access the microtubule and contribute to transport. However, the fluidity is expected to diminish the motors' ability to generate forces because the motors will slip in the plane of the membrane.

anchored to slowly diffusing regions or localized near boundaries where more motors accumulate with the help of diffusion [179,180].

These behaviors point toward a working model of the role of membrane fluidity in multimotor microtubule-based transport, in which membrane fluidity has two different roles (Fig. 7.2). First, due to motors slipping in the plane of the bilayer, membrane fluidity may reduce the ability of motors to generate force. The propensity of a motor to slip in the membrane is a function of the membrane viscosity, and so clustering of motors in gel-phase domains with higher viscosities could help to minimize this effect. Also, this behavior has more relevance for larger vesicles and organelles, because in small (<100 nm) vesicles any movement in the membrane will tend to move the motors toward the poles where their pulling forces are directed normal to the plane of the membrane and slipping no longer occurs. The second effect of membrane fluidity, which should enhance motor function, is clustering of motors near the microtubule. If motors are free to diffuse in the lipid bilayer, but upon microtubule binding are restricted to a small region of the vesicle surface, then the local motor concentration will be elevated near the microtubule. This increased local concentration will increase the motor rebinding rate, resulting in more motors pulling the vesicle than would occur if the motors were uniformly distributed on the surface and unable to diffuse. Interestingly, membrane viscosity works in opposite directions for these two phenomena. Increased membrane viscosity diminishes the loss in function due to motors slipping in the membrane, but it also diminishes the gain-of-function resulting from diffusion-based motor clustering.

Relating back to the important motor properties detailed in Fig. 7.1, we can pinpoint five potential effects of a fluid lipid bilayer. First, the number of motors, N_{motors}, is increased by motors diffusing in the membrane and being able to access the microtubule. Second, because motors can slip in the membrane, it is expected that the maximum stall force a motor can produce, F_{stall}, will be decreased. Third, if the motors are slipping in the membrane opposite to their direction of travel, the net velocity of the cargo relative to the microtubule will be diminished. Fourth, the slipping that reduces velocity and stall force also result in the motors being subject to smaller forces; this beneficial "shock-absorber" effect is expected to increase the run length of motors such as kinesin-2 that dissociate readily under load (i.e., have small F_{detach} values) [12,84]. Finally, the reattachment rate, $k_{reattach}$, is expected to be altered by the motor diffusing in the membrane; motors that have recently detached from the microtubule will reside near the microtubule, increasing $k_{reattach}$, but they will be able to diffuse away from the membrane, potentially diminishing k_{detach}. Clarifying these diffusional effects will require Brownian Dynamics models and other simulations that incorporate load-dependent motor kinetics, 2D diffusion dynamics in the lipid bilayer, 3D fluctuations of the motor domains as they search for the microtubule, and other details. Tight integration of modeling with experiments will be essential, with the models serving to generate experimentally testable hypotheses, and the experiments serving to validate the models.

7.8 Concluding thoughts

The study of bidirectional cargo transport along microtubules has a rich history and has a continuing relevance to biology and disease. Going forward, a major goal in the field is to integrate the rich understanding of motor properties at the single-molecule level into a model that can explain the diverse behaviors of intracellular cargo observed in cells. Even seemingly simple questions such as what triggers directional switching and why so many paused cargos are observed are difficult to definitively answer at present. And to understand the dysfunction in transport that occurs during neurodegenerative disease, it is crucial to have a better understanding of the mechanisms underlying normal transport as well as the various regulatory systems that control it. This is a multiscale problem spanning from nm-scale protein–protein interactions to transport dynamics at the mm scale and above. Progress will require a multidisciplinary attack that includes in vitro experiments, in vivo studies, and *in silico* modeling to bridge the scales.

References

[1] E.L. Holzbaur, Y.E. Goldman, Coordination of molecular motors: from in vitro assays to intracellular dynamics, Curr. Opin. Cell Biol. 22 (1) (2010) 4–13.

[2] S.E. Encalada, L.S. Goldstein, Biophysical challenges to axonal transport: motor-cargo deficiencies and neurodegeneration, Annu. Rev. Biophys. 43 (2014) 141–169.

[3] W.O. Hancock, Bidirectional cargo transport: moving beyond tug of war, Nat. Rev. Mol. Cell Biol. 15 (9) (2014) 615–628.

[4] A. Kunwar, et al., Mechanical stochastic tug-of-war models cannot explain bidirectional lipid-droplet transport, Proc. Natl. Acad. Sci. U.S.A. 108 (47) (2011) 18960–18965.

[5] G.T. Shubeita, et al., Consequences of motor copy number on the intracellular transport of kinesin-1-driven lipid droplets, Cell 135 (6) (2008) 1098–1107.

[6] L.A. Ligon, et al., A direct interaction between cytoplasmic dynein and kinesin I may coordinate motor activity, J. Biol. Chem. 279 (18) (2004) 19201–19208.

[7] V. Rodionov, et al., Switching between microtubule- and actin-based transport systems in melanophores is controlled by cAMP levels, Curr. Biol. 13 (21) (2003) 1837–1847.

[8] S.P. Gross, et al., A determinant for directionality of organelle transport in *Drosophila* embryos, Curr. Biol. 13 (19) (2003) 1660–1668.

[9] S.P. Gross, et al., Coordination of opposite-polarity microtubule motors, J. Cell Biol. 156 (4) (2002) 715–724.

[10] S.P. Gross, et al., Interactions and regulation of molecular motors in *Xenopus melanophores*, J. Cell Biol. 156 (5) (2002) 855–865.

[11] M.A. Welte, et al., Developmental regulation of vesicle transport in *Drosophila* embryos: forces and kinetics, Cell 92 (4) (1998) 547–557.

[12] H.W. Schroeder 3rd, et al., Force-dependent detachment of kinesin-2 biases track switching at cytoskeletal filament intersections, Biophys. J. 103 (1) (2012) 48–58.

[13] V.I. Rodionov, et al., Functional coordination of microtubule-based and actin-based motility in melanophores, Curr. Biol. 8 (3) (1998) 165–168.

[14] V.I. Rodionov, F.K. Gyoeva, V.I. Gelfand, Kinesin is responsible for centrifugal movement of pigment granules in melanophores, Proc. Natl. Acad. Sci. U.S.A. 88 (11) (1991) 4956–4960.

[15] M.M. Fu, E.L. Holzbaur, JIP1 regulates the directionality of APP axonal transport by coordinating kinesin and dynein motors, J. Cell Biol. 202 (3) (2013) 495–508.

[16] A.G. Hendricks, E.L. Holzbaur, Y.E. Goldman, Force measurements on cargoes in living cells reveal collective dynamics of microtubule motors, Proc. Natl. Acad. Sci. U.S.A. 109 (45) (2012) 18447–18452.

[17] J.P. Caviston, et al., Huntingtin coordinates the dynein-mediated dynamic positioning of endosomes and lysosomes, Mol. Biol. Cell 22 (4) (2011) 478–492.

[18] E. Perlson, et al., A switch in retrograde signaling from survival to stress in rapid-onset neurodegeneration, J. Neurosci. 29 (31) (2009) 9903–9917.

[19] J.P. Caviston, E.L. Holzbaur, Huntingtin as an essential integrator of intracellular vesicular trafficking, Trends Cell Biol. 19 (4) (2009) 147–155.

[20] V. Soppina, et al., Tug-of-war between dissimilar teams of microtubule motors regulates transport and fission of endosomes, Proc. Natl. Acad. Sci. U.S.A. 106 (46) (2009) 19381–19386.

[21] A.K. Rai, et al., Molecular adaptations allow dynein to generate large collective forces inside cells, Cell 152 (1–2) (2013) 172–182.

[22] A.K. Efremov, et al., Delineating cooperative responses of processive motors in living cells, Proc. Natl. Acad. Sci. U.S.A. 111 (3) (2014) E334–E343.

[23] A.G. Hendricks, et al., Motor coordination via a tug-of-war mechanism drives bidirectional vesicle transport, Curr. Biol. 20 (2010) 697–702.

[24] D.K. Jamison, et al., Two kinesins transport cargo primarily via the action of one motor: implications for intracellular transport, Biophys. J. 99 (9) (2010) 2967–2977.

[25] A.R. Rogers, et al., Negative interference dominates collective transport of kinesin motors in the absence of load, Phys. Chem. Chem. Phys. 11 (24) (2009) 4882–4889.

[26] M.R. Diehl, et al., Engineering cooperativity in biomotor-protein assemblies, Science 311 (5766) (2006) 1468–1471.

[27] N.D. Derr, et al., Tug-of-war in motor protein ensembles revealed with a programmable DNA origami scaffold, Science 338 (6107) (2012) 662–665.

[28] J. Beeg, et al., Transport of beads by several kinesin motors, Biophys. J. 94 (2) (2008) 532–541.

[29] S. Klumpp, R. Lipowsky, Cooperative cargo transport by several molecular motors, Proc. Natl. Acad. Sci. U.S.A. 102 (48) (2005) 17284–17289.

[30] C.B. Korn, et al., Stochastic simulations of cargo transport by processive molecular motors, J. Chem. Phys. 131 (24) (2009) 245107.

[31] M.J. Muller, S. Klumpp, R. Lipowsky, Tug-of-war as a cooperative mechanism for bidirectional cargo transport by molecular motors, Proc. Natl. Acad. Sci. U.S.A. 105 (12) (2008) 4609–4614.

[32] J.W. Driver, et al., Productive cooperation among processive motors depends inversely on their mechanochemical efficiency, Biophys. J. 101 (2) (2011) 386–395.

[33] J.W. Driver, et al., Coupling between motor proteins determines dynamic behaviors of motor protein assemblies, Phys. Chem. Chem. Phys. 12 (35) (2010) 10398–10405.

[34] D.K. Jamison, J.W. Driver, M.R. Diehl, Cooperative responses of multiple kinesins to variable and constant loads, J. Biol. Chem. 287 (5) (2011) 3357–3365.

[35] M.J. Muller, S. Klumpp, R. Lipowsky, Bidirectional transport by molecular motors: enhanced processivity and response to external forces, Biophys. J. 98 (11) (2010) 2610–2618.

[36] X. Pan, et al., Mechanism of transport of IFT particles in C. elegans cilia by the concerted action of kinesin-II and OSM-3 motors, J. Cell Biol. 174 (7) (2006) 1035–1045.

[37] A. Kunwar, et al., Stepping, strain gating, and an unexpected force-velocity curve for multiple-motor-based transport, Curr. Biol. 18 (16) (2008) 1173–1183.

[38] M. Badoual, F. Julicher, J. Prost, Bidirectional cooperative motion of molecular motors, Proc. Natl. Acad. Sci. U.S.A. 99 (10) (2002) 6696–6701.

[39] G. Arpag, et al., Transport by populations of fast and slow kinesins uncovers novel family-dependent motor characteristics important for in vivo function, Biophys. J. 107 (8) (2014) 1896–1904.

[40] R.P. Erickson, et al., How molecular motors are arranged on a cargo is important for vesicular transport, PLoS Comput. Biol. 7 (5) (2011) e1002032.

[41] K.J. De Vos, et al., Role of axonal transport in neurodegenerative diseases, Annu. Rev. Neurosci. 31 (2008) 151–173.

[42] G.B. Stokin, L.S. Goldstein, Linking molecular motors to Alzheimer's disease, J. Physiol. Paris 99 (2–3) (2006) 193–200.

[43] G.B. Stokin, L.S. Goldstein, Axonal transport and Alzheimer's disease, Annu. Rev. Biochem. 75 (2006) 607–627.

[44] G.B. Stokin, et al., Axonopathy and transport deficits early in the pathogenesis of Alzheimer's disease, Science 307 (5713) (2005) 1282–1288.

[45] S. Gunawardena, L.S. Goldstein, Disruption of axonal transport and neuronal viability by amyloid precursor protein mutations in Drosophila, Neuron 32 (3) (2001) 389–401.

[46] L.S. Goldstein, Kinesin molecular motors: transport pathways, receptors, and human disease, Proc. Natl. Acad. Sci. U.S.A. 98 (13) (2001) 6999–7003.

[47] E. Perlson, et al., Retrograde axonal transport: pathways to cell death? Trends Neurosci. 33 (7) (2010) 335–344.

[48] E. Chevalier-Larsen, E.L. Holzbaur, Axonal transport and neurodegenerative disease, Biochim. Biophys. Acta 1762 (11–12) (2006) 1094–1108.

[49] L.G. Bilsland, et al., Deficits in axonal transport precede ALS symptoms in vivo, Proc. Natl. Acad. Sci. U.S.A. 107 (47) (2010) 20523–20528.

[50] P.W. Baas, L. Qiang, Neuronal microtubules: when the MAP is the roadblock, Trends Cell Biol. 15 (4) (2005) 183–187.

[51] G.A. Morfini, et al., Pathogenic huntingtin inhibits fast axonal transport by activating JNK3 and phosphorylating kinesin, Nat. Neurosci. 12 (7) (2009) 864–871.

[52] S.P. Gross, et al., Dynein-mediated cargo transport in vivo. A switch controls travel distance, J. Cell Biol. 148 (5) (2000) 945–956.

[53] M.C. Tuma, et al., Heterotrimeric kinesin II is the microtubule motor protein responsible for pigment dispersion in Xenopus melanophores, J. Cell Biol. 143 (6) (1998) 1547–1558.

[54] S.L. Rogers, I.S. Tint, V.I. Gelfand, In vitro motility assay for melanophore pigment organelles, Methods Enzymol. 298 (1998) 361–372.

[55] A.R. Reilein, et al., Regulation of organelle movement in melanophores by protein kinase A (PKA), protein kinase C (PKC), and protein phosphatase 2A (PP2A), J. Cell Biol. 142 (3) (1998) 803–813.

[56] S.L. Rogers, et al., Regulated bidirectional motility of melanophore pigment granules along microtubules in vitro, Proc. Natl. Acad. Sci. U.S.A. 94 (8) (1997) 3720–3725.

[57] L.S. Goldstein, Z. Yang, Microtubule-based transport systems in neurons: the roles of kinesins and dyneins, Annu. Rev. Neurosci. 23 (2000) 39–71.

[58] D.G. Cole, et al., *Chlamydomonas* kinesin-II-dependent intraflagellar transport (IFT): IFT particles contain proteins required for ciliary assembly in *Caenorhabditis elegans* sensory neurons, J. Cell Biol. 141 (4) (1998) 993–1008.

[59] A. Rogers, et al., Construction and analyses of elastically coupled multiple-motor systems, Methods Enzymol. 540 (2014) 189–204.

[60] S.R. Norris, et al., A method for multiprotein assembly in cells reveals independent action of kinesins in complex, J. Cell Biol. 207 (3) (2014) 393–406.

[61] A. Kunwar, A. Mogilner, Robust transport by multiple motors with nonlinear force-velocity relations and stochastic load sharing, Phys. Biol. 7 (1) (2010) 16012.

[62] R.T. McLaughlin, M.R. Diehl, A.B. Kolomeisky, Collective dynamics of processive cytoskeletal motors, Soft Matter 12 (1) (2015) 14–21.

[63] L.C. Kapitein, et al., Probing intracellular motor protein activity using an inducible cargo trafficking assay, Biophys. J. 99 (7) (2010) 2143–2152.

[64] J.Y. Yi, et al., High-resolution imaging reveals indirect coordination of opposite motors and a role for LIS1 in high-load axonal transport, J. Cell Biol. 195 (2) (2011) 193–201.

[65] B.H. Blehm, et al., In vivo optical trapping indicates kinesin's stall force is reduced by dynein during intracellular transport, Proc. Natl. Acad. Sci. U.S.A. 110 (9) (2013) 3381–3386.

[66] D.J. Goldberg, Microinjection into an identified axon to study the mechanism of fast axonal transport, Proc. Natl. Acad. Sci. U.S.A. 79 (15) (1982) 4818–4822.

[67] S.T. Brady, K.K. Pfister, Kinesin interactions with membrane bounded organelles in vivo and in vitro, J. Cell Sci. Suppl. 14 (1991) 103–108.

[68] C.M. Waterman-Storer, et al., The interaction between cytoplasmic dynein and dynactin is required for fast axonal transport, Proc. Natl. Acad. Sci. U.S.A. 94 (22) (1997) 12180–12185.

[69] M. Martin, et al., Cytoplasmic dynein, the dynactin complex, and kinesin are interdependent and essential for fast axonal transport, Mol. Biol. Cell 10 (11) (1999) 3717–3728.

[70] R.V. Barkus, et al., Identification of an axonal kinesin-3 motor for fast anterograde vesicle transport that facilitates retrograde transport of neuropeptides, Mol. Biol. Cell 19 (1) (2008) 274–283.

[71] S.E. Encalada, et al., Stable kinesin and dynein assemblies drive the axonal transport of mammalian prion protein vesicles, Cell 144 (4) (2011) 551–565.

[72] A.J. Moughamian, E.L. Holzbaur, Dynactin is required for transport initiation from the distal axon, Neuron 74 (2) (2012) 331–343.

[73] M. Haghnia, et al., Dynactin is required for coordinated bidirectional motility, but not for dynein membrane attachment, Mol. Biol. Cell 18 (6) (2007) 2081–2089.

[74] S. Ally, et al., Opposite-polarity motors activate one another to trigger cargo transport in live cells, J. Cell Biol. 187 (7) (2009) 1071–1082.

[75] S.W. Deacon, et al., Dynactin is required for bidirectional organelle transport, J. Cell Biol. 160 (3) (2003) 297–301.

[76] A.D. Pilling, et al., Kinesin-1 and Dynein are the primary motors for fast transport of mitochondria in *Drosophila* motor axons, Mol. Biol. Cell 17 (4) (2006) 2057–2068.

[77] X. Wang, T.L. Schwarz, The mechanism of Ca^{2+}-dependent regulation of kinesin-mediated mitochondrial motility, Cell 136 (1) (2009) 163–174.

[78] W.M. Saxton, P.J. Hollenbeck, The axonal transport of mitochondria, J. Cell Sci. 125 (Pt 9) (2012) 2095–2104.

[79] A. Uchida, N.H. Alami, A. Brown, Tight functional coupling of kinesin-1A and dynein motors in the bidirectional transport of neurofilaments, Mol. Biol. Cell 20 (23) (2009) 4997–5006.

[80] S.C. Ling, et al., Transport of *Drosophila* fragile X mental retardation protein-containing ribonucleoprotein granules by kinesin-1 and cytoplasmic dynein, Proc. Natl. Acad. Sci. U.S.A. 101 (50) (2004) 17428–17433.

[81] C. Herold, et al., Long-range transport of giant vesicles along microtubule networks, Chemphyschem 13 (4) (2012) 1001–1006.

[82] M.J. Schnitzer, K. Visscher, S.M. Block, Force production by single kinesin motors, Nat. Cell Biol. 2 (10) (2000) 718–723.

[83] K. Svoboda, S.M. Block, Force and velocity measured for single kinesin molecules, Cell 77 (5) (1994) 773–784.

[84] J.O. Andreasson, et al., The mechanochemical cycle of mammalian kinesin-2 KIF3A/B under load, Curr. Biol. 25 (9) (2015) 1166–1175.

[85] T.M. Huckaba, et al., Kinesin-73 is a processive motor that localizes to Rab5-containing organelles, J. Biol. Chem. 286 (9) (2011) 7457–7467.

[86] M. Tomishige, D.R. Klopfenstein, R.D. Vale, Conversion of Unc104/KIF1A kinesin into a processive motor after dimerization, Science 297 (5590) (2002) 2263–2267.

[87] V. Belyy, et al., The mammalian dynein-dynactin complex is a strong opponent to kinesin in a tug-of-war competition, Nat. Cell Biol. 18 (9) (2016) 1018–1024.

[88] S. Shastry, W.O. Hancock, Interhead tension determines processivity across diverse N-terminal kinesins, Proc. Natl. Acad. Sci. U.S.A. 108 (39) (2011) 16253–16258.

[89] M.P. Nicholas, et al., Control of cytoplasmic dynein force production and processivity by its C-terminal domain, Nat. Commun. 6 (2015) 6206.

[90] R.J. McKenney, et al., Activation of cytoplasmic dynein motility by dynactin-cargo adapter complexes, Science 345 (6194) (2014) 337–341.

[91] K. Visscher, M.J. Schnitzer, S.M. Block, Single kinesin molecules studied with a molecular force clamp, Nature 400 (6740) (1999) 184–189.

[92] G.I. Bell, Models for the specific adhesion of cells to cells, Science 200 (4342) (1978) 618–627.

[93] E. Evans, K. Ritchie, Dynamic strength of molecular adhesion bonds, Biophys. J. 72 (4) (1997) 1541–1555.

[94] E. Evans, Probing the relation between force–lifetime–and chemistry in single molecular bonds, Annu. Rev. Biophys. Biomol. Struct. 30 (2001) 105–128.

[95] R. Merkel, et al., Energy landscapes of receptor-ligand bonds explored with dynamic force spectroscopy, Nature 397 (6714) (1999) 50–53.

[96] J. Howard, Mechanics of Motor Proteins and the Cytoskeleton, 1st. ed., Sinauer Associates, Inc., Sunderland, MA, 2001, 367.

[97] N. Hirokawa, R. Takemura, Kinesin superfamily proteins and their various functions and dynamics, Exp. Cell Res. 301 (1) (2004) 50–59.

[98] L.S. Goldstein, A.V. Philp, The road less traveled: emerging principles of kinesin motor utilization, Annu. Rev. Cell Dev. Biol. 15 (1999) 141–183.

[99] N. Hirokawa, Kinesin and dynein superfamily proteins and the mechanism of organelle transport, Science 279 (5350) (1998) 519–526.

[100] J.C. Hoeng, et al., High resolution crystal structure and in vivo function of a kinesin-2 homolog in *Giardia intestinalis*, Mol. Biol. Cell 19 (7) (2008) 3124–3137.

[101] J.R. Marszalek, L.S. Goldstein, Understanding the functions of kinesin-II, Biochim. Biophys. Acta 1496 (1) (2000) 142–150.

[102] G. Ou, et al., Functional coordination of intraflagellar transport motors, Nature 436 (7050) (2005) 583–587.

[103] J. Pan, W.J. Snell, Kinesin-II is required for flagellar sensory transduction during fertilization in *Chlamydomonas*, Mol. Biol. Cell 13 (4) (2002) 1417–1426.

[104] S. Nonaka, et al., Randomization of left-right asymmetry due to loss of nodal cilia generating leftward flow of extraembryonic fluid in mice lacking KIF3B motor protein, Cell 95 (6) (1998) 829–837.

[105] S. Takeda, et al., Left-right asymmetry and kinesin superfamily protein KIF3A: new insights in determination of laterality and mesoderm induction by kif3A$^{-/-}$ mice analysis, J. Cell Biol. 145 (4) (1999) 825–836.

[106] S. Loubery, et al., Different microtubule motors move early and late endocytic compartments, Traffic 9 (4) (2008) 492–509.

[107] C.L. Brown, et al., Kinesin-2 is a motor for late endosomes and lysosomes, Traffic 6 (12) (2005) 1114–1124.

[108] S. Takeda, et al., Kinesin superfamily protein 3 (KIF3) motor transports fodrin-associating vesicles important for neurite building, J. Cell Biol. 148 (6) (2000) 1255–1265.

[109] K. Haraguchi, et al., Role of the kinesin-2 family protein, KIF3, during mitosis, J. Biol. Chem. 281 (7) (2006) 4094–4099.

[110] T. Stauber, et al., A role for kinesin-2 in COPI-dependent recycling between the ER and the Golgi complex, Curr. Biol. 16 (22) (2006) 2245–2251.

[111] K. Ray, et al., Kinesin-II is required for axonal transport of choline acetyltransferase in *Drosophila*, J. Cell Biol. 147 (3) (1999) 507–518.

[112] J.R. Marszalek, et al., Situs inversus and embryonic ciliary morphogenesis defects in mouse mutants lacking the KIF3A subunit of kinesin-II, Proc. Natl. Acad. Sci. U.S.A. 96 (9) (1999) 5043–5048.

[113] F. Lin, et al., Kidney-specific inactivation of the KIF3A subunit of kinesin-II inhibits renal ciliogenesis and produces polycystic kidney disease, PNAS 100 (9) (2003) 5286–5291.

[114] G.J. Pazour, et al., *Chlamydomonas* IFT88 and its mouse homologue, polycystic kidney disease gene tg737, are required for assembly of cilia and flagella, J. Cell Biol. 151 (3) (2000) 709–718.

[115] Y. Okada, N. Hirokawa, A processive single-headed motor: kinesin superfamily protein KIF1A, Science 283 (5405) (1999) 1152–1157.

[116] D.R. Klopfenstein, et al., Role of phosphatidylinositol(4,5)bisphosphate organization in membrane transport by the Unc104 kinesin motor, Cell 109 (3) (2002) 347–358.

[117] V. Soppina, et al., Dimerization of mammalian kinesin-3 motors results in superprocessive motion, Proc. Natl. Acad. Sci. U.S.A. 111 (15) (2014) 5562–5567.

[118] V. Soppina, K.J. Verhey, The family-specific K-loop influences the microtubule on-rate but not the superprocessivity of kinesin-3 motors, Mol. Biol. Cell 25 (14) (2014) 2161–2170.

[119] J.W. Hammond, et al., Mammalian Kinesin-3 motors are dimeric in vivo and move by processive motility upon release of autoinhibition, PLoS Biol. 7 (3) (2009) e72.

[120] J.B. Riviere, et al., KIF1A, an axonal transporter of synaptic vesicles, is mutated in hereditary sensory and autonomic neuropathy type 2, Am. J. Hum. Genet. 89 (2) (2011) 219–230.

[121] N. Hirokawa, S. Niwa, Y. Tanaka, Molecular motors in neurons: transport mechanisms and roles in brain function, development, and disease, Neuron 68 (4) (2010) 610–638.

[122] A.P. Sagona, et al., PtdIns(3)P controls cytokinesis through KIF13A-mediated recruitment of FYVE-CENT to the midbody, Nat. Cell Biol. 12 (4) (2010) 362–371.

[123] J.Z. Torres, et al., The STARD9/Kif16a kinesin associates with mitotic microtubules and regulates spindle pole assembly, Cell 147 (6) (2011) 1309–1323.

[124] M. Samwer, et al., The nuclear F-actin interactome of *Xenopus* oocytes reveals an actin-bundling kinesin that is essential for meiotic cytokinesis, EMBO J. 32 (13) (2013) 1886–1902.

[125] H. Ueno, et al., KIF16B/Rab14 molecular motor complex is critical for early embryonic development by transporting FGF receptor, Dev. Cell 20 (1) (2011) 60–71.

[126] J.W. Tsai, et al., Kinesin 3 and cytoplasmic dynein mediate interkinetic nuclear migration in neural stem cells, Nat. Neurosci. 13 (12) (2010) 1463–1471.

[127] S.M. Ahmed, et al., KIF14 negatively regulates Rap1a-Radil signaling during breast cancer progression, J. Cell Biol. 199 (6) (2012) 951–967.

[128] C. Zhao, et al., Charcot-Marie-Tooth disease type 2A caused by mutation in a microtubule motor KIF1Bbeta, Cell 105 (5) (2001) 587–597.

[129] S. Klebe, et al., KIF1A missense mutations in SPG30, an autosomal recessive spastic paraplegia: distinct phenotypes according to the nature of the mutations, Eur. J. Hum. Genet. 20 (6) (2012) 645–649.

[130] Y. Erlich, et al., Exome sequencing and disease-network analysis of a single family implicate a mutation in KIF1A in hereditary spastic paraparesis, Genome Res. 21 (5) (2011) 658–664.

[131] G.Y. Chen, D.F. Arginteanu, W.O. Hancock, Processivity of the kinesin-2 KIF3A results from rear head gating and not front head gating, J. Biol. Chem. 290 (16) (2015) 10274–10294.

[132] Y. Tsuchizawa, et al., Cooperative force production by Unc104 forced dimers, in: Biophysical Society Annual Meeting, 2015. L3276-Pos.

[133] S.L. Reck-Peterson, et al., Single-molecule analysis of dynein processivity and stepping behavior, Cell 126 (2) (2006) 335–348.

[134] M.A. Schlager, et al., In vitro reconstitution of a highly processive recombinant human dynein complex, EMBO J. 33 (17) (2014) 1855–1868.

[135] L. Urnavicius, et al., The structure of the dynactin complex and its interaction with dynein, Science 347 (6229) (2015) 1441–1446.

[136] H. Schmidt, et al., Structure of human cytoplasmic dynein-2 primed for its power stroke, Nature 518 (7539) (2015) 435–438.

[137] W.B. Redwine, et al., Structural basis for microtubule binding and release by dynein, Science 337 (6101) (2012) 1532–1536.

[138] A. Gennerich, et al., Force-induced bidirectional stepping of cytoplasmic dynein, Cell 131 (5) (2007) 952–965.

[139] F.B. Cleary, et al., Tension on the linker gates the ATP-dependent release of dynein from microtubules, Nat. Commun. 5 (2014) 4587.

[140] S. Can, M.A. Dewitt, A. Yildiz, Bidirectional helical motility of cytoplasmic dynein around microtubules, Elife 3 (2014) e03205.

[141] V. Belyy, et al., Cytoplasmic dynein transports cargos via load-sharing between the heads, Nat. Commun. 5 (2014) 5544.

[142] M.A. DeWitt, et al., Cytoplasmic dynein moves through uncoordinated stepping of the AAA+ ring domains, Science 335 (6065) (2012) 221–225.

[143] W. Qiu, et al., Dynein achieves processive motion using both stochastic and coordinated stepping, Nat. Struct. Mol. Biol. 19 (2) (2012) 193–200.

[144] A. Gennerich, S.L. Reck-Peterson, Probing the force generation and stepping behavior of cytoplasmic Dynein, Methods Mol. Biol. 783 (2011) 63–80.

[145] C. Cho, S.L. Reck-Peterson, R.D. Vale, Regulatory ATPase sites of cytoplasmic dynein affect processivity and force generation, J. Biol. Chem. 283 (38) (2008) 25839–25845.

[146] L. Rao, et al., The yeast dynein Dyn2-Pac11 complex is a dynein dimerization/processivity factor: structural and single-molecule characterization, Mol. Biol. Cell 24 (15) (2013) 2362–2377.

[147] M.P. Nicholas, et al., Cytoplasmic dynein regulates its attachment to microtubules via nucleotide state-switched mechanosensing at multiple AAA domains, Proc. Natl. Acad. Sci. U.S.A. 112 (20) (2015) 6371–6376.

[148] R. Mallik, et al., Cytoplasmic dynein functions as a gear in response to load, Nature 427 (6975) (2004) 649–652.

[149] C. Leidel, et al., Measuring molecular motor forces in vivo: implications for tug-of-war models of bidirectional transport, Biophys. J. 103 (3) (2012) 492–500.

[150] N. Fukushima, et al., Post-translational modifications of tubulin in the nervous system, J. Neurochem. 109 (3) (2009) 683–693.

[151] N.E. LaPointe, et al., The amino terminus of tau inhibits kinesin-dependent axonal transport: implications for filament toxicity, J. Neurosci. Res. 87 (2) (2009) 440–451.

[152] N. Kaul, V. Soppina, K.J. Verhey, Effects of alpha-tubulin K40 acetylation and detyrosination on kinesin-1 motility in a purified system, Biophys. J. 106 (12) (2014) 2636–2643.

[153] N.A. Reed, et al., Microtubule acetylation promotes kinesin-1 binding and transport, Curr. Biol. 16 (21) (2006) 2166–2172.

[154] D. Wloga, J. Gaertig, Post-translational modifications of microtubules, J. Cell Sci. 123 (Pt 20) (2010) 3447–3455.

[155] K.J. Verhey, J. Gaertig, The tubulin code, Cell Cycle 6 (17) (2007) 2152–2160.

[156] J.W. Hammond, et al., Posttranslational modifications of tubulin and the polarized transport of kinesin-1 in neurons, Mol. Biol. Cell 21 (4) (2010) 572–583.

[157] D. Cai, et al., Single molecule imaging reveals differences in microtubule track selection between Kinesin motors, PLoS Biol. 7 (10) (2009) e1000216.

[158] J.W. Hammond, D. Cai, K.J. Verhey, Tubulin modifications and their cellular functions, Curr. Opin. Cell Biol. 20 (1) (2008) 71–76.

[159] A. Seitz, et al., Single-molecule investigation of the interference between kinesin, tau and MAP2c, EMBO J. 21 (18) (2002) 4896–4905.

[160] L.A. Lopez, M.P. Sheetz, Steric inhibition of cytoplasmic dynein and kinesin motility by MAP2, Cell Motil. Cytoskeleton 24 (1) (1993) 1–16.

[161] R. Dixit, et al., Differential regulation of dynein and kinesin motor proteins by tau, Science 319 (5866) (2008) 1086–1089.

[162] G.J. Hoeprich, et al., Kinesin's neck-linker determines its ability to navigate obstacles on the microtubule surface, Biophys. J. 106 (8) (2014) 1691–1700.

[163] J. Al-Bassam, et al., Analysis of the weak interactions of ADP-Unc104 and ADP-kinesin with microtubules and their inhibition by MAP2c, Cell Motil. Cytoskeleton 64 (5) (2007) 377–389.

[164] S. Dunn, et al., Differential trafficking of Kif5c on tyrosinated and detyrosinated microtubules in live cells, J. Cell Sci. 121 (Pt 7) (2008) 1085–1095.

[165] S. Lakamper, E. Meyhofer, The E-hook of tubulin interacts with kinesin's head to increase processivity and speed, Biophys. J. 89 (5) (2005) 3223–3234.

[166] I. Semenova, et al., Regulation of microtubule-based transport by MAP4, Mol. Biol. Cell 25 (20) (2014) 3119–3132.

[167] C. Leduc, et al., Cooperative extraction of membrane nanotubes by molecular motors, Proc. Natl. Acad. Sci. U.S.A. 101 (49) (2004) 17096–17101.

[168] C. Leduc, et al., Mechanism of membrane nanotube formation by molecular motors, Biochim. Biophys. Acta 1798 (7) (2010) 1418–1426.

[169] W.H. Roos, et al., Dynamic kinesin-1 clustering on microtubules due to mutually attractive interactions, Phys. Biol. 5 (4) (2008) 046004.

[170] O. Campas, et al., Coordination of kinesin motors pulling on fluid membranes, Biophys. J. 94 (12) (2008) 5009–5017.

[171] A. Roux, et al., A minimal system allowing tubulation with molecular motors pulling on giant liposomes, Proc. Natl. Acad. Sci. U.S.A. 99 (8) (2002) 5394–5399.

[172] P.M. Shaklee, et al., Bidirectional membrane tube dynamics driven by nonprocessive motors, Proc. Natl. Acad. Sci. U.S.A. 105 (23) (2008) 7993–7997.

[173] G. Koster, et al., Force barriers for membrane tube formation, Phys. Rev. Lett. 94 (6) (2005) 068101.

[174] G. Koster, et al., Membrane tube formation from giant vesicles by dynamic association of motor proteins, Proc. Natl. Acad. Sci. U.S.A. 100 (26) (2003) 15583–15588.

[175] N. Pollock, et al., Reconstitution of membrane transport powered by a novel dimeric kinesin motor of the Unc104/KIF1A family purified from *Dictyostelium*, J. Cell Biol. 147 (3) (1999) 493–506.

[176] D. Oriola, et al., Formation of helical membrane tubes around microtubules by single-headed kinesin KIF1A, Nat. Commun. 6 (2015) 8025.

[177] A. Rai, et al., Dynein clusters into lipid microdomains on phagosomes to drive rapid transport toward lysosomes, Cell 164 (4) (2016) 722–734.

[178] S.R. Nelson, K.M. Trybus, D.M. Warshaw, Motor coupling through lipid membranes enhances transport velocities for ensembles of myosin Va, Proc. Natl. Acad. Sci. U.S.A. 111 (38) (2014) E3986–E3995.

[179] S. Pyrpassopoulos, et al., Force generation by membrane-associated myosin-I, Sci. Rep. 6 (2016) 25524.

[180] M.J. Greenberg, et al., A perspective on the role of myosins as mechanosensors, Biophysical J. 110 (12) (2016) 2568–2576.

In this chapter

Chemical probes for dynein

Jonathan B. Steinman, Tarun M. Kapoor

The Rockefeller University, New York, NY, United States

Dynein, the first microtubule-based motor protein discovered, is responsible for most of the microtubule minus end–directed transport in eukaryotic cells. It is now clear that dynein-dependent transport is important for a wide range of cellular processes, including organelle positioning, mRNA transport, and cell division [1]. Dynein is a relatively fast motor, driving cargo transport at speeds of 1–10 μm/s [2]. Therefore, properly dissecting the function of this motor protein requires controlled perturbations of its activity on timescales of seconds to minutes. Such acute inhibition of protein function is possible with cell-permeable chemical inhibitors.

Dyneins can be divided into nine families comprised of seven axonemal dyneins and two classes of cytoplasmic dyneins [3]. Axonemal dyneins are localized to axoneme-containing organelles and drive ciliary and flagellar beating. Cytoplasmic dynein-2 is restricted to cilia and flagella, including the primary cilium, but does not participate in ciliary beating. Rather it drives motion of cargos along the axoneme toward the base of the cilium, a process termed retrograde intraflagellar transport. By contrast, cytoplasmic dynein-1 is located in the cytoplasm and participates in the transport of many types of cargo. In this chapter "dynein" refers to cytoplasmic dynein, unless indicated otherwise.

The ATP-hydrolyzing heavy chains of dynein belong to the AAA+ protein family (*ATPases associated with diverse cellular activities*), which is characterized by the AAA domain, a common structural motif for ATP binding [4]. These proteins typically function as oligomers (often hexamers), with ATP bound at the interface of two AAA domains. AAA+ enzymes couple nucleotide hydrolysis with conformational changes that lead to substrate remodeling or directional motion. In the case of dynein, all six AAA domains are in one single large polypeptide, and ATP hydrolysis leads to successive stepping of the motor protein along microtubule tracks.

The heavy chain of cytoplasmic dynein is an ~500 kDa polypeptide that in cells is believed to function as a homodimer bound to multiple accessory proteins. To move processively, the dynein complex (mass ~1.5 MDa) must bind dynactin, another megadalton-sized multiprotein complex [5,6]. At the N-terminus of the dynein heavy chain is an ~50- to 60-nm long "tail" that mediates dimerization and interacts with adapters necessary for cargo binding [7]. The C-terminus features the six AAA domains and the microtubule-binding domain (MTBD), which caps a coiled-coil stalk that extends ~15 nm from AAA4. A coiled-coil "buttress" emerging from AAA5 supports the stalk. In the dynein complex, ATP hydrolysis occurs mainly at two sites, AAA1 and AAA3, although AAA3 is active only in a subset of cytoplasmic dyneins [7] (ATP-binding sites are numbered such that AAA1 refers to the site between AAA+ domains 1 and 2). Of these sites, AAA1 is the major ATPase, whose activity can be controlled by other AAA sites and accessory proteins [7].

In this chapter, we review the recent advances in developing chemical inhibitors to probe dynein function. We highlight the ciliobrevins, the first selective cell-permeable small molecule inhibitors of dynein. We discuss the discovery of these compounds, their use, and how these chemical probes may be improved. We also discuss other strategies that have been developed to probe dynein function with fast temporal control.

8.1 General approach to inhibiting dynein

One simple approach to inhibiting dynein is to block its enzymatic activity using inhibitors that target its ATP-binding sites. The challenge of inhibiting the ATP-binding site is one of *selectivity*—how can we design or identify small molecules that bind this site in dynein but do not inhibit other nucleotide-binding enzymes? The structures of several AAA+ proteins have been solved and it is clear that their ATP-binding sites are remarkably similar. Root mean squared deviations between Cα atoms are in the range of 1.5–2.5 Å for residues that constitute the nucleotide-binding core [4]. Dynein's catalytically active nucleotide binding sites (AAA1 and AAA3) are similar to those of other AAA+ proteins in that ATP binds in a pocket containing a series of loops emanating from the top of several parallel β-sheets [4]. Polar residues such as the Walker A lysine coordinate negatively charged phosphate groups. The Walker B glutamate and aspartate orient the hydrolytic water molecule, while the "arginine finger," which extends from the adjacent AAA domain, stabilizes the terminal phosphate for nucleophilic attack by water. The adenine base forms hydrogen bonds with the peptide backbone of the "N-loop," and hydrophobic side chains of this motif contribute to the binding of adenine [8].

Although this general mode of interaction with ATP is conserved across AAA+ proteins, at the level of primary sequence these proteins have diverged to a greater degree. Variability in residue composition is tolerated adjacent to invariant residues, such as the Walker A lysine, the Walker

(A)

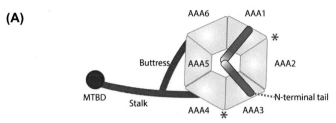

(B)

	Walker A	Walker B	Arg finger	N loop
cytoplasmic dynein 1 (AAA1)	GPaGtGKTe	wgcFDEfNrL	atVsRcGmV	qdkLvqtpl
cytoplasmic dynein 2 (AAA1)	GPaGtGKTe	wgcFDEfNrL	atIsRmGmI	askLvytpl
p97 (D2)	GPpGcGKTl	vlfFDElDsI	paIlRpGrL	wedIggled
NSF (D1)	GPpGcGKTl	iiiFDEiDaI	eaLlRpGrL	kmgIggldk

(C)

	Walker A	Walker B	Arg finger	N loop
cytoplasmic dynein 1 (AAA1)	GPAGTGKTE	WGCFDEFNRL	ATVSRcGMV	qdKLVqTPL
cytoplasmic dynein 2 (AAA1)	GPAGTGKTE	WGCFDEFNRL	ATISRmGMI	asKLVyTPL
axonemal dynein 5 (AAA1)	GPAGTGKTE	WGCFDEFNRL	ATVSRnGMV	tdRLViTPL

(D)

	Walker A	Walker B	Arg finger	N loop
cytoplasmic dynein 1 (AAA3)	GppGsgKtm	vLFcdEINLP	ALfnrCvLn	pdvVVptlD
cytoplasmic dynein 2 (AAA3)	GppGcgKgm	vLYlkDINLP	ALhkkCqVl	tlpVIqtpD
axonemal dynein 5 (AAA3)	GeqGtaKtv	tVFidDVNMP	ALisgCtId	gsiLVpnvD

Figure 8.1 Dynein, a motor protein in the AAA+ superfamily. (A) Schematic of the dynein heavy-chain. ATP hydrolyzing domains are indicated (asterisk). Gray: AAA domains. Blue: linker. Red: stalk, buttress, and microtubule-binding domain (MTBD). (B) Alignment of nucleotide-binding motifs in dynein-1 (AAA1), dynein-2 (AAA1), p97 (D2), and NSF (D1). Alignment of ATP-binding motifs for AAA1 (C) or AAA3 (D) in dynein-1, dynein-2, and axonemal dynein-5. In (B–D) identical residues across a given comparison are shown with bold capitals, similar residues are capitals, variable residues are lowercase. Key residues defining a motif are highlighted in red. Uniprot accession numbers: human cytoplasmic dynein 1 (Q14204), human cytoplasmic dynein-2 (Q8NCM8), human axonemal dynein-5 (Q8TE73), human p97 (P55072), human N-ethylmaleimide sensitive factor (P46459).

B aspartate/glutamate, and the arginine finger (see Fig. 8.1). Comparison of the ATP-binding pockets between dynein and other AAA+ enzymes or among human dynein isoforms (Fig. 8.1B–D) shows a high degree of similarity among residues that contact phosphate groups in ATP. By contrast, the residues positioned near the two hydrogen bonds that anchor the adenine base to the peptide backbone are more variable (N loop, Fig. 8.1B–D).

These comparisons suggest that it may be possible to identify small molecules that selectively target the ATP-binding site of one dynein isoform. In principle, inhibitor potency can be established via contacts with conserved residues and specificity can be achieved through interactions with non-conserved residues in dynein's ATP-binding pocket. However, high-resolution structural data for chemical inhibitor–dynein interactions will likely be needed to develop these chemical probes.

8.2 Nucleotide-mimetic inhibitors of dynein

Compounds that mimic ATP and the other nucleic acids are well-established enzyme inhibitors [9]. In the case of nucleotide mimetics, these compounds bind to enzymes in a manner similar to the endogenous nucleotide substrate but either lack a hydrolyzable phosphate moiety or have a slowly hydrolyzable phosphate-mimetic (Fig. 8.2).

8.2.1 Vanadate

Less than a decade after the discovery of dynein, researchers identified a "trace nucleotide contaminant" present in some batches of commercial ATP as an inhibitor of purified dynein [10,11]. Careful analysis identified this inhibitory contaminant as vanadate, an inorganic anion sometimes present in the animal tissue from which ATP was purified [12].

In general, vanadate exists as a tetra-coordinate $(VO_4)^{3-}$ anion but can also form a pentacoordinate complex. The relevant species for its inhibitory activity is ADP–vanadate, in which a terminal oxygen atom of ADP is bound to vanadium (Fig. 8.2). This molecule has a trigonal–bipyramidal geometry at vanadium, which mimics the transition state formed during nucleophilic attack by water on the gamma phosphate of ATP [13]. The stable substrate–vanadate complex is bound more tightly than the substrate or product [14]. Because this transition state is common to many enzymatic phosphate hydrolysis reactions, ADP–vanadate binds many ATP-binding enzymes and is generally accepted as a promiscuous

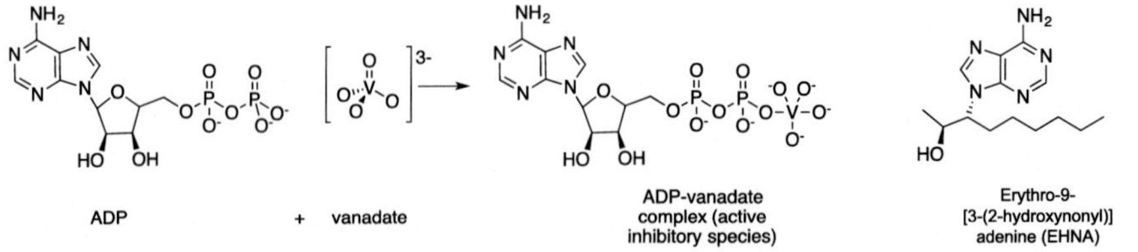

Figure 8.2 Nucleotide mimetic inhibitors of dynein. (A) Structure of the vanadate ion and its reaction with ADP to form the inhibitory species ADP–vanadate. (B) Structure of erythro-9-[3-(2-hydroxynonyl)]adenine (EHNA).

antagonist of enzymatic phosphate ester hydrolysis (including ATP) and phosphate transfer reactions [15].

The principles of inhibition by vanadate hold true in the case of dynein: the ion-ADP adduct inhibits (IC_{50}s usually 1–10 μM) many isoforms of dynein from different species in a nucleotide-dependent fashion. X-ray crystallography confirmed that dynein binds ADP–vanadate using the same Walker A and B residues that interact with the gamma phosphate of ATP [16].

Interestingly, although dynein has multiple sites capable of binding ATP, the ADP–vanadate complex is observed only at AAA1 in a crystal structure [16]. This may reflect selective binding of ADP–vanadate at the AAA1 site or simply the fact that ATP turnover—and thus ADP–vanadate binding—occurs most rapidly at this site [19]. Though it stands to reason that vanadate could inhibit any enzyme that hydrolyzes ATP, it does not bind to *all* nucleotide hydrolyzing proteins. Indeed, vanadate exhibits selective inhibition of dynein relative to other cytoskeletal motors such as myosin and kinesin, a property that made it a useful early probe for dynein [10,17]. A serendipitous discovery that irradiation of dynein–ADP–vanadate complexes with UV light leads to specific photocleavage at one site within the enzyme enabled annotation of dynein's structure and the commonalities between different dynein isoforms before this could be done by other approaches now commonly used [19]. Vanadate's hydrophilicity likely restricts its membrane permeability, which has limited its use as a probe in cellular settings.

8.2.2 Erythro-9-[3-(2-hydroxynonyl)]adenine

Erythro-9-[3-(2-hydroxynonyl)]adenine (EHNA) was identified as a dynein inhibitor by researchers studying sperm motility [20] (Fig. 8.2). It had previously been shown to inhibit protein carboxymethylase activity under certain conditions, which was thought to account for the observation that EHNA blocks sperm motility. However, even under conditions where carboxymethylase activity was only minimally inhibited, EHNA inhibited sperm motility. Further, it blocked motion of demembranated spermatozoa, and this inhibition could be reversed by addition of excess ATP, suggesting a direct effect on axonemal dynein, the ATPase that generated flagellar beating. Experiments on purified axonemal dynein confirmed direct inhibition of its ATPase activity [20].

EHNA is composed of an adenine base linked via N-9 (the site of linkage to ribose in adenosine/ATP) to an aliphatic alcohol. This compound is more hydrophobic than other nucleotide mimetics and is likely cell permeable, as it inhibits the motility of intact sperm.

The IC_{50} for EHNA inhibition of dynein varies from ~200 μM for dynein from sea urchin sperm to ~1 mM for dynein from rat sperm [20]. EHNA was shown to be a mixed-type inhibitor of dynein as it raises dynein's K_m and simultaneously lowers its V_{max} [21]. These data raise the possibility that EHNA may bind a site on dynein other than its ATP-binding pocket. However, additional biochemical or structural studies are needed to support this hypothesis. Based on its chemical structure and on a crystal structure in which EHNA is bound to adenosine deaminase, we posit that EHNA interacts with the ATP-binding pocket but can make additional contacts with proximal residues to achieve this complex mode of inhibition [22].

In vitro, EHNA inhibits dynein selectively in comparison to other motor proteins and ATPases [21]. As a result, EHNA has seen use as an in vitro probe for dynein function. For example, inhibition of MAP1C-driven microtubule gliding by EHNA contributed to the assignment of this minus end–directed motor as a cytoplasmic isoform of dynein [23].

As many cellular proteins bind adenosine-containing compounds, nucleotide analogues such as EHNA are unselective in cells. Indeed, this compound potently inhibits adenosine deaminase ($IC_{50} \sim 20\,nM$), and cyclic-GMP-stimulated phosphodiesterase II (PDE2, $IC_{50} \sim 800\,nM$) [22,24,25]. As a result, the utility of this compound to dissect dynein function in cells is limited.

8.3 Ciliobrevins: cell-permeable small molecule dynein inhibitors

The first dynein inhibitors not based on nucleotides or adenine were discovered in a cell-based screen for Hedgehog (Hh) pathway inhibitors. The approach that was used is referred to as "chemical genetics" as small molecule inhibitors, rather than genetic mutations, are identified that disrupt a cellular process of interest, which in this case was the Hh pathway [26]. Hh signaling controls development in multicellular organisms and was initially described as a regulator of polarity and segmentation in *Drosophila* [27,28]. A vast body of research has led to advanced models for this signaling pathway. Briefly, the binding of a soluble Hh ligand to its receptor Patched leads to the relief of inhibition of the seven-transmembrane protein Smoothened (Smo). Smoothened activity leads to the accumulation of activated Gli transcription factors. Gli then enters the nucleus and activates Hh responsive genes [29]. It has been established that Hh activity in vertebrates is dependent on the primary cilium [29,30].

The primary cilium is an antenna-shaped organelle that protrudes from vertebrate cells [31]. Transport of cargos along its axoneme, termed intraflagellar transport, is required for building and maintaining the cilium. Accumulation of Hh pathway components including Smo, Suppressor of Fused [Su(Fu)], and Gli within the cilium is required for the transcription of Hh responsive genes [32]. Dynein-2 is required for assembly of functional

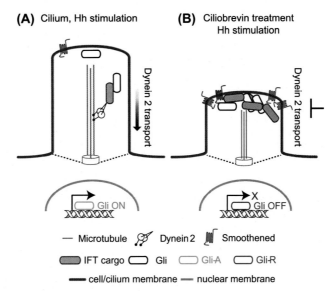

(A) Cilium, Hh stimulation   **(B)** Ciliobrevin treatment Hh stimulation

Dynein 2 transport

— Microtubule   Dynein 2   Smoothened

IFT cargo   Gli   Gli-A   Gli-R

cell/cilium membrane — nuclear membrane

Figure 8.3 Effect of ciliobrevin on the primary cilium and the Hh pathway. (A) Transport of intraflagellar transport cargos (purple) along the cilium is required for proper localization and activation of the Gli transcription factors. (B) Inhibition of dynein leads to shorter cilia, accumulation of IFT cargo and Gli proteins at the ciliary tip, and inhibition of the Hh pathway.

cilia and active transport within them as well as for active export of some ciliary factors [33]. In line with these roles, mutations in dynein-2 lead to both Hh pathway disruption and ciliary defects, such as ciliary shortening and accumulation of the intraflagellar transport component IFT88 [34,35] (Fig. 8.3).

Improper Hh pathway activation can cause cancer in humans. In particular, basal cell carcinoma, one of the most common human cancers, has been shown to be driven by activating mutations in the Hh pathway [36,37], and considerable effort has focused on discovering chemical inhibitors that block this pathway. With few exceptions, the compounds developed to disrupt the Hh pathway are inhibitors of Smo, and many bind to a common site on this protein [38]. These efforts have recently led to approved therapeutics [39]. One limitation of these clinical agents is the acquisition of resistance via mutations in Smo itself [40–42]. Furthermore, cancer-causing mutations can arise in other Hh pathway proteins, including Patched, Smo, Su(Fu), and Gli. As some of these proteins function downstream of Smo in the signaling cascade these cancers are unlikely to be responsive to drugs that inhibit Smo.

The chemical genetic screen in which ciliobrevin A was discovered was designed to uncover Hh inhibitors that act downstream of Smo [43]. For this

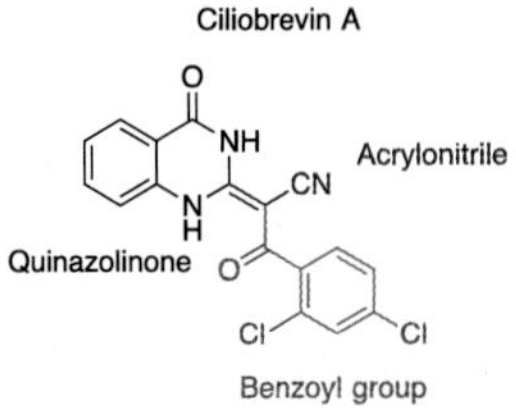

Figure 8.4 Chemical structure of ciliobrevin A, highlighting the components of its scaffold. Quinazolinone-black; acrylonitrile-blue; benzoyl group-red.

screen, the Hh pathway was activated using the Smo agonist (SAG), a small molecule that competes with other known Smo antagonists at a common binding site [44]. This innovation in the screen reduced the likelihood of discovering compounds that directly bind Smo. Of the ~1.2×10^5 compounds screened in this manner, four blocked Hh activity with IC_{50} values <10 µM.

Ciliobrevin A (or HPI-4, *H*edgehog *p*athway *i*nhibitor-4) was one of these four compounds. The chemical structure of this compound did not imply a mechanism of action, as its core structure, a benzoylacrylonitrile-substituted quinazolinone, differs from those usually observed among common inhibitor classes (e.g., kinase inhibitors) (Fig. 8.4). Analyses of reporter cell lines in which Hh signaling was constitutively active, including one expressing inhibitor-resistant allele of Smo, showed that the compound's action is "downstream" of Smo.

A clue to the target of this compound came from analysis of its effect on the primary cilium. Extended treatment with HPI-4 resulted in ciliary defects including cells with shortened or no primary cilia. This observation led to HPI-4 being named ciliobrevin [45]. Upon closer investigation, it was noted that ciliobrevin A treatment led to Gli2 accumulation at the distal tip of the cilium even in the absence of Hh pathway activation, and caused similar accumulation of IFT88, consistent with inhibition of retrograde intraflagellar transport [45]. Together, these data suggest that ciliobrevin A targets cytoplasmic dynein-2.

Confirmation that the ciliobrevins were dynein inhibitors came from biochemical experiments showing inhibition of microtubule gliding driven by dynein-1 purified from bovine brain and of the ATPase activity of the recombinant rat dynein-1 motor domain [45]. The half-maximal inhibitory concentration (IC_{50}) for dynein in vitro was ~30 µM. At the time this research was conducted, dynein-2 was not well characterized biochemically and could not be obtained in quantities sufficient for biochemical assays, but ciliobrevin A has since been shown to block ATP hydrolysis by dynein-2 at low ATP concentrations [46]. The ciliobrevins were the first small molecules that gave cellular phenotypes consistent with dynein inhibition and blocked dynein activity in vitro. As such they were the first selective cell-permeable small molecule probes of dynein function.

Two lines of evidence suggest that the ciliobrevins act at dynein's major ATPase site, AAA1 [45]. First, ciliobrevins inhibited dynein in an ATP-competitive manner. Second, ciliobrevins blocked vanadate-mediated photo-cleavage of recombinant dynein, which is consistent with direct displacement by ciliobrevin of ADP–vanadate at AAA1. We note, however, that direct competition cannot readily be distinguished from the effect of slow ADP–vanadate accumulation due to reduced ATPase activity. Furthermore, interpreting ATP-competition experiments is challenging for enzymes with multiple active sites. Additional work is needed to confirm the site(s) of ciliobrevin binding in dynein.

Analysis of other pathways known to interact with the Hh pathway demonstrated that ciliobrevin A does not inhibit signaling associated with protein kinase A, phosphatidylinositol 3-kinase (PI3K)/Akt, mitogen-activated protein kinase (MAPK), or the WNT pathway [43]. In vitro, ciliobrevin treatment at concentrations that fully inhibited dynein-driven microtubule gliding did not inhibit another microtubule-based motor protein (kinesin) or two other AAA+ proteins (p97, MCM2–7), indicating selective dynein inhibition. Additional studies will be required to determine the full selectivity profile of these first-generation ciliobrevins.

8.3.1 Other dynein inhibitors

In addition to the ciliobrevins, two compounds have been identified as modulators of dynein function. Both compounds described below have seen little use as probes for dynein.

Nordihydroguaiaretic acid (NDGA): This compound, a component of herbal medicines, exhibits pleiotropic activities in cells, which can be ascribed to its antioxidant properties or to its inhibition of lipoxygenase [47]. Some of the cellular effects of NDGA were noted to be independent of these properties, notably golgi complex disassembly and accumulation of ZW10 and EB1 at the centrosome [48]. These effects have been related to potentiation of interactions between dynactin and some dynein–dynactin cargos [49]. The mechanistic details of how NDGA alters dynein–dynactin cargo interactions are poorly understood and it is unclear if this effect is direct. Further biochemical characterization of the mechanism of action of NGDA is needed before it can be used as a probe of dynein function.

Purealin: This natural product was isolated from a species of sea sponge and shown to inhibit the beating of isolated flagella [50]. It was later shown to modestly (up to 50% at the highest concentration tested, 50 μM) inhibit the microtubule-stimulated ATPase activity of cytoplasmic dynein in vitro. This activity was shown to be ATP-uncompetitive [51].

8.3.2 Use of the ciliobrevins

We highlight a few examples of the use of ciliobrevins as probes of dynein biology. Other studies using ciliobrevins have been recently reviewed elsewhere [52].

Dynein-2 is required to build a primary cilium, and persistent inhibition of its function results in short or "stubby" cilia [29]. Acute blockade of dynein-2 can help dissect the role of dynein-driven retrograde intraflagellar transport in the intact cilium. In a cell-based assay where intraflagellar transport in a primary cilium could be directly observed using fluorescent tagging of IFT-88, treatment with ciliobrevin A or D blocked retrograde movement within 3 min [46,53]. The ability to rapidly inhibit dynein-2 allowed the uncoupling of intraflagellar transport from maintenance of ciliary architecture, which is largely unperturbed in the short-term inhibitor treatments. The somatostatin receptor (SSTR3), a cilium-based transmembrane protein, continued to diffuse following acute arrest of intraflagellar transport using ciliobrevin D, indicating that proper distribution of some ciliary signaling proteins is independent of dynein activity. This finding was corroborated for both SSTR3 and Smo by rapid depletion of cellular ATP [53] and was later supported by ultrahigh-resolution tracking of Smo movement [54].

The "immunological synapse" forms between antigen-presenting cells and T lymphocytes following recognition of a stimulatory antigen by the T cell receptor. In order for the T cell to mount a proper response at the immunological synapse, which includes the directed release of cytokines and cytotoxic substances toward the antigen presenting cell, T cells must rapidly rearrange and "polarize" components of their cytoplasm toward the synapse. Two studies used ciliobrevin D to examine the role of dynein in this dynamic process. In one study, treatment with ciliobrevin D, but not a related inactive control, reduced the polarization induced by T cell–antigen-presenting cell interactions. Addition of an inhibitor of nonmuscle myosin further reduced polarization, indicating a contribution from both dynein and myosin to this process [55]. In the other study, ciliobrevin D treatment slowed the velocity at which centrosomes translocated in the cytosol toward the newly formed immunological synapse, and caused centrosomes to pause farther away (~1.5 μm) from the synapse compared to treatments with a chemically related inactive control compound [56]. This study showed that dynein has a role in both rapid centrosome movement across the cytoplasm and in positioning the centrosome proximal to the immunological synapse. In both studies, the effect of ciliobrevin D on dynein was shown to be similar to that of blocking dynein function using other methods.

Ciliobrevins have also been used to study mechanisms of cell division. Longstanding models indicate that segregation of chromosomes during mitosis depends on microtubules connecting kinetochores to spindle poles. Two studies showed that this may not be absolutely required [57,58]. Using correlative light and electron microscopy, it was shown that, in the course of normal cell division, some chromosomes connected to the mitotic spindle by a microtubule bundle (K-fiber) that did not extend from chromosome to spindle pole [58]. Instead these K-fibers ended distal to the pole at an intersection with another microtubule fiber. Such chromosomes continue moving toward the appropriate spindle pole. Though this distal K-fiber attachment occurs in normal cells, it is difficult to detect. Chromosomes attached to the spindle in this manner can be generated by laser microsurgical ablation of a normal K-fiber, after which the free K-fiber stub rapidly (~30 s) connects with other microtubule fibers. Inhibition of dynein with a high dose of ciliobrevin (50 μM) disrupts spindle pole formation. However, an intermediate ciliobrevin dose in which spindles can form (10 μM), delays the resumption of chromosome movement after K-fiber severing [58]. One of these studies confirmed this finding by overexpressing a dynactin subunit to block dynein–cargo interaction [57]. Together, these findings show that dynein is required for the retention and movement of K-fibers that do not extend to the spindle pole.

Finally, ciliobrevins are active in tissue preparations (from mice and chickens, for example) [59,60]. In particular, ciliobrevin was shown to block intra-axonal movement within neurons from chicken egg ganglia and to impair nerve growth factor (NGF)-mediated regulation of their structure [60].

8.4 Other approaches that allow fast temporal control over dynein function

Recently, two new approaches have been designed to probe dynein function in cells. The first method utilizes chemically induced dimerization to rapidly recruit dynein complexes to a selected cargo [61,62]. Chemical inducers of dimerization (CIDs) were first developed by Schreiber and colleagues based on studies of natural products rapamycin and FK506 [63,64]. In particular, addition of rapamycin to cells can induce the dimerization of two proteins, FKBP12 and FRB (a domain of the mTOR kinase), within seconds [65]. This approach, which has been used in a wide-range of cellular contexts, has been adapted to control dynein function [65]. Specifically, a dynein cargo adapter, BicD2, was fused to FRB, and a dynein cargo protein with known localization (e.g., Pex3 for peroxisomal recruitment) was fused to FKBP12. Using this CID-based approach, researchers have studied the effect of rapid recruitment of dynein complexes to membrane-bound organelles such as mitochondria, peroxisomes, and endosomes or to the cell membrane [61,62,66,67]. The rapid recruitment and the ability to localize dynein to a specific cellular compartment allows analyses of the contribution of dynein to a given process. In one elegant example, rapamycin derivative–driven localization of dynein to the plasma membrane was used to show that cortical dynein can control the polarity of microtubules in developing axons [66].

The second method combines induced dimerization with optogenetics to enable reversible recruitment of dynein to a defined cargo [67]. In this system, illumination with blue light "uncages" a helix of a Light–Oxygen–Voltage domain (LovPEP), freeing it to bind an engineered PDZ domain (ePDZ) [68]. These small (<20 kDa) protein modules can be fused to proteins of interest and used to rapidly stimulate association upon exposure to the appropriate wavelength of light. In the case of the LovPEP–ePDZ system, association can be reversed (with rapid kinetics) upon turning off the excitation light. Further, use of precisely focused illumination makes it possible to recruit dynein complexes to a defined subcellular region, enabling spatial and temporal resolution of dynein's activity upon recruitment [67].

These approaches are exciting new developments in the field. However, one limitation of these methods is that the function of engineered protein constructs must be examined, rather than directly probing the activity of the endogenous motor protein, as can be possible with cell-permeable chemical inhibitors.

8.4.1 Outlook

Ciliobrevins provide a useful starting point to develop new chemical inhibitors with improved properties. Currently, there are two main limitations of ciliobrevins. First is their micromolar potency. This low potency makes it difficult to achieve target specificity in cellular contexts, as doses that fully suppress dynein activity may approach ~100 µM, a concentration at which the compound may interact with other proteins and suppress their activities. The second limitation of ciliobrevins is their low aqueous solubility (computed logarithm of the octanol:water partition coefficient [ClogP] ~4–5). This parameter is associated with decreased effective concentrations in cellular contexts and may account for the suppression of ciliobrevin activity by high-serum concentrations, such as those typically used for cell culture [69].

As a first step toward discovering improved dynein inhibitors, we and our collaborators generated and tested ~50 chemical derivatives of ciliobrevins [46]. These compounds were screened for inhibition of purified full-length human cytoplasmic dyneins 1 and 2 with an eye toward the development of selective dynein-2 antagonists. Dynein-2 selective inhibitors could block the Hh signaling pathway without inhibiting the many different cellular processes that

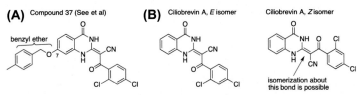

Figure 8.5 Structure of ciliobrevins. (A) Compound 37, a ciliobrevin derivative that selectively inhibits dynein-2. (B) The ciliobrevin scaffold may exist in two isomers about the indicated double bond.

depend on dynein-1 function. Most of the compounds synthesized differed from the parental compound at a single position, but left intact the cyanoacrylamide, quinazolinone, and dichlorobenzoyl groups present in ciliobrevins [46]. A set of derivatives typified by compound 37 (Fig. 8.5A) bearing aryl ethers at the 7-position of the compound retained activity toward dynein-2 while no longer inhibiting dynein-1, resulting in ~10- to >20-fold isoform selectivity under the in vitro assay conditions, which employed low ATP concentrations (<1 µM). It is noteworthy that inhibition of dynein under closer-to-physiologic ATP concentrations was not observed. Cell-based assays using these compounds demonstrated ~1.5- to 6-fold selective inhibition of dynein-2-dependent (e.g., intraflagellar transport, ciliogenesis) relative to dynein-1-associated processes (e.g., mitotic spindle assembly). It is possible that the diminished selectivity of these compounds in the cellular context is a result of the ATP-sensitive nature of their activity or may reflect reduced availability in cytosol due to the high hydrophobicity of these compounds (ClogP for compound 37: ~6).

Analysis of the ciliobrevin's chemical structure indicates that it can isomerize about its central double bond (Fig. 8.5B). If only a fraction of the ciliobrevins adopt the geometry that inhibits dynein, the overall potency will be proportionately reduced. If chemical modifications of the ciliobrevin scaffold changes the isomer preference of the acrylonitrile double bond, this potential variability in structure can confound structure–activity relationship analyses that would guide additional chemical modifications to improve properties. Therefore, an important next step is to restrict this potential isomerization, possibly by engineering in an additional cyclization of the core ciliobrevin scaffold that favors the conformation that inhibits dynein.

8.4.2 Selectively inhibiting a conserved site: lessons from the development of AAA+ inhibitors

The likelihood that potent and selective probes for dynein can be developed is indicated by the recent discoveries of chemical inhibitors for two other AAA+ proteins, VCP/p97 and midasin (Fig. 8.6).

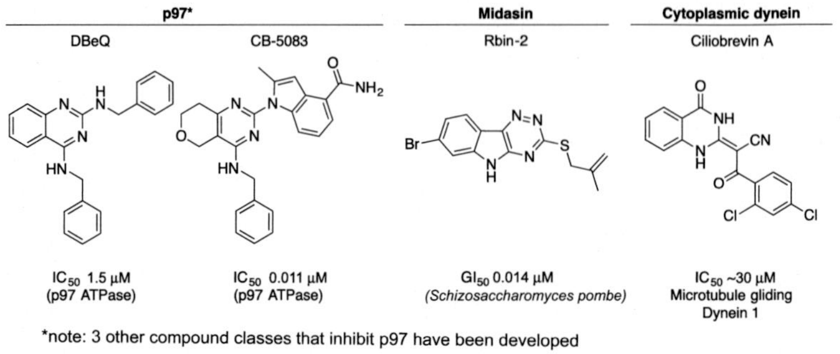

Dyneins

Figure 8.6 Inhibitors of AAA+ ATPases. Shown are the structures and inhibitory activities of: the first inhibitor of p97, DBeQ, and its derivative CB-5083, the midasin inhibitor Rbin-2 and the dynein inhibitor ciliobrevin A. * at least three other classes of compounds that inhibit p97 have been described.

CB-5083 has been developed as a potent (0.01 μM) and selective inhibitor of VCP/p97 and is the first AAA+ inhibitor to reach clinical trials for treatment of a human disease [70,71]. This compound was developed by iterative modification [72] and testing [73] of hundreds of derivatives of DBeQ (dibenzyl quinazoline-2,4-diamine), a chemical inhibitor of VCP/p97 and was identified via in vitro high-throughput screening [74,75]. In contrast to dynein, VCP/p97 can be generated in quantities sufficient for a large-scale initial screen and extensive biochemical follow-up work that helped in developing improved inhibitor analogues [76,77].

Midasin is an AAA+ enzyme required for the process of ribosome biogenesis. Like dynein, all six of its AAA+ subdomains are on a single large polypeptide of mass ~0.5 MDa. Rbin-1 (ribozinoindole-1), a cell-permeable chemical inhibitor, was identified from a cell-based high-throughput screen of ~10,000 compounds [78]. The target of this compound, which is a potent inhibitor of fission yeast *Schizosaccharomyces pombe* growth (GI$_{50}$ 0.14 μM), was identified by analyses of resistance-conferring mutations. In particular, multiple different point mutations, which clustered near the likely interface of the AAA3–AAA4 domains of midasin, were sufficient to confer resistance to Rbin-1 in cells. Testing of a series of analogues against both sensitive (wild-type) and resistant (midasin mutant) cell lines enabled the discovery of more potent analogues (Rbin-2, GI$_{50}$: 0.014 μM). Rbin-2 inhibited the ATPase activity of purified full-length recombinant midasin. Importantly, a mutation in midasin that conferred resistance to Rbins in cells was sufficient to suppress inhibition of ATPase activity, providing "gold standard" proof that midasin is Rbin's direct physiological target.

In summary, chemical inhibitors of dynein are likely to be powerful tools for dissecting this motor protein's functions across different cell-types and organisms. The work described here suggests that an initial foundation is in place for additional research that will lead to better chemical probes. It is also possible that these studies may lead to new therapeutics that selectively target individual isoforms of these complex cellular motor proteins.

Acknowledgments

T.M.K. was supported by NIH grant GM65933. J.B.S. was supported by NIH grant T32GM007739 to the Weill Cornell/Rockefeller/ Sloan-Kettering Tri-Institutional MD-PhD Program.

References

[1] V.J. Allan, Cytoplasmic dynein, Biochem. Soc. Trans. 39 (5) (2011) 1169–1178.

[2] C. Kural, H. Kim, S. Syed, G. Goshima, V.I. Gelfand, P.R. Selvin, Kinesin and dynein move a peroxisome in vivo: a tug-of-war or coordinated movement? Science 308 (5727) (2005) 1469–1472.

[3] B. Wickstead, K. Gull, Dyneins across eukaryotes: a comparative genomic analysis, Traffic 8 (12) (2007) 1708–1721.

[4] J.P. Erzberger, J.M. Berger, Evolutionary relationships and structural mechanisms of AAA+ proteins, Annu. Rev. Biophys. Biomol. Struct. 35 (2006) 93–114.

[5] L. Urnavicius, K. Zhang, A.G. Diamant, C. Motz, M.A. Schlager, M. Yu, et al., The structure of the dynactin complex and its interaction with dynein, Science 347 (6229) (2015) 1441–1446.

[6] T.A. Schroer, Dynactin, Annu. Rev. Cell Dev. Biol. 20 (2004) 759–779.

[7] A.P. Carter, Crystal clear insights into how the dynein motor moves, J. Cell Sci. 126 (Pt 3) (2013) 705–713.

[8] P.I. Hanson, S.W. Whiteheart, AAA+ proteins: have engine, will work, Nat. Rev. Mol. Cell Biol. 6 (7) (2005) 519–529.

[9] C. Simons, Nucleoside Mimetics: Their Chemistry and Biological Properties, Gordon and Breach Science Publishers, Amsterdam, Netherlands, 2001, 191 p.

[10] T. Kobayashi, T. Martensen, J. Nath, M. Flavin, Inhibition of dynein ATPase by vanadate, and its possible use as a probe for the role of dynein in cytoplasmic motility, Biochem. Biophys. Res. Commun. 81 (4) (1978) 1313–1318.

[11] I.R. Gibbons, M.P. Cosson, J.A. Evans, B.H. Gibbons, B. Houck, K.H. Martinson, et al., Potent inhibition of dynein adenosinetriphosphatase and of the motility of cilia and sperm flagella by vanadate, Proc. Natl. Acad. Sci. U.S.A. 75 (5) (1978) 2220–2224.

[12] L.C. Cantley Jr., L. Josephson, R. Warner, M. Yanagisawa, C. Lechene, G. Guidotti, Vanadate is a potent (Na,K)-ATPase inhibitor found in ATP derived from muscle, J. Biol. Chem. 252 (21) (1977) 7421–7423.

[13] R.L. VanEtten, P.P. Waymack, D.M. Rehkop, Letter: transition metal ion inhibition of enzyme-catalyzed phosphate ester displacement reactions, J. Am. Chem. Soc. 96 (21) (1974) 6782–6785.

[14] W.J. Ray Jr., J.M. Puvathingal, Characterization of a vanadate-based transition-state-analogue complex of phosphoglucomutase by kinetic and equilibrium binding studies. Mechanistic implications, Biochemistry 29 (11) (1990) 2790–2801.

[15] D.R. Davies, W.G. Hol, The power of vanadate in crystallographic investigations of phosphoryl transfer enzymes, FEBS Lett. 577 (3) (2004) 315–321.

[16] H. Schmidt, R. Zalyte, L. Urnavicius, A.P. Carter, Structure of human cytoplasmic dynein-2 primed for its power stroke, Nature 518 (7539) (2015) 435–438.

[17] T. Shimizu, Y.Y. Toyoshima, M. Edamatsu, R.D. Vale, Comparison of the motile and enzymatic properties of two microtubule minus-end-directed motors, ncd and cytoplasmic dynein, Biochemistry 34 (5) (1995) 1575–1582.

[18] A. Lee-Eiford, R.A. Ow, I.R. Gibbons, Specific cleavage of dynein heavy chains by ultraviolet irradiation in the presence of ATP and vanadate, J. Biol. Chem. 261 (5) (1986) 2337–2342.

[19] T. Kon, M. Nishiura, R. Ohkura, Y.Y. Toyoshima, K. Sutoh, Distinct functions of nucleotide-binding/hydrolysis sites in the four AAA modules of cytoplasmic dynein, Biochemistry 43 (35) (2004) 11266–11274.

[20] P. Bouchard, S.M. Penningroth, A. Cheung, C. Gagnon, C.W. Bardin, erythro-9-[3-(2-Hydroxynonyl)]adenine is an inhibitor of sperm motility that blocks dynein ATPase and protein carboxylmethylase activities, Proc. Natl. Acad. Sci. U.S.A. 78 (2) (1981) 1033–1036.

[21] S.M. Penningroth, A. Cheung, P. Bouchard, C. Gagnon, C.W. Bardin, Dynein ATPase is inhibited selectively in vitro by erythro-9-[3-2-(hydroxynonyl)]adenine, Biochem. Biophys. Res. Commun. 104 (1) (1982) 234–240.

[22] T. Kinoshita, T. Tada, I. Nakanishi, Conformational change of adenosine deaminase during ligand-exchange in a crystal, Biochem. Biophys. Res. Commun. 373 (1) (2008) 53–57.

[23] B.M. Paschal, R.B. Vallee, Retrograde transport by the microtubule-associated protein MAP 1C, Nature 330 (6144) (1987) 181–183.

[24] H.J. Schaeffer, C.F. Schwender, Enzyme inhibitors. 26. Bridging hydrophobic and hydrophilic regions on adenosine deaminase with some 9-(2-hydroxy-3-alkyl)adenines, J. Med. Chem. 17 (1) (1974) 6–8.

[25] T. Podzuweit, P. Nennstiel, A. Muller, Isozyme selective inhibition of cGMP-stimulated cyclic nucleotide phosphodiesterases by erythro-9-(2-hydroxy-3-nonyl) adenine, Cell Signal. 7 (7) (1995) 733–738.

[26] T.J. Mitchison, Towards a pharmacological genetics, Chem. Biol. 1 (1) (1994) 3–6.

[27] C. Nusslein-Volhard, E. Wieschaus, Mutations affecting segment number and polarity in *Drosophila*, Nature 287 (5785) (1980) 795–801.

[28] D. Huangfu, K.V. Anderson, Signaling from Smo to Ci/Gli: conservation and divergence of Hedgehog pathways from *Drosophila* to vertebrates, Dev. Camb. Engl. 133 (1) (2006) 3–14.

[29] D. Huangfu, K.V. Anderson, Cilia and Hedgehog responsiveness in the mouse, Proc. Natl. Acad. Sci. U.S.A. 102 (32) (2005) 11325–11330.

[30] D. Huangfu, A. Liu, A.S. Rakeman, N.S. Murcia, L. Niswander, K.V. Anderson, Hedgehog signaling in the mouse requires intraflagellar transport proteins, Nature 426 (6962) (2003) 83–87.

[31] S.C. Goetz, K.V. Anderson, The primary cilium: a signalling centre during vertebrate development, Nat. Rev. Genet. 11 (5) (2010) 331–344.

[32] H. Tukachinsky, L.V. Lopez, A. Salic, A mechanism for vertebrate Hedgehog signaling: recruitment to cilia and dissociation of SuFu-Gli protein complexes, J. Cell Biol. 191 (2) (2010) 415–428.

[33] Y. Hou, G.B. Witman, Dynein and intraflagellar transport, Exp. Cell Res. 334 (1) (2015) 26–34.

[34] P.J. Ocbina, J.T. Eggenschwiler, I. Moskowitz, K.V. Anderson, Complex interactions between genes controlling trafficking in primary cilia, Nat. Genet. 43 (6) (2011) 547–553.

[35] J. Kim, M. Kato, P.A. Beachy, Gli2 trafficking links Hedgehog-dependent activation of smoothened in the primary cilium to transcriptional activation in the nucleus, Proc. Natl. Acad. Sci. U.S.A. 106 (51) (2009) 21666–21671.

[36] L.L. Rubin, F.J. de Sauvage, Targeting the Hedgehog pathway in cancer, Nat. Rev. Drug Discov. 5 (12) (2006) 1026–1033.

[37] J. Xie, M. Murone, S.M. Luoh, A. Ryan, Q. Gu, C. Zhang, et al., Activating smoothened mutations in sporadic basal-cell carcinoma, Nature 391 (6662) (1998) 90–92.

[38] H.J. Sharpe, W. Wang, R.N. Hannoush, F.J. de Sauvage, Regulation of the oncoprotein smoothened by small molecules, Nat. Chem. Biol. 11 (4) (2015) 246–255.

[39] N. Basset-Seguin, H.J. Sharpe, F.J. de Sauvage, Efficacy of Hedgehog pathway inhibitors in basal cell carcinoma, Mol. Cancer Ther. 14 (3) (2015) 633–641.

[40] S.E. Gould, J.A. Low, J.C. Marsters Jr., K. Robarge, L.L. Rubin, F.J. de Sauvage, et al., Discovery and preclinical development of vismodegib, Expert Opin. Drug Discov. 9 (8) (2014) 969–984.

[41] R.L. Yauch, G.J. Dijkgraaf, B. Alicke, T. Januario, C.P. Ahn, T. Holcomb, et al., Smoothened mutation confers resistance to a Hedgehog pathway inhibitor in medulloblastoma, Science 326 (5952) (2009) 572–574.

[42] H.Q. Doan, S. Silapunt, M.R. Migden, Sonidegib, a novel smoothened inhibitor for the treatment of advanced basal cell carcinoma, Onco. Targets Ther. 9 (2016) 5671–5678.

[43] J.M. Hyman, A.J. Firestone, V.M. Heine, Y. Zhao, C.A. Ocasio, K. Han, et al., Small-molecule inhibitors reveal multiple strategies for Hedgehog pathway blockade, Proc. Natl. Acad. Sci. U.S.A. 106 (33) (2009) 14132–14137.

[44] J.K. Chen, J. Taipale, K.E. Young, T. Maiti, P.A. Beachy, Small molecule modulation of smoothened activity, Proc. Natl. Acad. Sci. U.S.A. 99 (22) (2002) 14071–14076.

[45] A.J. Firestone, J.S. Weinger, M. Maldonado, K. Barlan, L.D. Langston, M. O'Donnell, et al., Small-molecule inhibitors of the AAA+ ATPase motor cytoplasmic dynein, Nature 484 (7392) (2012) 125–129.

[46] S.K. See, S. Hoogendoorn, A.H. Chung, F. Ye, J.B. Steinman, T. Sakata-Kato, et al., Cytoplasmic dynein antagonists with improved potency and isoform selectivity, ACS Chem. Biol. 11 (1) (2016) 53–60.

[47] J.M. Lu, J. Nurko, S.M. Weakley, J. Jiang, P. Kougias, P.H. Lin, et al., Molecular mechanisms and clinical applications of nordihydroguaiaretic acid (NDGA) and its derivatives: an update, Med. Sci. Monit. 16 (5) (2010) Ra93–100.

[48] K. Arasaki, K. Tani, T. Yoshimori, D.J. Stephens, M. Tagaya, Nordihydroguaiaretic acid affects multiple dynein-dynactin functions in interphase and mitotic cells, Mol. Pharmacol. 71 (2) (2007) 454–460.

[49] J.K. Famulski, L.J. Vos, J.B. Rattner, G.K. Chan, Dynein/dynactin-mediated transport of kinetochore components off kinetochores and onto spindle poles induced by nordihydroguaiaretic acid, PLoS One 6 (1) (2011) e16494.

[50] Y.I. Fang, E. Yokota, I. Mabuchi, H. Nakamura, Y. Ohizumi, Purealin blocks the sliding movement of sea urchin flagellar axonemes by selective inhibition of half the ATPase activity of axonemal dyneins, Biochemistry 36 (50) (1997) 15561–15567.

[51] G. Zhu, F. Yang, R. Balachandran, P. Hook, R.B. Vallee, D.P. Curran, et al., Synthesis and biological evaluation of purealin and analogues as cytoplasmic dynein heavy chain inhibitors, J. Med. Chem. 49 (6) (2006) 2063–2076.

[52] D.H. Roossien, K.E. Miller, G. Gallo, Ciliobrevins as tools for studying dynein motor function, Front. Cell. Neurosci. 9 (2015) 252.

[53] F. Ye, D.K. Breslow, E.F. Koslover, A.J. Spakowitz, W.J. Nelson, M.V. Nachury, Single molecule imaging reveals a major role for diffusion in the exploration of ciliary space by signaling receptors, eLife 2 (2013) e00654.

[54] L. Milenkovic, L.E. Weiss, J. Yoon, T.L. Roth, Y.S. Su, S.J. Sahl, et al., Single-molecule imaging of Hedgehog pathway protein smoothened in primary cilia reveals binding events regulated by patched1, Proc. Natl. Acad. Sci. U.S.A. 112 (27) (2015) 8320–8325.

[55] X. Liu, T.M. Kapoor, J.K. Chen, M. Huse, Diacylglycerol promotes centrosome polarization in T cells via reciprocal localization of dynein and myosin II, Proc. Natl. Acad. Sci. U.S.A. 110 (29) (2013) 11976–11981.

[56] J. Yi, X. Wu, A.H. Chung, J.K. Chen, T.M. Kapoor, J.A. Hammer, Centrosome repositioning in T cells is biphasic and driven by microtubule end-on capture-shrinkage, J. Cell Biol. 202 (5) (2013) 779–792.

[57] M.W. Elting, C.L. Hueschen, D.B. Udy, S. Dumont, Force on spindle microtubule minus ends moves chromosomes, J. Cell Biol. 206 (2) (2014) 245–256.

[58] V. Sikirzhytski, V. Magidson, J.B. Steinman, J. He, M. Le Berre, I. Tikhonenko, et al., Direct kinetochore-spindle pole connections are not required for chromosome segregation, J. Cell Biol. 206 (2) (2014) 231–243.

[59] C.Y. Lee, H.F. Horn, C.L. Stewart, B. Burke, E. Bolcun-Filas, J.C. Schimenti, et al., Mechanism and regulation of rapid telomere prophase movements in mouse meiotic chromosomes, Cell Rep. 11 (4) (2015) 551–563.

[60] R. Sainath, G. Gallo, The dynein inhibitor ciliobrevin D inhibits the bidirectional transport of organelles along sensory axons and impairs NGF-mediated regulation of growth cones and axon branches, Dev. Neurobiol. 75 (7) (2015) 757–777.

[61] L.C. Kapitein, M.A. Schlager, W.A. van der Zwan, P.S. Wulf, N. Keijzer, C.C. Hoogenraad, Probing intracellular motor protein activity using an inducible cargo trafficking assay, Biophys. J. 99 (7) (2010) 2143–2152.

[62] C.C. Hoogenraad, P. Wulf, N. Schiefermeier, T. Stepanova, N. Galjart, J.V. Small, et al., Bicaudal D induces selective dynein-mediated microtubule minus end-directed transport, EMBO J. 22 (22) (2003) 6004–6015.

[63] S.L. Schreiber, Chemistry and biology of the immunophilins and their immunosuppressive ligands, Science 251 (4991) (1991) 283–287.

[64] R. Pollock, T. Clackson, Dimerizer-regulated gene expression, Curr. Opin. Biotechnol. 13 (5) (2002) 459–467.

[65] M. Putyrski, C. Schultz, Protein translocation as a tool: the current rapamycin story, FEBS Lett. 586 (15) (2012) 2097–2105.

[66] U. del Castillo, M. Winding, W. Lu, V.I. Gelfand, Interplay between kinesin-1 and cortical dynein during axonal outgrowth and microtubule organization in *Drosophila* neurons, eLife 4 (2015) e10140.

[67] P. van Bergeijk, M. Adrian, C.C. Hoogenraad, L.C. Kapitein, Optogenetic control of organelle transport and positioning, Nature 518 (7537) (2015) 111–114.

[68] D. Strickland, Y. Lin, E. Wagner, C.M. Hope, J. Zayner, C. Antoniou, et al., TULIPs: tunable, light-controlled interacting protein tags for cell biology, Nat. Methods 9 (4) (2012) 379–384.

[69] P.D. Leeson, B. Springthorpe, The influence of drug-like concepts on decision-making in medicinal chemistry, Nat. Rev. Drug Discov. 6 (11) (2007) 881–890.

[70] D.J. Anderson, R. Le Moigne, S. Djakovic, B. Kumar, J. Rice, S. Wong, et al., Targeting the AAA ATPase p97 as an approach to treat cancer through disruption of protein homeostasis, Cancer Cell 28 (5) (2015) 653–665.

[71] https://clinicaltrials.gov/ct2/show/NCT02223598.

[72] T.F. Chou, K. Li, K.J. Frankowski, F.J. Schoenen, R.J. Deshaies, Structure-activity relationship study reveals ML240 and ML241 as potent and selective inhibitors of p97 ATPase, ChemMedChem 8 (2) (2013) 297–312.

[73] T.F. Chou, S.L. Bulfer, C.C. Weihl, K. Li, L.G. Lis, M.A. Walters, et al., Specific inhibition of p97/VCP ATPase and kinetic analysis demonstrate interaction between D1 and D2 ATPase domains, J. Mol. Biol. 426 (15) (2014) 2886–2899.

[74] T.F. Chou, S.J. Brown, D. Minond, B.E. Nordin, K. Li, A.C. Jones, et al., Reversible inhibitor of p97, DBeQ, impairs both ubiquitin-dependent and autophagic protein clearance pathways, Proc. Natl. Acad. Sci. U.S.A. 108 (12) (2011) 4834–4839.

[75] P. Magnaghi, R. D'Alessio, B. Valsasina, N. Avanzi, S. Rizzi, D. Asa, et al., Covalent and allosteric inhibitors of the ATPase VCP/p97 induce cancer cell death, Nat. Chem. Biol. 9 (9) (2013) 548–556.

[76] C. Wojcik, M. Rowicka, A. Kudlicki, D. Nowis, E. McConnell, M. Kujawa, et al., Valosin-containing protein (p97) is a regulator of endoplasmic reticulum stress and of the degradation of N-end rule and ubiquitin-fusion degradation pathway substrates in mammalian cells, Mol. Biol. Cell 17 (11) (2006) 4606–4618.

[77] T.F. Chou, R.J. Deshaies, Quantitative cell-based protein degradation assays to identify and classify drugs that target the ubiquitin-proteasome system, J. Biol. Chem. 286 (19) (2011) 16546–16554.

[78] S.A. Kawashima, Z. Chen, Y. Aoi, A. Patgiri, Y. Kobayashi, P. Nurse, et al., Potent, reversible, and specific chemical inhibitors of eukaryotic ribosome biogenesis, Cell 167 (2) (2016) 512–524.e14.

In this chapter

Computational modeling of dynein activity and the generation of flagellar beating waveforms

Veikko F. Geyer[1,a], Pablo Sartori[2,a], Frank Jülicher[3], Jonathon Howard[1]

[1]Yale University, New Haven, CT, United States; [2]Institute for Advanced Study, Princeton, NJ, United States; [3]Max Planck Institute for the Physics of Complex Systems, Dresden, Germany

9.1 Introduction

The motility of the internal mechanical structure of cilia and flagella, which is the axoneme, can be reconstituted in a cell-free and membrane-free environment in buffers containing ATP [3]. This shows that neither intracellular chemical signals nor membrane potentials are required for the initiation or propagation of the flagellar beat. Instead, it is thought that the beat arises from a mechanical feedback circuit: the axonemal dyneins generate forces that bend the flagellum, and the stresses and/or strains that build up in the bent flagellum feed back on the activity of the dyneins. In principle, the feedback signals could include curvature, sliding forces, or transverse forces (that deform the circular cross section of the axoneme). Indeed, modeling studies show that these three mechanical signals can produce beating waveforms under certain conditions [6,7].

Models of the flagellar beat, which go back to the pioneering work of Machin [18], are mathematically complex (entailing fourth order partial differential equations) and computationally demanding (requiring advanced numerical solvers). Thus, the models are not accessible to cell biologists, structural biologists, or geneticists studying flagellar motility.

The aim of this chapter is to bridge the gap between theoretical and experimental biologists by providing a set of MATLAB (MathWorks, Natick, MA) routines that calculate solutions of the most common flagellar models, together with detailed explanations on how to use the code and how it works. The code is available at https://github.com/JHowardLab/AxonemeFits/. It is aimed at students with a minimal working knowledge of computing (using, e.g., MATLAB). Because many biology students now have training in computational techniques, we hope that this code will be usable in many biology labs to calculate flagellar beating waveforms.

---

[a] *Authors contributed equally to this work.*

Dyneins. https://doi.org/10.1016/B978-0-12-809470-9.00009-6

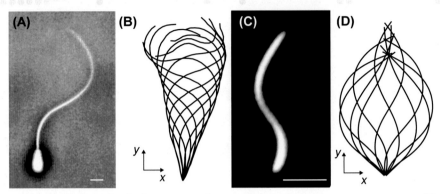

Figure 9.1 Snapshots and waveforms of a bull sperm and an isolated *Chlamydomonas* axoneme. (A) Snapshot of a bull sperm cell imaged with phase contrast microscopy with 4 ms exposure. The image contrast has been inverted. The scale bar is 5 μm. (B) Flagellar shapes of the bull sperm in (A) (one beat cycle). The time difference between adjacent shapes is 4 ms. The bending wave travels from the base (lower) to the tip (upper). (C) Snapshot of the symmetric beat of an isolated *Chlamydomonas* axoneme (*mbo2* mutant) imaged with darkfield microscopy with 1 ms exposure. The scale bar is 5 μm. (D) Shapes of the *Chlamydomonas* axoneme in (C) (one beat cycle). The time difference between adjacent shapes is 2 ms.

For simplicity, we consider only two-dimensional (i.e., planar) beats that are symmetric (the static component of the beat is zero [10]). Thus, the model applies in the cases of sperm (Fig. 9.1A and B), for axonemes isolated from *Chlamydomonas* cells carrying the *mbo2* mutation [24,22] (Fig. 9.1C and D) and for flagella of photoshocked *Chlamydomonas* cells. As a further simplification, we only study periodic solutions [10,22] and consider only sliding and curvature control models.

The routines presented here allow the reader to change parameters such as fluid friction, flagellar bending rigidity, sliding stiffness, motor force, and feedback strength, as well as the mechanical properties of the basal region of the flagellum. We hope to give the reader an intuitive understanding of how these parameters and conditions influence the beat. At the same time, the text guides the reader through the routines with some simple and biologically relevant examples. The hope is that we can demystify the modeling of the flagellar beat and allow the general reader who is interested in flagellar motility to explore the generation of flagellar waveforms.

9.2 Models for beat control in the flagellum

The earliest mechanical description of flagellar dynamics generated by internal force generators was developed by K.E. Machin in 1958 [18]. In his model, the flagellum was represented as a shearing beam whose movement was constrained to a plane. Using this mathematical description, Machin

Dyneins

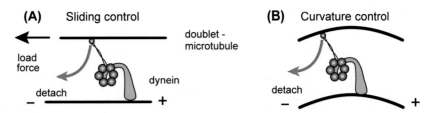

Figure 9.2 Schematic of motor control mechanisms in the flagellum. In *sliding control* (A) dynein detachment is enhanced by a tangential loading force. In *curvature control* (B) dyneins detach due to an increase in curvature. *Signs* indicate doublet polarity.

concluded that active bending moments must be generated along the entire length of the flagellum to produce bending waves that propagate all the way to the distal tip without decrease in wave amplitude. This prediction was subsequently confirmed by structural studies showing dynein arms all along the flagellum. Following the discovery that the doublet microtubules are essentially inextensible [23], and that they slide with respect to each other [26], the idea of a sliding filament model for the axoneme was born. Furthermore, the high frequency of the beat, up to 100 Hz, suggests that the coordination of the dynein motor activity in the bending wave is mediated by mechanical rather than chemical signals. For these reasons it is believed that a mechanical feedback mechanism is at the core of the flagellar beat.

Three molecular mechanisms for dynein coordination in the flagellum have been shown to generate beating patterns[1]. (1) In the sliding control mechanism (Fig. 9.2A), the dyneins detach in response to forces that act parallel to the long axis of the microtubule doublets and that oppose sliding [19]. (2) In the curvature control mechanism (Fig. 9.2B), the detachment of dynein is regulated by doublet curvature [5] (3). In the normal-force control mechanism, also called the geometric clutch, the detachment of dynein is regulated by transverse forces that act to separate adjacent doublets when they are curved [17,1]. Because the normal-force mechanism is related to curvature control [17,21], we will not consider it here.

The implementation of these models in the literature is mostly discussed for the two-dimensional case because twist, which would make the flagellar beat three-dimensional, is a second-order effect in the sliding filament model [12]. In this work we will only focus on how to obtain the generic beating patterns of a two-dimensional description of the beating flagellum.

[1] Additional regulatory mechanisms through radial spokes or the central pair apparatus [25] have received less attention by theorists, and will not be discussed here.

9.3 Theory

Here, we introduce the mechanical model of the flagellum, which is solved in the appended MATLAB program. This model was introduced in Ref. [9,8] and extended in Ref. [19] to include compliance of the axonemal base.

9.3.1 Planar description of the axonemal geometry

We model the axoneme, the mechanical apparatus inside the flagellum, as a pair of adjacent filaments in a plane, connected by motors and cross-linkers that keep the spacing a between the filaments fixed in the beating plane [11,21], as depicted in Fig. 9.3. Motors oriented in one direction (green) bend the filament in one direction while motors oriented in the opposite direction bend the filaments in the other direction.

The vector $\mathbf{r}(s)$ describes the position along the centerline (dashed line) where s is the arc-length ($s=0$ is the base and $s=L$ is the distal tip). Each point along the arc-length is characterized by its tangent angle $\psi(s)$ with respect to the horizontal, as shown in Fig. 9.3B. This position vector $\mathbf{r}(s)$ is related to the tangent angle by

$$r(s) = r(0) + \int_0^s \left(\cos\psi(s'), \sin\psi(s') \right) ds', \tag{9.1}$$

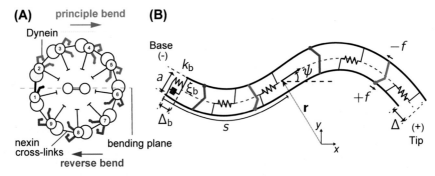

(A) principle bend **(B)**

Dynein

Base (-)

k_b

a

ξ_b

Δ_b

$-f$

ψ

$+f$

r

y

x

Δ (+)

Tip

nexin cross-links

bending plane

s

reverse bend

Figure 9.3 **Two-dimensional model of the axoneme.** (A) Cross section of an axoneme, as seen from the basal end looking toward the distal tip. The numbering of the doublets follows the convention for *Chlamydomonas* [13]. The green dyneins bend the axoneme such that the center of curvature is to the right. (B) Two-dimensional model of the axoneme, as seen in the bending plane, **xy**. The two filaments are constrained to have the spacing a. The point at arc-length s has position vector **r** and tangent angle ψ with respect to the horizontal axis of the lab frame **xy**. Dyneins on the upper filament (green) have microtubule-binding domains (MTBs) that walk along the lower filament toward the base and produce a (tensile) force density $+f$ on the lower filament. This force slides the lower filament toward the distal end. The dyneins on the opposite filament (blue) create sliding (and bending) forces in the opposite direction. The local sliding displacement is given by Δ, and the sliding at the base is Δ_b. The spring and dashpot at the base constitute the basal compliance, with stiffness k_b and friction coefficient ξ_b.

Dyneins

which intuitively corresponds to the centerline being formed by short segments $d\mathbf{r} = (\cos\psi(s), \sin\psi(s))\,ds$ of length and direction determined by the tangent angle. The sliding distance, Δ, between the filaments is

$$\Delta(s) = \Delta_b + a(\psi(s) - \psi(0)),\tag{9.2}$$

where Δ_b is the sliding displacement at the base. This equation describes how bending relates to sliding.

9.3.2 Dynamic equations of bending patterns

The dynamics of the flagellum are determined by the balance of hydrodynamic, bending, and motor forces, which are expressed as bending moments:

$$\xi_n\partial_t\psi(s,t) = -\kappa\partial_s^4\psi(s,t) + a\partial_s^2 f(s,t),\tag{9.3}$$

The term on the left is related to the hydrodynamic force; ξ_n is the (normal) friction coefficient, which characterizes the resistance experienced by the axoneme as it is moved through the fluid perpendicular to its long axis. The first term on the right is the force that resists bending of the axoneme characterized by the flexural rigidity κ. The last term arises from the sliding force f exerted by the molecular motors and passive cross-linkers. This equation is a partial differential equation, with derivatives in time denoted by ∂_t and derivatives in arclength denoted by ∂_s, with a superscript indicating the order of the derivative.

The sliding force along the length creates a force F_b at the base that deforms the compliant elements at the base (depicted as a spring and damper in Fig. 9.3), which restrict basal sliding.

$$F_b = -\int_0^L f(s)\,ds = k_b\Delta_b + \xi_b\partial_t\Delta_b,\tag{9.4}$$

where k_b is the spring constant of elements at the base and ξ_b is the corresponding friction coefficient.

Eqs. (9.3) and (9.4) can be solved if the sliding force $f(s,t)$ and boundary conditions are known. The boundary conditions are summarized in Table 9.1; the sliding force is discussed below.

Table 9.1 Boundary conditions

End	Free-free	Clamped-free
$s = 0$	$\kappa\partial_s\psi = aF_b$ $\kappa\partial_s^2\psi = af$	$\psi = 0$ $\partial_s\psi = 0$
$s = L$	$\partial_s\psi = 0$ $\kappa\partial_s^2\psi = af$	$\partial_s\psi = 0$ $\kappa\partial_s^2\psi = af$

For each condition we have four boundary conditions, two force and two torque boundary conditions (balance equations) at each end of the filament.

9.3.3 Oscillatory solutions and frequency representation

The beat of flagella is periodic in time, with a dominant frequency, called the beat frequency. We therefore look for periodic solutions to the equations. If the beat frequency is ν (with units cycles per second), the tangent angle can be written as

$$\psi(s, t) = \tilde{\psi}(s)\, e^{i\omega t} + \tilde{\psi}*(s)\, e^{-i\omega t} \tag{9.5}$$

where $\omega = 2\pi\nu$ is the angular frequency, $\tilde{\psi}$ is the (complex) first Fourier mode, and the star denotes complex conjugation. Analogous expressions can be written for all quantities that depend on time, and, in the following, tildes will denote the first Fourier mode, corresponding to the beat frequency. Eq. (9.5) corresponds to a sinusoidal wave with peak-to-peak amplitude $A(s) = 2\sqrt{\tilde{\psi}'^2(s) + \tilde{\psi}''^2(s)}$, and phase $\phi = \tan^{-1}\left(\tilde{\psi}'^2(s)/\tilde{\psi}''^2(s)\right)$. $\tilde{\psi}'(s)$ and $\tilde{\psi}''(s)$ correspond to the real (in phase, $\cos\omega t$) and imaginary (out of phase, $\sin\omega t$) components of the tangent angle. The advantage of using the representation in Eq. (9.5) is that $\tilde{\psi}(s)$ satisfies the ordinary differential equation, that is,

$$\xi_n i\omega\tilde{\psi} = -\kappa\partial_s^4\tilde{\psi} + a\partial_s^2\tilde{f}, \tag{9.6}$$

This equation is the Fourier transformed form of Eq. (9.3). The corresponding balance of sliding forces and basal force is

$$\int_0^L \tilde{f}(s)\, ds + \tilde{F}_b = 0, \tag{9.7}$$

where $\tilde{F}_b = \chi_b\tilde{\Delta}_b$ is the basal force, with $\chi_b = \chi_b' + i\chi_b''$ being the basal compliance (prime and double prime denote real and imaginary parts). For the case considered here, with elastic and viscous elements at the base, we have that $\chi_b' = k_b$ and $\chi_b'' = \omega\xi_b$.

9.3.3.1 Motor dynamics

To solve Eqs. (9.3) and (9.4) or, alternatively, Eqs. (9.6) and (9.7) for the oscillatory case, we need to choose one set of boundary conditions from Table 9.1 and provide an expression for the motor force. One approach is to provide an expression for the force at every point along the arc-length, $\tilde{f}(s)$. This will generate a solution for $\tilde{\psi}$. This approach was taken by Machin [18] who deduced the force $f(s)$ necessary to overcome the elastic and hydrodynamic resistance to bending.

We take an alternative approach by asking whether there are bending waves (i.e., solutions to Eq. 9.7) when the force is due to feedback from sliding or bending. In other words we do not specify the force explicitly, but rather ask whether we obtain solutions if the motor force is a function of the sliding force or the curvature. If there is a solution, then the function $\tilde{f}(s)$ is deduced from the solution. For simplicity, we assume a linear feedback model of the form

$$\tilde{f} = \chi\tilde{\Delta} + \beta\partial_s\tilde{\psi} \tag{9.8}$$

Table 9.2 Response coefficients for motor control models

Coefficient	Sliding control	Curvature[a]	Dynamic curvature[a]
(χ', χ'')	$(<0, <0)$	$(=0, =0)$	$(>0, =0)$
(β', β'')	$(=0, =0)$	$(>0, <0)$	$(=0, <0)$

The different motor control models constrain the range of values of the response coefficients. For example, for sliding control models, $\beta=0$ and the sign of χ distinguishes between passive and active sliding.
[a] The signs correspond to a particular direction of wave propagation, unlike for the signs for sliding control, which are intrinsic to the model.

where χ and β are the sliding and curvature response coefficients, respectively. The equation states that changes in sliding and/or curvature can affect the motor force. It is thus an expression of the mechanochemical feedback that underlies the flagellar dynamics and the formation of beating patterns. The response coefficients are complex numbers, and we can write them as $\chi = \chi' + i\chi''$ and $\beta = \beta' + i\beta''$, with χ', χ'', β', and β'', being real numbers. In the linear feedback model χ and β are constant in time and independent of arc-length. Thus, our linear feedback model has a simple form and depends at most on four parameters (and only two in the sliding control model where $\beta' = \beta'' = 0$).

The response coefficients χ and β are functions of frequency and stiffness as well as of microscopic parameters such as the motor density, the motor stall force, etc. (see Refs. [19,20], for examples). However, periodic beating patterns can be obtained without knowledge of all these details. It is for this reason that the beating patterns that emerge from this formalism are considered "generic." In Table 9.2 we provide some particular choices of χ and β that correspond to different types of motor feedback models (for example, $\beta' = \beta'' = 0$ is sliding control). Once we have found a solution for a particular choice of response coefficients, we can then ask if it matches the experimentally observed beating waveforms. If it does we say that such a feedback mechanism (sliding or curvature control) can explain the beat [19,22].

9.3.3.2 Solving the boundary value problem for time-periodic beats

For a particular choice of motor feedback from Table 9.2 and a set of boundary conditions from Table 9.1 we can solve Eqs. (9.6) and (9.7). To do so, we note that the general solution to Eq. (9.6) is given by

$$\tilde{\psi}(s) = \sum_{i=1}^{4} A_i e^{sk_i} \tag{9.9}$$

where the A_i are complex amplitudes and the k_i satisfy the characteristic polynomial

$$i\omega\xi_n = -\kappa k_i^4 + a\beta k_i^3 + a^2 k_i^2 \chi \tag{9.10}$$

obtained by substituting Eq. (9.9) into Eq. (9.6). The characteristic roots, k_i ($i=1,2,3,4$), are functions $k_i(\chi, \beta)$ of the complex response coefficients. These can be replaced in the expression of the general solution, Eq. (9.9). The basal force balance Eq. (9.7) becomes

$$\int_0^L \left[\chi \Delta_b + \sum_{i=1}^4 a\chi \left(A_i e^{sk_i} - A_i \right) + \beta k_i A_i e^{sk_i} \right] ds + \chi_b \Delta_b = 0$$
(9.11)

To obtain a particular beating pattern we still need to determine five unknowns: the four complex amplitudes A_i and the basal sliding $\tilde{\Delta}_b$. To do so, we use the global force balance and the four boundary conditions. Formally, the boundary value problem is a linear homogeneous system of equations on the five components of the vector, $\mathbf{v} = \{A_1, A_2, A_3, A_4, \Delta_b\}$. This system of equations is obtained by substituting the solution given by Eq. (9.9) into the four boundary conditions (Table 9.1) and also the global force balance Eq. (9.11) (see Appendix 1). The coefficients of this system compose the matrix $\hat{\mathbf{M}}$ that depends nonlinearly on χ_b as well as on χ and β (both explicitly through Eq. (9.11) and implicitly through $k_i(\chi, \beta)$). Because the system of equations is homogeneous, we can write it as

$$\hat{\mathbf{M}} \cdot \mathbf{v} = 0,$$
(9.12)

where $\mathbf{v}$ is the corresponding set of eigenvectors that satisfy the equation. This system will only have non-trivial solutions for the values of the parameters χ', χ'', β', β'', χ_b', χ_b'' for which the determinant is zero:

$$\det \hat{\mathbf{M}} = 0.$$
(9.13)

If the determinant is nonzero, then $\mathbf{v}=0$; and there will not be an emerging bending pattern. Thus, the determinant equation, Eq. (9.13), with separate real and imaginary parts, constitutes a pair of equations that constrain the values of the model's parameters, so that the solutions that exhibit bending waves are obtained. Fixing four of these parameters will determine the other two in the sense that they can take only a discrete set of values. For example, $\beta' = 0$, $\beta'' = 0$, $\chi_b' \to \infty$, $\chi_b'' \to \infty$ for sliding control with no sliding allowed at the base [8], will result in a discrete but infinite set of solutions $\{\chi_n', \chi_n''\}$, with $n=1$, 2, 3,..., for the determinant equations.

The logic of solving the model is therefore as follows. Choose four control parameters, choose one of the allowed values of the other two parameters (say the nth one) and compute the coefficients of the eigenvector $\mathbf{v}$ (from the boundary conditions and global force balance), then substitute into Eq. (9.9) to generate a waveform. Then we can ask: does this waveform match the experimentally observed waveform?

9.3.3.3 Numerical calculation of solutions

Along with this document, we provide a MATLAB program (for MATLAB 2016b) that performs the computations explained in the theory section. The purpose of this program is to: (1) allow the reader to calculate beating patterns, (2) explore different parameters and conditions without any required physics knowledge, and (3) provide functions that can easily be incorporated into more advanced programs used to, for example, fit observed beating patterns to experimental data.

The provided code contains a main program file called *compute_beats_examples.in.* It is used to specify the mechanical parameters, boundary conditions and motor models. It calls the following functions: (1) *parameters.m* that stores parameters from the user input, (2) *solspace.m* that solves the boundary value problem using the function *bcmat.m,* for given parameters and a specific choice of response coefficients, and (3) *beatmodes.m* that calculates the solutions. The function *plot_phasespace.m* visualizes solutions in the phase space (the parameter space, showing the determinant of $\hat{M}$, Eq. (9.13), for pairs of response coefficients, which we refer to as the determinant landscape). The function *plot_solution.m* displays a specific solution including detailed information about its properties.

In the following, we will go through the functions mentioned previously, explain their content, and present their applications using examples.

9.3.3.4 Parameter choice

We define the mechanical parameters of the axoneme, the motor control model, and the boundary conditions in *parameters.m.* This function accepts a structure called *input,* containing parameter inputs in random order as shown in Script-extract 1, lines 2–9. After running *parameters.m* with the parameter structure as an input, the parameters are stored in global variables (Script-extract 1, line 11). Two sets of example input parameters with their respective units are given in Table 9.3. If a parameter is not defined through an input, the default value defined within the function *parameters.m* is used (the default values represent the ones of Example 1, Table 9.3). The parameters are normalized as shown in Script-extract 1, lines 13–21 (see Appendix 2 and *parameters.m* for details) to facilitate subsequent computations. Throughout the program, normalized parameters are denoted with the extension *bar.* The structure also defines the boundary conditions as well as the motor model (Table 9.3). To choose a specific example presented in Table 9.3 select the respective example in section 0 (named "Choose a specific example") of the routine *compute_beats_examples.m.*

Table 9.3 Parameters, boundary conditions, and motor models

			Example 1: Bulls Sperm	Example 2: *Chlamydomonas mbo2*
	List of all available parameters			
	Parameter name			
	Length	L	58 µm	9 µm
	Frequency	ν	21 Hz	28 Hz
	Spacing	a	0.185 µm	0.06 µm
Mechanical parameters	Bending rigidity	κ	1700 pN/µm^2	400 pN/µm^2
	Basal stiffness	χ_b'	95000 pN/µm	2100 pN/µm
	Basal impedance	χ_b''	36154 pN/µm	2300 pN/µm
	Sliding stiffness[a]			
	Sliding friction[a] Normal friction	ξ_n	0.0034 pN s/µm^2	0.0034 pN s/µm^2
Boundary conditions	Boundaries		"Free-free" "Clamped-free"	"Clamped-free"	"Free-free"
Motor model	Motor		"Sliding curvature" "Dyn curvature"	"Sliding"	"Dyn curvature"

This table contains all available inputs of *parameter.m*. Example values are presented for bull sperm waveforms used in Ref. [19] and for *Chlamydomonas* waveforms used in Ref. [22]. The different values of the spacing (a) are due to different definitions in the two papers (diameter of the axoneme and center-to-center doublet spacing).

[a] *The sliding friction and sliding stiffness (with units of pN/µm^2) are input parameters for the curvature control model only, which is another option in the routine, which is not discussed in the text.*

Script-extract 1: Define the input parameters using *parameters.m*.

```
1  % Define a structure containing the input parameters
2  input = struct (...
3                 'Length',              58,...
4                 'Frequency',           21,...
5                 'Spacing',             0.185,...
6                 'BendingRigidity',     1700,...
7                 'NormalFriction',      0.0034,...
8                 'Boundaries',          'clamped-free'
9                 'Motor',               'sliding') ,...;
10 % Write the parameters to the global variable
11 parameters(input);
12
13 % Define dimensionless global variables used by the routines for calculations
14
15 % Normalized frequency ("sperm" number)
16 Sp_bar = 2*pi*Frequency*.NormalFriction*Length^4/BendingRigidity;
17 % Normalized Basal response coefficient
18 chib_bar = (BasalStiffness+i*BasalFriction)*(Length*Spacing^2/BendingRigidity)
      ;
19 % Normalized resonse coefficients
20 chi_bar=((a_0^2 *L^2)/kappa)*chi;  beta_bar =  ((a_0*L)/kappa)*beta;
```

9.3.3.5 Defining the matrix of coefficients

The first step in solving the boundary value problem is to construct the 5×5 matrix $\hat{M}$, as described in the text. This is done using the function *bcmat.m* (Script-extract 2), which takes as input the response coefficients χ and β (to determine the roots κ_i according to Eq. 9.10) and outputs the matrix $\hat{M}$. The exact definition is given in Script-extract 2, lines 9–15.

Script-extract 2: Solving the boundary value problem using *bcmat.m* here for free-free boundary conditions.

```
 1  % Define the complex response coefficients
 2  chi_bar  = response(1) + 1i*response(2);
 3  beta_bar = response(3) + 1i*response(4);
 4
 5  % Calculate the four roots to the characteristic polynomial
 6  r=[1 -beta_bar -chi_bar 0 1i*Sp_bar];
 7  k=roots(r);
 8
 9  % Generate matrix for corresponding boundary conditions
10  M1=[k(1),k(2),k(3),k(4),-chib_bar];
11  M2=[(k(1)-beta_bar)*k(1),(k(2)-beta_bar)*k(2),(k(3)-beta_bar)*k(3),(k(4)-
      beta_bar)*k(4),-chi_bar];
12  M3=[k(1)*exp(k(1)),k(2)*exp(k(2)),k(3)*exp(k(3)),k(4)*exp(k(4)),0];
13  M4=[(k(1)^2-chi_bar-beta_bar*k(1))*exp(k(1))+chi_bar,(k(2)^2-chi_bar-beta_bar*
      k(2))*exp(k(2))+chi_bar,(k(3)^2-chi_bar-beta_bar*k(3))*exp(k(3))+chi_bar(k
      (4)^2-chi_bar-beta_bar*k(4))*exp(k(4))+chi_bar,-chi_bar];
14  M5=[chi_bar*(exp(k(1))-1)/k(1)+beta_bar*exp(k(1))-(beta_bar+chi_bar)chi_bar*(
      exp(k(2))-1)/k(2)+beta_bar*exp(k(2))-(beta_bar+chi_bar)chi_bar*(exp(k(3))
      -1)/k(3)+beta_bar*exp(k(3))-(beta_bar+chi_bar)chi_bar*(exp(k(4))-1)/k(4)+
      beta_bar*exp(k(4))-(beta_bar+chi_bar),chi_bar+chib_bar];
15  M=[M1;M2;M3;M4;M5];
```

9.3.3.6 Exploring the phase space—the space of all solutions

To find all possible oscillatory solutions of the dynamic system, we explore the phase space of possible values of the response coefficients in a predefined range. This is done using the function *solspace.m*, which takes a parameter range for each response coefficients as input (Script-extract 3, lines 1–4). The function *solspace.m* uses the function *bcmat.m* to first calculate the matrix $\hat{M}$ for pairs of response coefficients in the predefined range. To find combinations of the response coefficients that satisfy the determinant equation Eq. (9.13) (for which the determinant equals zero and the system has nontrivial solutions) we first calculate the determinant of the matrix $\hat{M}$ for combinations of response coefficients and obtain the determinant landscape, Fig. 9.4. We then identify local minima in this landscape, which we refer to as seeds (Script-extract 3, lines 22–34). The function *solspace.m* outputs the determinant as a function of the response coefficients in the structure *space.xyz*, as well as the coordinates of all minima in the structure *space.seeds* (Script-extract 3). The function provides the option to also display the calculated determinant landscape and the identified minima, by giving the function input "plot" the value one (Script-extract 3, line 5). Those minima are shown in Fig. 9.4 (red dots) for Bull sperm flagella and *Chlamydomonas* axonemes using the parameter values of Table 9.3.

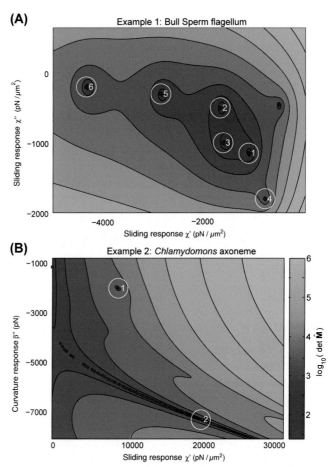

Figure 9.4 The phase space, determinant landscape for sets of response coefficients. The parameters, boundary conditions, and the motor model for Example 1(A) and Example 2(B) in Table 9.3 are used to calculate the determinant landscapes. The red dots represent the seeds, the minima in the landscape, the white circles and numbers identify solutions for which the determinant is zero (up to numerical error). Apparent solutions with relative error >0.1 are rejected (parameter *err_tol* in *compute_beats_examples.m*, Section 9.3.3.7). The figures are the output of the function *plot_phasespace.m*. Solutions are numbered by their proximity to the origin considering the normalized response coefficients.

Script-extract 3: Explore the phase space using *solspace.m*.

```
1   % Function input solspace.m : we explore the [chi1,chi2] space for sliding
        control
2   chi1=-400:1:0;   %(pN/ um^2)
3   chi2=-100:1:0;   %(pN/ um^2)
4   plot=0; % Plot option of the function (1=on/ 0=off)
5   space = solspace(chi1,chi2,plot);
6
7   % Function operations:
8           % define a function that calulates the determinant for a pair of
                response coefficients res(1), res(2)
9   determinant = @(res) det(bcmat([res(1),res(2),0,0]));
10
11          % compute the determinant over the range chi1, chi2
12  detdat = zeros(length(chi1_bar),length(chi2_bar));
13  for i=1:length(chi1_bar)
14          for j=1:length(chi2_bar)
15                  detdat(i,j)=abs(determinant([chi1_bar(i),chi2_bar(j)]));
16          end
17  end
18
19          % store the log of the determinant
20  [x,y]    = meshgrid(xrange,yrange);
21  z        = log10(detdat);
22
23          % calculate the minima of the determinant landscape
24  [~,~,~,imin] = extrema2(z');
25
26          % obtain seeds - find the points that are xy null
27  xmin=x(imin);  ymin=y(imin);
28  nonull=((xmin~=0)+(ymin~=0))~=0;
29  xmin = xmin(nonull);  ymin = ymin(nonull);
30  space.seeds = [xmin';ymin'];
31  space.xyz   = {x,y,z};
```

9.3.3.7 Calculate solutions

The function *beatmodes.m* solves Eq. (9.12) for pairs of response coefficients that satisfy the determinant equation (Eq. 9.13). It takes as input the seeds that are the preliminary pairs of response coefficients given through the structure *space.seeds*. These response coefficients are then refined through the *Mat. lab* function *fmincon* that finds the minimum of constrained nonlinear multivariable functions and are stored in the structure *sol.res* (Script-extract 4, line 10). Then, the matrix of coefficients $\hat{M}$ is calculated for the refined set of response coefficients. The matrix $\hat{M}$ is diagonalized using the *Mat.lab* function *svd*, which performs a singular value decomposition on $\hat{M}$ (Script-extract 4, line 13) to calculate the eigenvectors $v = \{A_1, A_2, A_3, A_4, \Delta_b\}$ (Script-extract 4, line 16), which are stored in the structure *sol. A*. The roots of the characteristic polynomial are calculated (Script-extract 4, line 16) and stored in *sol.k*. These are the complex amplitudes A_i and complex coefficients k_i, for $i=1,\ldots,5$ that define the shape by the angle ψ through the general solution Eq. (9.9), which is shown in Script-extract 5, lines 5–6. The variable *sol.err* represents the normalized error

ε that is associated with a solution. It is calculated by $\varepsilon = \sqrt{|\hat{\mathbf{M}} \cdot \mathbf{v}|^2} / \sqrt{|\mathbf{v}|^2}$.
The structure *sol* is the output of *beatmodes.m* (Script-extract 4, line 17). The function *beatmodes.m* can also be used to study solutions in an environment of a known solution by directly passing values to the function (Script-extract 4, line 3) and could be implemented into a fitting procedure if desired.

Script-extract 4: Calculate solutions using beatmodes.m.

```
1  % Function input:
2  solutions = beatmodes(space.seeds)
3  mysolution = beatmodes([chi1_bar;chi2_bar]))
4  % Function operations:
5  % Set options for the solver
6  options = optimset('Display','off','TolFun',1e-15,'TolX',1e-15);
7
8  % Solve the singular system using singular value decomposition
9  x0        = [seeds(1),seeds(2)];
10 sol(i).res = fmincon(revcond,x0,[[1,0];[0,1]],[0  0],[],[],[],[],[],options);
11 M          = bcmat([sol.res(1),sol.res(2),0,0]);
12 [~,~,V]    = svd(M,'econ');
13
14 % Assign amplitudes, modes, and errors
15 sol(i).A = V(:,size(M,2));
16 sol(i).k = roots([1,0,-(sol(i).res(1)+1i*sol(i).res(2)),0,1i*Sp]);
17 sol(i).err = sqrt(sum((abs(M*sol(i).A)).^2))/sqrt(sum((abs(sol(i).A)).^2));
```

9.3.3.8 Display of the phase space and specific solutions

To display the topography of the phase space and the calculated solutions use the function *plot_phasespace.m*. It takes as input the structures *space, solutions*. It also considers a value *err_tol* that allows to filter for displayed solutions based on the error calculated in *beatmodes.m* (Script-extract 4, line 17). Solutions are depicted with numbered, white circles; the seeds are shown as red dots. Through the refinement procedure, different seeds can converge to the same solution. Then the solution will only be displayed once. The function *plot_phasespace.m* outputs an index *(space.index)* that contains the numbers of the displayed solutions, ordered by their proximity to the origin of the phase space considering the normalized response coefficients. The user can now choose to study a specific solution by inputing the respective solution number input into *plot_solution.m*.

Script-extract 5: Calculate waveforms—sample a solution in arc-length "*s*" and time "t" using *beatmodes.m*.

```
1 % Sample a solution (tangent angle psi) in arc-length "s" and time "t"
2 ds  = 0.01; dt = 0.1;  % define arclength increment and time step
3 s   = 0:ds:1;          % define arclength vector
4 t   = 0:dt:1;          % define time vector
5 psi = (sol(i).A(1)*exp(sol(i).k(1)*s)+sol(i).A(2)*exp(sol(i).k(2)*s)...
6 +   sol(i).A(3)*exp(sol(i).k(3)*s)+sol(i).A(4)*exp(sol(i).k(4)*s));
7 psi = psi/(max([abs(psi)]));
```

To display a specific solution as a waveform in space and time, we sample the angle—in arc-length s and time t (Script-extract 5, lines 1–7). Since the magnitude of the angle is arbitrary in the linear theory presented here, we scale the angle to its maximum value (Script-extract 5, line 7). The function *plot_solution.m* displays the respective waveforms in different representations: We show the tangent angle (Script-extract 6, line 4) and the shape representation (Script-extract 6, lines 7–8) as time series that are time coded in rainbow colors. The waveform representation by amplitude and phase (Script-extract 6, lines 11 and 13) as well as by real and imaginary parts is given (Script-extract 6, lines 15–16).

Script-extract 6: Plot a solution using *plot_solution.m*

```
1  % Calulate the angle and x,y time series, amplitude and phase of the waveform
2  plot_solution(solutions,space.index,solution_number)
3     % Tangent angle time−series
4  psi = (s,real(psi'*exp(2*pi*1i*t)));
5
6     % x,y time series
7  x = ds*cumtrapz(cos(real(psi'*exp(2*pi*1i*t)))),
8  y = ds*cumtrapz(sin(real(psi'*exp(2*pi*1i*t))));
9
10    % Amplitude
11 Amplitude = (s,abs(psi));
12    % Phase
13 Phase = (s,unwrap(angle(psi)));
14    % Plot Real and Imaginary part of psi
15 real_psi = real(sol(i).psi);
16 imag_psi = imag(sol(i).psi);
```

The function outputs for the solutions that correspond to bull sperm and *Chlamydomonas* waveforms as shown in Fig. 9.5. The inputs of this function are the solutions (output of *beatmodes.m*), the index of the ordered and filtered solutions (*space.index*), and the number of the respective solution (*solution_number*) that should be displayed (Script-extract 6, line 2 and phase space plot Fig. 9.4).

9.4 Discussion

In this chapter we have presented MATLAB routines that calculate the beating waveforms of axonemes, the motile structure that generates the periodic motion of cilia and flagella. The reader can choose between two different regulatory mechanisms for dyneins, which are regulation by curvature and regulation by sliding displacement, and can vary the mechanical properties of the axoneme such as flagellar length, bending stiffness, and fluid friction. The key idea behind the models is that for specific compliant properties of the axonemal base (that determines the amount of sliding at the base), there are discrete values of the sliding control and/or curvature control coefficients

Dyneins

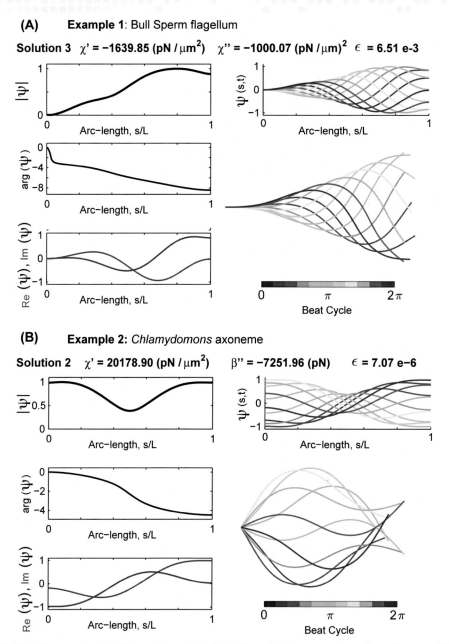

(A)   **Example 1**: Bull Sperm flagellum

Solution 3 $\chi' = -1639.85$ (pN / μm^2) $\chi'' = -1000.07$ (pN / μm)^2 $\epsilon = 6.51$ e-3

(B)   **Example 2**: *Chlamydomons* axoneme

Solution 2 $\chi' = 20178.90$ (pN / μm^2) $\beta'' = -7251.96$ (pN) $\epsilon = 7.07$ e-6

Figure 9.5 Waveform representations of a selected solution. For the parameters, boundary conditions, and the motor model (Table 9.3), a specific solution was identified in the phase space in Fig. 9.4. The waveforms of this solution are displayed using the function *plot_solution.m* as tangent angle (panel A; Script-extract 6, line 4), shape (panel B; Script-extract 6, lines 7–8), beat amplitude and phase (panel C and D; Script-extract 6, lines 11–13), and real and imaginary part of the angle ip (Script-extract 6, lines 15–16) for A bull sperm flagella parameters (Example 1), corresponding to the specific solution 3 shown in Fig. 9.4A and B *Chlamydomonas* axoneme parameters (Example 2), corresponding to the specific solution 2 shown in Fig. 9.4B. Both panels represent the standard output of the function *plot_solution.m*.

(χ and β, respectively) that give rise to periodic bending patterns. This is analogous to the different modes of vibration of strings and organ pipes, where there is a fundamental tone as well as higher harmonics. The routines identify these discrete solutions and calculate the corresponding waveforms. The reader can then compare these waveforms to the observed beats.

We present two specific examples of flagellar beats: intact bull sperm flagella [19] and isolated *Chlamydomonas* axonemes [22]. We show that the routines calculate the corresponding waveforms and that those are in good quantitative agreement with the observed waveforms.

Given a specific set of axonemal properties, each motor model provides a theoretical prediction for the phase between the motor force (i.e., the activity of the dynein motors on each side of the axoneme) and the curvature of the traveling bending wave (i.e., the shape), which can be compared to experiments. Thus, the model predicts where and when during the beat the motors are active. These predictions can be compared to structural studies, such as Cryo–electron microscopy [14,15].

The model solved in this chapter makes a number of simplifying assumptions. First, the model assumes a two-dimensional beat, in which the three-dimensional axoneme is replaced by a pair of filaments that bend in the plane [8,22]. The case of a three-dimensional axoneme, which has a small out-of-plane component has been explored recently in Ref. [21]. Second, the beats are symmetric, meaning that the time-averaged tangent angle is constant, i.e., the time-averaged curvature is zero. However, the beat patterns of wild-type *Chlamydomonas* are highly asymmetric [10], as are the waveforms of epithelial cilia. In addition, even the relative symmetric beating patterns of sperm usually have small asymmetries (that lead to circular or helical swimming paths). The model presented in this work can be extended to asymmetric beats [22].

The model has two major limitations. First, because the model is linear, the amplitude is not specified. Additional factors must limit the amplitude. One possible mechanism is that proteins like the N-DRC that cross-link the doublet microtubules cannot elongate more than about 100 nm and limit the amplitude of sliding [4]. This would add a nonlinear component to the sliding force [20,2]. A second limitation is that the model does not predict the frequency. Rather, the frequency is input into the model. According to the model, the frequency is selected implicitly: the curvature and/or sliding control parameters depend on microscopic properties of the motors such as their density, their force–velocity relations, their sensitivity to sliding forces and/or curvature, and the delays associated with the detachment of the motors from the microtubules. These properties lead to a frequency dependence of the curvature and sliding control parameters, and the beat frequency will be specified so that it matches the basal compliance.

The final question that the models raise, but do not answer, is about the molecular mechanisms by which curvature and/or sliding forces are detected and conveyed to the dyneins. Regulation by curvature, for example, could be achieved through axonemal twist, where a bend produced radial or transverse forces, which change the spacing between the doublet microtubules and could regulate the dyneins. This mechanism has been termed the t-force or geometric clutch mechanism [16]. For sliding control, one idea is that load force leads to motor detachment. However, this hypothesis has not been tested experimentally, for example, using single-molecule techniques. Thus, the molecular picture is incomplete.

In conclusion, the code presented here is able to recapitulate beating patterns of certain classes of cilia and flagella (i.e., that have symmetric, planar beats). The MATLAB routines also highlight open questions such as the mechanisms underlying frequency selectivity and beat amplitude.

A.1 Coefficient equations for the boundary value problem

The boundary value problem requires the global force balance equation, Eq. 9.11. After integrating, this equation becomes

$$\sum_{i=1}^{4} \left(a\chi \left(e^{Lk_i} - 1 \right) \right) / k_i + \beta \left(e^{Lk_i} - 1 \right) A_i + (L\chi + \chi_b) \Delta_b = 0 \tag{9.14}$$

In addition, the boundary conditions result in the following sets of equations:

- Free-free:

$$\sum_{i=1}^{4} \kappa k_i A_i - a\chi_b \tilde{\Delta}_b = 0$$

$$\sum_{i=1}^{4} (\kappa k_i - a\beta) k_i A_i - a\chi_b \tilde{\Delta}_b = 0$$

$$\sum_{i=1}^{4} k_i e^{k_i L} A_i = 0$$

$$\sum_{i=1}^{4} \left(\kappa k_i^2 e^{k_i L} - a^2 \chi \left(e^{k_i L} - 1 \right) - a\beta k_i e^{k_i L} \right) A_i - a\chi_b \tilde{\Delta}_b = 0 \tag{9.15}$$

- Clamped-free:

$$\sum_{i=1}^{4} A_i = 0$$

$$\sum_{i=1}^{4} \left(\kappa k_i^3 - a \left(a\chi k_i + \beta k_i^2 \right) \right) A_i = 0$$

$$\sum_{i=1}^{4} k_i e^{k_i L} A_i = 0$$

$$\sum_{i=1}^{4} \left(\kappa k_i^2 e^{k_i L} - a^2 \chi \left(e^{k_i L} - 1 \right) - a\beta k_i e^{k_i L} \right) A_i - a\chi_b \tilde{\Delta}_b = 0 \tag{9.16}$$

A.2 Parameter normalizations

The equations above can be rendered into dimensionless form using the following rescalings: $\bar{s} = s/L$, $\bar{\Delta} = \Delta/a_0$, $\bar{f} = a_0 L^2 f/\kappa$, $k_{\bar{b}} = a_0^2 L k_b/\kappa$, and $\bar{\xi}_b = a_0^2 L \omega \xi_b/\kappa$. This choice results in the additional rescalings $\bar{\chi} = a_0^2 L^2 \chi/\kappa$ and $\bar{\beta} = a_0 L \beta/\kappa$.

References

[1] P.V. Bayly, K.S. Wilson, Equations of interdoublet separation during flagella motion reveal mechanisms of wave propagation and instability, Biophys. J. 107 (7) (October 2014) 1756–1772.

[2] P.V. Bayly, K.S. Wilson, Analysis of unstable modes distinguishes mathematical models of flagellar motion, J. R. Soc. Interface 12 (106) (May 2015).

[3] M. Bessen, R.B. Fay, G.B. Witman, Calcium control of waveform in isolated flagellar axonemes of *Chlamydomonas*, J. Cell Biol. 86 (2) (August 1980) 446–455.

[4] R. Bower, D. Tritschler, K. Vanderwaal, C.A. Perrone, J. Mueller, L. Fox, W.S. Sale, M.E. Porter, The N-DRC forms a conserved biochemical complex that maintains outer doublet alignment and limits microtubule sliding in motile axonemes, Mol. Biol. Cell (February 2013).

[5] C.J. Brokaw, Bend propagation by a sliding filament model for flagella, J. Exp. Biol. 55 (2) (October 1971) 289–304.

[6] C.J. Brokaw, Computer simulation of flagellar movement. I. Demonstration of stable bend propagation and bend initiation by the sliding filament model, Biophys. J. 12 (5) (May 1972) 564–586.

[7] C.J. Brokaw, Computer simulation of flagellar movement X: doublet pair splitting and bend propagation modeled using stochastic dynein kinetics, Cytoskelet. (Hob. N.J.) 71 (4) (April 2014) 273–284.

[8] S. Camalet, F. Jülicher, Generic aspects of axonemal beating, New J. Phys. 2 (2000) 241–2423.

[9] S. Camalet, F. Jiilicher, J. Prost, Self-organized beating and swimming of internally driven filaments, Phys. Rev. Lett. (1999).

[10] V.F. Geyer, P. Sartori, B.M. Friedrich, F. Julicher, J. Howard, Independent control of the static and dynamic components of the *Chlamydomonas* flagellar beat, Curr. Biol. 26 (8) (April 2016) 1098–1103.

[11] A. Hilfinger, F. Jiilicher, The chirality of ciliary beats, Phys. Biol. 5 (1) (2008) 016003.

[12] M. Hines, J. J Blum, On the contribution of moment-bearing links to bending and twisting in a three-dimensional sliding filament model, Biophys. J. 46 (5) (November 1984) 559–565.

[13] H.J. Hoops, G.B. Witman, Outer doublet heterogeneity reveals structural polarity related to beat direction in *Chlamydomonas* flagella, J. Cell Biol. 97 (3) (September 1983) 902–908.

[14] T. Ishikawa, Cryo-electron tomography of motile cilia and flagella, Cilia 4 (1) (2015) 3.

[15] J. Lin, K. Okada, M. Raytchev, M.C. Smith, D. Nicastro, Structural mechanism of the dynein power stroke, Nat. Cell Biol. 16 (5) (May 2014) 479–485.

[16] C.B. Lindemann, A "geometric clutch" hypothesis to explain oscillations of the axoneme of cilia and flagella, J. Theor. Biol. 168 (2) (1994) 175–189.

[17] C.B. Lindemann, A model of flagellar and ciliary functioning which uses the forces transverse to the axoneme as the regulator of dynein activation, Cell Motil. Cytoskelet. 29 (2) (1994) 141–154.

[18] K.E. Machin, Wave propagation along flagella, J. Exp. Biol. 35 (4) (December 1958) 796–806.

[19] I.H. Riedel-Kruse, A. Hilfinger, J. Howard, F. Jiilicher, How molecular motors shape the flagellar beat, HFSP J. 1 (3) (September 2007) 192–208.

[20] P. Sartori, Effect of Curvature and Normal Forces on Motor Regulation of Cilia (Ph.D. thesis), September 2015.

[21] P. Sartori, V.F. Geyer, J. Howard, F. Jiilicher, Curvature regulation of the ciliary beat through axonemal twist, Phys. Rev. E 94 (4) (October 2016) 042426.

[22] P. Sartori, V.F. Geyer, A. Scholich, F. Julicher, J. Howard, Dynamic curvature regulation accounts for the symmetric and asymmetric beats of *Chlamydomonas* flagella, eLife 5 (May 2016).

[23] P. Satir, Studies on cilia : II. Examination of the distal region of the ciliary shaft and the role of the filaments in motility, J. Cell Biol. 26 (3) (September 1965) 805–834.

[24] R.A. Segal, B. Huang, Z. Ramanis, D.J. Luck, Mutant strains of *Chlamydomonas reinhardtii* that move backwards only, J. Cell Biol. 98 (6) (May 1984) 2026–2034.

[25] E.F. Smith, P. Yang, The radial spokes and central apparatus: Mechano-chemical transducers that regulate flagellar motility, Cell Motil. Cytoskelet. 57 (1) (2003) 8–17.

[26] K.E. Summers, I.R. Gibbons, Adenosine triphosphate-induced sliding of tubules in trypsin-treated flagella of sea-urchin sperm, Proc. Natl Acad. Sci. USA 68 (12) (December 1971) 3092–3096.

II

Dynein Dysfunction and Disease

In this chapter

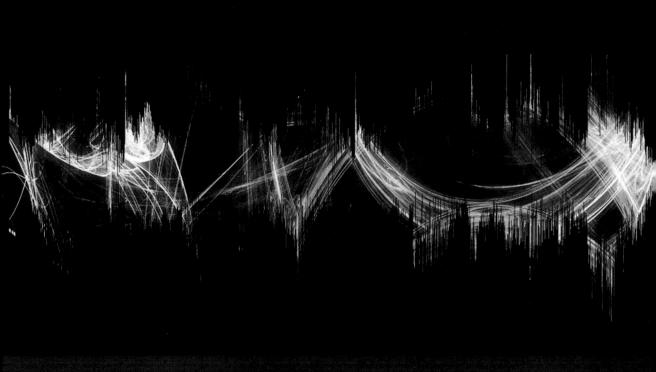

Impacts of virus-mediated manipulation of host Dynein

Miroslav P. Milev[1], Xaojian Yao[2], Lionel Berthoux[3],
Andrew J. Mouland[4,5]

[1]Concordia University, Montréal, QC, Canada; [2]University of Manitoba, Winnipeg, MB, Canada; [3]Université du Québec à Trois-Rivières, QC, Canada; [4]Lady Davis Institute at the Jewish General Hospital, Montréal, QC, Canada; [5]McGill University, Montréal, QC, Canada

10.1 Dynein and viral replication

The Dynein complex is a microtubule (MT)-associated protein complex that mediates retrograde transport of macromolecules in the cytoplasm [2]. It is a 1.6-MDa complex built around two copies of ATPase energy-generating subunits called Dynein heavy chains (DHCs). Two Dynein intermediate chains (DICs) and two light intermediate chains (LICs) bind directly to the DHCs. Three Dynein light chains (DLCs) serve as Dynein adapter proteins, such as DLC 1 (DYNLL1, LC8, DLC1), DLC Tctex-type 1 (DYNLT1), and p150[Glued], which have been implicated in cargo recruitment to Dynein complex during retrograde transport [3–6]. More details on Dynein function are found below and in accompanying chapters. Dynein involvement in virus replication has been well studied. Usually investigations have centered on major questions of viral capsid translocation in the cell. For neurotropic viruses such as herpesviruses that require transit through long distances in differentiated and extended cells this is intuitively obvious [7,8]. In this case, herpesvirus will interact on de *novo infection* and after gaining entry in the cell with Dynein subunits to transit toward the nucleus. The requirements for active transport are equally substantial for several other viruses such as human immunodeficiency virus type 1 (HIV-1) and recent investigations into other retroviruses such as murine leukemia virus (MLV), and reviewed extensively in Ref. [9–11]. Previous studies have implicated DYNLL1 and DYNLT1 proteins in different aspects of virus replication (reviewed in Ref. [12]). For example, DYNLL1 interacts with a rabies virus phosphoprotein and contributes to the viral gene expression [13–15]. Also, DYNLL1 interacts with the CA protein of bovine immunodeficiency virus (BIV) and contributes to retrograde transport [3]. DYNLT1 interacts with the CA protein of human papilloma virus type-16 (HPV-16) and contributes to the HPV-16 replication at an unknown replication step(s) [16]. Unlike the DYNLL1 and DYNLT1, the potential involvement of p150^{Glued} in the replication of viruses is currently not very clear. However, the component of the Dynein complex, Dynactin associates to p150^{Glued} [17] and has been implicated in viral replication and the retrograde transport of viruses [6,18,19].

10.2 Kinesins

The activity of kinesins co-opted by viruses is worthy of note. The tug-of-war that exists opposing motors, Kinesin and Dynein motors, is a precept that is generally accepted in the cell biology field. Morphological switches mediated by specific interactions with organelles and motor proteins will likely determine trafficking polarity in cells and the winner in the Dynein–Kinesin tug-of-war [8,20,21] especially since both motor proteins occupy intracytoplasmic vesicles, and virus capsids too, and cooperate to determine directionality [8,10,22]. This is clearly the case for retroviruses such as Mason–Pfizer Simian Virus (M-PSV) that interacts with the DLC (DYNLT1) Tctex-1 to target to intracellular viral assembly sites adjacent to the nucleus, at the microtubule organizing center (MTOC) [23,24]. In contrast to the directed capsid assembly of the lentivirus, HIV-1, at the plasma membrane, capsid assembly of M-PSV occurs near the MTOC. A morphogenetic switch, mediated by an interaction between de novo synthesized Gag and DYNLT1, allows for M-PSV to acquire bilipid envelope at the plasma membrane, likely by the activity of MT + end motors such as Kinesin [1,24]. Interestingly, a single point mutation in the Dynein-binding domain not only prevents DYNLT1 binding, but also changes the site of assembly to that exhibited by HIV-1, by targeting viral capsid assembly to the plasma membrane. The presence of viral proteins such as retroviral Gag as well as host proteins on endosomal membranes may constitute cargo that is available for directed transport on MTs toward viral assembly sites. Interactions of viral proteins with Dynein mediators such as Lissencephaly-1 (LIS-1) and other host proteins [25–27] could also provide information to switch from a minus-end to plus-end motor directed traffic. Curiously, the HIV-1 regulatory protein, Tat, enhances MT polymerization [28] and interacts with LIS-1 protein [29], whereas HIV-1 Rev, Vaccinia virus, and ASFV were reported to break down MTs [30–32]. Recent work demonstrates in fact that PKA phosphorylates cytoplasmic Dynein at a novel site in the Dynein light intermediate chain 1 (LIC1) that is essential for Dynein binding to the hexon capsid subunit and for virus motility [33,34]. This posttranslational modification mediates an adenovirus-derived switch that promotes late endosome/lysosome dispersal thereby inducing a tempered specific dispersal of late endosome/lysosomes. This was shown to be due to the disruption of the LIC1 interaction with the Rab7-interacting lysosomal protein (RILP). At the other end of the virus replication cycle, late events during HIV-1 replication for example are also characterized by virus-mediated disruptions in membrane trafficking. During viral egress, HIV-1 commandeers late endosome/lysosomes to promote outbound trafficking of viral components along with late endosome/lysosomes-associated factors such as mTORC1 and the viral RNA (Alessandro Cinti & A.J. Mouland, unpublished; [35,36]). Co-opting these membranes that bear characteristics of late endosome/lysosomes, containing LAMP-1 and other integral membrane markers, may have the obvious result of promoting correct targeting to virus assembly domains. Topologically, the eventual fusion of these vesicles at the plasma membrane should result in localization to the inner plasma membrane followed by outward, virus budding (see Ref. [1]). Interestingly, another recent study demonstrated that phagosome maturation was impaired in HIV-1–infected human macrophages. Testing pathogen clearance in these cells, it was found that the accessory Vpr gene product interacted with EB1 (plus-end MT tracking proteins), p150^{Glued}, and the DHC thereby blocking EB1 loading on Dynein motors on the plus ends of MTs that is necessary for phagosome maturation [9,37] (Fig. 10.1). These observations will necessarily have a major bearing on pathogenesis but as well as on the cell's ability to ward off other opportunistic infections, which is the case in mid- to late-stage infections. The effects of HIV-1 Vpr in this case might be reflected in patients as opportunistic coinfections take hold in largely untreated or drug-naive patients in the developing world [38]. Lastly, the RILP–Dynein interaction (Fig. 10.1) was found to be critical for HCV intracellular trafficking. Virus-mediated cleavage of RILP generates an N-terminally truncated protein that dissociates the Dynein motor from viral cargo over the course of infection. The resulting preference for outbound trafficking of viral cargo promotes the transition from the intracellular-predominant to the secretion-predominant phenotype that corresponds to the production of the cleaved RILP [39]. See Table 10.1.

10.3 Innate immunity, the Rabs, Rab7-interacting lysosomal protein, and vesicular transport

Following acute infection, the host cells mount a formidable innate response to infection that is not specific and usually leads to a burst of antiviral cytokine and interferon gene expression. This response includes the activity of well-described pattern recognition receptors and ensuing signaling to transcriptional activation of interferon stimulated genes, cytokines, and interferons. However, as immune-cell tropic viruses target T, B, and myeloid cell types, immunity is compromised [40]. Moreover, responses to acute infection in macrophages as well as other host cells such as dendritic cells appear to be compromised by a variety of mechanisms. One is a virus-mediated defect in viral antigen processing. By preventing autophagosome peptide generation and the presentation of antigen at the cell surface represents one way in which HIV-1, and likely other viruses, evade the host recognition [41]. Small Rab GTPases generally spike and define vesicles in the cell from early to late endosomes. Early endosomes will acidify following endocytosis and late endosomes that are generally actively directed to juxtanuclear positions carry Rab5 and then exchange with Rab7.

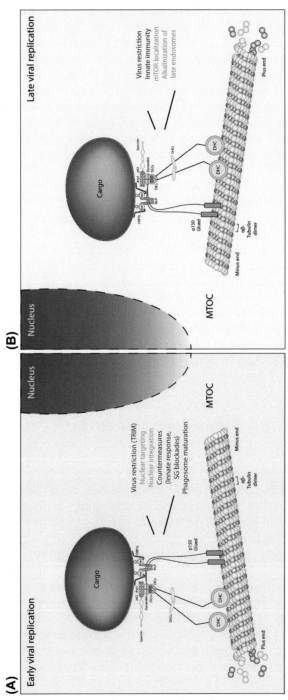

(A)

Early viral replication

Nucleus

MTOC

Cargo

Plus end

Minus end

α/β-
Tubulin
dimer

p150
Glued

Spectrin

DLC8

DHC

DHC

p62 Arp1

Dynamin

DLC1

CP/PIL

RILP

Virus restriction (TRIM)
Nuclear targeting
Nuclear integration
Countermeasures
(Innate response,
SG blockades)
Phagosome maturation

(B)

Nucleus

Late viral replication

Virus restriction
Innate immunity
mTOR localization
Alkalinization of
late endosomes

MTOC

Cargo

Minus end

Plus end

α/β-
Tubulin
dimer

p150
Glued

Spectrin

Arp1 p62

Dynamin
DLC1

DLC8

CP/PIL

RILP

DLC8

DHC

DHC

Figure 10.1 **Early and late virus replication steps are characterized by virus subversion of Dynein and associated factors.** During the replication cycle, viral particles (i.e., viral proteins and complexes) need to be transported within host cells. In the early (A) and late (B) (postreplication) stages, the trafficking of viral components relies on the host cell cytoskeleton and molecular motors. Viruses require minus-end–directed Dynein motor complex for retrograde transport on microtubules (from cell surface toward the nucleus) (A) and Dynein in addition to kinesins for the anterograde transport toward the plasma membrane (B). By interacting with different Dynein complex factors, viruses can positively (in green) or negatively (in red) regulate a number of cellular processes including but not limited to viral restriction, nuclear targeting and integration, autophagosome maturation, innate immunity, alkalinization, and localization of late endosome/lysosomes and mTOR positioning. Details on the specific viral–host (Dynein complex) interactions are discussed/provided in the text and listed in Table 10.1.

Dyneins

Table 10.1 Viruses that interact with Dynein components during replication to 2017

Virus[a]	Virus interaction domain	Dynein component	References
Adeno-Associated virus	Capsid	Dynein	[95]
Adenovirus	Capsid	Dynein light and light intermediate chain IC and LIC1	[95–101]
	FIP-1	TCTEL1	
African Swine Fever virus	P54 (13 amino acid motif)	DLC-8	[19,102]
Bovine Immunodeficiency virus	Capsid	DLC-8	[3]
Ebolavirus	VP35	DLC-8	[103]
Hantaan virus	Unknown	Unknown	[104]
Hepatitus B virus	Unknown	Unknown	[105]
Hepatitus C virus	Unknown	Unknown	[106].
Hepatitus E virus	Vp13	Dynein	[107]
Herpes Simplex Virus 1	VP26 capsid	RP3, Tctex-1	[6,8,48,108–112]
Equine Herpesvirus	Unknown	Unknown (acetylated tubulin)	[82]
Kaposi's Sarcoma Herpesvirus (HHV-8)	Unknown	Unknown (acetylated tubulin)	[83]
Mason-Pfizer Simian virus	Matrix	DYNLT1 (Tctex-1)	[23,24],
Bovine Immunodeficiency virus	Capsid	LC8	[3]
Human Foamy Virus	Capsid	LC8	[49]
Human immunodeficiency virus type 1	Preintegration complex	Dynein	[35,50,113]
Human papillomavirus (HPV)	MINOR CAPSID L2	DYNLT1 and DYNLT3	[114,115]
Poliovirus	Receptor CD155	Tctex-1	[116–119]
Rabies virus	P Phosphoprotein	DYNLL (LC8)	[13–15,120–122]
Sirevirus	Hopie Gag extension	DLC-8	[123]
Vaccinia virus	Unknown	Unknown	[32,124,125];
Rabies virus	polymerase L	DLC-1	[126]
Murine leukemia virus (MLV)	Preintegration complex	p50	[56,127]
Porcine circovirus (PCV2)	capsid (Cap)	DIC-1, Dynein light chain, DYNLL1	[60,128]
Ebola virus (EBOV)	VP35	Dynein light chain (LC8)	[103]
Human immunodeficiency virus type 1	Vpr	Dynein light chain protein, DYNLT1,	[129]

Continued

Table 10.1 Viruses that interact with Dynein components during replication to 2017—cont'd

Virus[a]	Virus interaction domain	Dynein component	References
Human immunodeficiency virus type 1	Vpr	EB1, p150[Glued], and DHC	[37]
Human immunodeficiency virus type 1	Integrase (IN)	Dynein light chain 1 (DYNLL1)	[78]
Hepatitus C virus	Unknown	RILP	[39]
Influenza A virus	Unknown	Dynein and dynactin	[130]
Foot-and-mouth disease virus (FMDV)	Nonstructural protein 3A	Dynactin 3	[131]
Mouse Polyomavirus (MPyV)	Unknown	Dynein	[132]
Rhesus rhadinovirus (RRV)	Unknown	Dynein	[133]
Pseudorabies virus (PRV)	Viral protein 1/2 (VP1/2)	DIC and p150[Glued]	[7]

[a]Highlighted columns indicate listings since first edition of this chapter (1). Most viruses are discussed in text.

Rab7-associated proteins such as RILP play particular roles in mediating the interaction of cargo with the Dynein motor via its association to p150[Glued]. RILP appears to be central to a number of processes involving intracellular trafficking of late endosome/lysosomes, membrane fusion events (e.g., endoplasmic reticulum (ER)–late endosome) and as described below, in immune responses and antigen presentation during virus infection. It remains a valuable tool in cell biology studies as the overexpression leads to the redistribution of late endosomes/lysosomes to the juxtanuclear region, the MTOC. This observed phenotype is evidence of a direct role in tethering the Dynein motor via the Dynein subunit p150[glued] and Rab7 expressed on late endosome/lysosomes. Expression of Dynein-interacting mutants leads to the release of late endosome/lysosomes with evidence that virus cargo is affected by RILP mutagenesis, to likely involve Dynein motor function. This was brought to the forefront of immunology in light of the involvement of Dynein in the presentation of HLA Class II antigen presentation. In multiple sclerosis for instance in which there is a susceptibility locus, late endosomal vesicle biogenesis, and HLA Class antigen presentation were shown to be dependent on CLEC16. CLEC16's interaction with RILP and the homotypic fusion and protein sorting–tethering complex [42] and perturbed Dynein retrograde transport led to a blockade of the recruitment of MHC Class II antigens to the perinuclear region. This is mentioned here to demonstrate that while a defect in the functioning of the host gene leads to human disease, MHC Class II antigen presentation is also crucial to mount a defense against HIV-1 and other viruses. Antigens derived from within infected cells, in contrast to extracellular delivery of virus-specific antigens from virally produced cells, are also shuttled intracellularly. Defects induced by virus infection lead to intracellular trafficking deficiencies and in this type of scenario, a role for autophagic vesicles was brought to light in dendritic cells to activate HIV-1 specific CD4+ T cells [41]. While autophagy was not critical to Ag presentation to CD4+ T cells, targeting of HIV-1 specific antigens to autophagosomes nevertheless enhances virus-specific T cell antigen presentation. The MT-associated dendritic cell protein L chain (LC3) played a key role in this event, likely propelled by the Dynein motor for recruitment to juxtanuclear membrane fusion and autophagy processes.

10.4 Dynein, viruses, and the innate immune response

Many viruses hijack the cellular machinery for their retrograde transport during the early stages of their replication cycle, as was reviewed previously by us and others [1,10,43,44]. In particular, multiple viruses subvert the MT network and MT-associated molecular motors; examples include adenoviruses [45,46], parvoviruses [47], herpes simplex virus [8,48], and retroviruses [49]. HIV-1 interactions with MTs and Dynein in the early infection stages have been relatively thoroughly investigated. A landmark live cell microscopy study that made use of fluorescently labeled individual viral particles yielded fascinating movies of HIV-1 traveling along MTs en route toward the MTOC [50]. Following MT-dependent transport, HIV-1 cores accumulate in the vicinity of the MTOC and then at nuclear pores [50–52].

Inspired by previous findings with other viruses, HIV-1 retrograde transport was proposed early on to be mediated by the Dynein motor, as evidenced by an accumulation of viral particles in the cell periphery following microinjection of anti-Dynein antibodies [50]. However, these results were descriptive in nature, offering no link between Dynein and MT functions and the infectivity of incoming HIV-1. Indeed, when functional studies were undertaken, they yielded rather conflicting results. In HeLa cells, for instance, none of several genetic or pharmacological approaches to interfere with Dynein or MTs had a significant impact on infectivity, even when they had a clear effect on HIV-1 intracellular movements [53–55]. Treatment with nocodazole and depletion of the DHC similarly did not affect the infectivity of another lentivirus, SIVmac, in human and nonhuman primate cells [53,54], and also had no effect on the infectivity of MLV, a gammaretro virus. We take these sobering observations as evidence that identifying the specific cytoskeleton and transporter subunit components involved is necessary to fully understand the mechanisms of retrograde transport for any given virus. Indeed, a specific DLC, DYNLRB2, was recently found to mediate MLV retrograde transport, and its expression levels correlated with MLV infectivity [56]. Likewise, a kinesin-1 adaptor, Fasciculation and Elongation Factor zeta 1 was recently found to be important for HIV-1 retrograde transport and infectivity [57]. In addition, not all MTs are alike, and HIV-1 seems to promote the formation of and then use "stable" MTs, characterized by a state of decreased polymerization/depolymerization, which renders them resistant to nocodazole (explaining why multiple investigators saw little to no effect of nocodazole on HIV-1 infectivity, and see below) [58]. Human herpes simplex virus 1 is transported on long distances toward the nucleus of sensory neurons through an association with MTs initiated by binding to protein complexes called "+TIPs," which comprise the Dynein interactor dynactin-1 [59]. The capsid protein of circoviruses interacts with the intermediate chain 1 of Dynein complexes for its retrograde transport [60]. Thus, viruses accomplish retrograde transport by interacting with Dynein motors, and possibly other motor complexes, using a variety of molecular mechanisms. Some viruses may modulate the dynamics and functions of MTs themselves, in addition to hijacking molecular motors.

The reliance of viruses on MT-dependent retrograde transport provides a weak spot that may be exploited by innate immune mechanisms to target these intracellular parasites. Indeed, several type 1 interferon-induced innate effectors act in the early postentry stages of viral infections [61]. One would expect some of these effectors to intercept viruses while they associate with MTs. This idea has been explored in the context of the inhibition of retroviruses by the tripartite motif (TRIM) protein family member TRIM5α. This E3 ubiquitin ligase forms cytoplasmic bodies in which inhibition-sensitive viruses are trapped, disassembled, and degraded [62]. Because these large cytoplasmic structures resemble aggresomes, their interactions with MTs were investigated soon after the discovery of their antiretroviral properties. It was found that indeed, TRIM5α cytoplasmic body subcellular localization is dependent on functional MTs [63]. Accordingly, the presence of functional MTs and Dynein motors was shown to be important for the inhibition of retroviruses by TRIM5α [54,55]. TRIM proteins form a large family of proteins of which many have antiviral properties [64] and, on the other hand, often show an affinity for MTs [65]. Therefore, it is entirely possible that TRIM5α is the tip of the iceberg and that more examples of molecular motors and MTs being essential to the antiviral roles of the TRIM proteins will be uncovered in the future.

10.5 IFITM3 and VAP-A

An additional restriction factor, that is also interferon-inducible, whose function depends on Dynein and Dynein-associated factors, is the interferon-inducible transmembrane protein 3 (IFITM3). IFITM3 has antiviral activity and depends on lipid metabolism in the cell. While many viruses depend on lipid scaffolds such as lipid rafts for assembly, viruses also depend on cholesterol dynamics for early and late infection. Interestingly, perturbation of cholesterol homeostasis can block viral entry steps (e.g., Ref. [66]). IFITM3 was shown to interact with VAP-A, a membrane-associated protein on the ER, on tight junctions and synaptic vesicles. Humans contain two VAP genes, VAP-A and VAP-B. VAP-A, as well as B, has an MSP domain at the N terminus, followed by a coiled coil region and ending in transmembrane domain that anchors the protein into the ER membrane. The MSP region has been shown to interact with FFAT domains commonly found on oxysterol binding proteins cholesterol sensor proteins, such as ORP1L. VAP-A has also been shown to interact with viral proteins from both HCV and Norwalk viruses and has been shown to modulate viral transport [67,68]. Prosser et al. determined that VAP-A creates a block in vesicular trafficking from the ER to the Golgi network and resolved the block by overexpressing VAP-A as well as mutants of this protein [69]. The overexpression also had dramatic effects on IFITM3 restriction activity, in that this blocked its interaction with ORP1L [70] thereby preventing the fusion of intraluminal virion-containing vesicles with endosomal membranes and thereby blocking virus release.

Interestingly, Rocha et al. [71] found that purified VAP-A was able to remove the C25 fragment of p150[glued] from preassembled, and purified, Rab7–RILP–ORP1L complexes on metal affinity beads. Through in vivo experiments they determined that late endosome/lysosomes were often in contact with the ER membrane and the conformation of ORP1L could dictate whether there were more or less membrane contact sites. When cholesterol is high, ORP1L takes on a conformation where the FFAT domain is sequestered and therefore cannot interact with ER-bound VAP-A. As a consequence, Dynein–p150[glued]–RILP promotes juxtanuclear localization of late endosome/lysosomes. In contrast, Dynein function is blocked when ORP1L is in a low cholesterol conformation with its FFAT domain exposed to interact with the MSP domain of the ER-associated, VAP-A. Importantly, HIV-1 subverts the cholesterol sensing mediated by ORP1L by maintaining peripheral localization of late endosomes/lysosomes not only impacting on HIV-1 protein and RNA localization but also on mTORC1 positioning and activity [71a] (Fig. 10.1).

As an integral component of late endosome/lysosomal membranes, the small GTPase, Rab7, which associates to RILP and the Dynein p150[Glued], was shown to play a critical role in viral restriction mediated by the bone marrow stromal antigen 2 or Tetherin host protein. Tetherin, a lipid raft-associated protein that is encoded by the *bst2* gene, is the last restriction factor discussed in this chapter. Tetherin, as the name implies promotes tethering of mature virus particles to the outer plasma membrane. Interestingly, HIV-1 and other viruses counter the innate activity of this host factor to allow for budding viruses to be released from the cell surface. While intense research into this factor has yielded almost a complete understanding of structure and mechanism of action, recent work has now identified a role of the Dynein effector, Rab7a, in enhancing Tetherin restriction activity in virus release and budding [34]. Similar to findings for HCV, this discovery suggests that endosomal sorting somehow influences the virus restriction activity of Tetherin, mediated by Dynein, thereby promoting both egress of late endosome/lysosomes and mature virus particle release from the cell surface, but this relationship will require further substantiation.

10.6 Dyneins and nuclear integration of viral DNA

On entry into the host cell, viruses translocate within the cytoplasm to sites of replication, to a perinuclear region and can use this as a scaffold to enter the nucleus in viruses that require a nuclear intermediate. In a manner identical to host cell vesicles or macromolecular complexes that are transported intracellularly, viral translocation is not achieved by passive diffusion but by energy-dependent, active transport mechanisms. Many viruses are transported along MTs and interact with various cellular cofactors including factors associated to the Dynein motor complex, as described herein. During the early stages of HIV-1 replication following acute infection for instance, the viral genomic RNA is reverse-transcribed into a complementary DNA and forms a preintegration complex, which undergoes intracytoplasmic retrograde transportation and nuclear import, and subsequently integrates into the host cell genome (reviewed in Ref. [72,73]). During these processes, different viral proteins interact with and utilize various cellular proteins for replication. Genome-wide si/shRNA screening and other functional studies have uncovered a large number of host proteins with putative roles in HIV-1 replication (reviewed in Ref. [74–76]). However, molecular events associated with HIV-1 retrograde transport in the cytoplasm are still not well understood. Interestingly, several studies have indicated that a functional Dynein complex or intact MT network is essential for efficient HIV-1 uncoating following entry [53,77,78] and retrograde transport toward the nucleus prior to nuclear integration [50,77,79].

Recent studies have demonstrated that the inhibition of DHC (DYNC1H1) by siRNA or the disruption of the intact MTs by nocodazole treatment delayed the uncoating process during HIV-1 infection [53,77]. This suggests that the Dynein complex and intact MT network facilitate the HIV-1 uncoating [53,77]. Another study showed that the DYNLL1-KD or the disruption of HIV-1 integrase/DYNLL1 interaction resulted in a significant loss of reverse transcription and an increase in the rate of HIV-1 uncoating [78]. However, it is still unclear whether it is the Dynein complex–associated DYNLL1 that promotes HIV-1 reverse transcription and uncoating. Interestingly, a previous study suggested that DYNLL1 may not be able to mediate cargo recruitment to the Dynein complex [80]. Therefore, further investigation is required to elucidate the roles of DYNLL1 and Dynein complex or MT network in HIV-1 replication.

The requirement of HIV integrase/DYNLL1 for the proper uncoating of HIV-1 and its reverse transcription has also been demonstrated [78]. However, the molecular mechanisms of this virus–host interaction that contributes to the proper uncoating of HIV-1 and reverse transcription will require further characterization. DYNLL1 has been shown to bind to a number of cellular proteins and facilitates protein complex formation [81]. Therefore, it can be speculated that cellular proteins may be recruited to the reverse transcription complex via DYNLL1 interaction, which helps HIV-1 RTC reorganization and/or stabilization, and consequently contributes to proper uncoating and/or efficient reverse transcription. Conversely, the disruption of HIV-1 integrase/DYNLL1 interaction could lead to aberrant uncoating and the formation of unstable RTC, resulting in low levels of HIV-1 reverse transcription. These findings provide evidence for a possible alternative mechanism by which HIV-1 integrase facilitates the proper uncoating and efficient viral reverse transcription.

10.7 Posttranslationally modified microtubules and Dynein

A major theme in virus-cell biology is the ability of the virus to commandeer factors to enable all aspects of the replication cycle. A variety of studies have focused attention on how viruses traffic on cytoskeletal elements. Notable interactions include motor protein translocation on MTs with cargo, including vesicles and viral components such as RNPs and incoming capsids. Indeed, these are critical to the establishment of infection. Nevertheless, recent work has uncovered that viruses will utilize more stable MTs that are modified posttranslationally, either by acetylation or tyrosination to confer enhanced stability and perhaps secure efficient viral targeting. This would contribute to the delivery of viral components to destinations in the cells during both ingress and egress. For example, resistance to the effects of the MT destabilizing agent nocodazole as mentioned above led to investigations on characterizing subpopulations of MTs that resisted nocodazole treatment. Indeed, several herpesviruses were found associated to acetylated MTs [82,83]. Recent evidence for retroviruses also demonstrates that there is a preference for acetylated and detyrosinated stable MTs early in infection [58], for which Dynein activity was shown to be critical [50]. As such, recent work on the specific roles of these posttranslationally modified MTs [84,85] highlights earlier work from several groups that demonstrate in vivo and in live cell experiments that retroviruses require late endosome/lysosomes for intracellular trafficking [35,36,86]. Future work focusing on the selectivity achieved by viruses for subpopulations of MTs during both ingress and egress should be informative.

10.8 Emerging viruses and co-opting of Dynein

Recent global threats to human health from emerging viruses have fueled intense efforts into the understanding of virus molecular and cellular biology. Emerging viruses include Ebola, severe acute and Middle East respiratory syndrome Coronavirus (SARS/MERS-CoV), Zika, and Chikungunya viruses, among others [87]. Despite new vaccines that are in the pipeline [88–90], scant new information on these viruses exist, especially in regard to how they commandeer host Dynein and associated machineries. For example, Ebola virus was shown to interact with the highly conserved 8 kDa cytoplasmic light chain (LC8), like many viruses ([91]; reviewed in Ref. [1]) but few details at the time were available. Current evidence indicates that this interaction indeed enhances RNA replication of this highly pathogenic virus [91]. Furthermore, Ebola replication was found to be sensitive Dynein as the expression of mutants of the cholesterol sensor, ORP1L, which acts as a regulator of Dynein motor binding to RILP, dramatically affected virus output [92]. In contrast, scant evidence for an association between SARS-CoV and Dynein motor complexes is available. Indeed, the SARS-CoV Envelope membrane–associated Envelope protein brought down the DHC, among other proteins; but little else is understood about this potential interaction [93]. Members of the flaviviridae including West Nile, HCV, Dengue, and Zika among others have each their own prevalence worldwide, the last two of which are considered as the current emerging threats to human health [94], in North America and elsewhere. There are little data on virus–host interactions between Zika and components of the Dynein machinery in mid-2017. Nevertheless, as Zika infects several cell types including neuronal cell lineages, and its study does not require high-level biosafety containment, a role for Dynein, as is the case for HCV [39], in executing virus-mediated programs will likely be revealed through current and intense further investigation.

Major outstanding questions

In light of the characterized interplay between viruses and molecular motor proteins such as Dynein and Kinesin, several questions remain, but are not necessarily readily answered. These include:

- Are virus–Dynein interactions druggable in selective and specific ways?
- Can the targeting of virus–Dynein interactions help in enhancing immune responses to infection?
- In what other ways do viruses commandeer intracellular trafficking machineries for early and late replication events?
- Is the selective commandeering of stable populations of microtubules a general feature of virus-mediated subversion of the host cell?

Acknowledgments

We would like to thank Shringar Rao for editorial input and Michael Sacher for critical contributions. This work was supported in part by grants from the Canadian Institutes of Health Research to X.Y., L.B., and A.J.M.

References

[1] A.J. Mouland, M.P. Milev, Role of dynein in viral pathogenesis, in: S.M. King (Ed.), Dynein: Structure, Biology and Disease, Elsevier, Inc., 2012, pp. 561–583.
[2] R. Mallik, S.P. Gross, Molecular motors: strategies to get along, Curr. Biol. 14 (22) (2004) R971–R982.

[3] Y. Su, W. Qiao, T. Guo, J. Tan, Z. Li, Y. Chen, et al., Microtubule-dependent retrograde transport of bovine immunodeficiency virus, Cell Microbiol. 12 (8) (2010) 1098–1107.

[4] C. Navarro, H. Puthalakath, J.M. Adams, A. Strasser, R. Lehmann, Egalitarian binds dynein light chain to establish oocyte polarity and maintain oocyte fate, Nat. Cell Biol. 6 (5) (2004) 427–435.

[5] K.W. Lo, J.M. Kogoy, K.K. Pfister, The DYNLT3 light chain directly links cytoplasmic dynein to a spindle checkpoint protein, Bub3, J. Biol. Chem. 282 (15) (2007) 11205–11212.

[6] K. Dohner, A. Wolfstein, U. Prank, C. Echeverri, D. Dujardin, R. Vallee, et al., Function of dynein and dynactin in herpes simplex virus capsid transport, Mol. Biol. Cell. 13 (8) (2002) 2795–2809.

[7] S.V. Zaichick, K.P. Bohannon, A. Hughes, P.J. Sollars, G.E. Pickard, G.A. Smith, The herpesvirus VP1/2 protein is an effector of dynein-mediated capsid transport and neuroinvasion, Cell Host Microbe 13 (2) (2013) 193–203.

[8] K. Radtke, D. Kieneke, A. Wolfstein, K. Michael, W. Steffen, T. Scholz, et al., Plus- and minus-end directed microtubule motors bind simultaneously to herpes simplex virus capsids using different inner tegument structures, PLoS Pathog. 6 (7) (2010) e1000991.

[9] R. Gaudin, B.C. de Alencar, N. Arhel, P. Benaroch, HIV trafficking in host cells: motors wanted!, Trends Cell Biol. 23 (12) (2013) 652–662.

[10] M.P. Dodding, M. Way, Coupling viruses to dynein and kinesin-1, EMBO J. 30 (17) (2011) 3527–3539.

[11] A. Slonska, R. Polowy, A. Golke, J. Cymerys, Role of cytoskeletal motor proteins in viral infection, Postepy Hig. Med. Dosw. (Online) 66 (2012) 810–817.

[12] J. Merino-Gracia, M.F. Garcia-Mayoral, I. Rodriguez-Crespo, The association of viral proteins with host cell dynein components during virus infection, FEBS J. 278 (17) (2011) 2997–3011.

[13] G.S. Tan, M.A. Preuss, J.C. Williams, M.J. Schnell, The dynein light chain 8 binding motif of rabies virus phosphoprotein promotes efficient viral transcription, Proc. Natl. Acad. Sci. U.S.A. 104 (17) (2007) 7229–7234.

[14] H. Raux, A. Flamand, D. Blondel, Interaction of the rabies virus P protein with the LC8 dynein light chain, J. Virol. 74 (21) (2000) 10212–10216.

[15] N. Poisson, E. Real, Y. Gaudin, M.C. Vaney, S. King, Y. Jacob, et al., Molecular basis for the interaction between rabies virus phosphoprotein P and the dynein light chain LC8: dissociation of dynein-binding properties and transcriptional functionality of P, J. Gen. Virol. 82 (Pt 11) (2001) 2691–2696.

[16] M.A. Schneider, G.A. Spoden, L. Florin, C. Lambert, Identification of the dynein light chains required for human papillomavirus infection, Cell Microbiol. 13 (1) (2011) 32–46.

[17] T.A. Schroer, Dynactin, Annu. Rev. Cell Dev. Biol. 20 (2004) 759–779.

[18] M.F. Engelke, C.J. Burckhardt, M.K. Morf, U.F. Greber, The dynactin complex enhances the speed of microtubule-dependent motions of adenovirus both towards and away from the nucleus, Viruses 3 (3) (2011) 233–253.

[19] C. Alonso, J. Miskin, B. Hernaez, P. Fernandez-Zapatero, L. Soto, C. Canto, et al., African swine fever virus protein p54 interacts with the microtubular motor complex through direct binding to light-chain dynein, J. Virol. 75 (20) (2001) 9819–9827.

[20] M.J. Muller, S. Klumpp, R. Lipowsky, Tug-of-war as a cooperative mechanism for bidirectional cargo transport by molecular motors, Proc. Natl. Acad. Sci. U.S.A. 105 (12) (2008) 4609–4614.

[21] C. Kural, H. Kim, S. Syed, G. Goshima, V.I. Gelfand, P.R. Selvin, Kinesin and dynein move a peroxisome in vivo: a tug-of-war or coordinated movement? Science 308 (5727) (2005) 1469–1472.

[22] N. Mizuno, S. Toba, M. Edamatsu, J. Watai-Nishii, N. Hirokawa, Y.Y. Toyoshima, et al., Dynein and kinesin share an overlapping microtubule-binding site, Embo J. 23 (13) (2004) 2459–2467.

[23] J.N. Sfakianos, R.A. LaCasse, E. Hunter, The M-PMV cytoplasmic targeting-retention signal directs nascent gag polypeptides to a pericentriolar region of the cell, Traffic 4 (10) (2003) 660–670.

[24] J. Vlach, J. Lipov, M. Rumlova, V. Veverka, J. Lang, P. Srb, et al., D-retrovirus morphogenetic switch driven by the targeting signal accessibility to Tctex-1 of dynein, Proc. Natl. Acad. Sci. U.S.A. 105 (30) (2008) 10565–10570.

[25] T. Torisawa, A. Nakayama, K. Furuta, M. Yamada, S. Hirotsune, Y.Y. Toyoshima, Functional dissection of LIS1 and NDEL1 towards understanding the molecular mechanism of cytoplasmic dynein regulation, J. Biol. Chem. 286 (3) (2010) 1959–1965.

[26] M. Yamada, S. Toba, T. Takitoh, Y. Yoshida, D. Mori, T. Nakamura, et al., mNUDC is required for plus-end-directed transport of cytoplasmic dynein and dynactins by kinesin-1, Embo J. 29 (3) (2010) 517–531.

[27] M. Yamada, S. Toba, Y. Yoshida, K. Haratani, D. Mori, Y. Yano, et al., LIS1 and NDEL1 coordinate the plus-end-directed transport of cytoplasmic dynein, Embo J. 27 (19) (2008) 2471–2483.

[28] J. de Mareuil, M. Carre, P. Barbier, G.R. Campbell, S. Lancelot, S. Opi, et al., HIV-1 Tat protein enhances microtubule polymerization, Retrovirology 2 (2005) 5.

[29] N. Epie, T. Ammosova, T. Sapir, Y. Voloshin, W.S. Lane, W. Turner, et al., HIV-1 Tat interacts with LIS1 protein, Retrovirology 2 (2005) 6.

[30] N.R. Watts, D.L. Sackett, R.D. Ward, M.W. Miller, P.T. Wingfield, S.S. Stahl, et al., HIV-1 rev depolymerizes microtubules to form stable bilayered rings, J. Cell Biol. 150 (2) (2000) 349–360.

[31] N. Jouvenet, T. Wileman, African swine fever virus infection disrupts centrosome assembly and function, J. Gen. Virol. 86 (Pt 3) (2005) 589–594.

[32] A. Ploubidou, V. Moreau, K. Ashman, I. Reckmann, C. Gonzalez, M. Way, Vaccinia virus infection disrupts microtubule organization and centrosome function, Embo J. 19 (15) (2000) 3932–3944.

[33] J. Scherer, R.B. Vallee, Conformational changes in the adenovirus hexon subunit responsible for regulating cytoplasmic dynein recruitment, J. Virol. 89 (2) (2015) 1013–1023.

[34] J. Scherer, J. Yi, R.B. Vallee, PKA-dependent dynein switching from lysosomes to adenovirus: a novel form of host-virus competition, J. Cell Biol. 205 (2) (2014) 163–177.

[35] M. Lehmann, M.P. Milev, L. Abrahamyan, X.J. Yao, N. Pante, A.J. Mouland, Intracellular transport of human immunodeficiency virus type 1 genomic RNA and viral production are dependent on dynein motor function and late endosome positioning, J. Biol. Chem. 284 (21) (2009) 14572–14585.

[36] D. Molle, C. Segura-Morales, G. Camus, C. Berlioz-Torrent, J. Kjems, E. Basyuk, et al., Endosomal trafficking of HIV-1 gag and genomic RNAs regulates viral egress, J. Biol. Chem. 284 (29) (2009) 19727–19743.

[37] A. Dumas, G. Le-Bury, F. Marie-Anais, F. Herit, J. Mazzolini, T. Guilbert, et al., The HIV-1 protein Vpr impairs phagosome maturation by controlling microtubule-dependent trafficking, J. Cell Biol. 211 (2) (2015) 359–372.

[38] A. Low, G. Gavriilidis, N. Larke, M.R. B-Lajoie, O. Drouin, J. Stover, et al., Incidence of opportunistic infections and the impact of antiretroviral therapy among HIV-infected adults in low- and middle-income countries: a systematic review and meta-analysis, Clin. Infect. Dis. 62 (12) (2016) 1595–1603.

[39] A.L. Wozniak, A. Long, K.N. Jones-Jamtgaard, S.A. Weinman, Hepatitis C virus promotes virion secretion through cleavage of the Rab7 adaptor protein RILP, Proc. Natl. Acad. Sci. U.S.A. 113 (44) (2016) 12484–12489.

[40] E.P. Scully, A. Lockhart, W. Garcia-Beltran, C.D. Palmer, C. Musante, E. Rosenberg, et al., Innate immune reconstitution with suppression of HIV-1, JCI Insight 1 (3) (2016) e85433.

[41] F.P. Blanchet, A. Moris, D.S. Nikolic, M. Lehmann, S. Cardinaud, R. Stalder, et al., Human immunodeficiency virus-1 inhibition of immunoamphisomes in dendritic cells impairs early innate and adaptive immune responses, Immunity 32 (5) (2010) 654–669.

[42] M.M. van Luijn, K.L. Kreft, M.L. Jongsma, S.W. Mes, A.F. Wierenga-Wolf, M. van Meurs, et al., Multiple sclerosis-associated CLEC16A controls HLA class II expression via late endosome biogenesis, Brain 138 (Pt 6) (2015) 1531–1547.

[43] K. Radtke, K. Dohner, B. Sodeik, Viral interactions with the cytoskeleton: a hitchhiker's guide to the cell, Cell Microbiol. 8 (3) (2006) 387–400.

[44] S. Cohen, S. Au, N. Pante, How viruses access the nucleus, Biochimica Biophys. Acta 1813 (9) (2011) 1634–1645.

[45] M. Suomalainen, M.Y. Nakano, S. Keller, K. Boucke, R.P. Stidwill, U.F. Greber, Microtubule-dependent plus- and minus end-directed motilities are competing processes for nuclear targeting of adenovirus, J. Cell Biol. 144 (4) (1999) 657–672.

[46] P.L. Leopold, G. Kreitzer, N. Miyazawa, S. Rempel, K.K. Pfister, E. Rodriguez-Boulan, et al., Dynein- and microtubule-mediated translocation of adenovirus serotype 5 occurs after endosomal lysis, Hum. Gene Ther. 11 (1) (2000) 151–165.

[47] P.J. Xiao, R.J. Samulski, Cytoplasmic trafficking, endosomal escape, and perinuclear accumulation of adeno-associated virus type 2 particles are facilitated by microtubule network, J. Virol. 86 (19) (2012) 10462–10473.

[48] A. Wolfstein, C.H. Nagel, K. Radtke, K. Dohner, V.J. Allan, B. Sodeik, The inner tegument promotes herpes simplex virus capsid motility along microtubules in vitro, Traffic 7 (2) (2006) 227–237.

[49] C. Petit, M.L. Giron, J. Tobaly-Tapiero, P. Bittoun, E. Real, Y. Jacob, et al., Targeting of incoming retroviral gag to the centrosome involves a direct interaction with the dynein light chain 8, J. Cell Sci. 116 (Pt 16) (2003) 3433–3442.

[50] D. McDonald, M.A. Vodicka, G. Lucero, T.M. Svitkina, G.G. Borisy, M. Emerman, et al., Visualization of the intracellular behavior of HIV in living cells, J. Cell Biol. 159 (3) (2002) 441–452.

[51] N. Arhel, A. Genovesio, K.A. Kim, S. Miko, E. Perret, J.C. Olivo-Marin, et al., Quantitative four-dimensional tracking of cytoplasmic and nuclear HIV-1 complexes, Nat. Methods 3 (10) (2006) 817–824.

[52] A. Zamborlini, J. Lehmann-Che, E. Clave, M.L. Giron, J. Tobaly-Tapiero, P. Roingeard, et al., Centrosomal pre-integration latency of HIV-1 in quiescent cells, Retrovirology 4 (2007) 63.

[53] P. Pawlica, L. Berthoux, Cytoplasmic dynein promotes HIV-1 uncoating, Viruses 6 (11) (2014) 4195–4211.

[54] P. Pawlica, C. Dufour, L. Berthoux, Inhibition of microtubules and dynein rescues human immunodeficiency virus type 1 from owl monkey TRIMCyp-mediated restriction in a cellular context-specific fashion, J. Gen. Virol. 96 (Pt 4) (2015) 874–886.

[55] P. Pawlica, V. Le Sage, N. Poccardi, M.J. Tremblay, A.J. Mouland, L. Berthoux, Functional evidence for the involvement of microtubules and dynein motor complexes in TRIM5α-mediated restriction of retroviruses, J. Virol. 88 (10) (2014) 5661–5676.

[56] T. Opazo, A. Garces, D. Tapia, F. Barraza, A. Bravo, T. Schwenke, et al., Functional evidence of the involvement of the dynein light chain DYNLRB2 in murine leukemia virus infection, J. Virol. 91 (10) (2017).

[57] V. Malikov, E.S. da Silva, V. Jovasevic, G. Bennett, D.A. de Souza Aranha Vieira, B. Schulte, et al., HIV-1 capsids bind and exploit the kinesin-1 adaptor FEZ1 for inward movement to the nucleus, Nat. Commun. 6 (2015) 6660.

[58] Y. Sabo, D. Walsh, D.S. Barry, S. Tinaztepe, K. de Los Santos, S.P. Goff, et al., HIV-1 induces the formation of stable microtubules to enhance early infection, Cell Host Microbe 14 (5) (2013) 535–546.

[59] V. Jovasevic, M.H. Naghavi, D. Walsh, Microtubule plus end-associated CLIP-170 initiates HSV-1 retrograde transport in primary human cells, J. Cell Biol. 211 (2) (2015) 323–337.

[60] J. Cao, C. Lin, H. Wang, L. Wang, N. Zhou, Y. Jin, et al., Circovirus transport proceeds via direct interaction of the cytoplasmic dynein IC1 subunit with the viral capsid protein, J. Virol. 89 (5) (2015) 2777–2791.

[61] N. Merindol, L. Berthoux, Restriction factors in HIV-1 disease progression, Curr. HIV Res. 13 (6) (2015) 448–461.

[62] Z. Lukic, S. Hausmann, S. Sebastian, J. Rucci, J. Sastri, S.L. Robia, et al., TRIM5α associates with proteasomal subunits in cells while in complex with HIV-1 virions, Retrovirology 8 (2011) 93.

[63] F. Diaz-Griffero, X. Li, H. Javanbakht, B. Song, S. Welikala, M. Stremlau, et al., Rapid turn-over and polyubiquitylation of the retroviral restriction factor TRIM5, Virology 349 (2) (2006) 300–315.

[64] R. Rajsbaum, A. Garcia-Sastre, G.A. Versteeg, TRIMmunity: the roles of the TRIM E3-ubiquitin ligase family in innate antiviral immunity, J. Mol. Biol. 426 (6) (2014) 1265–1284.

[65] T.C. Cox, The microtubule-associated C-I subfamily of TRIM proteins and the regulation of polarized cell responses, Adv. Exp. Med. Biol. 770 (2012) 105–118.

[66] M.K. Poh, G. Shui, X. Xie, P.Y. Shi, M.R. Wenk, F. Gu, U18666A, an intra-cellular cholesterol transport inhibitor, inhibits dengue virus entry and replication, Antiviral Res. 93 (1) (2012) 191–198.

[67] G. Xu, X. Xin, C. Zheng, GPS2 is required for the association of NS5A with VAP-A and hepatitis C virus replication, PLoS One 8 (11) (2013) e78195.

[68] K. Ettayebi, M.E. Hardy, Norwalk virus nonstructural protein p48 forms a complex with the SNARE regulator VAP-A and prevents cell surface expression of vesicular stomatitis virus G protein, J. Virol. 77 (21) (2003) 11790–11797.

[69] D.C. Prosser, D. Tran, P.Y. Gougeon, C. Verly, J.K. Ngsee, FFAT rescues VAPA-mediated inhibition of ER-to-Golgi transport and VAPB-mediated ER aggregation, J. Cell Sci. 121 (Pt 18) (2008) 3052–3061.

[70] S. Amini-Bavil-Olyaee, Y.J. Choi, J.H. Lee, M. Shi, I.C. Huang, M. Farzan, et al., The antiviral effector IFITM3 disrupts intracellular cholesterol homeostasis to block viral entry, Cell Host Microbe 13 (4) (2013) 452–464.

[71] N. Rocha, C. Kuijl, R. van der Kant, L. Janssen, D. Houben, H. Janssen, et al., Cholesterol sensor ORP1L contacts the ER protein VAP to control Rab7-RILP-p150 glued and late endosome positioning, J. Cell Biol. 185 (7) (2009) 1209–1225.

[71a] A. Cinti, V. Le Sage, MP. Milev, F. Valiente-Echeverría, C. Crossie, MJ. Miron, N. Panté, M. Olivier, AJ. Mouland. HIV-1 enhances mTORC1 activity and repositions lysosomes to the periphery by co-opting Rag GTPases. Sci Rep. 7 (1) (2017 Jul 14) 5515. doi: 10.1038/s41598-017-05410-0.

[72] K.D. Jayappa, Z. Ao, X. Yao, The HIV-1 passage from cytoplasm to nucleus: the process involving a complex exchange between the components of HIV-1 and cellular machinery to access nucleus and successful integration, Int. J. Biochem. Mol. Biol. 3 (1) (2012) 70–85.

[73] S. Nisole, A. Saib, Early steps of retrovirus replicative cycle, Retrovirology 1 (2004) 9.

[74] P.D. Bieniasz, An overview of intracellular interactions between immunodeficiency viruses and their hosts, Aids 26 (10) (2012) 1243–1254.

[75] K.H. Kok, T. Lei, D.Y. Jin, siRNA and shRNA screens advance key understanding of host factors required for HIV-1 replication, Retrovirology 6 (2009) 78.

[76] B.M. Friedrich, N. Dziuba, G. Li, M.A. Endsley, J.L. Murray, M.R. Ferguson, Host factors mediating HIV-1 replication, Virus Res. 161 (2) (2011) 101–114.

[77] Z. Lukic, A. Dharan, T. Fricke, F. Diaz-Griffero, E.M. Campbell, HIV-1 uncoating is facilitated by dynein and kinesin 1, J. Virol. 88 (23) (2014) 13613–13625.

[78] K.D. Jayappa, Z. Ao, X. Wang, A.J. Mouland, S. Shekhar, X. Yang, et al., Human immunodeficiency virus type 1 employs the cellular dynein light chain 1 protein for reverse transcription through interaction with its integrase protein, J. Virol. 89 (7) (2015) 3497–3511.

[79] N. Arhel, Revisiting HIV-1 uncoating, Retrovirology 7 (2010) 96.

[80] J.C. Williams, P.L. Roulhac, A.G. Roy, R.B. Vallee, M.C. Fitzgerald, W.A. Hendrickson, Structural and thermodynamic characterization of a cytoplasmic dynein light chain-intermediate chain complex, Proc. Natl. Acad. Sci. U.S.A. 104 (24) (2007) 10028–10033.

[81] P. Rapali, A. Szenes, L. Radnai, A. Bakos, G. Pal, L. Nyitray, DYNLL/LC8: a light chain sub-unit of the dynein motor complex and beyond, FEBS J. 278 (17) (2011) 2980–2996.

[82] A.R. Frampton Jr., H. Uchida, J. von Einem, W.F. Goins, P. Grandi, J.B. Cohen, et al., Equine herpesvirus type 1 (EHV-1) utilizes microtubules, dynein, and ROCK1 to productively infect cells, Vet. Microbiol. 141 (1–2) (2010) 12–21.

[83] P.P. Naranatt, H.H. Krishnan, M.S. Smith, B. Chandran, Kaposi's sarcoma-associated herpesvirus modulates microtubule dynamics via RhoA-GTP-diaphanous 2 signaling and utilizes the dynein motors to deliver its DNA to the nucleus, J. Virol. 79 (2) (2005) 1191–1206.

[84] J. Pu, C.M. Guardia, T. Keren-Kaplan, J.S. Bonifacino, Mechanisms and functions of lysosome positioning, J. Cell Sci. 129 (23) (2016) 4329–4339.

[85] C.M. Guardia, G.G. Farias, R. Jia, J. Pu, J.S. Bonifacino, BORC functions upstream of kinesins 1 and 3 to coordinate regional movement of lysosomes along different microtubule tracks, Cell Rep. 17 (8) (2016) 1950–1961.

[86] E. Basyuk, T. Galli, M. Mougel, J.M. Blanchard, M. Sitbon, E. Bertrand, Retroviral genomic RNAs are transported to the plasma membrane by endosomal vesicles, Dev. Cell 5 (1) (2003) 161–174.

[87] C. Scully, L.P. Samaranayake, Emerging and changing viral diseases in the new millennium, Oral Dis. 22 (3) (2016) 171–179.

[88] M.A. de La Vega, D. Stein, G.P. Kobinger, Ebolavirus evolution: past and present, PLoS Pathog. 11 (11) (2015) e1005221.

[89] A. Marzi, P.W. Hanley, E. Haddock, C. Martellaro, G. Kobinger, H. Feldmann, Efficacy of vesicular stomatitis virus-ebola virus postexposure treatment in rhesus macaques infected with Ebola virus Makona, J. Infect. Dis. 214 (suppl 3) (2016) S360–S366.

[90] P. Roques, K. Ljungberg, B.M. Kummerer, L. Gosse, N. Dereuddre-Bosquet, N. Tchitchek, et al., Attenuated and vectored vaccines protect nonhuman primates against Chikungunya virus, JCI Insight 2 (6) (2017) e83527.

[91] P. Luthra, D.S. Jordan, D.W. Leung, G.K. Amarasinghe, C.F. Basler, Ebola virus VP35 interaction with dynein LC8 regulates viral RNA synthesis, J. Virol. 89 (9) (2015) 5148–5153.

[92] R. van der Kant, A. Fish, L. Janssen, H. Janssen, S. Krom, N. Ho, et al., Late endosomal transport and tethering are coupled processes controlled by RILP and the cholesterol sensor ORP1L, J. Cell Sci. 126 (Pt 15) (2013) 3462–3474.

[93] E. Alvarez, M.L. DeDiego, J.L. Nieto-Torres, J.M. Jimenez-Guardeno, L. Marcos-Villar, L. Enjuanes, The envelope protein of severe acute respiratory syndrome coronavirus interacts with the non-structural protein 3 and is ubiquitinated, Virology 402 (2) (2010) 281–291.

[94] G. Ippolito, G. Rezza, Preface – emerging viruses: from early detection to intervention, Adv. Exp. Med. Biol. 972 (2017) 1–5.

[95] S.A. Kelkar, K.K. Pfister, R.G. Crystal, P.L. Leopold, Cytoplasmic dynein mediates adenovirus binding to microtubules, J. Virol. 78 (18) (2004) 10122–10132.

[96] M. Gazzola, C.J. Burckhardt, B. Bayati, M. Engelke, U.F. Greber, P. Koumoutsakos, A stochastic model for microtubule motors describes the in vivo cytoplasmic transport of human adenovirus, PLoS Comput. Biol. 5 (12) (2009) e1000623.

[97] Y. Liu, A. Shevchenko, A. Shevchenko, A.J. Berk, Adenovirus exploits the cellular aggresome response to accelerate inactivation of the MRN complex, J. Virol. 79 (22) (2005) 14004–14016.

[98] K.H. Bremner, J. Scherer, J. Yi, M. Vershinin, S.P. Gross, R.B. Vallee, Adenovirus transport via direct interaction of cytoplasmic dynein with the viral capsid hexon subunit, Cell Host Microbe 6 (6) (2009) 523–535.

[99] M. Suomalainen, M.Y. Nakano, K. Boucke, S. Keller, U.F. Greber, Adenovirus-activated PKA and p38/MAPK pathways boost microtubule-mediated nuclear targeting of virus, Embo J. 20 (6) (2001) 1310–1319.

[100] S.A. Lukashok, L. Tarassishin, Y. Li, M.S. Horwitz, An adenovirus inhibitor of tumor necrosis factor alpha-induced apoptosis complexes with dynein and a small GTPase, J. Virol. 74 (10) (2000) 4705–4709.

[101] H. Mabit, M.Y. Nakano, U. Prank, B. Saam, K. Dohner, B. Sodeik, et al., Intact microtubules support adenovirus and herpes simplex virus infections, J. Virol. 76 (19) (2002) 9962–9971.

[102] B. Hernaez, G. Diaz-Gil, M. Garcia-Gallo, J. Ignacio Quetglas, I. Rodriguez-Crespo, L. Dixon, et al., The African swine fever virus dynein-binding protein p54 induces infected cell apoptosis, FEBS Lett. 569 (1–3) (2004) 224–228.

[103] T. Kubota, M. Matsuoka, T.H. Chang, M. Bray, S. Jones, M. Tashiro, et al., Ebolavirus VP35 interacts with the cytoplasmic dynein light chain 8, J. Virol. 83 (13) (2009) 6952–6956.

[104] H.N. Ramanathan, D.H. Chung, S.J. Plane, E. Sztul, Y.K. Chu, M.C. Guttieri, et al., Dynein-dependent transport of the hantaan virus nucleocapsid protein to the endoplasmic reticulum-golgi intermediate compartment, J. Virol. 81 (16) (2007) 8634–8647.

[105] S. Kim, H.Y. Kim, S. Lee, S.W. Kim, S. Sohn, K. Kim, et al., Hepatitis B virus X protein induces perinuclear mitochondrial clustering in microtubule- and dynein-dependent manners, J. Virol. 81 (4) (2007) 1714–1726.

[106] S. Boulant, M.W. Douglas, L. Moody, A. Budkowska, P. Targett-Adams, J. McLauchlan, Hepatitis C virus core protein induces lipid droplet redistribution in a microtubule- and dynein-dependent manner, Traffic 9 (8) (2008) 1268–1282.

[107] H. Kannan, S. Fan, D. Patel, I. Bossis, Y.J. Zhang, The hepatitis E virus open reading frame 3 product interacts with microtubules and interferes with their dynamics, J. Virol. 83 (13) (2009) 6375–6382.

[108] K. Dohner, K. Radtke, S. Schmidt, B. Sodeik, Eclipse phase of herpes simplex virus type 1 infection: efficient dynein-mediated capsid transport without the small capsid protein VP26, J. Virol. 80 (16) (2006) 8211–8224.

[109] M.W. Douglas, R.J. Diefenbach, F.L. Homa, M. Miranda-Saksena, F.J. Rixon, V. Vittone, et al., Herpes simplex virus type 1 capsid protein VP26 interacts with dynein light chains RP3 and Tctex1 and plays a role in retrograde cellular transport, J. Biol. Chem. 279 (27) (2004) 28522–28530.

[110] B. Sodeik, M.W. Ebersold, A. Helenius, Microtubule-mediated transport of incoming herpes simplex virus 1 capsids to the nucleus, J. Cell Biol. 136 (5) (1997) 1007–1021.

[111] K.S. Topp, L.B. Meade, J.H. LaVail, Microtubule polarity in the peripheral processes of trigeminal ganglion cells: relevance for the retrograde transport of herpes simplex virus, J. Neurosci. 14 (1) (1994) 318–325.

[112] G.J. Ye, K.T. Vaughan, R.B. Vallee, B. Roizman, The herpes simplex virus 1 U(L)34 protein interacts with a cytoplasmic dynein intermediate chain and targets nuclear membrane, J. Virol. 74 (3) (2000) 1355–1363.

[113] K. Levesque, M. Halvorsen, L. Abrahamyan, L. Chatel-Chaix, V. Poupon, H. Gordon, et al., Trafficking of HIV-1 RNA is mediated by heterogeneous nuclear ribonucleoprotein A2 expression and impacts on viral assembly, Traffic 7 (9) (2006) 1177–1193.

[114] M.A. Schneider, G.A. Spoden, L. Florin, C. Lambert, Identification of the dynein light chains required for human papillomavirus infection, Cell Microbiol. (2010).

[115] L. Florin, K.A. Becker, C. Lambert, T. Nowak, C. Sapp, D. Strand, et al., Identification of a dynein interacting domain in the papillomavirus minor capsid protein l2, J. Virol. 80 (13) (2006) 6691–6696.

[116] S. Mueller, X. Cao, R. Welker, E. Wimmer, Interaction of the poliovirus receptor CD155 with the dynein light chain Tctex-1 and its implication for poliovirus pathogenesis, J. Biol. Chem. 277 (10) (2002) 7897–7904.

[117] A.A. Kondratova, N. Neznanov, R.V. Kondratov, A.V. Gudkov, Poliovirus protein 3A binds and inactivates LIS1, causing block of membrane protein trafficking and deregulation of cell division, Cell Cycle 4 (10) (2005) 1403–1410.

[118] S. Ohka, M. Sakai, S. Bohnert, H. Igarashi, K. Deinhardt, G. Schiavo, et al., Receptor-dependent and -independent axonal retrograde transport of poliovirus in motor neurons, J. Virol. 83 (10) (2009) 4995–5004.

[119] E. Gonzalez Duran, R.M. del Angel, J.S. Salas Benito, In vitro interaction of poliovirus with cytoplasmic dynein, Intervirology 50 (3) (2007) 214–218.

[120] P. Rasalingam, J.P. Rossiter, T. Mebatsion, A.C. Jackson, Comparative pathogenesis of the SAD-L16 strain of rabies virus and a mutant modifying the dynein light chain binding site of the rabies virus phosphoprotein in young mice, Virus Res. 111 (1) (2005) 55–60.

[121] Y. Jacob, H. Badrane, P.E. Ceccaldi, N. Tordo, Cytoplasmic dynein LC8 interacts with lyssavirus phosphoprotein, J. Virol. 74 (21) (2000) 10217–10222.

[122] T. Mebatsion, Extensive attenuation of rabies virus by simultaneously modifying the dynein light chain binding site in the P protein and replacing Arg333 in the G protein, J. Virol. 75 (23) (2001) 11496–11502.

[123] E.R. Havecker, X. Gao, D.F. Voytas, The Sireviruses, a plant-specific lineage of the Ty1/copia retrotransposons, interact with a family of proteins related to dynein light chain 8, Plant Physiol. 139 (2) (2005) 857–868.

[124] B.M. Ward, Visualization and characterization of the intracellular movement of vaccinia virus intracellular mature virions, J. Virol. 79 (8) (2005) 4755–4763.

[125] E. Herrero-Martinez, K.L. Roberts, M. Hollinshead, G.L. Smith, Vaccinia virus intracellular enveloped virions move to the cell periphery on microtubules in the absence of the A36R protein, J. Gen. Virol. 86 (Pt 11) (2005) 2961–2968.

[126] A. Bauer, T. Nolden, S. Nemitz, E. Perlson, S. Finke, A dynein light chain 1 binding motif in rabies virus polymerase L protein plays a role in microtubule reorganization and viral primary transcription, J. Virol. 89 (18) (2015) 9591–9600.

[127] R. Valle-Tenney, T. Opazo, J. Cancino, S.P. Goff, G. Arriagada, Dynein regulators are important for ecotropic murine leukemia virus infection, J. Virol. 90 (15) (2016) 6896–6905.

[128] S. Theerawatanasirikul, N. Phecharat, C. Prawettongsopon, W. Chaicumpa, P. Lekcharoensuk, Dynein light chain DYNLL1 subunit facilitates porcine circovirus type 2 intracellular transports along microtubules, Arch. Virol. 162 (3) (2017) 677–686.

[129] L. Caly, V.T. Kassouf, G.W. Moseley, R.J. Diefenbach, A.L. Cunningham, D.A. Jans, Fast track, dynein-dependent nuclear targeting of human immunodeficiency virus Vpr protein; impaired trafficking in a clinical isolate, Biochem. Biophys. Res. Commun. 470 (3) (2016) 735–740.

[130] I. Banerjee, Y. Miyake, S.P. Nobs, C. Schneider, P. Horvath, M. Kopf, et al., Influenza A virus uses the aggresome processing machinery for host cell entry, Science 346 (6208) (2014) 473–477.

[131] D.P. Gladue, V. O'Donnell, R. Baker-Bransetter, J.M. Pacheco, L.G. Holinka, J. Arzt, et al., Interaction of foot-and-mouth disease virus nonstructural protein 3A with host protein DCTN3 is important for viral virulence in cattle, J. Virol. 88 (5) (2014) 2737–2747.

[132] V. Zila, F. Difato, L. Klimova, S. Huerfano, J. Forstova, Involvement of microtubular network and its motors in productive endocytic trafficking of mouse polyomavirus, PLoS One 9 (5) (2014) e96922.

[133] W. Zhang, W. Greene, S.J. Gao, Microtubule- and dynein-dependent nuclear trafficking of rhesus rhadinovirus in rhesus fibroblasts, J. Virol. 86 (1) (2012) 599–604.

The use of mouse models to probe cytoplasmic dynein function

Marco Terenzio[1], Sandip Koley[1], Elizabeth M.C. Fisher[2], Mike Fainzilber[1]

[1]Weizmann Institute of Science, Rehovot, Israel; [2]University College London, London, United Kingdom

11.1 The rationale behind using a genetic approach to study the dynein complex

An outcome of evolutionary studies and full genome sequencing has been the realization that the mouse and human genomes share approximately 99% homology [1], hence mice seem appropriate model organisms to study physiological functions of proteins of interest. A particularly interesting and noteworthy biological phenomenon is active intracellular cargo transport, which is an essential cellular function on which cells rely to localize and/or shuttle proteins, lipids, RNA, and even organelles to and from specific subcellular locations. Depending on the cargo and its destination, large macromolecular complexes associated with two major groups of motor proteins, kinesins and dyneins, mediate trafficking along microtubules. This process is especially crucial in polarized cell types such as neurons, whose extreme size variations and unique morphology make them particularly reliant on molecular motors for the intracellular trafficking of essential cellular components [2]. As a consequence, perturbation of intracellular cargo transport often leads to neuropathologies [3].

Generally, kinesins are anterograde motors, i.e., they move toward the cell periphery, while dyneins traffic toward the center of the cell to the minus end of microtubules [2]. Dynein motility, particularly how it hydrolyses ATP and interacts with multiple cofactors to progress along microtubules, has been analyzed in detail [4]. In contrast, cargo recognition and sorting of the dynein complex is less well understood. One option for cargo loading onto the dynein complex is through association with several families of molecular motor adaptors (reviewed in Ref. [5]). These are often core components of much larger protein complexes, which include membrane-associated cargo receptors, scaffolding proteins, and signaling molecules (reviewed in Refs. [6–8]). Adaptor proteins can modulate organelle-specific responses to environmental stimuli [9] and are often linked to dynein through an accessory subunit.

Dynein motors are large multisubunit protein complexes of two major classes, axonemal dyneins involved in ciliary and flagellary movement, and cytoplasmic dyneins, involved in vesicular transport, cell migration, cell division, and maintenance of Golgi integrity [10,11]. Different subunit options exist of all the components of cytoplasmic dynein, including two heavy chains (HCs), intermediate chains (ICs) that provide direct binding to dynactin, four light intermediate chains (LICs), and several light chains (LCs) [12]. The dynactin complex mediates the binding of numerous cargoes to the main motor complex and facilitates its processivity along microtubules [13,14]. There are several families of dynein light chains (DLCs), LC7/road-block, LC8 and Tctex1/ rp3; the latter plays a role in the direct binding of cargoes to the dynein motor complex [15]. The nomenclature of the dynein genes is detailed in Box 11.1. The precise composition of the subunits that comprise the motor complex in different tissues and cell types is still remarkably unclear. The identity of the genes encoding the individual subunits of cytoplasmic dynein has been established [11,16,17], and the diverse patterns of expression of different subunits [18] might support the existence of distinct cytoplasmic dynein complexes in different tissues or even in different subcellular locations inside the cell [19]. Moreover, some subunits of cytoplasmic dynein have been described to have roles that are independent of the motor complex itself, such as the LC Tctex1, which was shown to mediate neurite outgrowth in hippocampal neurons by modulating actin dynamics and Rac1 activity [20,21]. Mouse mutants for Tctex1 display a complex phenotype and are discussed in detail elsewhere (see Chapter 16).

This complexity highlights the importance of dissecting the function of the various dynein subunits via use of mouse genetic models. Indeed a great opportunity is offered by the fact that mutant alleles have already been generated for most of the genes encoding dynein subunits (see Box 11.2 for an extensive description of currently available genetic mouse resources). Most of these mutants are the results of gene-trap technology (see Section 11.2.1) and have not been phenotyped to date. It seems also that no naturally occurring dynein mutants have been identified so far in the mouse [18], although human patients with mutations in *Dync1h1* exist and will be discussed later in this chapter. On the other hand a number of induced mutants have been mapped to *Dync1h1*, as well as a transgenic mutant *Dynll1* LC line [18]. The growing collection of dynein mutant mice currently listed in the Mouse Genome Informatics (MGI) database is described in Box 11.2 We will focus this chapter on published dynein mouse models as a paradigm for the targeted use of genetics to dissect the functions of the dynein complex (see Table 11.1).

Box 11.1 Dynein gene nomenclature

The naming of human and mouse genes and their mutations are standardized as approved by the Human Genome Organization Nomenclature Committee [79,80] and the International Committee on Standardized Nomenclature for Mice. The standard notation for mouse mutants is also available at http://www.informatics.jax.org/mgihome/nomen/index.shtml.

The cytoplasmic dynein complex 1 consists of up to 11 different subunits [79], each encoded by a single gene, whose names and abbreviation have now been standardized. Prior to this standardization, 15 different names could be found in the literature for dynein HC alone [80]. Fortunately now, following the abovementioned guidelines, it is possible to correctly assign a name for each dynein subunit as follows: specifically, each unique cytoplasmic dynein subunit starts with **Dyn** for dynein and **c** for cytoplasmic, followed by the specific dynein complex subtype, **1** or **2** (i.e., cytoplasmic dynein 2 would be **Dync2**). While *Dync1* and *Dync2* are used for protein that are thought to belong exclusively to cytoplasmic dynein protein complexes 1 or 2, respectively, DLCs are thought to be shared by different dynein complexes and therefore start with just **Dyn**, followed by a letter(s) denoting the size of said chain (i.e., **h** for the HC, **i** for the HC, **li** for the light intermediate chain, and **l** for the light chain (for example, cytoplasmic dynein 1 HC 1 would be written as **Dync1h1**). Finally, additional letters, **t**, **rb**, and **l**, are used to distinguish the three distinct LC families (for example, Cytoplasmic DLC Roadblock 1 is indicated as **Dynlrb1**). Note that we have referred in the text above to the mouse nomenclature; conventionally the gene names are written in italicized upper case for human genes (for example, *DYNC2H1*), italicized initial upper case and then lower-case letters for mouse (*Dync2h1*), and for proteins of both species everything is in upper case (DYNC2H1) [80].

Mutant strains are named by the name of the gene where the mutation takes place, followed by the name of the mutation itself in superscript. Some mutant strains have been historically named in a way that is descriptive of the corresponding phenotype, i.e., "Legs at odd angles," "Cramping 1," and "Sprawling" to cite a few, which are commonly abbreviated as "*Loa*," "*Cra1*" and "*Swl*," respectively. Therefore the full naming for the abovementioned mutations would be $Dync1h1^{Loa}$, $Dync1h1^{Cra1}$, and $Dync1h1^{Swl}$. When we need to discriminate between wild-type, heterozygote, and homozygote mice for a certain mutation, we use $Dync1h1^{+/+}$, $Dync1h1^{Loa/+}$ and $Dync1h1^{Loa/Loa}$, where "+" indicates the wild-type allele [18].

In this chapter we will also discuss an historical mouse model with a targeted mutation in the dynein HC gene *Dync1h1* [34]. For targeted mutations the nomenclature is slightly different than presented above: after the

Continued

Box 11.1 Dynein gene nomenclature—cont'd

name of the gene the superscript starts with "tm" (for targeted mutation), followed by a serial number from the laboratory that made the mouse and a code for that laboratory (in the case of the previously mentioned mutant the code is "Noh" for the laboratory of professor Nobutaka Hirokawa) [18]. Consequently, the full nomenclature of the dynein HC mutant in Ref. [34] is $Dync1h1^{tm1Noh}$.

Box 11.2 Mouse genetics resources

There are several useful resources online, which allow the gathering of information relative to individual mouse mutants, often organized in the form of databases that are freely accessible. The International Mouse Phenotype Consortium (IMPC) has a portal (http://www.mousephenotype.org) dedicated to the mouse mutants available or in the making by the consortium itself. The stated goal of this institution is to build the first truly comprehensive, functional catalogue of a mammalian genome, by generating a knockout mouse strain for every protein-coding gene. To do that, they are taking advantage of the embryonic stem cell resources generated by the International Knockout Mouse Consortium. Interestingly, the portal makes available the results of the systematic phenotyping performed by the consortium itself for each viable mouse strain generated. The IMPC website also contains information about a series of international programs aimed to the generation of gene-targeted animals including the European Conditional Mouse Mutagenesis Program. Other notable international programs are the North American Conditional Mouse Mutagenesis Project (http://www.norcomm.org/index.htm), and the American Knockout Mouse Project (https://www.komp.org).

Another excellent online resource is the MGI website, http://www.informatics.jax.org, which contains information of all mutants, regardless of the type of mutation, published in the literature. It also integrates the data generated by the IMPC. The goal of the MGI is to provide a platform of integrated genetic, genomic, and biological data to facilitate the study of mouse genetics. By searching for a gene, a large amount of information can be retrieved, such as its mapping in the genome, the protein domain encoded, the available clones and if there are available mouse knockout, gene-trapped or chemically and radiation-induced mutant alleles. Phenotypic information is also included wherever possible.

Table 11.1 Mouse mutants for cytoplasmic dynein

Allele name	Allele	Phenotype	References
Dync1h1^{tm1Noh}	Knockout of Dync1h1 first exon	Dync1h1^{tm1Noh} homozygous mice die at embryonic day 8.5; abnormal Golgi, endosomal and lysosomal distribution in cultured cells.	[34]
Dync1h1^{Loa}	Point mutant (F580Y) created by ENU mutagenesis	Dync1h1^{Loa} homozygous mice die 1 day after birth. Hind limb clasping, decreased number of motor neurons, severe sensory deficit. Abnormal Golgi distribution, axonal retrograde transport defects.	[31,43]
Dync1h1^{Cra1}	Point mutant (Y1055G) created by ENU mutagenesis	Dync1h1^{Cra1} homozygous mice die 1 day after birth. Hind limb clenching, decreased number of motor neurons. Abnormal Golgi distribution. Heterozygotes display progressive mitochondrial dysfunction in mutagenesis.	[31,41]
Dync1h1^{Swl}	Radiation mutant; 9-bp deletion from AA 1040 to1043 (GIVT to A)	Dync1h1^{Swl} homozygous mice die in utero before embryonic day 8.5. Heterozygous mice display an early onset sensory neuropathy.	[50,43]
Dync1h1^{N235Y}	Point mutant (N235Y) created by ENU mutagenesis	Dync1li1$^{N235Y/N235Y}$ mice exhibit increased anxiety-like behavior, altered gait and neuronal migration defects. DRG neurons have decreased dendritic branching. There are Golgi and endosomal abnormalities in cultured cells.	[63]
MMTV-DLC^{S88A}	Overexpression of the DLC^{S88A} mutant under the control of the murine MMTV promoter	Overexpression model of a DLC1 phosphomimetic mutant, which displayed mammary gland accelerated involution.	[68]
Dynll1$^{Gt(EUCE0287d04)}$ Hmgu	Gene-trapped mouse	Embryonic abnormalities due to abnormal length and morphology of the nodal cilia	[65]
Dynll1$^{tm1.1Jhh}$	Targeted deletion of the promoter and exons 1 and 2 of the Dynll1 gene	When deleted in B lymphoid cells, DYNLL1 was found essential for the development and expansion of MYC-driven lymphomas	[69]
Dync1i1-GFP	Insertion of a GFP tag and three FLAG tags before the stop codon in the middle of exon 17 of the endogenous Dync1i1 locus.	Knockin mouse utilized as a tool for biochemical characterization of dynein–dynactin and synaptic dynein complexes.	[27,70,71]

Table indicating the allele name, nature of the genetic mutation and known phenotype of the main mouse models available to study the function of cytoplasmic dynein.

11.2 Different approaches for mouse genetic studies

Mutants are commonly used in genetic studies to learn about the physiological function of genes and proteins. The model organism of choice for physiological studies of dynein is the mouse due to the availability of genetic resources, genome engineering tools, and behavioral and physiological tests built up over the past 100 years for this model [18].

11.2.1 Genotype-driven approach

The genotype-driven approach is the creation of animal models harboring specific mutations in the gene encoding the protein of interest. Correlating the genotype to emerging phenotypes in the genetically manipulated models provides new functional information and enables several levels of analyses, from behavior and physiology to cell biology and biochemistry. Two broad categories of this strategy exist, transgenic animals and animals carrying a gene-targeted mutation.

1. **Transgenic animal models**. These mice are created by the random insertion of a transgenic construct in the genome. The construct provides an extra copy (or copies) of a gene of interest (which can be human in origin), which is inserted randomly into the genome, thus likely resulting in gene overexpression (since the model will carry more than the endogenous two copies of the gene). The transgene's pattern of expression depends on the promoter to which it is adjacent, which may or may not be the one of the endogenous gene. Sequencewise, transgenes might be wild type or more often mutated forms of the gene of interest, which allow for specific functional studies.

   Because of its random nature, transgene insertion might disrupt an endogenous gene. This seemingly negative feature can be taken advantage of to create libraries of "gene-trap" mouse mutants, in which random endogenous genes are disrupted [22,23]. These mice are screened for interesting phenotypes and once such phenotype is found, the site of insertion, together with the disrupted gene, is identified. Most gene-trap mice are functional knockouts.

2. **Gene-targeted mouse models**. Since transgenic mice have the disadvantages of overexpression and random insertion of the transgenic gene, targeted approaches to mutagenesis were developed, in which the gene of interest is knocked out or otherwise altered within its endogenous locus. A variety of systems have been developed to achieve the desired specificity, mostly relying on homologous recombination, and more recently also on CRISPR/Cas9-mediated gene editing [24,25]. Further refined constructs for gene targeting allow for conditional or inducible mutations. Conditional mutants ensure that the intended genetic manipulation takes effect in

selected cell types, while inducible systems allow for translational control of a gene of interest that can be switched on and off at will [26]. The vast majority of targeting constructs are designed to disrupt or remove a particular gene, which results in the generation of a "knockout" mouse. Alternative approaches, however, have also been developed, whereby a sequence of interest, such as a novel protein domain, can be inserted in the targeted gene. These mouse models are commonly known as "knockin" mice. As an example, a *Dync1i1*-green fluorescent protein (GFP) knockin model [27] provides an excellent tool for in vivo proteomic studies of cytoplasmic dynein and will be discussed in depth in Section 11.5.2 of this chapter.

There are good arguments in favor of both transgenic and gene-targeted approaches, and the adoption of one over the other largely depends on the aim of the study. Regardless, both models have been extensively used and proved to be crucial for biological and medical research. For this reason a series of large-scale programs have been initiated to create mice harboring–targeted mutations for every known gene in the mouse genome. Most of these programs are adopting increasingly more sophisticated construct designs to create conditional mouse mutants in which the expression of a gene of interest can be spatially and temporally restricted. Examples of such programs and online resources for mouse genetics are discussed in Box 11.2 later in this chapter.

All the abovementioned genetic strategies, including transgenic, gene-trap, knockout, and knockin approaches might apply to the creation of mouse models to study cytoplasmic dynein function. We discuss below individual mouse mutants described in the literature and how they contributed to advance our knowledge of the dynein complex.

11.2.2 Phenotype-driven approach

The phenotype-driven approach, also known as "forward genetics" approach, is based on the identification of a phenotype of interest, for example, aberrant limb movements or gait patterns, followed by the characterization of the underlying unknown mutation using standard molecular genetics techniques [18]. The lack of prior assumptions about the gene or pathway involved can yield unexpected and interesting results in the form of new functions for known genes. This modus operandi is common in yeast and other model systems, including cell culture systems, where it has been exploited in large screens for the elucidation of entire molecular pathways [28]. In mammalian models the targeted gene approach is more commonly used than forward genetics, due to the amount of work (and expense) required to characterize unknown genotype–phenotype correlations in mammals.

Phenotype-driven approaches might yield mutations in unexpected parts of a gene that might help to assign new functions. Another advantage is that the mice are often affected by point mutations, which may result in subtle phenotypes and allow the study of genes that are lethal when knocked out. Large collections of such random mutants can be generated by utilizing the chemical mutagen *N*-ethyl-*N*-nitrosourea (ENU), a powerful alkylating agent that is highly mutagenic in mouse spermagonial germ cells [29]. The main advantage of ENU mutagenesis is that genetic lesions thus induced are mostly point mutations, hence any eventual phenotypes in mutant mice are most likely caused by single gene effects. This is due to an estimated mutation rate at gene loci of approximately 1 per 1000 genomes at an optimal dose of ENU [30]. In addition, because of the random nature of ENU mutagenesis, all DNA sequences, transcribed and nontranscribed, can be affected, which means that mutations in noncoding sequences can occur potentially disrupting splice sites or microRNAs [30].

ENU mutagenesis programs have been initiated internationally to create libraries of mice harboring random point mutations, which could potentially be linked to phenotype of interest for all areas of biological and medical research. These programs lead to the discovery of "Legs at odd angle" and "Cramping 1" [31] mouse mutants for dynein HC, which are the two most studied genetic models of cytoplasmic dynein function. Libraries of ENU mutants are typically too vast to keep and breed all resulting mutant mice; hence the libraries are stored as frozen embryo or sperm archives. DNA from these genetic banks can be screened for mutations in a specific gene of interest, offering a unique opportunity to perform genotype-driven discovery with high-throughput mutation detection protocols. If a suitable mutant is found, the mutation can be confirmed by DNA sequencing and the corresponding strain rederived to generate a mouse that can be phenotyped and further studied to gain further insights on the gene or protein of interest.

11.2.3 Genetic background matters

When working with mouse models, it is important to consider that while many genetic diseases and/or phenotypes observed both in mouse and humans are caused by one or few primary genetic variants, the background genetic milieu of each individual can have profound effects. Genetic modifiers have been well documented in model organisms, especially in mouse [32], where a particular mouse mutant crossed with two different inbred strains can have drastically different phenotypes. For example, a splicing mutation in the sodium channel gene *Scn8a* causes movement disorders on most genetic backgrounds, but when combined with a genetic modifier in the C57BL/6J mouse, the phenotype worsens to a lethal neurological disease [33]. Therefore even "simple" Mendelian diseases can be quite complex when genetic background comes into play. Mouse studies conducted on single genetic backgrounds can be informative for basic science, however, when findings are translated into the clinic, the complications of human genetic variability must be taken into consideration. Inbred mouse strains may amplify the visibility of a phenotype by reducing variability due to genetic background. The advantages of single inbred backgrounds for understanding the functions of a certain gene/protein must therefore be balanced with the possibility of differing physiological impacts of a mutation in an outbred background [32].

11.3 An allelic series of mutations in the cytoplasmic dynein heavy chain gene, *Dync1h1*

The most intense use of mouse genetics to study dynein is an allelic series of mutants (a series of mutations that map within the same gene) for *Dync1h1*, comprising a *Dync1h1* knockout mutant, two chemically induced mouse mutations and a radiation-induced mutant. The advantage of examining an allelic series is the level of information that we can infer regarding the function of the targeted gene. We will discuss these models as a paradigm to exemplify how mouse genetics can be used to probe cytoplasmic dynein function. The first mutant in this series was generated in 1998 [34] and is the first and so far only mouse harboring a targeted dynein HC mutation described in the literature. This mouse mutant is termed *Dync1h1*^{tm1Noh}; for a more in depth description on the rules governing mouse mutant nomenclature please refer to Box 11.1 in this chapter.

11.3.1 *Dync1h1*^{tm1Noh} knockout mouse

Yeast mutational studies in the early 1990s highlighted the role of cytoplasmic dynein in spindle orientation and anaphase chromosome segregation [35]. Other mutants generated in fungi showed abnormalities in nuclear positioning and movement [36,37] (see also Chapter 17 (vol. 1 of this book)), while Drosophila mutants were lethal [38], an indication that cytoplasmic dynein may be essential in multicellular organisms, while in yeast and fungi it is dispensable. The laboratory of Hirokawa at Tokyo University set out to generate a mammalian dynein mutant by generating a knockout mouse for dynein HC. Heterozygous *Dync1h1*^{tm1Noh} mice were reported to have no discernable phenotype; however, there were no homozygous mice in the litters. The growth of null blastocysts was indistinguishable from wild types, most likely due to residual maternal dynein HC1 mRNA in the null cells [34]. Homozygous blastocysts were implanted, but the embryos died very early at embryonic day 8.5 [34] when the investigators observed small decidua that contained completely resorbed embryos.

Cells cultured from wild type and homozygous blastocysts revealed that the Golgi complex was much smaller in size and fragmented throughout the cytoplasm in null cells, while wild-type cells retained normal Golgi morphology and distribution. Furthermore, endosomes and lysosomes labeled by uptake of extracellular fluorescein-isothiocyanate-labeled dextran had a uniform cytoplasmic distribution or remained close to the cell periphery in null blastocysts, whereas in wild-type cells they tended to distribute around the nucleus [34]. In contrast, endoplasmic reticulum and mitochondria morphology and distribution seemed unaffected by the dynein-targeted mutation. Interestingly, the dynactin complex, normally predominantly localized centrosomally, was found to be distributed throughout the cytoplasm in dynein null cells, though some centrosomal localization still remained, suggesting a role for dynein in dynactin centrosomal localization [34]. Indeed later studies confirmed a correlation between dynactin centrosomal localization and the mitotic cycle [39] and the role of dynein and the dynactin complexes in regulating centrosome integrity and positioning is now well established (see Chapters 19 and 20 (vol. 1 of this book)).

Overall the *Dync1h1*^{tm1Noh} mouse model proved instrumental in establishing dynein as an essential gene in mammals. It also sheds light on the cellular function of the dynein complex in maintaining Golgi homeostasis and integrity as well as proper endosomal and lysosomal distributions within the cell [34]. These findings were confirmed and expanded at the molecular level in successive studies [5].

11.3.2 *Dync1h1*^{Loa} and *Dync1h1*^{Cra1} N-ethyl-N-nitrosourea mutants

Both *Dync1h1*^{Loa} and *Dync1h1*^{Cra1} mutants originated from phenotype-driven ENU mutagenesis screens in mice (see Section 11.2.2 above). The first strain was generated at the UK Medical Research Council Mammalian Genetic Unit (MRC Mammalian Genetics Unit, Oxford, UK), while the latter resulted from work by Ingenium Pharmaceuticals (Munich, Germany). Both groups focused on mouse models with movement deficits. Although the two groups worked independently from each other, the mice were generated using the same approach and the mutants share a very similar phenotype. The *Dync1h1*^{Loa} strain was named "Legs at odd angles" due to a hind limb clasping phenotype displayed when suspended by the tail for more than 30 s [31] (Fig. 11.1A), while the *Dync1h1*^{Cra1} strain was called "Cramping 1" for similar reasons [31]. Clenching of the hind limb is normally interpreted as a genetic sign of neuromuscular damage, though heterozygotes for both mutant strains were reported to live up to at least 2 years, which is average in terms of life span for a laboratory mouse.

Since both *Dync1h1*^{Loa} and *Dync1h1*^{Cra1} strains were generated through ENU mutagenesis, they were each likely to carry a single point mutation that accounted for the phenotype; the site of mutation was eventually mapped in the *Dync1h1* gene independently by both groups with the use of standard genetic techniques [31]. Specifically, *Dync1h1*^{Loa} harbors a switch from

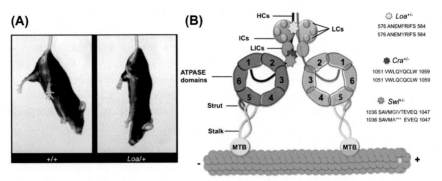

Figure 11.1 Mapping of dynein heavy chain (HC) mutants. (A) Photograph of two littermate mice suspended by the tail. Left—wild-type mouse, right one— *Dync1h1^Loa/+*. The Loa animal exhibits the typical cramping phenotype caused by the neuromuscular defect associated with the mutation. (B) Schematic representation of dynein (*HC*, heavy chain; *IC*, intermediate chain; *LC*, light chain; *LIC*, light intermediate chain; *MTB*, microtubule-binding domain), showing locations of the published dynein HC mutations (*colored stars*). Additionally, the position and the nature of the genetic alterations found in these dynein mutants are shown.

Phe to Tyr in position 580 of dynein HC; the mutated region of the HC polypeptide is responsible for both the homodimerization of the HC itself and its binding to the two ICs of the dynein complex [40]. Similarly, the causative point mutation in the *Dync1h1^Cra1* mouse was mapped as a Tyr to Cys switch in position 1055 of dynein HC, which once again lies in the HC homodimerization domain [40] (Fig. 11.1B). It is therefore unsurprising that both mice have a similar phenotype, although the *Dync1h1^Cra1* phenotype seems to be more severe.

Although heterozygous *Dync1h1^Loa* and *Dync1h1^Cra1* can live up to 2 years, homozygous mice show a more severe phenotype characterized by an inability to feed and move, which leads to death within 24 h of birth [31]. Consistent with the hind limb clasping phenotype, a decrease in the number of α motor neurons in the spinal cord anterior horn was determined by histopathological analysis in *Dync1h1^Loa* and *Dync1h1^Cra1* heterozygotes, which was paralleled by an altered composition of muscle fiber types in *Dync1h1^Cra1* heterozygotes. Homozygotes displayed once again a more severe phenotype, with a loss of 80% of the lower motor neurons in *Dync1h1^Cra1/Cra1* mice, and a 50% loss in *Dync1h1^Loa/Loa* mice when compared with wild-type littermate controls at embryonic day 18.5. In homozygous mice the loss of motor neurons was paralleled by enhanced apoptosis and the presence of intracellular perinuclear inclusion bodies in the surviving neurons, characterized by depositions of ubiquitin, SOD1, CDK5, and neurofilament [31]. Interestingly, similar depositions have been described as a feature of human amyotrophic lateral sclerosis (ALS), which is a lethal neuropathology characterized by progressive loss of motor neurons.

Dync1h1^{Loa} and *Dync1h1^{Cra1}* heterozygotes also display abnormalities in the development of facial motor neurons, and the branching and elongation of sensory neurons in the mutant hetero- and homozygous embryos was impaired [31]. Finally, *Dync1h1^{Cra1}* heterozygous mice display progressive mitochondrial dysfunction in muscle and develop hyperinsulinemia and hyperglycemia, which progress to glucose intolerance with age [41]. Indeed, *Dync1h1^{Cra1/Cra1}* and *Dync1h1^{Cra1/−}* fibroblasts exhibit mitochondrial morphological abnormalities, which are associated with the loss of mitofusin 1 [41].

Controversially, while *Dync1h1^{Loa/+}* mice (on a mixed genetic background) were described to develop a progressive age-dependent motor neuron loss [31], subsequent work failed to detect α-motor axon degeneration at any age point (albeit on different genetic backgrounds) [42,43]. Instead, a significant sensory axon deficit was reported, specifically from large proprioceptors. The deficit emerges early in life and most likely reflects developmental loss, which does not progress with age [42]. Although the mechanism underlying sensory neuron loss remains elusive, Cleveland and colleagues proposed deficits in neurotrophin-3 (NT-3) signaling as a likely candidate. Indeed NT-3 deletion leads to a selective loss of proprioceptive afferents and muscle spindles, accompanied by early postnatal death, similar to the phenotype of *Dync1h1^{Loa/Loa}* mice [42]. Further work on the *Dync1h1^{Loa/+}* mice showed a 52% reduction in hind limb grip strength, while the forelimb grip strength was unaffected, which correlated with a reduction of 86% in the number of muscle spindles in the hind limbs [43]. Also, while the number of cervical dorsal root ganglion (DRG) neurons was unaffected, *Dync1h1^{Loa/+}* mice displayed a 42% reduction in lumbar DRG neurons paralleled by a thinning of the lumbar segment of the dorsal roots [43]. Taken together, these data suggest that the movement disorder in *Dync1h1^{Loa/+}* mice [31,42] results from a sensory deficit rather than motor neuron loss. Similarly, Dupuis and colleagues reported that *Dync1h1^{Cra1}* heterozygous mice did not display any typical feature of motor neuron disease. Even in aged animals, *Dync1h1^{Cra1/+}* heterozygous neuromuscular junctions appeared to be normal and, in contrast to previous findings, no loss of motor neurons could be observed. *Dync1h1^{Cra1/+}* mice were described to develop instead an early onset sensory neuropathy, caused by proprioceptive sensory neuron loss, which the authors believe could be at least partially responsible for their characteristic motor phenotype [44].

In an attempt to describe the phenotype of *Dync1h1^{Loa}* and *Dync1h1^{Cra1}* mice at a cellular level, embryonic fibroblasts derived from both mutants were characterized. These cells displayed abnormalities in their Golgi distribution, similar to that previously reported in dynein HC knockout blastocysts, albeit more subtle. Specifically, while nuclear motility during cell division and steady state Golgi morphology and positioning were found not to be affected in these mouse mutants, the rebuilding of the pericentrosomal Golgi complex after nocodazole-induced dispersion was significantly impaired [31]. This finding suggests that the point mutations underlying $Dync1h1^{Loa}$ and $Dync1h1^{Cra1}$ strains affect cytoplasmic dynein performance only in situations of cellular stress.

Because cytoplasmic dynein is the most important retrograde motor in neurons, in addition to the abovementioned cellular phenotypes, the role of the $Dync1h1^{Loa}$ mutation in dynein-mediated retrograde axonal transport has also been extensively investigated. In the original study by Hafezparast and colleagues, the rate of retrograde axonal transport of $Dync1h1^{Loa/+}$ and $Dync1h1^{Loa/Loa}$ mice was assessed [31]. Heterozygous mice displayed identical transport kinetics to their wild-type counterparts, while in contrast the rate of axonal retrograde transport in homozygotes was severely impaired [31]. The assembly of the dynein complex in $Dync1h1^{Loa}$ mice was characterized, revealing enrichment of a lighter than usual subcomplex of dynein in the $Dync1h1^{Loa}$ mice. Moreover, while the association of mutant dynein to dynactin was reduced, the binding affinities for ICs, LIC1, and Tctex-1 were significantly increased in the mutant. Interestingly, kinesin LC 1 expression and its interaction with the dynein complex was also found to be perturbed, possibly as an effect of compensatory mechanisms in response to $Dync1h1^{Loa}$ retrograde transport defect [45]. Ore-McKenney and colleagues attempted to determine the molecular mechanisms responsible for the defect in axonal retrograde transport by purifying dynein complexes from $Dync1h1^{Loa/+}$ mutant mice and studying them in vitro with a combination of biochemical, single-molecule, and live-cell imaging techniques [46]. The authors discovered that mutant dynein displayed a marked reduction in motor run-length and significant alterations in motor domain coordination [46]. Finally, a recent study reported that the velocity of dynein-dependent retrograde transport of signaling endosomes induced by EGF and BDNF stimulation is significantly reduced in $Dync1h1^{Loa}$ embryonic fibroblasts and motor neurons [47]. Concomitantly, the number of the endosomes moving anterogradely was increased in mutant cells. As a consequence, mutant motor neurons, but not embryonic fibroblast, exhibited an abnormal ERK1/2 and c-Fos response to serum starvation-induced stress [47].

The $Dync1h1^{Loa/+}$ mouse was recently used to study mechanisms underlying the ability of a cell to sense its own length. The proposed model invokes motor-dependent transport to encode cell length in the frequency of an oscillating retrograde signal, which stems from a composite negative feedback loop between bidirectional motor-dependent signals [48]. The counterintuitive prediction from such model was that a decrease in either anterograde or retrograde motor complex should result in an increase in cell length. To test this prediction the study took advantage of reduced dynein levels in axons within the $Dync1h1^{Loa/+}$ sciatic nerve compared to wild-type littermates [48]. As the model suggested, the authors observed that neurite growth in cultured $Dync1h1^{Loa/+}$ adult sensory neurons reached greater lengths than wild-type littermates [48,49], and that $Dync1h1^{Loa/+}$ embryos displayed over 50% more forepaw innervation at embryonic day 11% and 30% more at embryonic day 12 [48].

Taken together, the evidence presented in this section highlights the utility of mouse genetics for the study of cytoplasmic dynein functions. Even relatively subtle point mutations, which map in a part of the *Dync1h1* gene that does not involve the ATPase domain, can be used to reveal critical functions of the dynein complex, most prominently in the nervous system. Indeed, these mouse models have been instrumental for probing the role of dynein in several neurological diseases including ALS, Charcot–Marie–Tooth disease (CMT), spinal muscular atrophy (SMA), and Huntington's disease [5]. For a detailed discussion of cytoplasmic dynein dysfunction in neurodegenerative disorders please see Chapter 10.

11.3.3 *Dync1h1^{Swl}* radiation-induced mutant

In the early 1970s a radiation-induced mutant mouse called the "Sprawling mouse" was generated at MRC Mammalian Genetics Unit; heterozygous mutant mice had a normal life span, but displayed an early onset sensory neuropathy and a loss of muscle spindles [50]. In 2007 Chen, Popko and colleagues from the University of Chicago, USA, narrowed down the genetic mutation underlying the mouse phenotype to a nine-base-pair deletion between residue 1040 and 1043 in the cargo binding domain of dynein HC; the deletion replaces four amino acids (glycine, isoleucine, and threonine) by a single alanine [43].

Phenotypically *Dync1h1$^{Swl/+}$* mice are characterized by the presence of hind limb clasping during tail suspension similar to *Dync1h1$^{Loa/+}$* and *Dync1h1$^{Cra1/+}$*, as soon as they reach 1 week of age. At 3–4 weeks of age, they develop an unsteady gait characterized by jerky and wobbly locomotion, while at rest their hind limbs are splayed and flexed forward [43]. Interestingly, while *Dync1h1$^{Swl/+}$* mice hind paws are incapable of gripping structures, forelimb locomotion does not seem to be impaired. Heterozygous mice have normal gestation periods, lactation ability, litter size, and life span and their phenotype does not appear to progress with age. In addition, no overt atrophy was observed in *Dync1h1$^{Swl/+}$* mice [43].

Further characterization of the mutant lumbar spinal cord revealed neither aberrant morphology nor decreased numbers of α-motor neurons (the large motor neurons) in *Dync1h1$^{Swl/+}$* animals at up to 2 years of age. Similarly, nociceptive sensory function in *Dync1h1$^{Swl/+}$* mice was assessed and found unaffected, while proprioceptive function, measured by H reflex, an electrophysiological readout of proprioceptive neuron activity, was shown to be severely impaired. Consistent with a proprioceptive defect in sensory nerves, hind limbs of *Dync1h1$^{Swl/+}$* mice displayed 88% fewer mature muscle spindles than the wild-type counterparts [43]. Immunostaining for DRG neuron subpopulation markers showed that proprioceptive sensory neurons were more severely compromised than the nociceptor subpopulation, with the number of proprioceptors reduced by 69%, while the number of nociceptive neurons was reduced by only 31% [43]. The authors proposed that lumbar proprioceptive DRG neuron loss occurs in *Dync1h1$^{Swl/+}$* mice concomitantly with muscle spindle degeneration during late embryonic development [43].

11.4 *DYNC1H1* mutations in humans

Clinical supporting evidence for the findings reported from mouse dynein mutants has become available only recently (Fig. 11.2). Because of the number of described dynein human mutations we will limit our discussion to few well-characterized cases. The first human patient harboring a *DYNC1H1* mutation (*DYNC1H1^H3822P*) was described in 2010, presenting with developmental delay, hypotonia, and brain malformation [51]. Two years later a second study identified the *DYNC1H1^H3822P* mutation and a novel *DYNC1H1^E1518KP* mutation in two patients with severe intellectual disability and variable neuronal migration defects [52]. These findings provide supporting evidence in humans for the well-established role of dynein in neuronal migration [5].

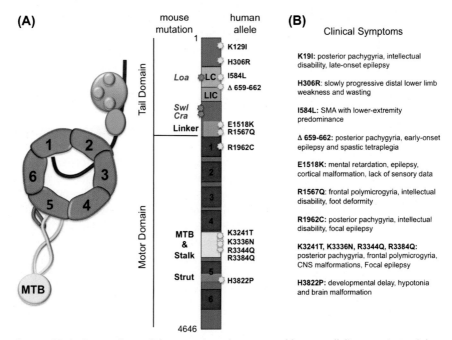

Figure 11.2 Comparison of the mapping of mouse and human allelic mutations of the dynein cytoplasmic complex. (A) Diagram of the cytoplasmic dynein motor complex including the heavy chain (HC) (in blue) and its associated subunits (*LC*, light chain; *LIC*, light intermediate chain; *MTB*, microtubule-binding domain). The positions on the dynein HC of the three mouse mutations (Loa, Cra and Swl) and the main human mutations discussed in Section 11.4 of this chapter are indicated (*colored stars*). (B) The most prominent clinical symptoms exhibited by affected patients are described for each human allelic mutation reported in A.

Additional *DYNC1H1* mutations have been identified in disorders of cortical development, which include lissencephaly, pachygyria, polymicrogyria, and microcephaly, and are often associated with severe intellectual disability and epilepsy. Several mutations (*DYNC1H1^{K3241T}*, *DYNC1HK^{K3336N}*, *DYNC1H1^{R3344Q}*, *DYNC1H1^{R3384Q}*) cluster in the stalk region and microtubule-binding domain of dynein HC and have been shown to decrease the binding of the motor to microtubules [53]. Another three mutations are mapped in different areas of the HC, *DYNC1H1^{R1962C}* lies in the ATPase domain, *DYNC1H1^{R1567Q}* is located in proximity of the HC linker region, *DYNC1H1^{K129I}* at the N-terminus of the HC tail domain and the deletion *DYNC1H1$^{d659-662}$* map within the binding sites of the HC to the ICs and LICs [53]. These, together with the previously described mutations suggest that mutations in *DYNC1H1* underlie a broad neuronal phenotypic spectrum in humans, ranging from peripheral neuropathies to cerebral malformations. Indeed, *DYNC1H1* could be an important locus for mutations underlying disorders in cortical development.

Interestingly, another *DYNC1H1* mutant, *DYNC1H1^{H306R}*, was described in CMT disease, which is an inherited neuromuscular disorder categorized as a chronic motor and sensory neuropathy with distal muscle weakness and atrophy, possibly associated with sensory loss [54]. Significantly, the mutation maps at a highly conserved residue within the dynein HC homodimerization domain and patient symptoms are broadly similar to the phenotypes observed in *Dync1h1$^{Loa/+}$*, *Dync1h1$^{Cra1/+}$*, and *Dync1h1$^{Swl1/+}$* mice. *DYNC1H1^{H306R}* patients display delayed motor functions and/or an abnormal gait and early onset progressive distal lower limb weakness [54]. The same mutation was also described in another neuropathy, neurogenic muscular atrophy, by sequencing of three affected individuals [55]. In addition, investigating patient families affected by SMA, a developmental and degenerative disorder affecting motor neurons or their axons, three more missense mutations in the *DYNC1H1* tail domain were identified [55]. Interestingly, the authors of the latter study demonstrated that purified dynein complexes from patients with the *DYNC1H1^{I585L}* mutation displayed altered stability and function [56]. Finally, a recent study of 16 families affected by SMA reported 30 mutations in the tail and motor domains of *DYNC1H1*, 10 of which were novel. The patients were characterized by congenital or childhood-onset lower limb wasting and weakness, which was frequently associated with cognitive impairment of variable severity. Patient lower limb muscle displayed signs of denervation in MRI [57]. Another recent study screened a database of 1024 sequencing samples of motor neuron and related diseases and found 13 patients carrying both novel variants and known single nucleotide variations in the human *DYNC1H1* gene implicated in SMA with lower extremity dominance , intellectual disability with neuronal migration defects, malformations of cortical development, and CMT disease, type 2O [58]. Interestingly, given the controversy about the mouse HC mutations in motor neuron loss, these human mutations clusters both in the dynein microtubule-binding domain and in the region where the mouse *Loa*, *Cra*, and *Swl* mutants are mapped [58].

The prior information obtained from mouse mutants was of great value in analyzing the above-described examples of *DYNC1H1* mutations in human patients, with significant similarities between patient symptoms and mouse phenotypes resulting from motor, sensory, and cortical migration defects in *Dync1h1^Loa/+^*, *Dync1h1^Cra1/+^*, and *Dync1h1^Swl1/+^* mice (Fig. 11.2).

11.5 Dynein light chain and intermediate chain mutants

In addition to the above-described allelic series of mutations in dynein HC, three additional dynein mouse models have been reported, a mutant of dynein LIC 1, three for DLC 1 and a knockin mouse for dynein IC-1.

11.5.1 *Dync1li1^N235Y^* mutant mouse

Two LIC subunits, DYNC1LI1 and DYNC1LI2, are present in vertebrates, but only one type of LIC is incorporate in each given dynein complex; both LICs are believed to play an important role in cargo transport and in the stability of the HC [12]. DYNC1LI1 and DYNC1LI2 have been implicated in intracellular trafficking [12], endoplasmic reticulum export [59], lysosomal localization and morphology [60] and axonal retrograde transport [61]. Several proteins involved in membrane trafficking have been described to interact with DYNC1LIs, including Rab4a and FIP3, a Rab11-family interacting protein [12]. Additionally, DYNC1LI1/2 are required for many cytoplasmic dynein 1 functions during mitosis [12]. While DYNC1LI1/2 have similar roles, pericentrin binds specifically to DYNC1LI1 and Par3 to DYNC1LI2 [12].

An archive of genomic DNAs from mice with random point mutations generated by the MRC Mammalian Genetics Unit, UK, in an ENU-driven mutagenesis program [62] was screened for exon 5 of *Dync1li1* and an A to T mutant at base pair 795 was identified, which causes an asparagine to tyrosine change at the highly conserved residue 235 [63]. The line was subsequently rederived by IVF on a C57BL/6J background. In behavioral tests *Dync1li1^N235Y/N235Y^* mice displayed increased anxiety-like behavior and altered gait while on a morphological level *Dync1li1^N235Y/N235Y^* were characterized by neuronal migration defects affecting the cortex and prefrontal cortex [63]. Cortical neurons from *Dync1li1^N235Y/N235Y^* mice exhibited increased dendrite length while DRG neurons from the same mice showed an increased number of dendrite branches [63], which might correlate with Drosophila studies suggesting that the cellular machinery controlling dendrite branching is transported in endosomes and Golgi outposts [64,65]. Indeed, the *Dync1li1^N235Y^* mutation was found to reduce the trafficking of Golgi fragments and endosomes [63].

When compared to the HC mutants, the $Dync1li1^{N235Y}$ mouse model might be able to shed light on the role of specific subunits, versus the entire cytoplasmic dynein complex. Indeed, both the HC (*Loa, Cra,* and *Swl*) and $Dync1li1^{N235Y}$ mutants display defects in Golgi reassembly and EGF trafficking [63,31]; however, the behavioral phenotype of the latter is milder (less obvious gait and motor defects). In addition, Dync1h1$^{Loa/+}$ mice exhibit a severe sensory neuron deficit, which is not observed in the $Dync1li1^{N235Y/N235Y}$ mutant.

11.5.2 *Dynll1* mutant mice

DLC 1 (also known as DYNLL1, LC8, and PIN) is ubiquitously expressed and participates in a variety of essential intracellular events. DLC1 transition from monomer to dimer is crucial for its function, and this process is regulated by p21-activated kinase 1 (Pak1) phosphorylation of Ser^{88} in DLC1 [66,67]. Regulation of DLC1 dimerization promotes mammalian cell survival by modulating its interaction with the proapoptotic protein Bim [66,67]. Song and colleagues engineered two transgenic mouse models, a DLC1 wild-type overexpressor, and a mouse that overexpresses DLC1-S88A (Ser to Alanine mutation of the phosphor-site), to study the role of DLC1-Ser^{88} phosphorylation in vivo. The expression of the transgene was driven by the murine mammary tumor virus (MMTV) promoter in the mammary epithelium [68]. The authors proceeded to evaluate the effect of the DLC1-S88A mutant on apoptosis during involution of the mammary glands. In accordance with the literature, overexpression of DLC1-S8A in mammary glands accelerated involution, while mammary glands from MMTV-DLC1 mice looked similar to that of their wild-type counterpart [68]. Moreover, mammary glands from MMTV-DLC1 mice retained a percentage of adipocytes similar to wild-type mice, while a significant increase in the proportion of adipose tissue could be observed in glands from the MMTV-DLC1-S88A mice [68]. Overall, the use of this mouse model was instrumental to show how DLC1-Ser^{88} phosphorylation is critical for the role of DLC1 and overexpression of DLC1-S88A mutation can accelerate mammary gland involution.

Another two *Dynll1* mouse models are present in the literature, the $Dynll1^{Gt(EUCE0287d04)Hmgu}$ and $Dynll1^{tm1.1Jhh}$ mice. This first was generated by gene-trap and resulted in highly significant downregulation of *Dynll1* expression, consistent with it being a functional null allele in embryonic tissue [65]. At embryonic day 13.5, $Dynll1^{Gt(EUCE0287d04)Hmgu/Gt(EUCE0287d04)Hmgu}$ embryos were affected by oedema, exencephaly and coloboma, miss-patterned or no lungs, high incidence of interrupted aortic arch and common outflow tract development. In addition, high incidence of reversed heart and stomach situs was present [65]. These phenotypes are attributed to the observation that nodal cilia from $Dynll1^{Gt(EUCE0287d04)Hmgu/Gt(EUCE0287d04)Hmgu}$ embryos were shorter than wild-type controls and presented morphological abnormality such as a high incidence of bulges around the base [65].

The *Dynll1*[tm1.1Jhh] mouse model is the result of the targeted deletion of a region comprising the promoter and exons 1 and 2 of the *Dynll1* gene [69]. Homozygous floxed mice were born at the expected Mendelian ratios and were indistinguishable from wild-type littermates. The authors crossed these mice with mice bearing the Mbi-Cre to drive *Dynll1* deletion in B lymphoid cells and demonstrate that DYNLL1 is essential for the development and expansion of MYC-driven lymphomas [69].

11.5.3 *Dync1i1*-green fluorescent protein knockin mouse

Despite the importance of cytoplasmic dynein in neuronal homeostasis, not many genetic tools are available for studying this protein in neurons. For this reason Zhang and colleagues generated a knockin mouse that enables visualization of endogenous fluorescently labeled dynein in neurons. The knockin construct was also engineered to allow biochemical isolation of neuronal dynein complexes under various physiological, pathological, and pathogenic conditions to identify dynein binding partners.

Dynein IC-1 (*Dync1i1* gene) is mainly expressed in the nervous tissue and was therefore an ideal target for engineering a mouse with a neuronal-specific GFP tagged dynein. A knockin strategy was used to insert a GFP-tag and three FLAG tags before the stop codon in the middle of exon 17 of the endogenous *Dync1i1* locus. This strategy results in the expression of an IC-1-GFP-3xFLAG fusion protein under the control of the *Dync1i1* endogenous promoter [27]. The IC-1-GFP-3xFLAG fusion protein is integrated into the endogenous dynein complex, retaining dynactin interaction capacity. Furthermore its expression levels increased with the maturation of hippocampal and cortical neurons in culture, as for native protein. In addition the GFP fusion protein could be successfully visualized in cultured hippocampal neurons and the ability of the FLAG tag to pull down the dynein complex from brain tissue was confirmed by mass spectrometry [27]. Indeed, GFP-labeled dynein complexes purified from the brain tissue of *Dync1i1* GFP mice were employed to investigate the formation and motility of dynein–dynactin cocomplexes in an in vitro system. The cytoskeleton-associated protein-glycine-rich domain of dynactin was found to function as a dynamic tether by increasing the recruitment of dynein onto microtubules and decreasing the likelihood of its detachment from them [70]. This role of dynactin might be required in regions of the cell with a low microtubule density or for the maintenance of progressive cargo motility, which is crucial for long-range axonal transport [70].

Dync1i1 GFP mice were further utilized in a recent study to characterize synaptic dynein complexes via differential proteomic analysis of pull down experiments from synaptosomes of an ALS mouse model [71]. As a result, dynein was found to contribute to the localization of Staufen1, an

RNA-binding protein required for the transport of neuronal RNAs along axons and synapses [71].

In conclusion, this mouse represents an example of how genetic models can be generated to create tools for the community to facilitate the study of cytoplasmic dynein at multiple levels. Because defects in cytoplasmic dynein function are implicated in diverse neuronal disorders, including the brain development disorder lissencephaly, and neuronal diseases such as ALS, SMA, CMT disease, and Perry syndrome, tools such as this represents an invaluable resource.

11.6 Dynactin mutant mice

We have discussed a series of mouse models for cytoplasmic dynein and their contribution to the functional study of the dynein complex. Other mutants are present in the literature for proteins that, though not belonging to cytoplasmic dynein per se, are its binding partners or adaptors and can also be used to investigate its physiology.

The dynactin protein complex is an essential cofactor of cytoplasmic dynein [72]. The dynactin large subunit is called p150[Glued] and missense mutations in the *Dctn1* gene encoding p150[Glued] have been linked to neurodegenerative diseases. In particular, the C59S mutation of p150[Glued] has been associated with a slowly progressive, autosomal dominant form of lower motor neuron disease [73]. Consequently a *Dctn1* knockout and a *Dctn1* G59S knockin mouse have been generated [74]. Both of these models proved to be embryonic lethal in homozygotes, however, Dnct1[+/−] mice showed no overt phenotype, while heterozygous *Dctn1* G59S knockin mice developed a late-onset, slowly progressive motor neuron disease characterized by abnormal accumulation of neurofilaments, loss of motor neurons, and gait abnormalities [74]. One year later another group published a knockin model expressing the human G59S mutant of p150[Glued] [75]. This transgenic mouse shows similarities with the previous one and exhibits motor neuron degeneration associated with impairment of intracellular vesicular trafficking, axonal swelling, and degeneration of the axonal termini [75]. Interestingly, a recent study has shown that genetic background influences the severity of the phenotype in human *Dctn1* G59S knockin mice [76]. A more severe phenotype with earlier disease onset and decreased survival was observed when the transgene was bred onto the SJL background compared to a C57BL/6 background [76].

Interestingly, despite p150[Glued] having a dynein-binding site, in vitro studies have shown that purified dynein and dynactin form a stable complex only in the presence of the cargo adaptor Bicaudal-D2 (BICD2), a protein associated with transport of vesicles, mRNA, and nuclei [9,72]. A transgenic mouse with neuronal-specific expression of BICD2 N-terminal fragment (BICD2-N), which strongly binds to the dynein–dynactin complex [72], was generated by Teuling et al. [77]. Predictably, overexpression of BICD2-N in motor neurons affected the dynein complex and resulted in the accumulation of dynein and dynactin in the cell body and Golgi fragmentation. Axonal retrograde trafficking was also affected. However, the authors reported that *Bicd2*-N mice did not develop motor abnormalities and did not present signs of motor neuron degeneration [77]. A *Bicd2* knockout mouse was published in 2014 [78]. Interestingly, late embryonic and newborn *Bicd2*$^{-/-}$ mice were indistinguishable from their wild-type littermates, but developed later on an enlarged dome-shaped skull, weight loss, and died at postnatal day 30 [78]. At 3 weeks of age *Bicd2*$^{-/-}$ mice displayed altered locomotion, abnormal wide gait, hind limb clasping, and developed severe hydrocephalus [78]. Crossing *Bicd2*$^{-/-}$ mice with a GFAP-Cre line resulted in the disruption of the laminar organization of cerebral cortex and cerebellum due to impaired radial neuronal migration [78].

11.7 Conclusions

The cytoplasmic dynein HC allelic series that we have explored in this chapter has helped the community to establish the importance of dynein for the homeostasis of the Golgi apparatus, the endoplasmic reticulum and the endocytic pathway at a cellular level and in a physiological context, as well as pointing out the crucial role of the dynein complex in the nervous system, both in developing and ageing. Interestingly, sensory and specifically proprioceptive neurons seem to be particularly reliant on cytoplasmic dynein. Moreover these findings are supported by human patient mutations, strengthening the links between mouse models and human physiology. The MGI website (see Box 11.2) lists a large number of available mouse lines harboring mutations in all cytoplasmic dynein subunits, including mutants of the cytoplasmic ICs (*Dync1i2 and Dync2li1*), the light chain LC8-type 1 and 2 (*Dynll1 and Dynll2*), and the LC roadblock-type 1 and 2 (*Dynlrb1 and Dynlrb2*). Several mutants are also available for axonemal dyneins, which are also crucial motor complexes (see Chapters 4–11 (vol. 1 of this book)). Most of these mutant strains are the result of chemical mutagenesis or gene-trap approaches. This wealth of models should facilitate future efforts to characterize dynein physiological functions.

Acknowledgments

Our research on these topics is generously supported by the European Research Council (Neurogrowth), the Dr. Miriam and Sheldon G. Adelson Medical Research Foundation, and the Israel Science Foundation. M.F. is the incumbent of the Chaya Professorial Chair in Molecular Neuroscience at the Weizmann Institute of Science. The authors have no conflicting financial interests.

References

[1] R.H. Waterston, K. Lindblad-Toh, E. Birney, J. Rogers, J.F. Abril, P. Agarwal, R. Agarwala, R. Ainscough, M. Alexandersson, P. An, S.E. Antonarakis, J. Attwood, R. Baertsch, J. Bailey, K. Barlow, S. Beck, E. Berry, B. Birren, T. Bloom, P. Bork, M. Botcherby, N. Bray, M.R. Brent, D.G. Brown, S.D. Brown, C. Bult, J. Burton, J. Butler, R.D. Campbell, P. Carninci, S. Cawley, F. Chiaromonte, A.T. Chinwalla, D.M. Church, M. Clamp, C. Clee, F.S. Collins, L.L. Cook, R.R. Copley, A. Coulson, O. Couronne, J. Cuff, V. Curwen, T. Cutts, M. Daly, R. David, J. Davies, K.D. Delehaunty, J. Deri, E.T. Dermitzakis, C. Dewey, N.J. Dickens, M. Diekhans, S. Dodge, I. Dubchak, D.M. Dunn, S.R. Eddy, L. Elnitski, R.D. Emes, P. Eswara, E. Eyras, A. Felsenfeld, G.A. Fewell, P. Flicek, K. Foley, W.N. Frankel, L.A. Fulton, R.S. Fulton, T.S. Furey, D. Gage, R.A. Gibbs, G. Glusman, S. Gnerre, N. Goldman, L. Goodstadt, D. Grafham, T.A. Graves, E.D. Green, S. Gregory, R. Guigo, M. Guyer, R.C. Hardison, D. Haussler, Y. Hayashizaki, L.W. Hillier, A. Hinrichs, W. Hlavina, T. Holzer, F. Hsu, A. Hua, T. Hubbard, A. Hunt, I. Jackson, D.B. Jaffe, L.S. Johnson, M. Jones, T.A. Jones, A. Joy, M. Kamal, E.K. Karlsson, D. Karolchik, A. Kasprzyk, J. Kawai, E. Keibler, C. Kells, W.J. Kent, A. Kirby, D.L. Kolbe, I. Korf, R.S. Kucherlapati, E.J. Kulbokas, D. Kulp, T. Landers, J.P. Leger, S. Leonard, I. Letunic, R. Levine, J. Li, M. Li, C. Lloyd, S. Lucas, B. Ma, D.R. Maglott, E.R. Mardis, L. Matthews, E. Mauceli, J.H. Mayer, M. McCarthy, W.R. McCombie, S. McLaren, K. McLay, J.D. McPherson, J. Meldrim, B. Meredith, J.P. Mesirov, W. Miller, T.L. Miner, E. Mongin, K.T. Montgomery, M. Morgan, R. Mott, J.C. Mullikin, D.M. Muzny, W.E. Nash, J.O. Nelson, M.N. Nhan, R. Nicol, Z. Ning, C. Nusbaum, M.J. O'Connor, Y. Okazaki, K. Oliver, E.O. Larty, L. Pachter, G. Parra, K.H. Pepin, J. Peterson, P. Pevzner, R. Plumb, C.S. Pohl, A. Poliakov, T.C. Ponce, C.P. Ponting, S. Potter, M. Quail, A. Reymond, B.A. Roe, K.M. Roskin, E.M. Rubin, A.G. Rust, R. Santos, V. Sapojnikov, B. Schultz, J. Schultz, M.S. Schwartz, S. Schwartz, C. Scott, S. Seaman, S. Searle, T. Sharpe, A. Sheridan, R. Shownkeen, S. Sims, J.B. Singer, G. Slater, A. Smit, D.R. Smith, B. Spencer, A. Stabenau, N.S. Strange-Thomann, C. Sugnet, M. Suyama, G. Tesler, J. Thompson, D. Torrents, E. Trevaskis, J. Tromp, C. Ucla, A.U. Vidal, J.P. Vinson, A.C. von Niederhausern, C.M. Wade, M. Wall, R.J. Weber, R.B. Weiss, M.C. Wendl, A.P. West, K. Wetterstrand, R. Wheeler, S. Whelan, J. Wierzbowski, D. Willey, S. Williams, R.K. Wilson, E. Winter, K.C. Worley, D. Wyman, S. Yang, S.P. Yang, E.M. Zdobnov, M.C. Zody, E.S. Lander, C. Mouse Genome Sequencing, Initial sequencing and comparative analysis of the mouse genome, Nature 420 (2002) 520–562, http://dx.doi.org/10.1038/nature01262.

[2] K.L. Gibbs, L. Greensmith, G. Schiavo, Regulation of axonal transport by protein kinases, Trends Biochem. Sci. 40 (2015) 597–610, http://dx.doi.org/10.1016/j.tibs.2015.08.003.

[3] A.M. Schreij, E.A. Fon, P.S. McPherson, Endocytic membrane trafficking and neurodegenerative disease, Cell. Mol. Life Sci. 73 (2016) 1529–1545, http://dx.doi.org/10.1007/s00018-015-2105-x.

[4] A.P. Carter, A.G. Diamant, L. Urnavicius, How dynein and dynactin transport cargos: a structural perspective, Curr. Opin. Struct. Biol. 37 (2016) 62–70, http://dx.doi.org/10.1016/j.sbi.2015.12.003.

[5] G. Schiavo, L. Greensmith, M. Hafezparast, E.M. Fisher, Cytoplasmic dynein heavy chain: the servant of many masters, Trends Neurosci. 36 (2013) 641–651, http://dx.doi.org/10.1016/j.tins.2013.08.001.

[6] M.A. Schlager, C.C. Hoogenraad, Basic mechanisms for recognition and transport of synaptic cargos, Mol. Brain 2 (2009) 25, http://dx.doi.org/10.1186/1756-6606-2-25.

[7] A. Akhmanova, J.A. Hammer, Linking molecular motors to membrane cargo, Curr. Opin. Cell Biol. 22 (2010) 479–487, http://dx.doi.org/10.1016/j.ceb.2010.04.008.

[8] M.V. Hinckelmann, D. Zala, F. Saudou, Releasing the brake: restoring fast axonal transport in neurodegenerative disorders, Trends Cell Biol. 23 (2013) 634–643, http://dx.doi.org/10.1016/j.tcb.2013.08.007.

[9] C.C. Hoogenraad, A. Akhmanova, Bicaudal D family of motor adaptors: linking dynein motility to cargo binding, Trends Cell Biol. 26 (5) (2016) 327–340, http://dx.doi.org/10.1016/j.tcb.2016.01.001. 1–14.

[10] N. Hirokawa, S. Niwa, Y. Tanaka, Molecular motors in neurons: transport mechanisms and roles in brain function, development, and disease, Neuron 68 (2010) 610–638, http://dx.doi.org/10.1016/j.neuron.2010.09.039.

[11] P. Höök, R.B. Vallee, The dynein family at a glance, J. Cell Sci. 119 (2006) 4369–4371, http://dx.doi.org/10.1242/jcs.03176.

[12] V.J. Allan, Cytoplasmic dynein, Biochem. Soc. Trans. 39 (2011) 1169–1178, http://dx.doi.org/10.1042/BST0391169.

[13] S.J. King, T.A. Schroer, Dynactin increases the processivity of the cytoplasmic dynein motor, Nat. Cell Biol. 2 (2000) 20–24, http://dx.doi.org/10.1038/71338.

[14] J.D. Lane, M.A.S. Vergnolle, P.G. Woodman, V.J. Allan, Apoptotic cleavage of cytoplasmic dynein intermediate chain and p150^{Glued} stops dynein-dependent membrane motility, J. Cell Biol. 153 (2001) 1415–1426, http://dx.doi.org/10.1083/jcb.153.7.1415.

[15] A.W. Tai, J.Z. Chuang, C. Bode, U. Wolfrum, C.H. Sung, Rhodopsin's carboxy-terminal cytoplasmic tail acts as a membrane receptor for cytoplasmic dynein by binding to the dynein light chain Tctex-1, Cell 97 (1999) 877–887, http://dx.doi.org/10.1016/S0092-8674(00)80800-4.

[16] K.T. Vaughan, R.B. Vallee, Cytoplasmic dynein binds dynactin through a direct interaction between the intermediate chains and p150, J. Cell Biol. 131 (1995) 1507–1516.

[17] B. Wickstead, K. Gull, Dyneins across eukaryotes: a comparative genomic analysis, Traffic 8 (2007) 1708–1721, http://dx.doi.org/10.1111/j.1600-0854.2007.00646.x.

[18] A. Kuta, M. Hafezparast, G. Schiavo, E.M.C. Fisher, Genetic Insights Into Mammalian Cytoplasmic Dynein Function provided by Novel Mutations in the Mouse, Elsevier, 2012http://dx.doi.org/10.1016/B978-0-12-382004-4.10018-4.

[19] K.J. Palmer, H. Hughes, D.J. Stephens, Specificity of cytoplasmic dynein subunits in discrete membrane-trafficking steps, Mol. Biol. Cell 20 (2009) 2885–2899, http://dx.doi.org/10.1091/mbc.E08-12-1160.

[20] J.Z. Chuang, T.Y. Yeh, F. Bollati, C. Conde, F. Canavosio, A. Caceres, C.H. Sung, The dynein light chain tctex-1 has a dynein-independent role in actin remodeling during neurite outgrowth, Dev. Cell 9 (2005) 75–86, http://dx.doi.org/10.1016/j.devcel.2005.04.003.

[21] C. Conde, C. Arias, M. Robin, A. Li, M. Saito, J.-Z. Chuang, A.C. Nairn, C.-H. Sung, A. Cáceres, Evidence for the involvement of Lfc and Tctex-1 in axon formation, J. Neurosci. 30 (2010) 6793–6800, http://dx.doi.org/10.1523/JNEUROSCI.5420-09.2010.

[22] T. Lee, C. Shah, E.Y. Xu, Gene trap mutagenesis: a functional genomics approach towards reproductive research, Mol. Hum. Reprod. 13 (2007) 771–779, http://dx.doi.org/10.1093/molehr/gam069.

[23] C. Guan, C. Ye, X. Yang, J. Gao, A review of current large-scale mouse knockout efforts, Genesis 48 (2010) 73–85, http://dx.doi.org/10.1002/dvg.20594.

[24] Y. Ma, L. Zhang, X. Huang, Genome modification by CRISPR/Cas9, FEBS J. 281 (2014) 5186–5193, http://dx.doi.org/10.1111/febs.13110.

[25] R. Peng, G. Lin, J. Li, Potential pitfalls of CRISPR/Cas9-mediated genome editing, FEBS J. 283 (2015) 1218–1231, http://dx.doi.org/10.1111/febs.13586.

[26] L. van der Weyden, J.K. White, D.J. Adams, D.W. Logan, The mouse genetics toolkit: revealing function and mechanism, Genome Biol. 12 (2011) 224, http://dx.doi.org/10.1186/gb-2011-12-6-224.

[27] J. Zhang, A.E. Twelvetrees, J.E. Lazarus, K.R. Blasier, X. Yao, N.A. Inamdar, E.L.F. Holzbaur, K.K. Pfister, X. Xiang, Establishing a novel knock-in mouse line for studying neuronal cytoplasmic dynein under normal and pathologic conditions, Cytoskeleton 70 (2013) 215–227, http://dx.doi.org/10.1002/cm.21102.

[28] A.A. Duina, M.E. Miller, J.B. Keeney, Budding yeast for budding geneticists: a primer on the *Saccharomyces cerevisiae* model system, Genetics 197 (2014) 33–48, http://dx.doi.org/10.1534/genetics.114.163188.

[29] S. Hitotsumachi, D.A. Carpenter, W.L. Russell, Dose-repetition increases the mutagenic effectiveness of N-ethyl-N-nitrosourea in mouse spermatogonia, Proc. Natl. Acad. Sci. U.S.A. 82 (1985) 6619–6621, http://dx.doi.org/10.1073/pnas.82.19.6619.

[30] P.M. Nolan, A. Hugill, R.D. Cox, ENU mutagenesis in the mouse: application to human genetic disease, Brief. Funct. Genomic. Proteomic. 1 (2002) 278–289, http://dx.doi.org/10.1093/bfgp/1.3.278.

[31] M. Hafezparast, R. Klocke, C. Ruhrberg, A. Marquardt, A. Ahmad-Annuar, S. Bowen, G. Lalli, A.S. Witherden, H. Hummerich, S. Nicholson, P.J. Morgan, R. Oozageer, J.V. Priestley, S. Averill, V.R. King, S. Ball, J. Peters, T. Toda, A. Yamamoto, Y. Hiraoka, M. Augustin, D. Korthaus, S. Wattler, P. Wabnitz, C. Dickneite, S. Lampel, F. Boehme, G. Peraus, A. Popp, M. Rudelius, J. Schlegel, H. Fuchs, M. Hrabe de Angelis, G. Schiavo, D.T. Shima, A.P. Russ, G. Stumm, J.E. Martin, E.M.C. Fisher, Mutations in dynein link motor neuron degeneration to defects in retrograde transport, Science 300 (2003) 808–812, http://dx.doi.org/10.1126/science.1083129.

[32] C.Y. Chow, Bringing genetic background into focus, Nat. Rev. Genet. 17 (2016) 63–64, http://dx.doi.org/10.1038/nrg.2015.9.

[33] D.A. Buchner, M. Trudeau, M.H. Meisler, SCNM1, a putative RNA splicing factor that modifies disease severity in mice, Science 301 (2003) 967–969, http://dx.doi.org/10.1126/science.1086187.

[34] A. Harada, Y. Takei, Y. Kanai, Y. Tanaka, S. Nonaka, N. Hirokawa, Golgi vesiculation and lysosome dispersion in cells lacking cytoplasmic dynein, J. Cell Biol. 141 (1998) 51–59, http://dx.doi.org/10.1083/jcb.141.1.51.

[35] W. Saunders, D. Koshland, D. Eshel, I.R. Gibbons, M.A. Hoyt, *Saccharomyces cerevisiae* kinesin- and dynein-related protein required for anaphase chromosome segregation, J. Cell Biol. 128 (1995) 617–624, http://dx.doi.org/10.1083/jcb.128.4.617.

[36] M. Plamann, P.F. Minke, J.H. Tinsley, K.S. Bruno, Cytoplasmic dynein and actin-related protein Arp1 are required for normal nuclear distribution in filamentous fungi, J. Cell Biol. 127 (1994) 139–149, http://dx.doi.org/10.1083/jcb.127.1.139.

[37] X. Xiang, S.M. Beckwith, N.R. Morris, Cytoplasmic dynein is involved in nuclear migration in Aspergillus nidulans, Proc. Natl. Acad. Sci. U.S.A. 91 (1994) 2100–2104. http://www.ncbi.nlm.nih.gov/pubmed/8134356.

[38] J. Gepner, M. Li, S. Ludmann, C. Kortas, K. Boylan, S.J. Iyadurai, M. McGrail, T.S. Hays, Cytoplasmic dynein function is essential in *Drosophila melanogaster*, Genetics 142 (1996) 865–878. http://www.ncbi.nlm.nih.gov/pubmed/8849893.

[39] T.-Y. Chen, J.-S. Syu, T.-Y. Han, H.-L. Cheng, F.-I. Lu, C.-Y. Wang, Cell cycle-dependent localization of dynactin subunit p150glued at centrosome, J. Cell. Biochem. 116 (2015) 2049–2060, http://dx.doi.org/10.1002/jcb.25160.

[40] S.H. Tynan, M.A. Gee, R.B. Vallee, Distinct but overlapping sites within the cytoplasmic dynein heavy chain for dimerization and for intermediate chain and light intermediate chain binding, J. Biol. Chem. 275 (2000) 32769–32774, http://dx.doi.org/10.1074/jbc.M001537200.

[41] J. Eschbach, J. Sinniger, J. Bouitbir, A. Fergani, A.-I. Schlagowski, J. Zoll, B. Geny, F. René, Y. Larmet, V. Marion, R.H. Baloh, M.B. Harms, M.E. Shy, N. Messadeq, P. Weydt, J.-P. Loeffler, A.C. Ludolph, L. Dupuis, Dynein mutations associated with hereditary motor neuropathies impair mitochondrial morphology and function with age, Neurobiol. Dis. 58 (2013) 220–230, http://dx.doi.org/10.1016/j.nbd.2013.05.015.

[42] H.S. Ilieva, K. Yamanaka, S. Malkmus, O. Kakinohana, T. Yaksh, M. Marsala, D.W. Cleveland, Mutant dynein (Loa) triggers proprioceptive axon loss that extends survival only in the SOD1 ALS model with highest motor neuron death, Proc. Natl. Acad. Sci. U.S.A. 105 (2008) 12599–12604, http://dx.doi.org/10.1073/pnas.0805422105.

[43] X.-J. Chen, E.N. Levedakou, K.J. Millen, R.L. Wollmann, B. Soliven, B. Popko, Proprioceptive sensory neuropathy in mice with a mutation in the cytoplasmic Dynein heavy chain 1 gene, J. Neurosci. 27 (2007) 14515–14524, http://dx.doi.org/10.1523/JNEUROSCI.4338-07.2007.

[44] L. Dupuis, A. Fergani, K.E. Braunstein, J. Eschbach, N. Holl, F. Rene, J.L. Gonzalez De Aguilar, B. Zoerner, B. Schwalenstocker, A.C. Ludolph, J.P. Loeffler, Mice with a mutation in the dynein heavy chain 1 gene display sensory neuropathy but lack motor neuron disease, Exp. Neurol. 215 (2009) 146–152, http://dx.doi.org/10.1016/j.expneurol.2008.09.019.

[45] W. Deng, C. Garrett, B. Dombert, V. Soura, G. Banks, E.M.C. Fisher, M.P. Van Der Brug, M. Hafezparast, Neurodegenerative mutation in cytoplasmic dynein alters its organization and dynein-dynactin and dynein-kinesin interactions, J. Biol. Chem. 285 (2010) 39922–39934, http://dx.doi.org/10.1074/jbc.M110.178087.

[46] K.M. Ori-McKenney, J. Xu, S.P. Gross, R.B. Vallee, A cytoplasmic dynein tail mutation impairs motor processivity, Nat. Cell Biol. 12 (2010) 1228–1234, http://dx.doi.org/10.1038/ncb2127.

[47] C.A. Garrett, M. Barri, A. Kuta, V. Soura, W. Deng, E.M.C. Fisher, G. Schiavo, M. Hafezparast, DYNC1H1 mutation alters transport kinetics and ERK1/2-cFos signalling in a mouse model of distal spinal muscular atrophy, Brain 137 (2014) 1883–1893, http://dx.doi.org/10.1093/brain/awu097.

[48] I. Rishal, N. Kam, R.B. Perry, V. Shinder, E.M. Fisher, G. Schiavo, M. Fainzilber, A motor-driven mechanism for cell-length sensing, Cell Rep. 1 (2012) 608–616, http://dx.doi.org/10.1016/j.celrep.2012.05.013.

[49] P.B. Perry, I. Rishal, E. Doron-Mandel, A.L. Kalinski, K.F. Medzihradszky, M. Terenzio, S. Alber, S. Koley, A. Lin, M. Rozenbaum, D. Yudin, P.K. Sahoo, C. Gomes, V. Shinder, W. Geraisy, E.A. Huebner, C.J. Woolf, A. Yaron, A.L. Burlingame, J.L. Twiss, M. Fainzilber, Nucleolin-mediated RNA localization regulates article nucleolin-mediated RNA localization regulates neuron growth and cycling cell size, Cell Rep. 16 (2016) 1–13, http://dx.doi.org/10.1016/j.celrep.2016.07.005.

[50] L.W. Duchen, A dominant hereditary sensory disorder in the mouse with deficiency of muscle spindles: the mutant sprawling, J. Physiol. 237 (1974) 10P–11P. http://www.ncbi.nlm.nih.gov/pubmed/4274920.

[51] L.E. Vissers, J. de Ligt, C. Gilissen, I. Janssen, M. Steehouwer, P. de Vries, B. van Lier, P. Arts, N. Wieskamp, M. del Rosario, B.W. van Bon, A. Hoischen, B.B. de Vries, H.G. Brunner, J.A. Veltman, A de novo paradigm for mental retardation, Nat. Genet. 42 (2010) 1109–1112, http://dx.doi.org/10.1038/ng.712.

[52] M.H. Willemsen, L.E.L. Vissers, M.A. Willemsen, B.W. van Bon, T. Kroes, J. de Ligt, B.B. de Vries, J. Schoots, D. Lugtenberg, B.C.J. Hamel, H. van Bokhoven, H.G. Brunner, J.A. Veltman, T. Kleefstra, Mutations in DYNC1H1 cause severe intellectual disability with neuronal migration defects, J. Med. Genet. 49 (2012) 179–183, http://dx.doi.org/10.1136/jmedgenet-2011-100542.

[53] K. Poirier, N. Lebrun, L. Broix, G. Tian, Y. Saillour, C. Boscheron, E. Parrini, S. Valence, B. Saint Pierre, M. Oger, D. Lacombe, D. Geneviève, E. Fontana, F. Darra, C. Cances, M. Barth, D. Bonneau, B.D. Bernadina, S. N'guyen, C. Gitiaux, P. Parent, V. des Portes, J.M. Pedespan, V. Legrez, L. Castelnau-Ptakine, P. Nitschke, T. Hieu, C. Masson, D. Zelenika, A. Andrieux, F. Francis, R. Guerrini, N.J. Cowan, N. Bahi-Buisson, J. Chelly, Mutations in TUBG1, DYNC1H1, KIF5C and KIF2A cause malformations of cortical development and microcephaly, Nat. Genet. 45 (2013) 639–647, http://dx.doi.org/10.1038/ng.2613.

[54] M.N. Weedon, R. Hastings, R. Caswell, W. Xie, K. Paszkiewicz, T. Antoniadi, M. Williams, C. King, L. Greenhalgh, R. Newbury-Ecob, S. Ellard, Exome sequencing identifies a DYNC1H1 mutation in a large pedigree with dominant axonal Charcot-Marie-Tooth disease, Am. J. Hum. Genet. 89 (2011) 308–312, http://dx.doi.org/10.1016/j.ajhg.2011.07.002.

[55] Y. Tsurusaki, S. Saitoh, K. Tomizawa, A. Sudo, N. Asahina, H. Shiraishi, J. Ito, H. Tanaka, H. Doi, H. Saitsu, N. Miyake, N. Matsumoto, A DYNC1H1 mutation causes a dominant spinal muscular atrophy with lower extremity predominance, Neurogenetics 13 (2012) 327–332, http://dx.doi.org/10.1007/s10048-012-0337-6.

[56] M.B. Harms, K.M. Ori-McKenney, M. Scoto, E.P. Tuck, S. Bell, D. Ma, S. Masi, P. Allred, M. Al-Lozi, M.M. Reilly, L.J. Miller, A. Jani-Acsadi, A. Pestronk, M.E. Shy, F. Muntoni, R.B. Vallee, R.H. Baloh, Mutations in the tail domain of DYNC1H1 cause dominant spinal muscular atrophy, Neurology 78 (2012) 1714–1720, http://dx.doi.org/10.1212/WNL.0b013e3182556c05.

[57] M. Scoto, A.M. Rossor, M.B. Harms, S. Cirak, M. Calissano, S. Robb, A.Y. Manzur, A. Martínez Arroyo, A. Rodriguez Sanz, S. Mansour, P. Fallon, I. Hadjikoumi, A. Klein, M. Yang, M. De Visser, W.C. Overweg-Plandsoen, F. Baas, J.P. Taylor, M. Benatar, A.M. Connolly, M.T. Al-Lozi, J. Nixon, C.G. de Goede, A.R. Foley, C. Mcwilliam, M. Pitt, C. Sewry, R. Phadke, M. Hafezparast, W.K. Chong, E. Mercuri, R.H. Baloh, M.M. Reilly, F. Muntoni, Novel mutations expand the clinical spectrum of DYNC1H1 -associated spinal muscular atrophy, Neurology 84 (2015) 668–679.

[58] A.V. Strickland, M. Schabhüttl, H. Offenbacher, M. Synofzik, N.S. Hauser, M. Brunner-Krainz, U. Gruber-Sedlmayr, S.A. Moore, R. Windhager, B. Bender, M. Harms, S. Klebe, P. Young, M. Kennerson, A.S. Garcia, M.A. Gonzalez, S. Züchner, R. Schule, M.E. Shy, M. Auer-Grumbach, Mutation screen reveals novel variants and expands the phenotypes associated with DYNC1H1, J. Neurol. 9 (2015) 2124–2134, http://dx.doi.org/10.1007/s00415-015-7727-2.

[59] S. Kong, X. Du, X. Du, C. Peng, Y. Wu, H. Li, X. Jin, L. Hou, K. Deng, T. Xu, W. Tao, Dlic1 deficiency impairs ciliogenesis of photoreceptors by destabilizing dynein, Cell Res. 23 (2013) 835–850, http://dx.doi.org/10.1038/cr.2013.59.

[60] S.C. Tan, J. Scherer, R.B. Vallee, Recruitment of dynein to late endosomes and lysosomes through light intermediate chains, Mol. Biol. Cell 22 (2011) 467–477, http://dx.doi.org/10.1091/mbc.E10-02-0129.

[61] S.P. Koushika, A.M. Schaefer, R. Vincent, J.H. Willis, B. Bowerman, M.L. Nonet, Mutations in *Caenorhabditis elegans* cytoplasmic dynein components reveal specificity of neuronal retrograde cargo, J. Neurosci. 24 (2004) 3907–3916, http://dx.doi.org/10.1523/JNEUROSCI.5039-03.2004.

[62] P.M. Nolan, J. Peters, M. Strivens, D. Rogers, J. Hagan, N. Spurr, I.C. Gray, L. Vizor, D. Brooker, E. Whitehill, R. Washbourne, T. Hough, S. Greenaway, M. Hewitt, X. Liu, S. McCormack, K. Pickford, R. Selley, C. Wells, Z. Tymowska-Lalanne, P. Roby, P. Glenister, C. Thornton, C. Thaung, J.A. Stevenson, R. Arkell, P. Mburu, R. Hardisty, A. Kiernan, A. Erven, K.P. Steel, S. Voegeling, J.L. Guenet, C. Nickols, R. Sadri, M. Nasse, a Isaacs, K. Davies, M. Browne, E.M. Fisher, J. Martin, S. Rastan, S.D. Brown, J. Hunter, A systematic, genome-wide, phenotype-driven mutagenesis programme for gene function studies in the mouse, Nat. Genet. 25 (2000) 440–443, http://dx.doi.org/10.1038/78140.

[63] G.T. Banks, M.A. Haas, S. Line, H.L. Shepherd, M. Alqatari, S. Stewart, I. Rishal, A. Philpott, B. Kalmar, A. Kuta, M. Groves, N. Parkinson, A. Acevedo-Arozena, S. Brandner, D. Bannerman, L. Greensmith, M. Hafezparast, M. Koltzenburg, R. Deacon, M. Fainzilber, E.M. Fisher, Behavioral and other phenotypes in a cytoplasmic dynein light intermediate chain 1 mutant mouse, J. Neurosci. 31 (2011) 5483–5494, http://dx.doi.org/10.1523/JNEUROSCI.5244-10.2011.

[64] D. Satoh, D. Sato, T. Tsuyama, M. Saito, H. Ohkura, M.M. Rolls, F. Ishikawa, T. Uemura, Spatial control of branching within dendritic arbors by dynein-dependent transport of Rab5-endosomes, Nat. Cell Biol. 10 (2008) 1164–1171, http://dx.doi.org/10.1038/ncb1776.

[65] P. Goggolidou, J.L. Stevens, F. Agueci, J. Keynton, G. Wheway, D.T. Grimes, S.H. Patel, H. Hilton, S.K. Morthorst, A. Di Paolo, D.J. Williams, J. Sanderson, S.V. Khoronenkova, N. Powles-Glover, A. Ermakov, C.T. Esapa, R. Romero, G.L. Dianov, J. Briscoe, C.A. Johnson, L.B. Pedersen, D.P. Norris, ATMIN is a transcriptional regulator of both lung morphogenesis and ciliogenesis, Development 141 (2014) 3966–3977, http://dx.doi.org/10.1242/dev.107755.

[66] H. Puthalakath, D.C. Huang, L.A. O'Reilly, S.M. King, A. Strasser, The proapoptotic activity of the Bcl-2 family member Bim is regulated by interaction with the dynein motor complex, Mol. Cell 3 (1999) 287–296, http://dx.doi.org/10.1016/S1097-2765(00)80456-6.

[67] R.K. Vadlamudi, R. Bagheri-Yarmand, Z. Yang, S. Balasenthil, D. Nguyen, A.A. Sahin, P. Den Hollander, R. Kumar, Dynein light chain 1, a p21-activated kinase 1-interacting substrate, promotes cancerous phenotypes, Cancer Cell 5 (2004) 575–585, http://dx.doi.org/10.1016/j.ccr.2004.05.022.

[68] C. Song, W. Wen, S.K. Rayala, M. Chen, J. Ma, M. Zhang, R. Kumar, Serine 88 phosphorylation of the 8-kDa dynein light chain 1 is a molecular switch for its dimerization status and functions, J. Biol. Chem. 283 (2008) 4004–4013, http://dx.doi.org/10.1074/jbc.M704512200.

[69] D.M. Wong, L. Li, S. Jurado, A. King, R. Bamford, M. Wall, M.K. Walia, G.L. Kelly, C.R. Walkley, D.M. Tarlinton, A. Strasser, J. Heierhorst, The transcription factor ASCIZ and its target DYNLL1 are essential for the development and expansion of MYC-driven B cell lymphoma, Cell Rep. 14 (2016) 1488–1499, http://dx.doi.org/10.1016/j.celrep.2016.01.012.

[70] S. Ayloo, J.E. Lazarus, A. Dodda, M. Tokito, E.M. Ostap, E.L.F. Holzbaur, Dynactin functions as both a dynamic tether and brake during dynein-driven motility, Nat. Commun. 5 (2014) 4807, http://dx.doi.org/10.1038/ncomms5807.

[71] N. Gershoni-Emek, A. Mazza, M. Chein, T. Gradus-Pery, X. Xiang, K.W. Li, R. Sharan, E. Perlson, Proteomic analysis of dynein-interacting proteins in amyotrophic lateral sclerosis synaptosomes reveals alterations in the RNA-binding protein staufen1, Mol. Cell. Proteomics (2015)http://dx.doi.org/10.1074/mcp.M115.049965.

[72] L. Urnavicius, K. Zhang, A.G. Diamant, C. Motz, M.A. Schlager, M. Yu, N.A. Patel, C. V Robinson, A.P. Carter, The structure of the dynactin complex and its interaction with dynein, Science 347 (2015) 1441–1446, http://dx.doi.org/10.1126/science.aaa4080.

[73] I. Puls, C. Jonnakuty, B.H. LaMonte, E.L.F. Holzbaur, M. Tokito, E. Mann, M.K. Floeter, K. Bidus, D. Drayna, S.J. Oh, R.H. Brown, C.L. Ludlow, K.H. Fischbeck, Mutant dynactin in motor neuron disease, Nat. Genet. 33 (2003) 455–456, http://dx.doi.org/10.1038/ng1123.

[74] C. Lai, X. Lin, J. Chandran, H. Shim, W.-J. Yang, H. Cai, The G59S mutation in p150[glued] causes dysfunction of dynactin in mice, J. Neurosci. 51 (2007) 13982–13990, http://dx.doi.org/10.1523/JNEUROSCI.4226-07.2007.

[75] F.M. Laird, M.H. Farah, S. Ackerley, A. Hoke, N. Maragakis, J.D. Rothstein, J. Griffin, D.L. Price, L.J. Martin, P.C. Wong, Motor neuron disease occurring in a mutant dynactin mouse model is characterized by defects in vesicular trafficking, J. Neurosci. 28 (2008) 1997–2005, http://dx.doi.org/10.1523/JNEUROSCI.4231-07.2008.

[76] T.D. Heiman-Patterson, E.P. Blankenhorn, R.B. Sher, J. Jiang, P. Welsh, M.C. Dixon, J.I. Jeffrey, P. Wong, G.A. Cox, G.M. Alexander, Genetic background effects on disease onset and lifespan of the mutant dynactin p150[glued] mouse model of motor neuron disease, PLoS One 10 (2015) e0117848, http://dx.doi.org/10.1371/journal.pone.0117848.

[77] E. Teuling, V. van Dis, P.S. Wulf, E.D. Haasdijk, A. Akhmanova, C.C. Hoogenraad, D. Jaarsma, A novel mouse model with impaired dynein/dynactin function develops amyotrophic lateral sclerosis (ALS)-like features in motor neurons and improves lifespan in SOD1-ALS mice, Hum. Mol. Genet. 17 (2008) 2849–2862, http://dx.doi.org/10.1093/hmg/ddn182.

[78] D. Jaarsma, R. van den Berg, P.S. Wulf, S. van Erp, N. Keijzer, M.A. Schlager, E. de Graaff, C.I. De Zeeuw, R.J. Pasterkamp, A. Akhmanova, C.C. Hoogenraad, A role for Bicaudal-D2 in radial cerebellar granule cell migration, Nat. Commun. 5 (2014) 3411, http://dx.doi.org/10.1038/ncomms4411.

[79] K. Kevin Pfister, E.M. Fisher, I.R. Gibbons, T.S. Hays, E.L. Holzbaur, J. Richard McIntosh, M.E. Porter, T.A. Schroer, K.T. Vaughan, G.B. Witman, S.M. King, R.B. Vallee, Cytoplasmic dynein nomenclature, J. Cell Biol. 171 (2005) 411–413, http://dx.doi.org/10.1083/jcb.200508078.

[80] K.K. Pfister, P.R. Shah, H. Hummerich, A. Russ, J. Cotton, A.A. Annuar, S.M. King, E.M. Fisher, Genetic analysis of the cytoplasmic dynein subunit families, PLoS Genet. 2 (2006) 11–26, http://dx.doi.org/10.1371/journal.pgen.0020001.

In this chapter

Cytoplasmic dynein and its regulators in neocortical development and disease

David J. Doobin, Richard B. Vallee
Columbia University, New York, NY, United States

12.1 Neocortical development

The neocortex is a specialized structure of the mammalian central nervous system made up of six distinct neuronal layers located in the superficial portion of the forebrain [1] (Fig. 12.1). Eighty percent of the neurons here are excitatory and project axons to other regions of the neocortex or deep brain structures such as the thalamus, basal ganglia, cerebellum, hindbrain, or nuclei within the spinal cord [2]. Therefore, the neocortex plays a paramount role in coordinating the activity of neurons throughout the central nervous system.

Development of the rodent neocortex is quite similar to that of the human neocortex (Fig. 12.1), with the most noticeable difference being between the proliferative zones containing the stem cells, leading to the gross differences in the extent and size of the gyrations observed in the adult human brain [3]. Development of the neocortex begins during embryogenesis with the formation and closure of the neural tube, which begins as a single layer of pseudostratified neuroepithelial cells [4] (Fig. 12.2). These cells are highly elongated, with an apical process that contacts the lumen of the neural tube (eventually the ventricular surface), and a basal process that extends to the extracellular matrix at the outer surface of the neural tube. The neuroepithelial cells exhibit an unusual form of proliferative behavior involving interkinetic nuclear migration (INM), in which the nucleus oscillates between the apical and basal surfaces of the neural tube in a cell cycle-dependent manner [5–7]. Mitosis occurs exclusively at the ventricular surface. During G1 the nucleus moves basally, undergoes S phase at the apex of migration, and migrates apically back toward the ventricular surface during G2 (Fig. 12.2). Notably, the neuroepithelial cells enter mitosis only after returning to the ventricular surface [7].

The neuroepithelial cells undergo multiple rounds of INM, dividing symmetrically to produce two progenitors, leading to exponential expansion of the progenitor pool. By embryonic day 9 in mice, roughly corresponding to gestation

Dyneins. https://doi.org/10.1016/B978-0-12-809470-9.00012-6

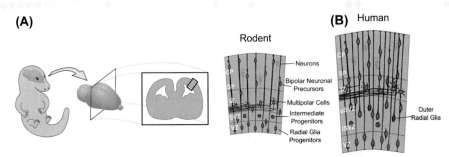

Figure 12.1 Comparison of rodent and human neocortical development. Schematic illustrating the location of the neocortex within the developing rodent and human embryos. There are striking differences in the proliferative regions containing radial glia progenitors (RGPs). (A) In the rodent neocortex all RGP cell bodies are located within the ventricular zone (VZ), with some transient amplifying cells in the subventricular zone (SVZ). The intermediate zone (IZ) contains horizontally oriented axons, while the subplate (SP) and cortical plate (CP) represent the expanding neuronal layers in which the bipolar migrating neurons accumulate. (B) In the developing human neocortex, the SVZ is greatly expanded into an inner SVZ (iSVZ) and outer SVZ (oSVZ). It is here that the cell bodies of the specialized outer radial glia (oRG) are found, which help greatly amplify the number of neurons generated in the primate brain. Otherwise the zones of the developing neocortex are quite similar to those in the rodent neocortex.

week 6 in humans, the first asymmetric neuroepithelial cell divisions occur [8]. In this case one of the progeny retains its stem cell potential and its elongated morphology. The other daughter cell retracts its apical and basal processes and migrates as a postmitotic neuronal precursor toward the basal (outer) surface of the neural tube [4,8]. The arriving cells establish a transient embryonic structure known as the preplate, which consists of Cajal–Retzius cells, which secrete reelin (RLN) to establish a chemoattractant gradient that guides neuronal migration, and the earliest neurons of the neocortex that send pioneering axons to target deep subcortical regions [2,9].

Following the period of preplate formation neuroepithelial cells are referred to as radial glia progenitors (RGPs). They retain their connections to the apical and basal surfaces, which become the respective ventricular and pial surfaces of the developing neocortex. The RGP nuclei continue to exhibit INM, though nuclear migration within the cell is now restricted to the ventricular zone (VZ), defined as the portion of the neocortex below the preplate [4,8]. As with most zones in embryonic development, the VZ is a transient region of great importance during development, though it disappears in the adult brain. RGPs have been found to give rise to all of the cells that populate the adult neocortex, with the exception of the microglia, which arise from the bone marrow [8] The basal processes of the RGP continue to lengthen throughout embryonic brain development and serve as tracks along which postmitotic neuronal precursors migrate toward their final destinations within the neocortex [6,10].

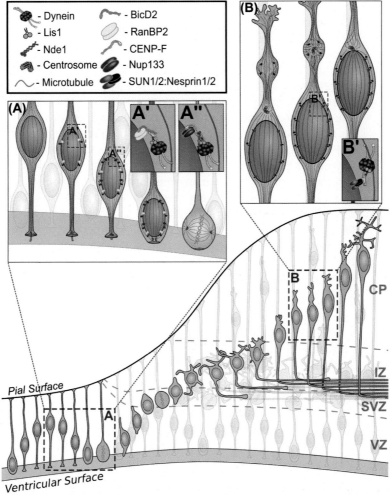

Figure 12.2 Various roles for the dynein complex during neocortical development. The neocortex begins as a single layer of neuroepithelial progenitors that eventually divide asymmetrically, creating a postmitotic neuronal precursor and a RGP. The nuclei of neuroepithelial cells and, the RGPs they become, oscillate in synchrony with cell cycle progression, restricting mitosis to the ventricular surface, in a behavior termed INM. The RGP nuclei only undergo INM within the VZ of the developing brain. The progeny of asymmetric RGP divisions may divide a couple of times in the SVZ as intermediate progenitors, and they then migrate into the IZ where they adopt a multipolar morphology. Here they emit a nascent axon, and retract the remaining neurites with the exception of the neurite oriented most basally, which becomes the leading process. These neuronal precursors then migrate along basal radial glia fibers to their final destinations within the CP, which develops into the neocortex. (A) During INM there are two separate pathways by which dynein is recruited to the nuclear envelope. The first involves the nucleoporin RanBP2 binding the dynein adaptor BicD2 (*A′*), while the second involves Nup133 recruiting CENP-F to bring in Nde1, which binds the dynein complex (*A″*). Lis1 is present in both pathways. (B) During bipolar migration of the neuronal precursors dynein is required to move the centrosome and nucleus, where the SUN/Nesprin complex is thought to help anchor dynein to the nuclear envelope of the migrating bipolar cell (*B′*).

As the RGP cells continue to divide, the progeny of asymmetric divisions leave the VZ and migrate toward the outer pial surface of the developing brain [10](Fig. 12.2). There are a number of behaviors these cells undergo, however, while en route. The precursors first pass through the subventricular zone (SVZ), where they can divide further as transient amplifying cells termed intermediate/basal progenitors (but it is unclear if all RGP progeny must pass through an intermediate progenitor stage) [11]. Following these divisions the neuronal precursors continue to migrate outward and enter the intermediate zone (IZ). Here, in the lower IZ, they encounter a dense arrangement of axons oriented orthogonally to the RGP basal fibers (Fig. 12.2). The neuronal precursors adopt a multipolar morphology at this point, with several processes extending transiently in all directions, and they remain in this state anywhere from 24 to 48 hours [12]. One of the processes outgrows the others and becomes the nascent axon, which then typically extends in the medial or apical direction and grows along the axon tracts established by the earlier arriving neurons [13,14]. A second polarization event follows as another process thickens to become the leading migratory process [15], and the remaining transient processes retract. All together these changes are referred to as the multipolar-to-bipolar transition.

The migratory process attaches to the basal fibers of the RGPs and helps guide the neuronal precursors as they migrate radially through the IZ and into the developing cortical plate (CP), which will become the neocortex during postnatal development (Fig. 12.2) [16,17]. As the CP expands, the six neuronal layers are established in an inside-out manner, with earlier born neuronal precursors populating the deeper layers and later born precursors contributing to the more superficial layers [10]. Whole cell migration is sometimes referred to as "locomotion," to contrast with another form of migration—translocation—in which the migratory process attaches to the pial surface of the brain, and then shortens to bring the nucleus closer to this region [18].

As a consequence of radial migration, neurons generated from individual RGPs are arranged in columnar radial units, which preferentially form microcircuits with each other [19–21]. Additionally, roughly 20% of neurons in the neocortex are inhibitory interneurons, which arise from RGPs located on the ventral surface of the rostral lateral ventricle [22,23]. These interneuron precursors must migrate tangentially as they make their way to the neocortex [24], where they then incorporate into the circuitry and provide important modulatory signals.

Neocortical development in higher order mammals, such as primates and humans, differs from that of rodents most noticeably in the complexity of the proliferative regions [3,25]. The primate SVZ, which in rodents contains the transit amplifying intermediate progenitors, is greatly expanded and separated into the inner and outer SVZ (Fig. 12.1). The outer SVZ was recently found to contain a novel and distinct progenitor cell population known as basal radial glia (also termed outer radial glia [oRGs]) [26,27]. oRGs are derived from the

RGPs in the VZ and retain the capacity to divide symmetrically or asymmetrically and to create different fate lineages [28]. The oRGs have a unique morphology, with a basal process connecting to the pial surface, but no apical process. The presence of the oRGs is thought to greatly expand the proliferative capacity of the stem cell population of the developing neocortex, and contribute to increased neuronal production, leading to the complex gyri and sulci found in humans and other primates.

The extent to which oRGs retain the unusual subcellular behavior of RGPs remains to be fully explored. So far it has been observed that oRG nuclei exhibit a form of blebbistatin-inhibited, myosin-driven basal movement termed mitotic somal translocation, which occurs immediately prior to mitosis [29]. The basal processes of the oRGs also contribute as guides for the neuronal precursors as they migrate through the CP in the later stages of primate neocortical development, though during early development the basal processes of RGPs in the VZ are still required [30].

12.2 The role of dynein in radial glia progenitors and interkinetic nuclear migration

Cytoplasmic dynein is involved in a number of independent aspects of neurogenesis and migration during neocortical development. One crucial role for dynein and its regulatory proteins is during INM in RGPs [15]. Throughout this process the RGP centrosome remains at the apical terminus, within the end foot of the cell, where it serves as the microtubule-organizing center [31]. This creates a highly uniform microtubule network, with microtubule minus ends located at the apical pole of the RGPs, and 93% of plus ends extending up toward the basal process [31]. The nucleus is able to utilize this arrangement of microtubules and to recruit microtubule motors at different stages of the cell cycle to drive nuclear migration in the basal and apical directions. During G1 the unconventional kinesin-3 KIF1A was found to be responsible for basal movement of the nucleus [31,32], and it is not until G2 that dynein is recruited to the nuclear envelope to drive apical INM [33].

We have found that there are two separate and sequential pathways through which dynein is recruited to the nuclear envelope during apical INM (Fig. 12.2A) [7]. The first pathway is anchored by the nucleoporin RanBP2, which recruits the dynein adaptor BicD2 to the nuclear envelope, which in turn recruits Lis1 and dynein, and likely dynactin as well [7,33]. Early G2 dynein recruitment is responsible for the initial stage of apical INM, but as the nucleus approaches the ventricle, it is thought that the biophysical forces needed to overcome the increased crowding are greater [34]. Therefore more dynein is required at the nuclear envelope to complete apical INM. For this purpose a second pathway of dynein recruitment becomes active, in which the nucleoporin Nup133 binds CENP-F, which is sequestered within the nucleus until late G2 [33,35,36]. The nuclear envelope-associated CENP-F then recruits the dynein regulatory proteins Nde1 and Lis1, which are thought to recruit and activate enough dynein at the nuclear envelope to allow the RGP nucleus to complete apical INM, reach the ventricular surface, and enter mitosis (Fig. 12.2A) [7].

These two pathways for dynein recruitment also function in cultured nonneuronal cells, including HeLa, in which dynein may facilitate nuclear envelope breakdown and initiate mitotic spindle formation, though nuclear envelope dynein is not required for entry into mitosis [33,36–38]. In the RGPs, however, dynein recruitment is required for apical INM, and apical INM is, in turn, critical for entry into mitosis at the ventricular surface. This conclusion is supported by in utero electroporation of shRNAs directed against several of the dynein recruitment factors [7,15]. When there is insufficient dynein at the nuclear envelope, the RGP nuclei are blocked from reaching the ventricular surface, leading to a great reduction in the mitotic index and proliferative capacity of the progenitor pool [7,15,31,33,39]. When the late pathway of dynein recruitment to the nuclear envelope is interfered with—such as with knockdown of *Nup133*, *CENP-F*, or *Nde1*—overexpression of BicD2 will bring enough dynein to the nuclear envelope of RGPs to allow for the completion of apical INM [7,39].

Because apical INM only occurs during G2, we speculated that recruitment of dynein to the nuclear envelope might depend on a cell cycle kinase. Specifically, the increase in CyclinB levels during G2 results in gradual activation of cyclin-dependent kinase-1 (Cdk1) [33]. We found the early G2 pathway of dynein recruitment to be activated by Cdk1 phosphorylation of RanBP2 at four sites, which promotes binding of BicD2 and the additional components of the dynein complex [33]. This mechanism is conserved in RGPs and HeLa cells [33]. Pharmacological inhibition of Cdk1 in HeLa cells blocks CENP-F from ever leaving the nucleus, suggesting a role for Cdk1 in regulating the late-G2 pathway of dynein recruitment [33]. In addition to functioning during apical INM, Lis1 has also been shown to contribute to prophase nuclear envelope invagination in RGPs [40]. Finally multiple of these dynein regulatory proteins are also critical for mitotic spindle formation and mitotic progression in RGPs [41,42].

A similar pattern of INM has been found in other developing tissues and has been investigated in detail in the progenitor cells of the developing zebrafish retina, *Drosophila* imaginal discs, and *Cnidarian* pseudostratified ectoderm. In these tissues basal and apical INM have been found to rely heavily on myosin II [43–45]. Myosin II inhibition was found to interfere with the basal INM in analysis of cell cohorts in mouse, at least during earlier neocortical development [46]. Using live imaging, our lab found no effect of myosin II in INM using blebbistatin and myosin IIb RNAi. Whether the observed differences in motor protein use are tissue, species, or developmental stage dependent is unknown [31]. We do note a role for dynactin during INM in zebrafish retinal epithelium, suggesting that the dynein pathway is also active in this system [47].

12.3 Roles for the dynein pathway in postmitotic neuronal precursors

Following mitosis, neuronal precursors migrate into the IZ, where dynein is required as they assume a multipolar morphology and begin axonogenesis (Fig. 12.2) [10]. The multipolar-to-bipolar transition is a complex process, dependent on changes in transcription [48], centrosome reorientation [13,49], proper outgrowth of the nascent axon [50], and N-cadherin exocytosis [51,52]. Major cytoskeletal changes must occur, including the retraction of auxiliary neurites and the formation of an enlarged process on the pial-directed side of the multipolar cell, which becomes the leading process [53]. Knockdown of dynein or a variety of its accessory proteins, including *Lis1*, *Nde1*, *Ndel1*, *CENP-F*, and *BicD2*, impairs the multipolar-to-bipolar transition [7,39]. The underlying mechanism for these effects is unclear, though they suggest a role for dynein in the cytoskeletal rearrangements that accompany this aspect of neuronal morphogenesis.

During the multipolar stage neuronal precursors extend and retract a variety of neurites in a very dynamic manner, one of which later becomes the axon [12]. Axonal growth then occurs and involves both extension and guidance of the axon by the exploratory movements of the growth cone [54]. Neuronal precursors can progress through the multipolar-to-bipolar transition even if axonogenesis is blocked [50], though the majority of neuronal precursors extend a defined axon prior to beginning bipolar radial migration [55,56].

Axonogenesis, itself, involves proteins in the dynein pathway, as determined by in vitro analysis of primary cultured neurons and in vivo analysis of the multipolar-to-bipolar transition in neuronal precursors. Dynein, dynactin, and Lis1 all advance into the leading edge of the growth cone, followed by microtubule extension. *Lis1* RNAi in multipolar cells within the IZ causes axons to grow at a slower rate and with a more branched growth cone than normal, though the axons are still oriented correctly in the IZ [15]. Work in cultured dorsal root ganglia and hippocampal neurons suggests that Lis1 and dynein are critical for growth cone remodeling and allowing microtubules to enter into the growth cone [57]. Additionally dynein and its regulatory proteins are thought to facilitate microtubule sliding to promote axon outgrowth by anchoring to stabilized F-actin in the cortex [57–59]. Dynein- and kinesin-dependent vesicular transport are also critical components of axonal outgrowth and maintenance [60], and this aspect of dynein function is discussed at length in Chapter 13.

Following nascent axon specification and outgrowth, the multipolar cell in the lower IZ must assume a bipolar morphology and migrate toward its final location in the CP. As the leading process forms at the end of the multipolar-to-bipolar transition, there is an accumulation of growing microtubules into the leading process, which may help with its enlargement and stability [49]. The centrosome must reorient itself from the base of the growing axon to the base of the leading process [49]. Once the bipolar neuronal precursor forms, it must migrate in the pial direction along the basal processes of RGP fibers. The gap junction proteins connexin 26 and connexin 43 have been reported to be critical for the attachment of neuronal precursors to the radial glia [61], just as astrotactin 1 is necessary for cerebellar granular cells to migrate along Bergman glial fibers in the cerebellum [62].

During bipolar migration dynein is required at multiple stages of the saltatory migration cycle (Fig. 12.2B). First, a swelling forms within the leading process distal to the nucleus and centrosome, with the centrosome serving as the microtubule-organizing center. Dynein is enriched in the swelling prior to movement of the centrosome or nucleus [53,63,64]. The centrosome advances into the swelling, likely a consequence of the anchored dynein in the swelling pulling on the microtubule network to draw the centrosome closer [53]. The microtubule array advances with the centrosome, leaving the nucleus surrounded by an ordered array of microtubules. The nucleus then advances by microtubule minus end-directed transport toward the centrosome. This behavior suggests that dynein must also be localized to the nuclear envelope. Although nuclear envelope dynein has not been visualized as it has during INM [7], dynein localization within the entire cell body region of migrating bipolar cells suggests that there is a role for nuclear envelope anchored dynein [53,65].

Interference with dynein or its regulatory proteins impedes the migration of the bipolar neuronal precursors and offers insight into a further dynein requirement during neocortical development. Strong knockdown of *Lis1* blocks the movement of both the centrosome and nucleus into the leading process, whereas weak knockdown of *Lis1* or dynein heavy chain allows for slower than usual centrosome excursions into the leading process without permitting any apparent nucleokinesis [53]. Intriguingly, RNAi for *Ndel1*—which does not cause as strong of an arrest at the multipolar stage as shRNA to its paralog *Nde1*—results in a similar bipolar arrest, wherein the centrosome can enter the leading process but the nucleus is unable to follow [39].

Similar to RGPs, migrating bipolar cells could in principle employ nuclear envelope anchored dynein to accomplish nucleokinesis. In fact, the SUN domain proteins SUN1 and SUN2, which span the inner nuclear membrane, and the KASH domain proteins Nesprin1 and Nesprin2, which span the outer nuclear membrane, play roles in tethering dynein to the nuclear envelope during neuronal migration [65]. SUN1/2 and Nesprin2 are all required for bipolar migration in the neocortex, with Nesprin1 not essential for neuronal migration but functionally complementary to Nesprin2. Colocalization and immunoprecipitation studies have suggested that dynactin participates in the Nesprin complexes during bipolar migration [65], but further work into the dynein regulatory proteins involved in this process is needed.

In contrast to the lack of a role for myosin II during INM in embryonic rat brain [31], myosin II has been found to contribute to nuclear advance in migrating bipolar neuronal precursors in several studies [53,64,65]. The myosin inhibitor blebbistatin impeded nucleokinesis, though centrosome movement into the leading process was unaffected [53]. The same effect was observed following in utero electroporation of myosin IIb shRNA [53]. Thus, the dynein complex plays a critical role in centrosomal advance into the leading process and subsequent nucleokinesis, with myosin II assisting during nucleokinesis. Once the migrating neuronal precursors reach their final destination in the correct layer of the neocortex, the leading process develops many elaborate protrusions and the process of dendrite formation begins [66].

12.4 Overview of malformations of cortical development associated with dynein mutations

The multiple requirements for the dynein complex during neocortical development illustrate the many opportunities for errors that can potentially impact brain development, and in the past few years there has been a great increase in the identification of human mutations in the dynein heavy chain and in dynein regulatory proteins associated with malformations of cortical development (MCD). The clinical classifications for MCDs are quite complex and depend largely on neuroradiological analysis [67], and a variety of genetic mutations have been associated with each type of MCD [68]. Here the focus will be given to MCDs caused by mutations in genes encoding dynein and its regulatory proteins. We give particular attention to mutations in *LIS1* and *DYNC1H1*, encoding dynein heavy chain, which are associated with lissencephaly (smooth brain lacking gyrations) [69,70], polymicrogyria (many smaller gyrations), pachygyria (fewer gyrations overall), and microcephaly (gross reduction in total brain size) [71], and mutations in *NDE1* associated with severe cases of microcephaly [72,73]. Although many mutations in genes encoding dynein, dynactin, and *BICD2* have been associated with neurological disorders such as Charcot–Marie–Tooth, spinal muscular atrophy, and amyotrophic lateral sclerosis [74–77], these diseases are predominantly peripheral neuropathies and likely involve roles for dynein in axonal transport [78] and perhaps other neuronal functions.

12.5 Lissencephaly associated with *LIS1* mutations

One of the first pieces of evidence that cytoplasmic dynein might be important during neocortical development came from patients with lissencephaly, an MCD characterized by a smooth brain lacking gyri and sulci [69]. The causative gene, *LIS1*, was identified [69], and then shortly thereafter found to be homologous to *Aspergillus NudF,* in the cytoplasmic dynein pathway [79]. The severity of lissencephaly varies depending on which gene mutation a patient carries, and there can be variable penetrance among patients with the same genetic alterations [80,81]. Clinical symptoms are often severe and include intellectual disability, epilepsy, and failure to thrive. On pathological examination the lissencephalic cortex has an oversimplified pattern of gyri and sulci macroscopically, and a novel four-layered cortex microscopically (very severe forms of lissencephaly only have two layers) [70]. Mutations in doublecortin (*DCX*; which is involved in the regulation of kinesin KIF1A [32,82]), tubulin A (*TUBA1A*), *RLN*, and other genes have been linked to lissencephaly [68], though the first discovered and most common genetic causes of lissencephaly are deletions and mutations in *LIS1*.

Human mutations in *LIS1* (also known as *PAFAH1B1* for an additional role as a subunit of platelet activating factor acetylhydrolase [83]) can range from heterozygous deletions of *LIS1* and its adjacent gene *YWHAE*, resulting in the Miller–Dieker form of lissencephaly, or intragenic deletions, duplications, substitutions, or insertions within *LIS1*, resulting in isolated lissencephaly sequence. Rare instances of *LIS1* mutations are associated with subcortical band heterotopia, which is where there is a normal six-layered neocortex with an ectopic band of neurons closer to the ventricle, and thought of as an MCD on the same pathological spectrum as lissencephaly but with less severe pathological impairments [69,70,81,84].

Given the critical role of Lis1 during neocortical development it becomes apparent how mutations in *LIS1* could contribute to lissencephaly [15,31,53,85,86]. The organization of the four-layered lissencephalic neocortex suggests that the position of the Cajal–Retzius cells is preserved in layer 1, but the other five layers are essentially inverted from the normal neocortical organization and organized into three distinct layers, and located in a region normally occupied only by axonal tracts [70]. This suggests that postmitotic neurons in LIS1-deficient patients migrate into the IZ, but cannot successfully complete bipolar neuronal migration to reach the CP, and they build up in successive layers opposite to the normal inside-out organization of the neocortex [87]. Patients with *LIS1* mutations often exhibit agenesis of the corpus callosum as well. Although this is a gross defect that can result from many causes, it may reflect the role for LIS1 in proper axon elongation and guidance during corticogenesis [80]. Mouse models have provided evidence that neocortical development is highly sensitive to Lis1 dosage [86,88], which may explain why the effects of *LIS1* mutations are not more severe, such as resulting in blocked INM leading to microcephaly. We also suspect that this is due to the heterozygous nature of the human mutations, with residual amounts of LIS1 being sufficient for neurogenesis but impeding neuronal migration and cortical lamination.

12.6 Malformations of cortical development associated with *DYNC1H1* mutations

Recently mutations in the cytoplasmic dynein 1 heavy chain gene, *DYNC1H1*, have also been implicated in MCDs, with features of polymicrogyria, pachygyria, and microcephaly (Fig. 12.3A) [71,75,89–92]. Nearly all patients

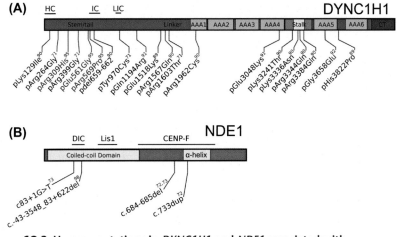

Figure 12.3 Human mutations in *DYNC1H1* and *NDE1* associated with malformations in cortical development. Over the past few years there has been a large increase in the number of human mutations associated with MCDs identified in the *DYNC1H1* and *NDE1* genes. (A) Most of the *DYNC1H1* mutations lead to single amino acid substitutions, with the exception of one 4-amino acid deletion, and are denoted by the amino acid changes on the bar diagram. The scale of the diagram is proportional to the amino acid position in the protein. All patients are heterozygous for given mutations. Superscript reflects the reference of the original studies that uncovered the mutation. *Blue text* indicates that the mutation was found by deep sequencing to be somatic mosaicism. (B) Mutations in *NDE1* associated with microcephaly. Nomenclature is that given to the mutations in the original studies. Scale of diagram is proportional to protein size. Not included is a schematic of Lis1 and the known associated human mutations, in part because of the large volume of documented mutations in *LIS1*, and the complexity of many of these mutations.

with *DYNC1H1* mutations have microcephaly, though in a much less severe form than microcephalies associated with *NDE1* mutations (see Section 12.7 below). Patients with polymicrogyria have multiple gyri that are smaller than normal, and the mutations in *DYNC1H1* were the first recorded cases of polymicrogyria with pachygyria, which refers to fewer gyri overall [90]. These patients can also have agenesis of the corpus callosum and basal ganglia malformations [90]. The dynein mutations are in all cases heterozygous de novo missense mutations (and one known deletion of four amino acids), the majority of which occur in the stalk, linker, or microtubule-binding domains of the protein (Fig. 12.3A).

The patients with *DYNC1H1* mutations typically present with intellectual disability or attention deficit hyperactivity disorder (ADHD) [71], and a subset of the patients also have epilepsy [89]. Unlike *LIS1* mutations that predominantly affect the posterior neocortex, *DYNC1H1* mutations tend to affect the anterior neocortex [71,80]. Because dynein is required for the proliferation of RGPs in the VZ, the formation and extension of the axon, and the postmitotic migration of neuronal precursors [7,53,93], it is readily apparent how dynein heavy chain mutations might affect brain development.

DYNC1H1 mutations have classically been associated with peripheral neuropathies, but recent reports introduce clinical cases of *DYNC1H1* mutations leading to peripheral neuropathies and MCDs that were possibly overlooked earlier [71,89,91]. Interestingly, MCDs coinciding with peripheral neuropathies are predicted from legs at odd angles (*Loa*) mouse model [93]. *Loa* heterozygous mice exhibit clear retrograde axonal transport defects coupled with neurodegeneration [94], and *Loa* homozygous recessive mice have profound impairments in bipolar neuronal migration and axonal extension [93].

Symptomatically, patients with *DYNC1H1* mutations leading to MCDs and peripheral neuropathies have mild cognitive impairments with or without ADHD and provide a functional clinical link between how mutations in *DYNC1H1* can effect neocortical development plus peripheral nerve health [71,89,91]. Additionally, an extensive recent study used deep-sequencing technology to investigate MCDs of unknown genetic origins, and found a substantial number of patients that had somatic mutations in genes already known to cause MCDs when mutated in the germ line [95], with *LIS1* and *DYNC1H1* mutations among those found. Thus, over the past few years there has been an increasingly apparent role of *DYNC1H1* mutations in causing MCDs, particularly those with atypical neuroradiology [71].

Mutations in *DYNC1H1* and the genes encoding other regulatory proteins, such as BicD2 and dynactin should remain high on a clinical index of suspicion for undiagnosed cases of MCDs, especially in patients with peripheral neuropathies linked to *BICD2* and *p150* mutations [74,76,96]. Additionally, the reported substitution mutations on cytoplasmic dynein could offer new insight into the biophysical and biochemical function of the protein with great relevancy to human health.

12.7 *NDE1* mutations and the pathogenesis of severe microcephaly

Though *DYNC1H1* mutations are associated with microcephaly, another dynein regulatory protein has recently been found to be associated with some of the most severe genetic forms of this condition encountered clinically. Mutations in *NDE1* have been associated with multiple unrelated cases of severe microcephaly, often accompanied by lissencephaly [72,73], fetal brain disruption (collapsed skull with prominent scalp rugae) [97], or hydranencephaly (gross underdevelopment of the neocortex accompanied by prominent expansion of the ventricles) [98]. The lissencephaly in these patients can be so severe that only the Sylvian fissure—one of the most prominent and conserved neuroanatomical structures—remains. The mutations in *NDE1* involve premature terminations [72,73], or deletions of entire chromosomal regions containing the *NDE1*-locus [97]. All patients have homozygous recessive mutations (Fig. 12.3B).

NDE1-associated microcephaly patients all exhibit extreme microcephaly, with reductions in neocortical size more severe than other genetic forms of microcephaly, such as those associated with mutations in *ASPM* or *WDR62* [99,100]. Ventriculomegaly, which is the abnormal expansion of the ventricular space, is also seen in all patients with *NDE1* mutations. Some patients with *NDE1*-mediated microcephaly have epilepsy, whereas others have such drastic underdevelopment of neuroanatomical structures that no seizures occur [72,73,97,98]. All patients have severe cognitive impairment and meet only minimal developmental milestones. Some of the patients develop persistent hypothermia, which suggests brainstem abnormalities, though all other organs develop normally [97], and all patients are immunocompetent.

Nde1 is highly expressed in the developing neocortex in comparison to other embryonic tissue, and its level then drops significantly after birth [101]. Within the developing neocortex *Nde1* expression is highest in the VZ and SVZ, which helps to explain why aberrations in Nde1 function impair neurogenesis [102]. *Nde1* knockout mice displayed altered RGP mitotic spindle alignment and reduced cortical thickness, though they did not recapitulate the severity of the disease seen in human patients [41,103].

Conditional knockout of the *Nde1* paralog *Ndel1* resulted in neuronal migration defects with no impairments in RGP proliferation [88]. There have been no recorded cases of human mutations in *NDEL1* leading to MCDs, likely because all *NDE1* mutations are penetrant only when homozygous recessive, and *Ndel1* knockout mice are embryonic lethal at the peri-implantation stage, implying that any patients with *NDEL1* mutations may die during early embryonic development [104]. It was not until recently that acute knockdown of both *Nde1* and *Ndel1* in the developing rodent brain revealed the cellular mechanisms by which these severe cases of microlissencephaly may occur [39].

While most genetic causes of microcephaly impact cell cycle progression of RGPs in either DNA-repair or mitosis [67], the severity of *NDE1*-mediated microcephaly suggested particularly severe impediments to cell cycle progression. RNAi for *Nde1* blocked a substantial number of RGPs during apical INM, consistent with the role for the protein in the late-G2 pathway of dynein recruitment to the nuclear envelope (Figs. 12.2A and 12.4A) [39]. Whereas RNAi for the *Nde1* paralog *Ndel1* had no discernable effect on RGP cell cycle progression, knockdown of both genes caused a marked increase in RGPs arrested further from the ventricular surface than RNAi for *Nde1* (Fig. 12.4A) [39]. Surprisingly, the majority of the RGP cells subjected to *Nde1/Ndel1* RNAi were positive for the G1 marker CyclinD1. Since previous work in cultured nonneuronal cells had implicated Nde1 in the initiation of primary cilium resorption, which is linked to G1-to-S progression, we tested whether primary cilium length was impacted by *Nde1/Ndel1* double knockdown. We found these structures to be overelongated in the *Nde1/Ndel1* double knockdown, where they were doubled in length. When ciliogenesis was completely blocked in these cells, the G1/S-phase block was rescued, indicating a causative role of the primary cilia in the G1-S transition in RGPs [39].

An additional unexpected finding regarded the G2-M transition. We found that the *Nde1* RNAi-mediated apical INM arrest could be rescued not only by expression of RNAi-insensitive *Nde1* but also by expression of *Ndel1*. However, *Ndel1* expression could not rescue entry into mitosis in RGP cells lacking *Nde1*. Instead, these cells accumulated at the ventricular surface with intact nuclear envelopes, uncondensed DNA, and centrosomes still sequestered to the base of the apical end foot. To ensure that this arrest at the G2-M transition was not an artifact of *Ndel1* overexpression, and indeed represented a unique role for *Nde1*, we overexpressed *BicD2* to recruit additional dynein to the nuclear envelope in RGP cells treated with *Nde1* RNAi. Though expression of *BicD2* was able to rescue apical INM and mitotic entry in RGPs treated with *CENP-F* RNAi [7], it could not rescue mitotic entry in RGPs treated with *Nde1* RNAi (Fig. 12.4A) [39]. This reinforced a unique premitotic role of Nde1 in RGP cells, distinct from its involvement in dynein recruitment to the nuclear envelope. These results revealed three critical roles for Nde1 in RGP cell cycle progression (Fig. 12.4B), in addition

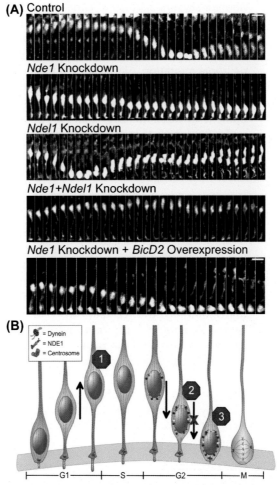

Figure 12.4 Multiple distinct defects in cell cycle progression account for the severity of *NDE1*-associated microcephaly. (A) Time-lapse images of radial glia progenitor (RGP) cells expressing shRNA to *Nde1* or *Ndel1* in tandem with a cytoplasmic GFP fluorescent marker in an E19 rat brain slice [39]. In comparison with control and *Ndel1* knockdown conditions, which complete apical INM and mitosis normally, *Nde1* and *Nde1/Ndel1* double knockdown conditions impede the movement of the RGP nucleus. *BicD2* overexpression, with *Nde1* knockdown, rescues apical INM but not entry into mitosis. Each panel is separated by 30 min. Scale bar is 10 μm. (B) Cartoon illustration of the three nonmitotic requirements for Nde1 in the RGP cell cycle: (1) at the primary cilia during the G1-to-S transition, (2) during apical nuclear migration in G2, and (3) at the G2-to-M transition.

to its known roles during mitosis [105–107], which may all contribute to the increased severity of NDE1 microcephaly in comparison to microcephaly resulting from other genes, which have less impact on RGP cell cycle progression.

Additionally, knockdown of either *Nde1* or *Ndel1* impaired postmitotic neuronal migration and caused a large accumulation of postmitotic neuronal precursors in the multipolar stage, with nearly no cells reaching the CP [39]. Interestingly, both proteins could functionally compensate for each other during postmitotic neuronal migration. Since Nde1 is required for postmitotic neuronal migration, it becomes apparent why most patients with *NDE1*-associated microcephaly exhibit altered lamination, as well as lissencephaly. The drastic underdevelopment of other parts of the central nervous system in *NDE1* microcephaly patients suggest that NDE1 may be critical in RGPs outside of the neocortex and highlights the need to further study the development of structures such as the basal ganglia and brainstem.

12.8 Summary

In conclusion, dynein and its regulatory proteins are critical for a variety of aspects of neurogenesis and neuronal migration during neocortical development. Mutations in these genes lead to a variety of cortical malformations with clinical symptoms ranging from ADHD to crippling seizures and severe intellectual disability. Considerable progress has been made toward understanding the cellular basis of these defects, but many questions remain regarding basic mechanisms, as well as the distinct effects of specific mutations on these proteins. For example, mutations in both *DYNC1H1* and *NDE1* cause microcephaly, though *NDE1*-associated microcephaly is far more severe, likely due to homozygous recessive *NDE1* mutations in patients compared to heterozygous substitution mutations in *DYNC1H1* patients. Similarly, all of the patient mutations discussed lead to cortical lamination defects, reflecting the role of the dynein complex in bipolar radial migration.

Our expanding knowledge of how adaptor proteins regulate the dynein motor complex and increase its ability to carry cargo along microtubules, coupled with the knowledge of the cell biology underlying neocortical development, should help clinicians and researchers provide improved diagnoses for patients with neurological disorders, and will fuel the research into potential therapeutic options. As the role of proper neocortical organization becomes apparent for diseases such as autism [108], and there becomes a greater appreciation for the role of somatic mutations in causing focal areas of malformed neocortex [109], a greater understanding of the roles of dynein and its regulatory proteins during brain development will be crucial.

Acknowledgments

We would like to thank Aurelie Carabalona, Tiago Dantas, and João Goncalves for their critical reading of the manuscript. Supported by NIH HD40182 to R.B.V, and NIH NINDS F30NS095577 to D.J.D.

References

[1] M.F. Glasser, T.S. Coalson, E.C. Robinson, C.D. Hacker, J. Harwell, E. Yacoub, et al., A multi-modal parcellation of human cerebral cortex, Nature 536 (2016) 171–178, http://dx.doi.org/10.1038/nature18933.

[2] O. Marín, U. Müller, Lineage origins of GABAergic versus glutamatergic neurons in the neocortex, Curr. Opin. Neurobiol. 26 (2014) 132–141, http://dx.doi.org/10.1016/j.conb. 2014.01.015.

[3] E. Lewitus, I. Kelava, W.B. Huttner, Conical expansion of the outer subventricular zone and the role of neocortical folding in evolution and development, Front. Hum. Neurosci. 7 (2013) 424, http://dx.doi.org/10.3389/fnhum.2013.00424.

[4] J.T. Paridaen, W.B. Huttner, Neurogenesis during development of the vertebrate central nervous system, EMBO Rep. 15 (2014) 351–364, http://dx.doi.org/10.1002/embr.201438447.

[5] F.C. Sauer, Mitosis in the neural tube, J. Comp. Neurol. 62 (1935) 377–405, http://dx.doi.org/10.1002/cne.900620207.

[6] S.C. Noctor, A.C. Flint, T.A. Weissman, R.S. Dammerman, A.R. Kriegstein, Neurons derived from radial glial cells establish radial units in neocortex, Nature 409 (2001) 714–720, http://dx.doi.org/10.1038/35055553.

[7] D.J.-K. Hu, A.D. Baffet, T. Nayak, A. Akhmanova, V. Doye, R.B. Vallee, Dynein recruitment to nuclear pores activates apical nuclear migration and mitotic entry in brain progenitor cells, Cell 154 (2013) 1300–1313, http://dx.doi.org/10.1016/j.cell.2013.08.024.

[8] A. Kriegstein, A. Alvarez-Buylla, The glial nature of embryonic and adult neural stem cells, Annu. Rev. Neurosci. 32 (2009) 149–184, http://dx.doi.org/10.1146/annurev. neuro.051508.135600.

[9] E.-E. Govek, M.E. Hatten, L. Van Aelst, The role of Rho GTPase proteins in CNS neuronal migration, Dev. Neurobiol. 71 (2011) 528–553, http://dx.doi.org/10.1002/dneu.20850.

[10] S.C. Noctor, V. Martínez-Cerdeño, L. Ivic, A.R. Kriegstein, Cortical neurons arise in symmetric and asymmetric division zones and migrate through specific phases, Nat. Neurosci. 7 (2004) 136–144, http://dx.doi.org/10.1038/nn1172.

[11] W. Haubensak, A. Attardo, W. Denk, W.B. Huttner, Neurons arise in the basal neuroepithelium of the early mammalian telencephalon: a major site of neurogenesis, Proc. Natl. Acad. Sci. USA 101 (2004) 3196–3201, http://dx.doi.org/10.1073/pnas.0308600100.

[12] H. Tabata, K. Nakajima, Multipolar migration: the third mode of radial neuronal migration in the developing cerebral cortex, J. Neurosci. 23 (2003) 9996–10001.

[13] F.C. de Anda, K. Meletis, X. Ge, D. Rei, L.H. Tsai, Centrosome motility is essential for initial axon formation in the neocortex, J. Neurosci. 30 (2010) 10391–10406, http://dx.doi.org/10.1523/JNEUROSCI.0381-10.2010.

[14] F. Polleux, W. Snider, Initiating and growing an axon, Cold Spring Harb. Perspect. Biol. 2 (2010) a001925, http://dx.doi.org/10.1101/cshperspect.a001925.

[15] J.W. Tsai, Y. Chen, A.R. Kriegstein, R.B. Vallee, LIS1 RNA interference blocks neural stem cell division, morphogenesis, and motility at multiple stages, J. Cell Biol. 170 (2005) 935–945, http://dx.doi.org/10.1083/jcb.200505166.

[16] P. Rakic, Mode of cell migration to the superficial layers of fetal monkey neocortex, J. Comp. Neurol. 145 (1972) 61–83, http://dx.doi.org/10.1002/cne.901450105.

[17] R.B. Vallee, G.E. Seale, J.-W. Tsai, Emerging roles for myosin II and cytoplasmic dynein in migrating neurons and growth cones, Trends Cell Biol. 19 (2009) 347–355, http://dx.doi.org/10.1016/j.tcb.2009.03.009.

[18] O. Marín, M. Valiente, X. Ge, L.-H. Tsai, Guiding neuronal cell migrations, Cold Spring Harb. Perspect. Biol. 2 (2010) a001834, http://dx.doi.org/10.1101/cshperspect.a001834.

[19] P. Rakic, Specification of cerebral cortical areas, Science 241 (1988) 170–176.

[20] P. Gao, M.P. Postiglione, T.G. Krieger, L. Hernandez, C. Wang, Z. Han, et al., Deterministic progenitor behavior and unitary production of neurons in the neocortex, Cell 159 (2014) 775–788, http://dx.doi.org/10.1016/j.cell.2014.10.027.

[21] S. He, Z. Li, S. Ge, Y.-C. Yu, S.-H. Shi, Inside-out radial migration facilitates lineage-dependent neocortical microcircuit assembly, Neuron 86 (2015) 1159–1166, http://dx.doi.org/10.1016/j.neuron.2015.05.002.

[22] C.C. Harwell, L.C. Fuentealba, A. Gonzalez-Cerrillo, P.R.L. Parker, C.C. Gertz, E. Mazzola, et al., Wide dispersion and diversity of clonally related inhibitory interneurons, Neuron 87 (2015) 999–1007, http://dx.doi.org/10.1016/j.neuron.2015.07.030.

[23] C. Mayer, X.H. Jaglin, L.V. Cobbs, R.C. Bandler, C. Streicher, C.L. Cepko, et al., Clonally related forebrain interneurons disperse broadly across both functional areas and structural boundaries, Neuron 87 (2015) 989–998, http://dx.doi.org/10.1016/j.neuron.2015.07.011.

[24] F. Polleux, K.L. Whitford, P.A. Dijkhuizen, T. Vitalis, A. Ghosh, Control of cortical interneuron migration by neurotrophins and PI3-kinase signaling, Development (2002) 3147–3160.

[25] B.E. LaMonica, J.H. Lui, X. Wang, A.R. Kriegstein, OSVZ progenitors in the human cortex: an updated perspective on neurodevelopmental disease, Curr. Opin. Neurobiol. 22 (2012) 747–753, http://dx.doi.org/10.1016/j.conb.2012.03.006.

[26] D.V. Hansen, J.H. Lui, P.R.L. Parker, A.R. Kriegstein, Neurogenic radial glia in the outer subventricular zone of human neocortex, Nature 464 (2010) 554–561, http://dx.doi.org/10.1038/nature08845.

[27] S.A. Fietz, I. Kelava, J. Vogt, M. Wilsch-Bräuninger, D. Stenzel, J.L. Fish, et al., OSVZ progenitors of human and ferret neocortex are epithelial-like and expand by integrin signaling, Nat. Neurosci. 13 (2010) 690–699, http://dx.doi.org/10.1038/nn.2553.

[28] B.E. LaMonica, J.H. Lui, D.V. Hansen, A.R. Kriegstein, Mitotic spindle orientation predicts outer radial glial cell generation in human neocortex, Nat. Commun. 4 (2013) 1665–1711, http://dx.doi.org/10.1038/ncomms2647.

[29] B.E.L. Ostrem, J.H. Lui, C.C. Gertz, A.R. Kriegstein, Control of outer radial glial stem cell mitosis in the human brain, Cell Rep. 8 (2014) 656–664, http://dx.doi.org/10.1016/j.celrep.2014.06.058.

[30] T.J. Nowakowski, A.A. Pollen, C. Sandoval-Espinosa, A.R. Kriegstein, Transformation of the radial glia scaffold demarcates two stages of human cerebral cortex development, Neuron 91 (2016) 1219–1227, http://dx.doi.org/10.1016/j.neuron.2016.09.005.

[31] J.-W. Tsai, W.-N. Lian, S. Kemal, A.R. Kriegstein, R.B. Vallee, Kinesin 3 and cytoplasmic dynein mediate interkinetic nuclear migration in neural stem cells, Nat. Publ. Group 13 (2010) 1463–1471, http://dx.doi.org/10.1038/nn.2665.

[32] A. Carabalona, D.J.-K. Hu, R.B. Vallee, KIF1A inhibition immortalizes brain stem cells but blocks BDNF-mediated neuronal migration, Nat. Neurosci. 19 (2016) 253–262, http://dx.doi.org/10.1038/nn.4213.

[33] A.D. Baffet, D.J. Hu, R.B. Vallee, Cdk1 activates pre-mitotic nuclear envelope dynein recruitment and apical nuclear migration in neural stem cells, Dev. Cell 33 (2015) 703–716, http://dx.doi.org/10.1016/j.devcel.2015.04.022.

[34] T. Miyata, M. Okamoto, T. Shinoda, A. Kawaguchi, Interkinetic nuclear migration generates and opposes ventricular-zone crowding: insight into tissue mechanics, Front. Cell Neurosci. 8 (2014) 473, http://dx.doi.org/10.3389/fncel.2014.00473.

[35] M. Zuccolo, A. Alves, V. Galy, S. Bolhy, E. Formstecher, V. Racine, et al., The human Nup107–160 nuclear pore subcomplex contributes to proper kinetochore functions, EMBO J. 26 (2007) 1853–1864.

[36] S. Bolhy, I. Bouhlel, E. Dultz, T. Nayak, M. Zuccolo, X. Gatti, et al., A Nup133-dependent NPC-anchored network tethers centrosomes to the nuclear envelope in prophase, J. Cell Biol. 192 (2011) 855–871, http://dx.doi.org/10.1083/jcb.201007118.

[37] D. Splinter, M.E. Tanenbaum, A. Lindqvist, D. Jaarsma, A. Flotho, K.L. Yu, et al., Bicaudal D2, dynein, and kinesin-1 associate with nuclear pore complexes and regulate centrosome and nuclear positioning during mitotic entry, PLoS Biol. 8 (2010) e1000350, http://dx.doi.org/10.1371/journal.pbio.1000350.

[38] D. Splinter, D.S. Razafsky, M.A. Schlager, A. Serra-Marques, I. Grigoriev, J. Demmers, et al., BICD2, dynactin, and LIS1 cooperate in regulating dynein recruitment to cellular structures, Mol. Biol. Cell 23 (2012) 4226–4241, http://dx.doi.org/10.1091/mbc.E12-03-0210.

[39] D.J. Doobin, S. Kemal, T.J. Dantas, R.B. Vallee, Severe NDE1-mediated microcephaly results from neural progenitor cell cycle arrests at multiple specific stages, Nat. Commun. 7 (2016) 12551, http://dx.doi.org/10.1038/ncomms12551.

[40] S. Hebbar, M.T. Mesngon, A.M. Guillotte, B. Desai, R. Ayala, D.S. Smith, Lis1 and Ndel1 influence the timing of nuclear envelope breakdown in neural stem cells, J. Cell Biol. 182 (2008) 1063–1071, http://dx.doi.org/10.1083/jcb.200803071.

[41] Y. Feng, C.A. Walsh, Mitotic spindle regulation by Nde1 controls cerebral cortical size, Neuron 44 (2004) 279–293, http://dx.doi.org/10.1016/j.neuron.2004.09.023.

[42] J. Yingling, Y.H. Youn, D. Darling, K. Toyo-oka, T. Pramparo, S. Hirotsune, et al., Neuroepithelial stem cell proliferation requires LIS1 for precise spindle orientation and symmetric division, Cell 132 (2008) 474–486, http://dx.doi.org/10.1016/j.cell.2008.01.026.

[43] C. Norden, S. Young, B.A. Link, W.A. Harris, Actomyosin is the main driver of interkinetic nuclear migration in the retina, Cell 138 (2009) 1195–1208, http://dx.doi.org/10.1016/j.cell.2009.06.032.

[44] P.J. Strzyz, H.O. Lee, J. Sidhaye, I.P. Weber, L.C. Leung, C. Norden, Interkinetic nuclear migration is centrosome independent and ensures apical cell division to maintain tissue integrity, Dev. Cell 32 (2015) 203–219, http://dx.doi.org/10.1016/j.devcel.2014.12.001.

[45] E.J. Meyer, A. Ikmi, M.C. Gibson, Interkinetic nuclear migration is a broadly conserved feature of cell division in pseudostratified epithelia, Curr. Biol. 21 (2011) 485–491, http://dx.doi.org/10.1016/j.cub.2011.02.002.

[46] J. Schenk, M. Wilsch-Bräuninger, F. Calegari, W.B. Huttner, Myosin II is required for interkinetic nuclear migration of neural progenitors, Proc. Natl. Acad. Sci. USA 106 (2009) 16487–16492, http://dx.doi.org/10.1073/pnas.0908928106.

[47] F. Del Bene, A.M. Wehman, B.A. Link, H. Baier, Regulation of neurogenesis by interkinetic nuclear migration through an apical-basal notch gradient, Cell 134 (2008) 1055–1065, http://dx.doi.org/10.1016/j.cell.2008.07.017.

[48] C. Ohtaka-Maruyama, S. Hirai, A. Miwa, J.I.-T. Heng, H. Shitara, R. Ishii, et al., RP58 regulates the multipolar-bipolar transition of newborn neurons in the developing cerebral cortex, Cell Rep. 3 (2013) 458–471, http://dx.doi.org/10.1016/j.celrep.2013.01.012.

[49] A. Sakakibara, T. Sato, R. Ando, N. Noguchi, M. Masaoka, T. Miyata, Dynamics of centrosome translocation and microtubule organization in neocortical neurons during distinct modes of polarization, Cereb. Cortex (2013)http://dx.doi.org/10.1093/cercor/bhs411.

[50] A.P. Barnes, B.N. Lilley, Y.A. Pan, L.J. Plummer, A.W. Powell, A.N. Raines, et al., LKB1 and SAD kinases define a pathway required for the polarization of cortical neurons, Cell 129 (2007) 549–563, http://dx.doi.org/10.1016/j.cell.2007.03.025.

[51] Y. Jossin, J.A. Cooper, Reelin, Rap1 and N-cadherin orient the migration of multipolar neurons in the developing neocortex, Nat. Neurosci. 14 (2011) 697–703, http://dx.doi.org/10.1038/nn.2816.

[52] G. La Fata, A. Gärtner, N. Domínguez-Iturza, T. Dresselaers, J. Dawitz, R.B. Poorthuis, et al., FMRP regulates multipolar to bipolar transition affecting neuronal migration and cortical circuitry, Nat. Neurosci. 17 (2014) 1693–1700, http://dx.doi.org/10.1038/nn.3870.

[53] J.-W. Tsai, K.H. Bremner, R.B. Vallee, Dual subcellular roles for LIS1 and dynein in radial neuronal migration in live brain tissue, Nat. Neurosci. 10 (2007) 970–979, http://dx.doi.org/10.1038/nn1934.

[54] T.L. Lewis, J. Courchet, F. Polleux, Cell biology in neuroscience: cellular and molecular mechanisms underlying axon formation, growth, and branching, J. Cell Biol. 202 (2013) 837–848, http://dx.doi.org/10.1083/jcb.201305098.

[55] Y. Hatanaka, K. Yamauchi, Excitatory cortical neurons with multipolar shape establish neuronal polarity by forming a tangentially oriented axon in the intermediate zone, Cereb. Cortex 23 (2012) 105–113, http://dx.doi.org/10.1093/cercor/bhr383.

[56] T. Namba, Y. Kibe, Y. Funahashi, S. Nakamuta, T. Takano, T. Ueno, et al., Pioneering axons regulate neuronal polarization in the developing cerebral cortex, Neuron 81 (2014) 814–829, http://dx.doi.org/10.1016/j.neuron.2013.12.015.

[57] P.W. Grabham, G.E. Seale, M. Bennecib, D.J. Goldberg, R.B. Vallee, Cytoplasmic dynein and LIS1 are required for microtubule advance during growth cone remodeling and fast axonal outgrowth, J. Neurosci. 27 (2007) 5823–5834, http://dx.doi.org/10.1523/JNEUROSCI.1135-07.2007.

[58] A. Prokop, The intricate relationship between microtubules and their associated motor proteins during axon growth and maintenance, Neural Dev. 8 (2013) 17, http://dx.doi.org/10.1186/1749-8104-8-17.

[59] P.W. Baas, C. Vidya Nadar, K.A. Myers, Axonal transport of microtubules: the long and short of it, Traffic 7 (2006) 490–498, http://dx.doi.org/10.1111/j.1600-0854.2006.00392.x.

[60] A.E. Twelvetrees, S. Pernigo, A. Sanger, P. Guedes-Dias, G. Schiavo, R.A. Steiner, et al., The dynamic localization of cytoplasmic dynein in neurons is driven by Kinesin-1, Neuron (2016) 1–17, http://dx.doi.org/10.1016/j.neuron.2016.04.046.

[61] L.A.B. Elias, D.D. Wang, A.R. Kriegstein, Gap junction adhesion is necessary for radial migration in the neocortex, Nature 448 (2007) 901–907, http://dx.doi.org/10.1038/nature06063.

[62] C. Zheng, N. Heintz, M.E. Hatten, CNS gene encoding astrotactin, which supports neuronal migration along glial fibers, Science 272 (1996) 417–419.

[63] D.J. Solecki, L. Model, J. Gaetz, T.M. Kapoor, M.E. Hatten, Par6α signaling controls glial-guided neuronal migration, Nat. Neurosci. 7 (2004) 1195–1203, http://dx.doi.org/10.1038/nn1332.

[64] D.J. Solecki, N. Trivedi, E.-E. Govek, R.A. Kerekes, S.S. Gleason, M.E. Hatten, Myosin II motors and F-actin dynamics drive the coordinated movement of the centrosome and soma during CNS glial-guided neuronal migration, Neuron 63 (2009) 63–80, http://dx.doi.org/10.1016/j.neuron.2009.05.028.

[65] X. Zhang, K. Lei, X. Yuan, X. Wu, Y. Zhuang, T. Xu, et al., SUN1/2 and Syne/Nesprin-1/2 complexes connect centrosome to the nucleus during neurogenesis and neuronal migration in mice, Neuron 64 (2009) 173–187, http://dx.doi.org/10.1016/j.neuron.2009.08.018.

[66] R. Hand, D. Bortone, P. Mattar, L. Nguyen, J.I.-T. Heng, S. Guerrier, et al., Phosphorylation of Neurogenin2 specifies the migration properties and the dendritic morphology of pyramidal neurons in the neocortex, Neuron 48 (2005) 45–62, http://dx.doi.org/10.1016/j.neuron.2005.08.032.

[67] A.J. Barkovich, R. Guerrini, R.I. Kuzniecky, G.D. Jackson, W.B. Dobyns, A developmental and genetic classification for malformations of cortical development: update 2012, Brain 135 (2012) 1348–1369, http://dx.doi.org/10.1093/brain/aws019.

[68] R. Guerrini, W.B. Dobyns, Malformations of cortical development: clinical features and genetic causes, Lancet Neurol. 13 (2014) 710–726, http://dx.doi.org/10.1016/S1474-4422(14)70040-7.

[69] O. Reiner, R. Carrozzo, Y. Shen, M. Wehnert, F. Faustinella, W.B. Dobyns, et al., Isolation of a Miller-Dieker lissencephaly gene containing G protein beta-subunit-like repeats, Nature 364 (1993) 717–721, http://dx.doi.org/10.1038/364717a0.

[70] O. Reiner, T. Sapir, LIS1 functions in normal development and disease, Curr. Opin. Neurobiol. 23 (2013) 951–956, http://dx.doi.org/10.1016/j.conb.2013.08.001.

[71] M. Scoto, A.M. Rossor, M.B. Harms, S. Cirak, M. Calissano, S. Robb, et al., Novel mutations expand the clinical spectrum of DYNC1H1-associated spinal muscular atrophy, Neurology 84 (2015) 668–679, http://dx.doi.org/10.1212/WNL.0000000000001269.

[72] F.S. Alkuraya, X. Cai, C. Emery, G.H. Mochida, M.S. Al-Dosari, J.M. Felie, et al., Human mutations in NDE1 cause extreme microcephaly with lissencephaly, Am. J. Hum. Genet. 88 (2011) 536–547, http://dx.doi.org/10.1016/j.ajhg.2011.04.003.

[73] M. Bakircioglu, O.P. Carvalho, M. Khurshid, J.J. Cox, B. Tuysuz, T. Barak, et al., The essential role of centrosomal NDE1 in human cerebral cortex neurogenesis, Am. J. Hum. Genet. 88 (2011) 523–535, http://dx.doi.org/10.1016/j.ajhg.2011.03.019.

[74] C. Munch, R. Sedlmeier, T. Meyer, V. Homberg, A.D. Sperfeld, A. Kurt, et al., Point mutations of the p150 subunit of dynactin (DCTN1) gene in ALS, Neurology 63 (2004) 724–726.

[75] M.B. Harms, K.M. Ori-McKenney, M. Scoto, E.P. Tuck, S. Bell, D Ma, et al., Mutations in the tail domain of DYNC1H1 cause dominant spinal muscular atrophy, Neurology 78 (2012) 1714–1720, http://dx.doi.org/10.1212/WNL.0b013e3182556c05.

[76] E.C. Oates, A.M. Rossor, M. Hafezparast, M. Gonzalez, F. Speziani, D.G. MacArthur, et al., Mutations in BICD2 cause dominant congenital spinal muscular atrophy and hereditary Spastic paraplegia, Am. J. Hum. Genet. 92 (2013) 1–9, http://dx.doi.org/10.1016/j.ajhg.2013.04.018.

[77] J. Lipka, M. Kujipers, J. Jaworski, C.C. Hoogenraad, Mutations in cytoplasmic dynein and its regulators cause malformations of cortical development and neurodegenerative diseases, Biochem. Soc. Trans. 41 (2013) 1605–1612, http://dx.doi.org/10.1042/BST20130188.

[78] M. Hafezparast, R. Klocke, C. Ruhrberg, A. Marquardt, A. Ahmad-Annuar, S. Bowen, et al., Mutations in dynein link motor neuron degeneration to defects in retrograde transport, Science 300 (2003) 808–812, http://dx.doi.org/10.1126/science.1083129.

[79] X. Xiang, A.H. Osmani, S.A. Osmani, M. Xin, N.R. Morris, NudF, a nuclear migration gene in Aspergillus nidulans, is similar to the human LIS-1 gene required for neuronal migration, Mol. Biol. Cell 6 (1995) 297–310.

[80] R. Guerrini, W.B. Dobyns, A.J. Barkovich, Abnormal development of the human cerebral cortex: genetics, functional consequences and treatment options, Trends Neurosci. 31 (2008) 154–162, http://dx.doi.org/10.1016/j.tins.2007.12.004.

[81] E.V. Haverfield, A.J. Whited, K.S. Petras, W.B. Dobyns, S. Das, Intragenic deletions and duplications of the LIS1 and DCX genes: a major disease-causing mechanism in lissencephaly and subcortical band heterotopia, Eur. J. Hum. Genet. 17 (2009) 911–918, http://dx.doi.org/10.1038/ejhg.2008.213.

[82] J.S. Liu, C.R. Schubert, X. Fu, F.J. Fourniol, J.K. Jaiswal, A. Houdusse, et al., Molecular basis for specific regulation of neuronal Kinesin-3 motors by doublecortin family proteins, Mol. Cell 47 (2012) 707–721, http://dx.doi.org/10.1016/j.molcel.2012.06.025.

[83] M. Hattori, H. Adachi, M. Tsujimoto, H. Arai, K. Inoue, Miller-Dieker lissencephaly gene encodes a subunit of brain platelet-activating factor acetylhydrolase [corrected], Nature 370 (1994) 216–218, http://dx.doi.org/10.1038/370216a0.

[84] D.T. Pilz, N. Matsumoto, S. Minnerath, P. Mills, J.G. Gleeson, K.M. Allen, et al., LIS1 and XLIS (DCX) mutations cause most classical lissencephaly, but different patterns of malformation, Hum. Mol. Genet. 7 (1998) 2029–2037.

[85] S. Hippenmeyer, Y.H. Youn, H.M. Moon, K. Miyamichi, H. Zong, A. Wynshaw-Boris, et al., Genetic mosaic dissection of Lis1 and Ndel1 in neuronal migration, Neuron 68 (2010) 695–709, http://dx.doi.org/10.1016/j.neuron.2010.09.027.

[86] W. Bi, T. Sapir, O.A. Shchelochkov, F. Zhang, M.A. Withers, J.V. Hunter, et al., Increased LIS1 expression affects human and mouse brain development, Nat. Genet. 41 (2009) 168–177, http://dx.doi.org/10.1038/ng.302.

[87] T. Saito, S. Hanai, S. Takashima, E. Nakagawa, S. Okazaki, T. Inoue, et al., Neocortical layer formation of human developing brains and lissencephalies: consideration of layer-specific marker expression, Cereb. Cortex 21 (2011) 588–596, http://dx.doi.org/10.1093/cercor/bhq125.

[88] Y.H. Youn, T. Pramparo, S. Hirotsune, A. Wynshaw-Boris, Distinct dose-dependent cortical neuronal migration and neurite extension defects in Lis1 and Ndel1 mutant mice, J. Neurosci. 29 (2009) 15520–15530, http://dx.doi.org/10.1523/JNEUROSCI.4630-09.2009.

[89] M.H. Willemsen, L.E.L. Vissers, M.A.A.P. Willemsen, B.W.M. van Bon, T. Kroes, J. de Ligt, et al., Mutations in DYNC1H1 cause severe intellectual disability with neuronal migration defects, J. Med. Genet. 49 (2012) 179–183, http://dx.doi.org/10.1136/jmedgenet-2011-100542.

[90] K. Poirier, N. Lebrun, L. Broix, G. Tian, Y. Saillour, C.E.C. Boscheron, et al., Mutations in TUBG1, DYNC1H1, KIF5C and KIF2A cause malformations of cortical development and microcephaly, Nat. Genet. 45 (2013) 639–647, http://dx.doi.org/10.1038/ng.2613.

[91] C. Fiorillo, F. Moro, J. Yi, S. Weil, G. Brisca, G. Astrea, et al., Novel dynein DYNC1H1 neck and motor domain mutations link distal spinal muscular atrophy and abnormal cortical development, Hum. Mutat. 35 (2014) 298–302, http://dx.doi.org/10.1002/humu.22491.

[92] J. Hertecant, M. Komara, A. Nagi, J. Suleiman, L. Al-Gazali, B.R. Ali, A novel de novo mutation in DYNC1H1 gene underlying malformation of cortical development and cataract, Meta Gene 9 (2016) 124–127, http://dx.doi.org/10.1016/j.mgene.2016.05.004.

[93] K.M. Ori-McKenney, R.B. Vallee, Neuronal migration defects in the Loa dynein mutant mouse, Neural Dev. 6 (2011) 26, http://dx.doi.org/10.1186/1749-8104-6-26.

[94] K.M. Ori-McKenney, J. Xu, S.P. Gross, R.B. Vallee, A cytoplasmic dynein tail mutation impairs motor processivity, Nat. Cell Biol. 12 (2010) 1228–1234, http://dx.doi.org/10.1038/ncb2127.

[95] S.S. Jamuar, A.-T.N. Lam, M. Kircher, A.M. D'Gama, J. Wang, B.J. Barry, et al., Somatic mutations in cerebral cortical malformations, N. Engl. J. Med. 371 (2014) 733–743, http://dx.doi.org/10.1056/NEJMoa1314432.

[96] K. Neveling, L.A. Martinez-Carrera, I. Hölker, A. Heister, A. Verrips, S.M. Hosseini-Barkooie, et al., Mutations in BICD2, which encodes a Golgin and important motor adaptor, cause Congenital Autosomal-dominant spinal muscular atrophy, Am. J. Hum. Genet. 92 (2013) 1–9, http://dx.doi.org/10.1016/j.ajhg.2013.04.011.

[97] A.R. Paciorkowski, K. Keppler-Noreuil, L. Robinson, C. Sullivan, S. Sajan, S.L. Christian, et al., Deletion 16p13.11 uncovers NDE1 mutations on the non-deleted homolog and extends the spectrum of severe microcephaly to include fetal brain disruption, Am. J. Med. Genet. A 161A (2013) 1523–1530, http://dx.doi.org/10.1002/ajmg.a.35969.

[98] A. Guven, A. Gunduz, T.M. Bozoglu, C. Yalcinkaya, A. Tolun, Novel NDE1 homozygous mutation resulting in microhydranencephaly and not microlyssencephaly, Neurogenetics 13 (2012) 189–194, http://dx.doi.org/10.1007/s10048-012-0326-9.

[99] A.K. Nicholas, M. Khurshid, J. Désir, O.P. Carvalho, J.J. Cox, G. Thornton, et al., WDR62 is associated with the spindle pole and is mutated in human microcephaly, Nat. Genet. 42 (2010) 1010–1014, http://dx.doi.org/10.1038/ng.682.

[100] J. Bond, E. Roberts, G.H. Mochida, D.J. Hampshire, S. Scott, J.M. Askham, et al., ASPM is a major determinant of cerebral cortical size, Nat. Genet. 32 (2002) 316–320, http://dx.doi.org/10.1038/ng995.

[101] N.J. Bradshaw, W. Hennah, D.C. Soares, NDE1 and NDEL1: twin neurodevelopmental proteins with similar 'nature' but different 'nurture', Biomol. Concepts 4 (2013) 447–464, http://dx.doi.org/10.1515/bmc-2013-0023.

[102] Y. Feng, E.C. Olson, P.T. Stukenberg, L.A. Flanagan, M.W. Kirschner, C.A. Walsh, LIS1 regulates CNS lamination by interacting with mNudE, a central component of the centrosome, Neuron 28 (2000) 665–679.

[103] S.L. Houlihan, Y. Feng, The scaffold protein Nde1 safeguards the brain genome during S phase of early neural progenitor differentiation, eLife 3 (2014) e03297, http://dx.doi.org/10.7554/eLife.03297.

[104] S. Sasaki, D. Mori, K. Toyo-oka, A. Chen, L. Garrett-Beal, M. Muramatsu, et al., Complete loss of Ndel1 results in neuronal migration defects and early embryonic lethality, Mol. Cell Biol. 25 (2005) 7812–7827, http://dx.doi.org/10.1128/MCB.25.17.7812-7827.2005.

[105] S.A. Stehman, Y. Chen, R.J. McKenney, R.B. Vallee, NudE and NudEL are required for mitotic progression and are involved in dynein recruitment to kinetochores, J. Cell Biol. 178 (2007) 583–594, http://dx.doi.org/10.1083/jcb.200610112.

[106] Y. Liang, W. Yu, Y. Li, L. Yu, Q. Zhang, F. Wang, et al., Nudel modulates kinetochore association and function of cytoplasmic dynein in M phase, Mol. Biol. Cell 18 (2007) 2656–2666, http://dx.doi.org/10.1091/mbc.E06-04-0345.

[107] M.A.S. Vergnolle, S.S. Taylor, Cenp-F links kinetochores to Ndel1/Nde1/Lis1/dynein microtubule motor complexes, Curr. Biol. 17 (2007) 1173–1179, http://dx.doi.org/10.1016/j.cub.2007.05.077.

[108] O. Reiner, E. Karzbrun, A. Kshirsagar, K. Kaibuchi, Regulation of neuronal migration, an emerging topic in autism spectrum disorders, J. Neurochem. 136 (2015) 440–456, http://dx.doi.org/10.1111/jnc.13403.

[109] A. Poduri, G.D. Evrony, X. Cai, C.A. Walsh, Somatic mutation, genomic variation, and neurological disease, Science 341 (2013) 1237758, http://dx.doi.org/10.1126/science.1237758.

In this chapter

Cytoplasmic dynein dysfunction and neurodegenerative disease

Armen J. Moughamian[1,2], Erika L.F. Holzbaur[1]

[1]University of Pennsylvania Perelman School of Medicine, Philadelphia, PA, United States;
[2]University of California, San Francisco, CA, United States

13.1 Introduction

Active, directed transport of organelles and other cellular cargo along cytoplasmic microtubules is a characteristic feature of mammalian cells. This transport, driven by the microtubule motor proteins dynein and kinesin, is required for normal intracellular trafficking in both the biosynthetic and degradative pathways. Long-distance transport along microtubules is especially important in large cells, which cannot rely on diffusion to efficiently distribute proteins or organelles. Microtubule-based transport is also critical in polarized cells, where the directed trafficking of newly synthesized components to specific cellular destinations is required to maintain distinct functional domains.

Neurons are both large cells, with processes that can extend up to a meter in length, and highly polarized with distinct somal, axonal, and dendritic compartments. Thus, these cells are particularly reliant on microtubule-based transport (Fig. 13.1) and uniquely vulnerable to defects in cytoplasmic dynein function. Mutations in cytoplasmic dynein or dynein effectors, such as dynactin, BICD2, and Lis1 result in neurodegenerative and neurodevelopmental disease. Here we review the critical functions for cytoplasmic dynein in neurons and discuss the effects that mutations in dynein or its binding partners have on neuronal function in both mouse models and affected human patients. We also discuss neurodegenerative diseases in which dynein dysfunction is not the proximal cause but may be a key contributor to development of the pathological state, including amyotrophic lateral sclerosis (ALS) and Huntington's disease (HD).

Dyneins. https://doi.org/10.1016/B978-0-12-809470-9.00013-8

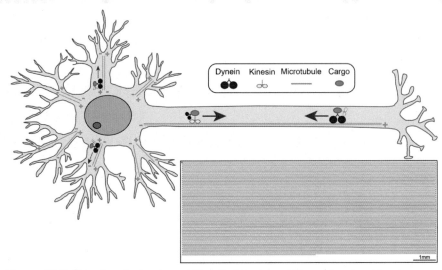

Figure 13.1 Cytoplasmic dynein and kinesin drive long-distance transport in neurons. Neurons are polarized cells with dendritic and axonal projections emanating from the cell body, or soma. Microtubules in the axon are uniformly oriented with plus ends outward. The plus end-directed microtubule motor, kinesin, transports cargos from the cell body to the distal axon and synapse, while the minus end-directed microtubule motor, cytoplasmic dynein, transports cargos from the axon back to the cell body. Microtubules are organized with mixed polarity in dendrites of mammalian neurons, making the analysis of trafficking more complex in this compartment. However, dynein is known to drive cargo both into and out of dendrites. Most motile cargos in neurons have both dynein and kinesin motors stably attached; bidirectional transport is regulated by interactions with effectors and scaffolding proteins. (Inset) This to-scale representation of an α motor neuron illustrates the vast distances that microtubule motors must traverse to maintain the distal axon and synapse. An α motor neuron projecting from the lumbar spinal cord to a muscle in the foot has a typical cell body diameter of 50 mm and an axon length of approximately 1 m.

13.2 Cytoplasmic dynein function in neurons

13.2.1 Molecular motors drive intracellular trafficking in neurons

In neurons, the microtubule array is uniformly polarized within the soma (cell body) and the axon, with microtubule minus ends oriented toward the cell center and plus ends directed outward toward the cell periphery [1]. In dendrites, microtubule organization is more complex (Fig. 13.1). Microtubules are oriented with a mixed polarity within dendrites of mammalian neurons and may have a completed inverted polarity (plus ends in and minus ends out) in the dendrites of neurons from invertebrates such as Drosophila or *Caenorhabditis elegans* [1,2].

The motor proteins that drive long-distance transport in neurons, kinesins, and dynein move in opposite directions along microtubules. The canonical kinesin motor kinesin-1, as well as kinesin-2 and kinesin-3 motors that also actively move cargo in neurons, all move toward microtubule plus ends. These motors drive the anterograde motility of newly synthesized material from the soma out along the axon [2].

In contrast to the multiple forms of kinesin that drive plus end-directed motility, cytoplasmic dynein is the major long-distance minus end-directed motor for vesicles and organelles. Thus, dynein drives the motility of a diverse array of cellular cargos both within the soma and along the axon and dendrites [2]. For example, within the soma dynein drives vesicles moving from the endoplasmic reticulum to the Golgi complex during the synthesis of membrane-associated and secreted proteins [3] and also is involved in early endosome formation, motility, maturation, and sorting [4]. In axons, following the uptake of growth factors such as NGF, dynein drives the long-distance movement of the resulting signaling endosomes back to the nucleus to affect changes in gene expression [5], so defects in dynein function will lead to neurotrophic deprivation. Dynein also drives the minus end-directed motility of degradative organelles, including both lysosomes [6] and autophagosomes [7,8], so defects in dynein function can lead to altered proteostasis and the buildup of aggregated proteins or dysfunctional organelles [2]. The multiple essential functions of dynein in neurons are described in more detail below.

13.2.2 Axonal transport

Perhaps the best-studied role of dynein in neurons is driving the robust retrograde motility of vesicles and organelles along the axon. This activity was first demonstrated in classic nerve ligation assays by Hirokawa and coworkers [9,10], in which dynein-associated organelles were observed to accumulate both proximally and distally to the site of ligature, while kinesin-associated organelles accumulated on the proximal side only. This result was interpreted as evidence that dynein motors, synthesized in the cytoplasm, are passively transported outward along the axon by kinesin-1. Activation of dynein motors in the distal axon then drives cargo transport back to the cell soma. In contrast, kinesins may take a one-way trip, driving anterograde transport outward along the axon followed by degradation at the distal terminal, or alternatively may be recycled back to the soma via diffusion for reuse [11].

Classic studies on axonal transport identified two major forms of motility: a fast phase and a slower phase. Fast axonal transport rates are reported as ~100–400 mm/day or ~1 μm/s and slow axonal transport rates are ~0.2–10 mm/day [2]. Fast axonal transport has been studied with an array of approaches, including live-cell imaging of cargos including lysosomal and mitochondrial markers, synaptic protein cargos, amyloid precursor protein (APP), fluorescently labeled RABs, NGF-bound quantum dots, and tetanus toxin [2]. These studies have provided many new insights into the mechanisms involved. Some of the observed transport is bidirectional, with cargos such as lysosomes and mitochondria alternating in movement in either the anterograde or retrograde direction [12,13]. In contrast, other dynein cargos, such as NGF-associated signaling endosomes and autophagosomes, show pronounced unidirectional motility [14,15]. Of note, translocation velocities are motor- and cargo-specific, as some cargos such as APP move significantly faster than other cargos such as mitochondria [16].

Multiple members of the kinesin superfamily drive fast anterograde transport along the axon, including kinesin-1, kinesin-2, and kinesin-3 [2]. Kinesin-1, or conventional kinesin, drives the motility of a number of different cargos including vesicles, organelles, proteins, and RNA granules. Kinesin-1 and kinesin-2 are both implicated in late endosome–lysosome transport [12]; kinesin-2 also drives the motility of specific cargos including fodrin-positive plasma membrane precursors [17]. Kinesin-3 is essential for the transport of synaptic vesicle precursors [18,19] and dense core vesicles [20]. In contrast, cytoplasmic dynein remains the only known motor for fast retrograde motility of cargos along the axon including signaling endosomes and autophagosomes [15,21,22].

While the motors driving fast axonal transport were identified relatively rapidly [9,10], the mechanisms driving the much slower transport of proteins, such as neurofilaments, remained less clear for many years. Insight into the process came from a breakthrough study demonstrating that the net slow movement of neurofilaments results from short bursts of rapid movement punctuated by prolonged pauses [23]. Kinesin-1 and dynein were identified as the motors for the bidirectional slow transport of neurofilaments along the axon [24–26]. As the slow transport of soluble axonal proteins, such as synapsin, is also dependent on the mechanisms driving fast vesicular transport [27], it is now thought that the same motors drive both fast and slow motility, with the key difference being an increased frequency of pausing during transport for cargos moving at net slow transport rates.

Dynein itself is a component of slow axonal transport, as metabolic labeling studies indicate that newly synthesize dynein motors are transported outward along the axon at velocities of 1–20 mm/day [28,29]. Mechanistic studies have shown that the anterograde transport of dynein is mediated by the direct binding of kinesin-1 to the neuron-specific isoform of dynein intermediate chain (DIC) [30]. Dynein and its activator dynactin accumulate in the distal axon, as a pool of motors available for initiation of retrograde transport back to the cell body [30,31]. The initiation of transport from the distal axon is tightly regulated [32] and is specifically disrupted by mutations in dynactin causative for Perry syndrome [31], as detailed below.

13.2.3 Dendritic trafficking

The uniform polarity of microtubule organization in axons makes it relatively easy to assess the mechanisms driving transport. In contrast, the more complex, mixed polarity of microtubules in dendrites has slowed the analysis of transport in this compartment. Targeted recruitment studies suggest that while kinesin-1 transports cargo into the axon, dynein is required for the polarized targeting of cargos both into and out of dendrites [33,33a]. Further complicating the issue of trafficking in dendrites are studies that demonstrate a critical role for dynein in the initial establishment of the mixed microtubule polarity characteristic of this compartment [34]. While the underlying mechanisms involved remain to be determined, this represents an additional role for dynein required for normal neuronal organization and function.

13.2.4 Dynein in neurodevelopment

13.2.4.1 Trophic factor signaling

Dynein is required for the trafficking of neurotrophic signals from the axon terminal to the nucleus to elicit changes in gene expression. Trophic factors such as NGF and brain-derived neurotrophic factor (BDNF) bind to receptors including TrkA and TrkB and are internalized to form signaling endosomes [35]. Dynein then drives the long-distance motility of signaling endosomes back toward the nucleus [5,36,37], a process that has been effectively tracked in live neurons using quantum dots [14]. The trafficking of BDNF bound to TrkB from dendrites to the soma is dynein-dependent [33a,38]. Defects in signaling endosome transport have been observed in neurons cultured from dynein mutant mice [39], consistent with an essential role for dynein in this transport. Trophic factor deprivation may contribute to both the developmental and degenerative phenotypes observed in mouse models and human patients, reviewed below.

13.2.4.2 Neuronal migration

In addition to the roles for dynein in mature neurons discussed above, dynein is required for nuclear migration in the developing brain, reviewed elsewhere in this volume. Consistent with this critical role, mutations in dynein can cause neurodevelopmental defects including cognitive dysfunction. Mutations in dynein regulatory proteins LIS1 and NDE1 also cause defective brain development. LIS1 mutations lead to a severe developmental disease called lissencephaly [40], characterized by a "smooth brain" as the disrupted migration of neurons during development leads to misformation of the cortex. NDE1 interacts with both dynein and LIS1, and mutations in NDE1 have also been shown to cause cortical malformation including severe microcephaly [41,42].

13.2.5 Dynein function in degradative pathways in the neuron

Neurons in the adult brain are predominantly postmitotic cells with very long lifetimes (>75 years). Thus neurons are particularly reliant on efficient degradative processes to remove misfolded or aggregated proteins or dysfunctional organelles that can accumulate during aging. Of note, the major neurodegenerative diseases, including ALS, HD, Alzheimer's disease, and Parkinson's disease (PD) are characterized pathologically by the accumulation of protein aggregates and organelles with abnormal morphology. These observations suggest that neurons may be uniquely vulnerable to even minor dysfunction of degradative pathways including lysosomal and proteasomal pathways, aggresome formation, and autophagic clearance of protein aggregates and organelles. Importantly, dynein function is implicated in all of these pathways, and defects in dynein function within the degradative pathway have been suggested to contribute to the pathology observed in neurodegeneration [43].

13.2.5.1 Lysosomes

In conjunction with kinesin-1 and kinesin-2, dynein is known to drive the bidirectional transport of lysosomes [12]. While it has not been directly demonstrated that defects in lysosomal motility lead to defects in protein degradation, it remains a compelling hypothesis that the bidirectional movement of lysosomes throughout both axonal and dendritic processes enhances the efficient clearance of misfolded or aging proteins or small protein aggregates. The continuous bidirectional motility of lysosomes may thus serve as a surveillance mechanism allowing for efficient clearance of protein aggregates from extended neuronal processes that can range up to 1 m in length. Consistent with this hypothesis, it has been shown that subtle defects in Rab7 function are sufficient to cause one form of Charcot–Marie–Tooth (CMT) disease (type 2B) [44,45]. Rab7 has been shown to regulate the association of dynein with the lysosomal compartment [46], so a defect in Rab7-dependent regulation may potentially affect either the localization or motility of lysosomes in affected patients.

13.2.5.2 Proteasomes

Significant links between motor activity and proteosomal function have not been established. However, a recent study suggested that the active depletion of proteasomes from the distal axon is driven by dynein, and that this depletion enhances protein stability and promotes axon outgrowth for neurons in culture [47]. The role of dynein in the regulation of proteasome distribution in mature neurons, and how this might contribute to the regulation of proteostasis during aging has not yet been addressed.

13.2.5.3 Aggresomes

In cells in culture, proteotoxic stress can induce aggresome formation. Aggresomes are accumulations of filamentous proteins and entrapped organelles in inclusions near the nucleus [48]. Dynein drives the inward movement of the components of this structure [49], potentially through a link mediated by the microtubule-associated deacetylase HDAC6 [50]. However, it remains unclear whether aggresomes have a role in regulating proteostasis in neurons in vivo or instead are a response of cells in culture to proteotoxic insult induced by overexpression of misfolded or mutant proteins.

13.2.5.4 Autophagy

One of the major pathways used by the neuron to remove aggregated proteins and old or damaged organelles is macroautophagy, referred to here as autophagy. Neurons are dependent on autophagy, as neuron-specific knockout of either of two essential autophagy genes, Atg5 and Atg7, leads to degeneration in the cerebellar and cerebral cortices [51,52].

Studies in primary neurons have characterized a robust, constitutive pathway for autophagy in neurons that is spatially specific [15,53]. Autophagosomes are generated in a constitutive process at the distal end of the axon and initially undergo localized bidirectional transport [15]. The recruitment of the scaffolding protein JIP1, which binds to and coordinates the activities of kinesin and dynein motors, leads to a dramatic switch to unidirectional, highly processive dynein-driven motility [54]. Following this JIP1-dependent switch, autophagosomes are rapidly transported along the axon toward the soma over distances reaching hundreds of microns. Dynein-driven autophagosome motility has been proposed to mediate efficient fusion with lysosomes [7]; consistent with this proposal, failure to recruit JIP1 and thus effectively regulate kinesin and dynein motor activity leads to impaired acidification of the autophagosome, likely due to inefficient fusion with lysosomes [54]. Impairment in dynein loading leads to the distal accumulation of autophagosomes in neurites and synaptic termini [55], and mutations in dynein impair clearance of protein aggregates in Drosophila [43].

13.2.6 Dynein is an essential motor in neurons

In summary, cytoplasmic dynein plays multiple essential roles in neurons, with required functions in neuronal migration, neurotrophic signaling, and multiple intracellular trafficking pathways, both biosynthetic and degradative. The large, polarized morphology, and postmitotic nature of neurons make these cells uniquely vulnerable to even subtle defects in dynein function. Thus, mutations in cytoplasmic dynein would be expected to have deleterious effects on neuronal function, as observed in both animal models and human patients. Mutations in dynein or in associated proteins, such as dynactin and BICD2, cause neuron-specific phenotypes including developmental defects and degenerative conditions involving the peripheral and central nervous system, as discussed below.

13.3 Dynein dysfunction in mice

13.3.1 Mouse models of hereditary motor neuropathy VIIB

The first mouse model to examine the effects of dynein dysfunction on neuro-degenerative disease was a transgenic line engineered to overexpress the dyna-mitin subunit of dynactin primarily in motor neurons. Dynamitin overexpression destabilizes dynactin, causing dynactin dysfunction. Transgenic mice overex-pressing dynamitin develop a late-onset, slowly progressive degeneration of motor neurons [56]. Strikingly, the pathology observed in this model resembled the late-onset slowly degenerative phenotype reported in human patients with a G59S mutation in the DCTN1 gene described one year later [57]. The G59S mutation in DCTN1 disrupts the folding of the conserved cytoskeletal-associ-ated protein, glycine-rich (CAP-Gly) domain found at the amino terminus of the encoded polypeptide, p150[Glued] subunit, the largest subunit of the dynactin complex. This mutation abrogates interactions of dynactin with both microtu-bules and plus end-binding proteins, thus inhibiting dynein–dynactin function both in primary neurons and in a Drosophila model [31,57–59].

Subsequently, the human G59S mutation in DCTN1 was modeled in the mouse, using both knockin and transgenic approaches [60–62]. There is some variability in the phenotypes observed in the resulting lines. In homozygous knockin mice, expression of the G59S mutation was embryonic lethal [60]. Heterozygous knockin mice and one of the transgenic models developed slowly progressive motor neuron degeneration, consistent with the disease course in humans [60,62]. A second transgenic line displayed more marked degeneration that was fatal within ~10 months [61]; the enhanced degeneration observed in this model may reflect higher expression levels of the mutant dynactin sub-unit. Pronounced defects in axonal transport were not observed [62], but all three models displayed destabilization of the neuromuscular junction. Marked changes in membranous organelles within the cell body were also noted, most notably the proliferation of tertiary lysosomes and lipofuscin granules [62].

These studies indicate that disruption of axonal transport is not the sole pathogenic mechanism caused by mutations in the dynein–dynactin path-way. Consistently, synaptic defects have been observed in models of dynein/dynactin dysfunction. Destabilization of the presynaptic terminal was first observed in Drosophila with the Glued (GL[1]) mutation, which causes a trun-cation of the p150[Glued] polypeptide, which acts as a dominant negative inhibitor of dynein–dynactin function [63,64]. Presynaptic defects have sub-sequently been observed in both fly and mouse models of dynein-pathway dysfunction [60–62,65]. These observations suggest that dynein and dyn-actin have an additional role in the maintenance of synaptic stability that extends beyond the well-characterized role of the motor in minus end-directed transport along the axon.

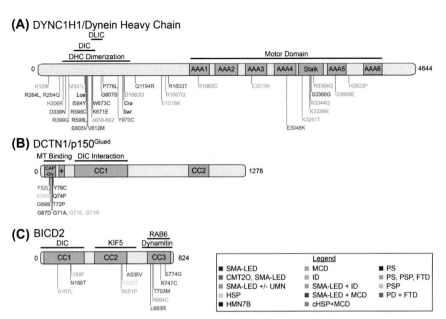

Figure 13.2 Mutations in dynein and dynein-binding proteins cause neurodevelopmental and neurodegenerative disease. (A) Cytoplasmic dynein heavy chain, encoded by the DYNC1H1 gene in humans, is a large polypeptide with an N-terminal dimerization domain, six tandem AAA domains and a microtubule-binding stalk domain. Binding determinants for dynein intermediate chain (IC) and light intermediate chain localize to the N-terminus. More than 30 pathogenic mutations, both missense and deletion, have been identified to date within the human DYNC1H1 gene. The *Loa*, *Cra1*, and *Swl* mutations identified in mouse (identified in black font) map to the N-terminal region. (B) The DCTN1 gene encodes p150^Glued, the largest subunit of the dynactin complex. There is a highly conserved CAP-Gly domain at the N-terminus of p150^Glued that binds to both microtubules and plus end tracking proteins including EB1 and EB3, followed by two coiled coil domains (CC1 and CC2); CC1 is the binding site for the dynein intermediate chain. (C) BICD2 is a dynein effector, characterized primarily by coiled coil domains. Binding sites for dynein IC, kinesin-1 (KIF5), Rab6, and the dynamitin subunit of dynactin have been identified. (A–C) Mutations in the dynein pathway result in three broad categories of disease: motor neuron disease, parkinsonism, and alterations in cortical development. These categories can overlap and certain mutations result in a combination of diseases (i.e.: SMA-LED and ID), this is represented by a+sign. Mutations are color-coded based on disease with similar colors representing a similar category of disease. Diseases identified to date include SMA-LED (spinal muscular atrophy-lower extremity predominant); (Charcot–Marie–Tooth disease type 2O (CMT2O), upper motor neuron disease (UMN), hereditary spastic paraplegia (HSP), hereditary motor neuropathy, type 7B (HMN7B), malformations in cortical development (MCD), intellectual disability (ID), complex hereditary spastic paraplegia (cHSP), (Perry syndrome (PS), progressive supranuclear palsy (PSP), frontotemporal dementia (FTD).

13.3.2 Legs-at-odd-angles, cramping-1, and sprawling mice

Three lines of mice with mutations in dynein heavy chain (HC) have been characterized in some detail. Two lines, Legs-at-odd-angles (*Loa*) and Cramping-1 (*Cra1*), have point mutations in the N-terminal tail domain of cytoplasmic dynein HC. Sprawling (Swl) mice have a three-amino-acid residue deletion, also in the tail domain of dynein HC [66,67] (Fig. 13.2). As the tail domain of dynein mediates both HC dimerization and the association between dynein HC and dynein IC, these mutations are predicted to affect the integrity of the dynein holoprotein complex. Biochemical and biophysical characterization of dynein purified from both heterozygous and homozygous Loa mice further supports this interpretation [68].

All three mutations are lethal when homozygous. *Swl/Swl* mice die during embryogenesis [69] and homozygous *Loa* and *Cra1* mice die within 24 h of birth, demonstrating an inability to move or feed [66]. The most detailed phenotypic analyses have been performed on heterozygous mice from all three lines, which show a characteristic clasping or clenching of the hind limbs when suspended by their tails [66,67].

Both the *Loa* and *Cra1* heterozygous phenotypes were initially interpreted as age-dependent motor neuron degeneration [66]. Characterization of the *Swl/+* mouse as primarily a sensory defect due to the abnormal developmental of proprioceptive neurons [67] led to some debate as to the neuronal populations most affected by mutations in cytoplasmic dynein HC (motor or sensory), and whether the defects should be considered developmental or degenerative [70,71].

Further studies on the *Loa/+* mouse reported the slowly progressive loss of motor neurons, but not proprioceptive neurons, in the trigeminal nerve [72], leading to the proposal that the heterozygous Loa mutation differentially affects specific neuronal subsets. Further extending the types of neurons affected by dynein mutations, in vivo brain imaging of *Cra1 mice* revealed significant deficits in striatal neuron morphology, including striatal atrophy and altered dopamine signaling [73].

Detailed longitudinal studies on *Cra1/+* mice revealed early onset weakness in both forelimbs and hind limbs that is stable with aging, but no underlying loss of either motor or sensory neurons [74]. The most striking effects observed in this study were found at neuromuscular junctions, which showed morphological alterations and decreased innervation in comparison to control mice, suggesting that dynein is required for the development and/or maintenance of synapses in the mouse. These data parallel previous observations in Drosophila and in mice expressing mutant dynactin that the motor complex is required for synaptic stabilization [60–62,65,75].

Dynein also functions in dendritic transport [33,33a,34]. Careful characterization of the Loa mouse [72] revealed altered dendritic morphology of large trigeminal motor neurons, and a depletion of organelles such as lysosomes, multivesicular bodies, and polyribosomes from proximal dendrites. Further, abnormal, vesiculated mitochondria were observed in dendrites from Loa mice [72].

As cytoplasmic dynein HC is encoded by a single gene whose expression is required for the normal function of all neurons, it is likely that all three lines (*Loa*, *Cra1*, *Swl*) exhibit pleiotropic defects, both developmental and degenerative. Multiple pathways are affected, including axonal and dendritic trafficking and synaptic stability. Neuronal subpopulations may be differentially sensitive to these defects based on variables such as requirement for continuous trophic factor support, maintenance of extended morphologies such as meter-long axons or extensive axonal or dendritic arbors, or sensitivity to defects in degradative pathway function such as deficits in autophagy.

13.4 Mutations in dynein and in dynein effectors result in a spectrum of neurodevelopmental and neurodegenerative disease in humans

The disparate phenotypes described in dynein mutant mice caused some initial controversy, but accurately foreshadowed the wide range of neurological disorders that have now been observed in patients with mutations in dynein, dynein effectors including dynactin and BICD2, and dynein-associated proteins huntingtin and Rab7. The numerous clinical phenotypes include motor and sensory neuropathy, upper and lower motor neuron dysfunction, parkinsonism, oculomotility abnormalities, depression, childhood-onset intellectual disability, and adult-onset cognitive decline. In the last several years more than 15 unique pathogenic mutations have been discovered within the cytoplasmic dynein HC gene, a least 10 mutations within the DCTN1 gene encoding the dynactin subunit p150^{Glued}, and at least 11 mutations within the BICD2 gene (Fig. 13.2). These mutations result in a range of neurodevelopmental and neurodegenerative phenotypes. Despite the clinical variability, the syndromes can be divided into three broad categories: motor neuron disease, intellectual disability, and Perry syndrome (a form of atypical parkinsonism). Although there is some overlap among these syndromes, these categories suggest multiple pathogenic consequences due to dynein pathway dysfunction.

The broad spectrum of mutations and clinical manifestations defies a straightforward genotype–phenotype analysis, but rather requires a mechanistic understanding of how a specific mutation disrupts dynein function. For example, in many cases mutations causing malformations of cortical development or spinal muscular atrophy can be found clustered within discrete domains of the primary structure of the dynein HC (Fig. 13.2). This clustering suggests that mutations affecting the motor domain, and in particular the microtubule-binding stalk, lead to malformations of cortical development. In contrast, mutations in the tail domain, predicted to affect complex integrity, lead to spinal muscular atrophy characterized by muscle weakness, and atrophy leading to gait abnormalities, with or without sensory deficits. However, we must note that some patients exhibit evidence of both phenotypes, suggesting that it is unlikely for a mutation to specifically disrupt a single dynein function without altering other functions.

To date, mutations have only been identified within the cytoplasmic dynein HC (DYNC1H1), and not in the associated intermediate, light intermediate, and light chains. This may reflect the very large size of the DYNC1H1 gene, or the high number of residues critical for enzymatic function and binding interactions both within the complex and with key interactors such as the microtubule. Similarly, the only pathogenic mutations to date within the multisubunit dynactin complex are found within the DCTN1 gene encoding the largest subunit, p150^{Glued}. As discussed below, pathogenic mutations have also been identified in dynein effectors, including BICD2, Huntingtin, and Rab7. Mutations in binding partners Lis1 and Nde1 are also highly pathogenic but are described in more detail elsewhere in this volume.

13.4.1 Dynein mutations can lead to cortical malformation

One of the first mutations to be identified in cytoplasmic dynein was a de novo missense mutation, H3822P, in the DYNC1H HC (Fig. 13.2) [76]. The affected patient exhibited intellectual disability, developmental delays, and an altered gait. Willemsen and colleagues [77] described another novel de novo mutation, E1518K, in a patient with severe intellectual disability due to neuronal migration defects and cortical malformation. This patient also exhibited inability to walk and progressive swallowing difficulties, indicating that defective dynein function could affect both brain development and motor function.

In a broader study of mutations leading to malformations of cortical development [78], a familial missense mutation in cytoplasmic dynein HC was identified (K3241T), as well as additional de novo mutations, including one deletion of four residues (659–662), and the missense mutations: K129I, R1567Q, R1962C, K3241T, K3336N, R3344Q, and R3384Q. Examination of patient brains by MRI revealed a wide range of malformations. Of note, one patient exhibited peripheral neuropathy [78]. Based on these observations, the authors proposed that DYNC1H1 should be considered as a principal locus for cases of lissencephaly, pachygyria, and polymicrogyria not explained by mutations in other known genes such as Lis1 [78]. Given the key role for dynein and its effector Lis1 in neuronal migration and cortical development, reviewed elsewhere in this volume, it is not surprising that dynein mutations would have such striking effects on neurodevelopment.

13.4.2 Dynein mutations cause CMT-2O and SMA-LED

Shortly after the initial description of dynein mutations with a phenotype of intellectual disability, an analysis of a large pedigree with dominant axonal CMT disease, type 2 determined that the causative mutation was a H306R missense mutation in DYNC1H1 (Fig. 13.2) [79]. Affected individuals were generally characterized by delayed motor milestones, distal limb weakness and atrophy, and abnormal gain. Some patients also displayed variability in proprioception and fine touch. This variant has now been termed CMT-2O. A R598C mutation was also identified as a cause of CMT (also known as axonal hereditary motor and sensory neuropathy or HMSN) in a multigenerational Australian family. Disease in this family is characterized by early onset and slowly progressive weakness that is primarily distal, along with atrophy of lower limbs and a waddling gait with mild sensory abnormalities.

Patients with dominant spinal muscular atrophy-lower extremity predominant (SMA-LED) exhibit a somewhat similar phenotype, with early onset (congenital or childhood) proximal leg weakness with muscle atrophy. Harms and colleagues [80] identified three novel missense mutations in DYNC1H1 as causative for SMA-LED, including I584L, K671E, and Y970C. This study also established variability within the phenotype, as one patient demonstrated mild cognitive delay.

This theme of variable clinical presentation, even when patients express the same mutation, continued with a report that the same H306R mutation reported to cause CMT-2O results in SMA-LED in an affected family, with clinical observations including early childhood onset of proximal leg weakness, mildly delayed motor development, unstable gait, and lower limb atrophy, but no sensory deficits [81].

There have now been many reports on DYNC1H1 mutations causative for SMA-LED, including R264L [82], Q1194R and E3048K [83], G807S [84], R598C [85], E2616K, R1603T, Y970C, W673C, V612M, E603V, R598L, Ile584L, M581L, R399G, D399G, D338N, and R264G [86], and S3360G [87]. Most of these patients display core symptoms of early onset proximal leg weakness, delayed motor milestones, and a waddling gait. In most cases, the defects are reported to remain relatively stable over time, while in other cases slowly progressive worsening is reported, with weakness extending to upper limbs [82,84].

Many studies further extend the phenotypic abnormalities reported. Some but not all patients display cognitive impairment. MRI studies of these patients reveal underlying structural malformations in the brain [86] consistent with defects in neuronal migration. These malformations may also result in a variable degree of behavioral impairment. Other studies report pes cavus in patients with a P776L mutation [88], and formation of bilateral cataracts during infancy in a patient with the de novo mutation G3658E [89]. Perhaps the most complex case reported to date involves a patient with a D1062G mutation that causes complex hereditary spastic paraplegia (HSP) that has a progressively worsening gait and a progression to upper motor neuron disease, along with cognitive deficits, behavioral disturbances, and polymicrogyria. This patient also suffered from epilepsy and cataracts [87].

To date, limited cellular and biochemical studies have been performed exploring the underlying defects caused by these mutations. Analysis of fibroblasts derived from a patient with an I584L mutation causative for SMA-LED suggested reduced stability of the dynein complex, consistent with the localization of the mutation to the tail domain, which is required for assembly of the dynein holocomplex [80]. This is similar to observations from the Loa mouse [68]; the Loa mutation is similarly localized within the primary sequence of the cytoplasmic dynein HC. In contrast, Fiorillo et al. [83] report that the mutations Q1994R and E3048Q do not affect dynein complex stability, but instead affect dynein function. While organelles are normally distributed in patient fibroblasts, they note a delay in Golgi reassembly following microtubule depolymerization with nocodazole followed by drug washout. A delay but not a complete block is consistent with a minor impairment of dynein function.

In summary, mutations in the cytoplasmic dynein HC result in a wide range of neurodevelopmental and occasionally degenerative deficits, affecting cognition, behavior, motor function, sensory function, skeletal development (scoliosis and pes cavus), seizures, and early cataract development. Some clinical manifestations are relatively common, such as a waddling gait, but it is clear that there is a spectrum of deficits that can result from dominant disruptions of dynein function. Given the requirement for dynein in the early stages of embryogenesis, it is likely that more severe mutations would be lethal.

13.4.3 Dynactin mutations cause two distinct forms of neurodegeneration

Mutations in dynein effectors have also been implicated in human disease. One of first of these mutations to be identified was a G59S mutation in the p150Glued subunit of dynactin, encoded by the DCTN1 gene [57]. The G59S mutation has also been recently found in two unrelated families of Korean descent [90]. Dynactin is a highly conserved accessory or activator complex for dynein and required for most dynein-mediated functions, including vesicle trafficking and axonal transport. The p150Glued subunit of dynactin binds to dynein via a direct association with dynein DIC [91,92]. p150[Glued] also mediates the direct association of dynactin with microtubules, via a highly conserved N-terminal CAP-Gly domain [93]. Most dynactin mutations identified to date are localized within this CAP-Gly domain.

Patients with the G59S mutation display an autosomal dominant motor neuron-specific degenerative disease, hereditary motor neuropathy VIIB (HMN7B), also known as distal spinal and bulbar muscular atrophy [57,94]. Biochemical and cellular approaches indicate that the mutant protein can misfold and has a tendency to aggregate at higher expression levels. In nonneuronal cells, the misfolded protein is efficiently cleared. However, the misfolded polypeptide accumulates in aggregates in motor neurons in affected patients; dynein is also recruited to these aggregates [58,94].

Both loss of dynein/dynactin activity as well as the toxic accumulation of aggregated proteins are likely to contribute to disease in patients with HMN7B. As noted above, this mutation has been studied extensively in mouse models. Dynactin mutations have also been implicated in ALS [95,96], although it remains unclear if these specific mutations are pathogenic or, alternatively, are allelic variants.

Mutations in the DCTN1 gene encoding the p150[Glued] subunit of dynactin can also cause a distinct and lethal form of neurodegeneration known as Perry syndrome [97]. Perry syndrome is an autosomal dominant disorder in which parkinsonism is accompanied by depression, weight loss, and hypoventilation. Most cases of Perry syndrome are caused by mutations in DCTN1 (Fig. 13.2), although a mutation in the microtubule-associated protein tau gene has been shown to cause a similar syndrome of parkinsonism, depression, weight loss, and hypoventilation [98]. Perry syndrome-associated mutations in DCTN1 include G67D, G71R, G71E, G71A, T72P, Q74P, and Y78C [97,99,100], all of which map to the highly conserved CAP-Gly domain at the amino terminus of p150[Glued].

The identification of additional patients affected by Perry syndrome has demonstrated that the clinical presentation is more variable than previously thought, even within a single family. For example, the G71E mutation has been found in patients diagnosed with a progressive supranuclear palsy (PSP) or behavioral-variant frontotemporal dementia [101]. Additional manifestations in these patients include vertical gaze slowness, dystonia, axial rigidity, a dysexecutive syndrome and midbrain atrophy. The K56R and G71R mutations in DCTN1 have been associated with PSP [102,103], while the F52L mutation causes both parkinsonism and frontotemporal atrophy [104]. Taken together, Perry syndrome is an atypical form of parkinsonism that not only encompasses core features of parkinsonism, hypoventilation, depression, and weight loss but can also cause ocular manifestations and cognitive decline.

Intriguingly all of the Perry syndrome mutations within DCTN1 are clustered within the CAP-Gly domain at the N-terminus of p150^{Glued}, only a few amino acid residues away from the G59S mutation that causes late-onset, slowly progressive motor neuron degeneration [57]. Both structural and cellular studies have provided insight into the differential effects of these mutations. Structural studies indicate that the G59S mutation maps to the core of the folded CAP-Gly domain [57], and biochemical analysis suggests that the substitution of the larger serine residue destabilizes the domain, leading to protein misfolding and aggregation in cells [58] and in knockin and transgenic mouse models [60–62]. In contrast, the Perry-associated mutations map to an exposed surface loop within the CAP-Gly domain, involved in binding to microtubules and to microtubule plus end-binding proteins (EB1 and EB3) or EEY motif containing proteins (CLIP-170 and alpha-tubulin). Biochemical studies indicate that the Perry mutations disrupt the binding of dynactin to microtubules and to EB1/EB3, causing a loss-of-function phenotype [31]. Cellular studies in primary neurons and in vivo studies in a Drosophila model indicate that loss of CAP-Gly function leads to a specific disruption of retrograde axonal transport [31,59]. Surprisingly, transport is unaffected along most of the axon, indicating that the CAP-Gly domain is dispensable for long-distance processive movement of cargos. However, CAP-Gly function is specifically required for efficient transport initiation in the distal axon [150], for a range of cargos including late endosomes and lysosomes, mitochondria, and signaling endosomes [105]. Still to be determined is why some neuronal subtypes are preferentially affected by these mutations, but the answer may lie in the remarkable morphology of midbrain projection neurons, which are characterized by extensive axonal arbors [106]. Maintaining these arbors over many decades may require highly efficient transport from the distal axon, making these neurons more susceptible to haploinsufficiency than other neuronal subtypes.

13.4.4 Mutations in the dynein effector BIDC2 also cause human disease

Dynein motor function is enhanced by the binding of effectors, including proteins from the BICD, TRAK, and Hook families [107–113]. Structural studies suggest that these activators function by enhancing the formation of a ternary complex composed of the adaptor, dynein, and dynactin [114]. These activators also include domains mediating targeting to specific organelles, including the Golgi (BICD proteins) and early endosomes (Hook1).

Therefore, mutations in dynein effectors are also likely to cause disease in humans. To date, 14 distinct mutations have been identified in BICD2; these pathogenic variants include both autosomal dominant and de novo mutations [82,83,115–119]. These mutations are distributed throughout the primary sequence, affecting both the dynein-binding and Rab6-binding domains. Affected patients exhibit a spectrum of phenotypes that overall are remarkably similar to those observed for patients with mutations in the cytoplasmic dynein HC gene DYNC1H1, discussed above. In most patients, symptoms include early onset weakness and atrophy of the lower extremities, delayed motor milestones, a waddling gait, atrophy of the shoulder girdle, and contractures of the foot, consistent with a diagnosis of SMA-LED. Disease progression is typically slow, and most patients do not exhibit either sensory deficits or significant cognitive impairment. However, mutations in BICD2 have also been identified in one family with HSP and in three families with a combination of SMA and upper motor neuron features consistent with HSP [116]. A patient with a de novo BICD2 mutation was diagnosed with cerebellar hypoplasia by MRI [120], but, in general, mutations in BCID2 do not result in cognitive dysfunction suggesting that brain development proceeds normally.

Together, these genetic and clinical findings are consistent with the known role of BICD2 as a key dynein effector, but one that is potentially involved in a subset of dynein functions within the cell.

13.4.5 Rab7

Mutations in Rab7 may affect dynein-mediated organelle transport by affecting motor–cargo associations. Rab7 is a small GTPase involved in the regulation of late endosomes and lysosomes. A mutation in Rab7 was identified as the cause of CMT disease type 2B, a disorder characterized primarily by the length-dependent degeneration of sensory neurons [121]. Detailed structural and biochemical studies have shown that the mutation causes a relatively mild misregulation of Rab7 function [45]. Rab7 has been shown to be key to the recruitment of dynein to late endosomes and lysosomes, via a mechanism that includes RLIP, ORPIL, and βIII spectrin, so it is possible that this misregulation of Rab7 activity affects the motility of lysosomes along extended axonal processes [46].

13.5 Dynein dysfunction in the pathogenesis of neurodegenerative disease: ALS, HD, and PD

The neurodegenerative diseases discussed above are caused directly by mutations in dynein-associated proteins. However, it is likely that the overall number of patients affected directly by mutations in this pathway is likely to remain low. As there is no genetic or functional redundancy for most aspects of the cytoplasmic dynein pathway, mutations must be relatively mild for affected individuals to survive development. In affected patients with known mutations in DCTN1, for example, dynein–dynactin function is only mildly compromised and in most cases the deficits do not become apparent until one or more decades have passed.

However, dysfunction of the dynein pathway has been implicated in the pathogenesis of a much broader range of neurodegenerative disease. Trafficking and transport defects may be at the heart of pathogenesis in a number of major diseases. For example, specific deficits in axonal transport have been reported in animal models of ALS, HD, and PD (reviewed in Ref. [122]). Here, we review the available evidence suggesting that alterations in dynein function contribute to the development of three of the major neurodegenerative diseases, ALS, HD, and PD.

13.5.1 Amyotrophic lateral sclerosis

In ALS, defects in fast axonal transport have been noted early in the disease process in multiple models, both in vitro and in vivo [123–129]. A common observation in many of these studies is the observation of more frequent pausing of cargos during translocation along the axon. One mechanism for this increased pausing involves activation of the stress-induced kinase CDK5 [128]. Phosphorylation of the dynein-binding protein NDEL1 by CDK5 enhances the formation of a high affinity dynein-LIS1-NDEL1 complex [130,131]. The binding of Lis1 to the dynein motor domain is thought to block the ATP-dependent release of the dynein motor from its microtubule track [132]. Thus, stress-induced CDK activity leads to increased pausing of cargos moving along the axon. Of note, blocking CDK5 activation is sufficient to rescue transport deficits in neurons expressing mutant SOD1 [128].

However, it is not just the processivity of cargo transport along the axon that is affected in models of ALS. Pronounced changes in the nature of the cargos being transported, particularly in the retrograde direction, have been observed, including marked changes in key signaling molecules [123]. Thus, while motor dysregulation likely contributes to neurodegeneration, pathological changes in dynein cargo may be just as, or even more significant in the onset of pathogenesis. One example of how the nature of cargo transported by dynein is changed due to alterations in the cellular environment occurs during axonal injury. Following an axonal injury, the continuous retrograde flow of neurotrophins transported by dynein may cease [5]. Instead, dynein begins to transport injury and stress signals to the nucleus [133]. A similar process may occur in neurodegeneration, with axonal transport mediating the transport of stress signals from the cell periphery to the nucleus [122,123].

13.5.2 Huntington's disease

Expansion of a polyglutamine repeat in the N-terminus of huntingtin, from ≤35 repeats in normal individuals to >35 in affected individuals, is the sole cause of HD. HD is a fatal neurodegenerative disorder that primarily affects medium spiny neurons in the basal ganglia and is characterized by chorea, behavioral change, and cognitive decline. Pathogenic expansion of the normal polyglutamine sequence leads to increased aggregation of the huntingtin protein in both cell body and nucleus and induces an array of cellular changes including alterations in gene expression [134].

It remains unclear whether the autosomal dominant disease is caused entirely by a toxic gain of function, or whether loss of normal huntingtin function contributes. Part of this uncertainty stems from the fact that the normal function of huntingtin is not well understood. However, multiple studies have implicated huntingtin in vesicle and organelle trafficking within the neuron. Huntingtin is known to bind directly to cytoplasmic dynein [135] and indirectly to both dynactin and kinesin-1 via the adaptor protein HAP1 [136–139]. Accumulating data indicate that huntingtin serves as a scaffolding protein involved in the regulation of bidirectional vesicle transport along microtubules [140,141]. Huntingtin regulates dynein motor function on multiple organelles including: mitochondria, BDNF- and TrkB-containing vesicles and autophagosomes [38,141–144].

While it is not clear how much this interaction contributes to the pathogenesis of HD, recent work has shown that the degeneration of neurons induced by depletion of huntingtin cannot be rescued by a construct that lacks the dynein-binding domain [145]. Further, depletion of either huntingtin or its binding partner HAP1 in neurons leads to a disruption of dynein-driven autophagosome transport along axons [144]. Expression of mutant huntingtin had a dominant negative effect on autophagosome function, leading to the accumulation of undigested mitochondrial fragments within the lumen of axonal autophagosomes.

In dendrites, both huntingtin and dynein colocalize with TrkB, the receptor for the neurotrophic factor BDNF [38]. The trafficking of BDNF-bound TrkB from dendrites to the cell body, and the resultant signaling to affect gene expression were altered in neurons from a mouse model of HD [38], further supporting a key role for the Huntington–dynein interaction in the regulation of intracellular trafficking. Further work is required to determine how these defects in tropic factor signaling, autophagic clearance of protein aggregates and aging organelles contribute to the pathophysiology of HD.

Data suggesting a key role for dynein in the striatal neurons that degenerate during HD comes from in vivo brain imaging of the striatum of Cra/+ mice, which express a point mutation in cytoplasmic dynein HC (Fig. 13.2A).

Both striatal atrophy and altered dopamine signaling were noted [73]. Further, axonal transport defects have been observed in multiple models of HD [146,147]. Proposed mechanisms include defects in the uptake or delivery of the essential neurotrophic factor BDNF [38,143], and/or defects in the autophagy machinery [144] that may lead to accumulation of mutant huntingtin (reviewed in Ref. [148]). Further work is required to determine to what extent defects in dynein-dependent trafficking functions contribute to HD.

13.5.3 Parkinson's disease

The most direct link between dynein and PD to date is the observation that an atypical form of parkinsonism, known as Perry syndrome, is caused by mutations in dynactin [97]. Perry syndrome is a rapidly progressive, lethal form of parkinsonism characterized by depression, weight loss, insomnia, and hypoventilation [97,149]. At the cellular level, detailed mechanistic studies indicate that the Perry mutations have a striking effect on the initiation of retrograde axonal transport from the distal axon [31,59]. The initiation of dynein-mediated retrograde transport is a stepwise regulated mechanism involving the microtubule plus end tracking proteins EB3 and CLIP-170, which act together to recruit dynein and dynactin to the microtubule [32,105,150]. Recruitment of the motor complex is further enhanced by the locally high concentration of tyrosinated microtubules found in the distal axon [150]. Defects in transport initiation may lead to ineffective trophic factor signaling or reduced clearance of aggregated proteins via autophagy. The cell type specificity of the disease remains to be explained, but recent work suggests that some Perry-associated mutations may affect additional neuronal subtypes [104], consistent with the essential role of dynactin in neurons.

Genome-wide association studies have not directly implicated mutations in dynein as a cause of familial PD. However, of the 24 genetic loci implicated in PD to date, 11 are involved in the degradative pathway, including autophagy and lysosomal degradation [150]. These include the kinase LRRK2, implicated in the regulation of autophagy [151], and also linked to the dynein-dependent positioning of lysosomes [152] and Golgi outposts in dendrites [153]. Further work will be required to dissect the role of dynein-mediated organelle dynamics in the pathogenesis of PD.

13.6 Conclusions

The identification of cytoplasmic dynein as an essential minus end-directed motor in higher eukaryotes has led to a much deeper understanding of the role of active transport in the neuron. Many of the key pathways involved have been illuminated, but many important questions remain.

The significance of these questions to neurological disease is becoming increasingly clear with the identification of human mutations in dynein and dynein-associated proteins, and the increasing evidence for dynein-pathway dysfunction in diseases such as ALS and HD. The broad range of clinical presentations observed in these diseases may stem directly from the diverse roles of dynein in the cell. Further, analysis of pathological mutations in dynein and dynein effectors provides an invaluable tool to dissecting the varied and vital functions of dynein in the cell. This is now a very rapidly progressing field, where basic research is informing clinical discoveries and in turn clinical observations are accelerating our understanding of basic cellular processes.

References

[1] L.C. Kapitein, C.C. Hoogenraad, Building the neuronal microtubule cytoskeleton, Neuron 87 (3) (2015) 492–506.

[2] S. Maday, A.E. Twelvetrees, A.J. Moughamian, E.L. Holzbaur, Axonal transport: cargo-specific mechanisms of motility and regulation, Neuron 84 (2) (2014) 292–309.

[3] J.F. Presley, N.B. Cole, T.A. Schroer, K. Hirschberg, K.J. Zaal, J. Lippincott-Schwartz, ER-to-Golgi transport visualized in living cells, Nature 389 (6646) (1997) 81–85.

[4] O.J. Driskell, A. Mironov, V.J. Allan, P.G. Woodman, Dynein is required for receptor sorting and the morphogenesis of early endosomes, Nat. Cell Biol. 9 (1) (2007) 113–120.

[5] H.M. Heerssen, M.F. Pazyra, R.A. Segal, Dynein motors transport activated Trks to promote survival of target-dependent neurons, Nat. Neurosci. 7 (6) (2004) 596–604.

[6] A. Harada, Y. Takei, Y. Kanai, Y. Tanaka, S. Nonaka, N. Hirokawa, Golgi vesiculation and lysosome dispersion in cells lacking cytoplasmic dynein, J. Cell Biol. 141 (1) (1998) 51–59.

[7] S. Kimura, T. Noda, T. Yoshimori, Dynein-dependent movement of autophagosomes mediates efficient encounters with lysosomes, Cell Struct. Funct. 33 (1) (2008) 109–122.

[8] S. Maday, E.L. Holzbaur, Autophagosome assembly and cargo capture in the distal axon, Autophagy 8 (5) (2012) 858–860.

[9] N. Hirokawa, R. Sato-Yoshitake, T. Yoshida, T. Kawashima, Brain dynein (MAP1C) localizes on both anterogradely and retrogradely transported membranous organelles in vivo, J. Cell Biol. 111 (3) (1990) 1027–1037.

[10] N. Hirokawa, R. Sato-Yoshitake, N. Kobayashi, K.K. Pfister, G.S. Bloom, S.T. Brady, Kinesin associates with anterogradely transported membranous organelles in vivo, J. Cell Biol. 114 (2) (1991) 295–302.

[11] T.L. Blasius, N. Reed, B.M. Slepchenko, K.J. Verhey, Recycling of kinesin-1 motors by diffusion after transport, PLoS One 8 (9) (2013) e76081.

[12] A.G. Hendricks, E. Perlson, J.L. Ross, H.W. Schroeder 3rd, M. Tokito, E.L. Holzbaur, Motor coordination via a tug-of-war mechanism drives bidirectional vesicle transport, Curr. Biol. 20 (8) (2010) 697–702.

[13] A.D. Pilling, D. Horiuchi, C.M. Lively, W.M. Saxton, Kinesin-1 and Dynein are the primary motors for fast transport of mitochondria in *Drosophila* motor axons, Mol. Biol. Cell 17 (4) (2006) 2057–2068.

[14] B. Cui, C. Wu, L. Chen, A. Ramirez, E.L. Bearer, W.P. Li, et al., One at a time, live tracking of NGF axonal transport using quantum dots, Proc. Natl. Acad. Sci. USA 104 (34) (2007) 13666–13671.

[15] S. Maday, K.E. Wallace, E.L. Holzbaur, Autophagosomes initiate distally and mature during transport toward the cell soma in primary neurons, J. Cell Biol. 196 (4) (2012) 407–417.

[16] M.M. Fu, E.L. Holzbaur, JIP1 regulates the directionality of APP axonal transport by coordinating kinesin and dynein motors, J. Cell Biol. 202 (3) (2013) 495–508.

[17] S. Takeda, H. Yamazaki, D.H. Seog, Y. Kanai, S. Terada, N. Hirokawa, Kinesin superfamily protein 3 (KIF3) motor transports fodrin-associating vesicles important for neurite building, J. Cell Biol. 148 (6) (2000) 1255–1265.

[18] D.H. Hall, E.M. Hedgecock, Kinesin-related gene unc-104 is required for axonal transport of synaptic vesicles in *C. elegans*, Cell 65 (5) (1991) 837–847.

[19] Y. Okada, H. Yamazaki, Y. Sekine-Aizawa, N. Hirokawa, The neuron-specific kinesin superfamily protein KIF1A is a unique monomeric motor for anterograde axonal transport of synaptic vesicle precursors, Cell 81 (5) (1995) 769–780.

[20] K.Y. Lo, A. Kuzmin, S.M. Unger, J.D. Petersen, M.A. Silverman, KIF1A is the primary anterograde motor protein required for the axonal transport of dense-core vesicles in cultured hippocampal neurons, Neurosci. Lett. 491 (3) (2011) 168–173.

[21] C. Wu, A. Ramirez, B. Cui, J. Ding, J.D. Delcroix, J.S. Valletta, et al., A functional dynein-microtubule network is required for NGF signaling through the Rap1/MAPK pathway, Traffic 8 (11) (2007) 1503–1520.

[22] J. Ha, K.W. Lo, K.R. Myers, T.M. Carr, M.K. Humsi, B.A. Rasoul, et al., A neuron-specific cytoplasmic dynein isoform preferentially transports TrkB signaling endosomes, J. Cell Biol. 181 (6) (2008) 1027–1039.

[23] L. Wang, C.L. Ho, D. Sun, R.K. Liem, A. Brown, Rapid movement of axonal neurofilaments interrupted by prolonged pauses, Nat. Cell Biol. 2 (3) (2000) 137–141.

[24] O.I. Wagner, J. Ascano, M. Tokito, J.F. Leterrier, P.A. Janmey, E.L. Holzbaur, The interaction of neurofilaments with the microtubule motor cytoplasmic dynein, Mol. Biol. Cell 15 (11) (2004) 5092–5100.

[25] Y. He, F. Francis, K.A. Myers, W. Yu, M.M. Black, P.W. Baas, Role of cytoplasmic dynein in the axonal transport of microtubules and neurofilaments, J. Cell Biol. 168 (5) (2005) 697–703.

[26] A. Uchida, N.H. Alami, A. Brown, Tight functional coupling of kinesin-1A and dynein motors in the bidirectional transport of neurofilaments, Mol. Biol. Cell 20 (23) (2009) 4997–5006.

[27] Y. Tang, D. Scott, U. Das, D. Gitler, A. Ganguly, S. Roy, Fast vesicle transport is required for the slow axonal transport of synapsin, J. Neurosci. 33 (39) (2013) 15362–15375.

[28] J.F. Dillman 3rd, L.P. Dabney, S. Karki, B.M. Paschal, E.L. Holzbaur, K.K. Pfister, Functional analysis of dynactin and cytoplasmic dynein in slow axonal transport, J. Neurosci. 16 (21) (1996) 6742–6752.

[29] J.F. Dillman 3rd, L.P. Dabney, K.K. Pfister, Cytoplasmic dynein is associated with slow axonal transport, Proc. Natl. Acad. Sci. USA 93 (1) (1996) 141–144.

[30] A.E. Twelvetrees, S. Pernigo, A. Sanger, P. Guedes-Dias, G. Schiavo, R.A. Steiner, et al., The dynamic localization of cytoplasmic dynein in neurons is driven by Kinesin-1, Neuron 90 (5) (2016) 1000–1015.

[31] A.J. Moughamian, E.L. Holzbaur, Dynactin is required for transport initiation from the distal axon, Neuron 74 (2) (2012) 331–343.

[32] J.J. Nirschl, M.M. Magiera, J.E. Lazarus, C. Janke, E.L. Holzbaur, Alpha-tubulin tyrosination and CLIP-170 phosphorylation regulate the initiation of dynein-driven transport in neurons, Cell Rep. 14 (11) (2016) 2637–2652.

[33] L.C. Kapitein, M.A. Schlager, M. Kuijpers, P.S. Wulf, M. van Spronsen, F.C. MacKintosh, et al., Mixed microtubules steer dynein-driven cargo transport into dendrites, Curr. Biol. 20 (4) (2010) 290–299.

[33a] S. Ayloo, P. Guedes-Dias, A.E. Ghiretti, E.L. Holzbaur, Dynein efficiently navigates the dendritic cytoskeleton to drive the retrograde trafficking of BDNF/TrkB signaling endosomes, Mol. Biol. Cell 28 (19), (2017) 2543–2554.

[34] Y. Zheng, J. Wildonger, B. Ye, Y. Zhang, A. Kita, S.H. Younger, et al., Dynein is required for polarized dendritic transport and uniform microtubule orientation in axons, Nat. Cell Biol. 10 (10) (2008) 1172–1180.

[35] A.W. Harrington, D.D. Ginty, Long-distance retrograde neurotrophic factor signalling in neurons, Nat. Rev. Neurosci. 14 (3) (2013) 177–187.

[36] H. Yano, F.S. Lee, H. Kong, J. Chuang, J. Arevalo, P. Perez, et al., Association of Trk neurotrophin receptors with components of the cytoplasmic dynein motor, J. Neurosci. 21 (3) (2001) RC125.

[37] A. Bhattacharyya, F.L. Watson, S.L. Pomeroy, Y.Z. Zhang, C.D. Stiles, R.A. Segal, High-resolution imaging demonstrates dynein-based vesicular transport of activated Trk receptors, J. Neurobiol. 51 (4) (2002) 302–312.

[38] G. Liot, D. Zala, P. Pla, G. Mottet, M. Piel, F. Saudou, Mutant Huntingtin alters retrograde transport of TrkB receptors in striatal dendrites, J. Neurosci. 33 (15) (2013) 6298–6309.

[39] C.A. Garrett, M. Barri, A. Kuta, V. Soura, W. Deng, E.M. Fisher, et al., DYNC1H1 mutation alters transport kinetics and ERK1/2-cFos signalling in a mouse model of distal spinal muscular atrophy, Brain 137 (Pt 7) (2014) 1883–1893.

[40] O. Reiner, T. Sapir, LIS1 functions in normal development and disease, Curr. Opin. Neurobiol. 23 (6) (2013) 951–956.

[41] F.S. Alkuraya, X. Cai, C. Emery, G.H. Mochida, M.S. Al-Dosari, J.M. Felie, et al., Human mutations in NDE1 cause extreme microcephaly with lissencephaly [corrected], Am. J. Hum. Genet. 88 (5) (2011) 536–547.

[42] A.R. Paciorkowski, K. Keppler-Noreuil, L. Robinson, C. Sullivan, S. Sajan, S.L. Christian, et al., Deletion 16p13.11 uncovers NDE1 mutations on the non-deleted homolog and extends the spectrum of severe microcephaly to include fetal brain disruption, Am. J. Med. Genet. A 161A (7) (2013) 1523–1530.

[43] B. Ravikumar, A. Acevedo-Arozena, S. Imarisio, Z. Berger, C. Vacher, C.J. O'Kane, et al., Dynein mutations impair autophagic clearance of aggregate-prone proteins, Nat. Genet. 37 (7) (2005) 771–776.

[44] L. Cogli, F. Piro, C. Bucci, Rab7 and the CMT2B disease, Biochem. Soc. Trans. 37 (Pt 5) (2009) 1027–1031.

[45] B.A. McCray, E. Skordalakes, J.P. Taylor, Disease mutations in Rab7 result in unregulated nucleotide exchange and inappropriate activation, Hum. Mol. Genet. 19 (6) (2010) 1033–1047.

[46] I. Jordens, M. Fernandez-Borja, M. Marsman, S. Dusseljee, L. Janssen, J. Calafat, et al., The Rab7 effector protein RILP controls lysosomal transport by inducing the recruitment of dynein-dynactin motors, Curr. Biol. 11 (21) (2001) 1680–1685.

[47] M.T. Hsu, C.L. Guo, A.Y. Liou, T.Y. Chang, M.C. Ng, B.I. Florea, et al., Stage-dependent axon transport of proteasomes contributes to axon development, Dev. Cell 35 (4) (2015) 418–431.

[48] J.A. Olzmann, L. Li, L.S. Chin, Aggresome formation and neurodegenerative diseases: therapeutic implications, Curr. Med. Chem. 15 (1) (2008) 47–60.

[49] J.A. Johnston, M.E. Illing, R.R. Kopito, Cytoplasmic dynein/dynactin mediates the assembly of aggresomes, Cell Motil. Cytoskelet. 53 (1) (2002) 26–38.

[50] Y. Kawaguchi, J.J. Kovacs, A. McLaurin, J.M. Vance, A. Ito, T.P. Yao, The deacetylase HDAC6 regulates aggresome formation and cell viability in response to misfolded protein stress, Cell 115 (6) (2003) 727–738.

[51] T. Hara, K. Nakamura, M. Matsui, A. Yamamoto, Y. Nakahara, R. Suzuki-Migishima, et al., Suppression of basal autophagy in neural cells causes neurodegenerative disease in mice, Nature 441 (7095) (2006) 885–889.

[52] M. Komatsu, S. Waguri, T. Chiba, S. Murata, J. Iwata, I. Tanida, et al., Loss of autophagy in the central nervous system causes neurodegeneration in mice, Nature 441 (7095) (2006) 880–884.

[53] S. Maday, E.L. Holzbaur, Autophagosome biogenesis in primary neurons follows an ordered and spatially regulated pathway, Dev. Cell 30 (1) (2014) 71–85.

[54] M.-M. Fu, J.J. Nirschl, E.L.F. Holzbaur, LC3 binding to the scaffolding protein JIP1 regulates processive dynein-driven transport of autophagosomes, Dev. Cell 29 (5) (2014) 577–590.

[55] X.T. Cheng, B. Zhou, M.Y. Lin, Q. Cai, Z.H. Sheng, Axonal autophagosomes recruit dynein for retrograde transport through fusion with late endosomes, J. Cell Biol. 209 (3) (2015) 377–386.

[56] B.H. LaMonte, K.E. Wallace, B.A. Holloway, S.S. Shelly, J. Ascano, M. Tokito, et al., Disruption of dynein/dynactin inhibits axonal transport in motor neurons causing late-onset progressive degeneration, Neuron 34 (5) (2002) 715–727.

[57] I. Puls, C. Jonnakuty, B.H. LaMonte, E.L. Holzbaur, M. Tokito, E. Mann, et al., Mutant dynactin in motor neuron disease, Nat. Genet. 33 (4) (2003) 455–456.

[58] J.R. Levy, C.J. Sumner, J.P. Caviston, M.K. Tokito, S. Ranganathan, L.A. Ligon, et al., A motor neuron disease-associated mutation in p150Glued perturbs dynactin function and induces protein aggregation, J. Cell Biol. 172 (5) (2006) 733–745.

[59] T.E. Lloyd, J. Machamer, K. O'Hara, J.H. Kim, S.E. Collins, M.Y. Wong, et al., The p150(Glued) CAP-Gly domain regulates initiation of retrograde transport at synaptic termini, Neuron 74 (2) (2012) 344–360.

[60] C. Lai, X. Lin, J. Chandran, H. Shim, W.J. Yang, H. Cai, The G59S mutation in p150(glued) causes dysfunction of dynactin in mice, J. Neurosci. 27 (51) (2007) 13982–13990.

[61] F.M. Laird, M.H. Farah, S. Ackerley, A. Hoke, N. Maragakis, J.D. Rothstein, et al., Motor neuron disease occurring in a mutant dynactin mouse model is characterized by defects in vesicular trafficking, J. Neurosci. 28 (9) (2008) 1997–2005.

[62] E.S. Chevalier-Larsen, K.E. Wallace, C.R. Pennise, E.L. Holzbaur, Lysosomal proliferation and distal degeneration in motor neurons expressing the G59S mutation in the p150Glued subunit of dynactin, Hum. Mol. Genet. 17 (13) (2008) 1946–1955.

[63] R.D. Allen, J. Metuzals, I. Tasaki, S.T. Brady, S.P. Gilbert, Fast axonal transport in squid giant axon, Science 218 (4577) (1982) 1127–1129.

[64] C.M. Waterman-Storer, E.L. Holzbaur, The product of the *Drosophila* gene, Glued, is the functional homologue of the p150Glued component of the vertebrate dynactin complex, J. Biol. Chem. 271 (2) (1996) 1153–1159.

[65] B.A. Eaton, R.D. Fetter, G.W. Davis, Dynactin is necessary for synapse stabilization, Neuron 34 (5) (2002) 729–741.

[66] M. Hafezparast, R. Klocke, C. Ruhrberg, A. Marquardt, A. Ahmad-Annuar, S. Bowen, et al., Mutations in dynein link motor neuron degeneration to defects in retrograde transport, Science 300 (5620) (2003) 808–812.

[67] X.J. Chen, E.N. Levedakou, K.J. Millen, R.L. Wollmann, B. Soliven, B. Popko, Proprioceptive sensory neuropathy in mice with a mutation in the cytoplasmic Dynein heavy chain 1 gene, J. Neurosci. 27 (52) (2007) 14515–14524.

[68] K.M. Ori-McKenney, J. Xu, S.P. Gross, R.B. Vallee, A cytoplasmic dynein tail mutation impairs motor processivity, Nat. Cell Biol. 12 (12) (2010) 1228–1234.

[69] J. Zhao, Y. Wang, H. Xu, Y. Fu, T. Qian, D. Bo, et al., Dync1h1 mutation causes proprioceptive sensory neuron loss and impaired retrograde axonal transport of dorsal root ganglion neurons, CNS Neurosci. Ther. 22 (7) (2016) 593–601.

[70] H.S. Ilieva, K. Yamanaka, S. Malkmus, O. Kakinohana, T. Yaksh, M. Marsala, et al., Mutant dynein (Loa) triggers proprioceptive axon loss that extends survival only in the SOD1 ALS model with highest motor neuron death, Proc. Natl. Acad. Sci. USA 105 (34) (2008) 12599–12604.

[71] L. Dupuis, A. Fergani, K.E. Braunstein, J. Eschbach, N. Holl, F. Rene, et al., Mice with a mutation in the dynein heavy chain 1 gene display sensory neuropathy but lack motor neuron disease, Exp. Neurol. 215 (1) (2009) 146–152.

[72] L.M. Wiggins, A. Kuta, J.C. Stevens, E.M. Fisher, C.S. von Bartheld, A novel phenotype for the dynein heavy chain mutation Loa: altered dendritic morphology, organelle density, and reduced numbers of trigeminal motoneurons, J. Comp. Neurol. 520 (12) (2012) 2757–2773.

[73] K.E. Braunstein, J. Eschbach, K. Rona-Voros, R. Soylu, E. Mikrouli, Y. Larmet, et al., A point mutation in the dynein heavy chain gene leads to striatal atrophy and compromises neurite outgrowth of striatal neurons, Hum. Mol. Genet. 19 (22) (2010) 4385–4398.

[74] S.L. Courchesne, M.F. Pazyra-Murphy, D.J. Lee, R.A. Segal, Neuromuscular junction defects in mice with mutation of dynein heavy chain 1, PLoS One 6 (2) (2011) e16753.

[75] M.J. Allen, X. Shan, P. Caruccio, S.J. Froggett, K.G. Moffat, R.K. Murphey, Targeted expression of truncated glued disrupts giant fiber synapse formation in *Drosophila*, J. Neurosci. 19 (21) (1999) 9374–9384.

[76] L.E. Vissers, J. de Ligt, C. Gilissen, I. Janssen, M. Steehouwer, P. de Vries, et al., A de novo paradigm for mental retardation, Nat. Genet. 42 (12) (2010) 1109–1112.

[77] M.H. Willemsen, L.E. Vissers, M.A. Willemsen, B.W. van Bon, T. Kroes, J. de Ligt, et al., Mutations in DYNC1H1 cause severe intellectual disability with neuronal migration defects, J. Med. Genet. 49 (3) (2012) 179–183.

[78] K. Poirier, N. Lebrun, L. Broix, G. Tian, Y. Saillour, C. Boscheron, et al., Mutations in TUBG1, DYNC1H1, KIF5C and KIF2A cause malformations of cortical development and microcephaly, Nat. Genet. 45 (6) (2013) 639–647.

[79] M.N. Weedon, R. Hastings, R. Caswell, W. Xie, K. Paszkiewicz, T. Antoniadi, et al., Exome sequencing identifies a DYNC1H1 mutation in a large pedigree with dominant axonal Charcot-Marie-Tooth disease, Am. J. Hum. Genet. 89 (2) (2011) 308–312.

[80] M.B. Harms, K.M. Ori-McKenney, M. Scoto, E.P. Tuck, S. Bell, D. Ma, et al., Mutations in the tail domain of DYNC1H1 cause dominant spinal muscular atrophy, Neurology 78 (22) (2012) 1714–1720.

[81] Y. Tsurusaki, S. Saitoh, K. Tomizawa, A. Sudo, N. Asahina, H. Shiraishi, et al., A DYNC1H1 mutation causes a dominant spinal muscular atrophy with lower extremity predominance, Neurogenetics 13 (4) (2012) 327–332.

[82] K. Peeters, I. Litvinenko, B. Asselbergh, L. Almeida-Souza, T. Chamova, T. Geuens, et al., Molecular defects in the motor adaptor BICD2 cause proximal spinal muscular atrophy with autosomal-dominant inheritance, Am. J. Hum. Genet. 92 (6) (2013) 955–964.

[83] C. Fiorillo, F. Moro, J. Yi, S. Weil, G. Brisca, G. Astrea, et al., Novel dynein DYNC1H1 neck and motor domain mutations link distal spinal muscular atrophy and abnormal cortical development, Hum. Mutation 35 (3) (2014) 298–302.

[84] Q. Niu, X. Wang, M. Shi, Q. Jin, A novel DYNC1H1 mutation causing spinal muscular atrophy with lower extremity predominance, Neurol. Genet. 1 (2) (2015) e20.

[85] J. Punetha, S. Monges, M.E. Franchi, E.P. Hoffman, S. Cirak, C. Tesi-Rocha, Exome sequencing identifies DYNC1H1 variant associated with vertebral abnormality and spinal muscular atrophy with lower extremity predominance, Pediatr. Neurol. 52 (2) (2015) 239–244.

[86] M. Scoto, A.M. Rossor, M.B. Harms, S. Cirak, M. Calissano, S. Robb, et al., Novel mutations expand the clinical spectrum of DYNC1H1-associated spinal muscular atrophy, Neurology 84 (7) (2015) 668–679.

[87] A.V. Strickland, M. Schabhuttl, H. Offenbacher, M. Synofzik, N.S. Hauser, M. Brunner-Krainz, et al., Mutation screen reveals novel variants and expands the phenotypes associated with DYNC1H1, J. Neurol. 262 (9) (2015) 2124–2134.

[88] D. Ding, Z. Chen, K. Li, Z. Long, W. Ye, Z. Tang, et al., Identification of a de novo DYNC1H1 mutation via WES according to published guidelines, Sci. Rep. 6 (2016) 20423.

[89] J. Hertecant, M. Komara, A. Nagi, J. Suleiman, L. Al-Gazali, B.R. Ali, A novel de novo mutation in DYNC1H1 gene underlying malformation of cortical development and cataract, Meta Gene 9 (2016) 124–127.

[90] S.H. Hwang, E.J. Kim, Y.B. Hong, J. Joo, S.M. Kim, S.H. Nam, et al., Distal hereditary motor neuropathy type 7B with Dynactin 1 mutation, Mol. Med. Rep. 14 (4) (2016) 3362–3368.

[91] S. Karki, E.L. Holzbaur, Affinity chromatography demonstrates a direct binding between cytoplasmic dynein and the dynactin complex, J. Biol. Chem. 270 (48) (1995) 28806–28811.

[92] K.T. Vaughan, R.B. Vallee, Cytoplasmic dynein binds dynactin through a direct interaction between the intermediate chains and p150Glued, J. Cell Biol. 131 (6 Pt 1) (1995) 1507–1516.

[93] C.M. Waterman-Storer, S. Karki, E.L. Holzbaur, The p150Glued component of the dynactin complex binds to both microtubules and the actin-related protein centractin (Arp-1), Proc. Natl. Acad. Sci. USA 92 (5) (1995) 1634–1638.

[94] I. Puls, S.J. Oh, C.J. Sumner, K.E. Wallace, M.K. Floeter, E.A. Mann, et al., Distal spinal and bulbar muscular atrophy caused by dynactin mutation, Ann. Neurol. 57 (5) (2005) 687–694.

[95] C. Munch, R. Sedlmeier, T. Meyer, V. Homberg, A.D. Sperfeld, A. Kurt, et al., Point mutations of the p150 subunit of dynactin (DCTN1) gene in ALS, Neurology 63 (4) (2004) 724–726.

[96] C. Munch, A. Rosenbohm, A.D. Sperfeld, I. Uttner, S. Reske, B.J. Krause, et al., Heterozygous R1101K mutation of the DCTN1 gene in a family with ALS and FTD, Ann. Neurol. 58 (5) (2005) 777–780.

[97] M.J. Farrer, M.M. Hulihan, J.M. Kachergus, J.C. Dachsel, A.J. Stoessl, L.L. Grantier, et al., DCTN1 mutations in Perry syndrome, Nat. Genet. 41 (2) (2009) 163–165.

[98] M. Omoto, S. Suzuki, T. Ikeuchi, T. Ishihara, T. Kobayashi, Y. Tsuboi, et al., Autosomal dominant tauopathy with parkinsonism and central hypoventilation, Neurology 78 (10) (2012) 762–764.

[99] E.J. Chung, J.H. Hwang, M.J. Lee, J.H. Hong, K.H. Ji, W.K. Yoo, et al., Expansion of the clinicopathological and mutational spectrum of Perry syndrome, Parkinsonism Relat. Disord. 20 (4) (2014) 388–393.

[100] P. Tacik, F.C. Fiesel, S. Fujioka, O.A. Ross, F. Pretelt, C. Castaneda Cardona, et al., Three families with Perry syndrome from distinct parts of the world, Parkinsonism Relat. Disord. 20 (8) (2014) 884–888.

[101] P. Caroppo, I. Le Ber, F. Clot, S. Rivaud-Pechoux, A. Camuzat, A. De Septenville, et al., DCTN1 mutation analysis in families with progressive supranuclear palsy-like phenotypes, JAMA Neurol. 71 (2) (2014) 208–215.

[102] E.K. Gustavsson, J. Trinh, I. Guella, C. Szu-Tu, J. Khinda, C.H. Lin, et al., DCTN1 p.K56R in progressive supranuclear palsy, Parkinsonism Relat. Disord. 28 (2016) 56–61.

[103] V. Newsway, M. Fish, J.D. Rohrer, E. Majounie, N. Williams, M. Hack, et al., Perry syndrome due to the DCTN1 G71R mutation: a distinctive levodopa responsive disorder with behavioral syndrome, vertical gaze palsy, and respiratory failure, Mov. Disord. 25 (6) (2010) 767–770.

[104] E. Araki, Y. Tsuboi, J. Daechsel, A. Milnerwood, C. Vilarino-Guell, N. Fujii, et al., A novel DCTN1 mutation with late-onset parkinsonism and frontotemporal atrophy, Mov. Disord. 29 (9) (2014) 1201–1204.

[105] A.J. Moughamian, G.E. Osborn, J.E. Lazarus, S. Maday, E.L. Holzbaur, Ordered recruitment of dynactin to the microtubule plus-end is required for efficient initiation of retrograde axonal transport, J. Neurosci. 33 (32) (2013) 13190–13203.

[106] W. Matsuda, T. Furuta, K.C. Nakamura, H. Hioki, F. Fujiyama, R. Arai, et al., Single nigrostriatal dopaminergic neurons form widely spread and highly dense axonal arborizations in the neostriatum, J. Neurosci. 29 (2) (2009) 444–453.

[107] R.J. McKenney, W. Huynh, M.E. Tanenbaum, G. Bhabha, R.D. Vale, Activation of cytoplasmic dynein motility by dynactin-cargo adapter complexes, Science 345 (6194) (2014) 337–341.

[108] C.C. Hoogenraad, A. Akhmanova, Bicaudal D family of motor adaptors: linking dynein motility to cargo binding, Trends Cell Biol. 26 (5) (2016) 327–340.

[109] M. van Spronsen, M. Mikhaylova, J. Lipka, M.A. Schlager, D.J. van den Heuvel, M. Kuijpers, et al., TRAK/Milton motor-adaptor proteins steer mitochondrial trafficking to axons and dendrites, Neuron 77 (3) (2013) 485–502.

[110] J. Zhang, R. Qiu, H.N. Arst Jr., M.A. Penalva, X. Xiang, HookA is a novel dynein-early endosome linker critical for cargo movement in vivo, J. Cell Biol. 204 (6) (2014) 1009–1026.

[111] E. Bielska, M. Schuster, Y. Roger, A. Berepiki, D.M. Soanes, N.J. Talbot, et al., Hook is an adapter that coordinates kinesin-3 and dynein cargo attachment on early endosomes, J. Cell Biol. 204 (6) (2014) 989–1007.

[112] M.A. Olenick, M. Tokito, M. Boczkowska, R. Dominguez, E.L. Holzbaur, Hook adaptors induce unidirectional processive motility by enhancing the dynein-dynactin interaction, J. Biol. Chem. 291 (35) (2016) 18239–18251.

[113] C.M. Schroeder, R.D. Vale, Assembly and activation of dynein-dynactin by the cargo adaptor protein Hook3, J. Cell Biol. 214 (3) (2016) 309–318.

[114] L. Urnavicius, K. Zhang, A.G. Diamant, C. Motz, M.A. Schlager, M. Yu, et al., The structure of the dynactin complex and its interaction with dynein, Science 347 (6229) (2015) 1441–1446.

[115] K. Neveling, L.A. Martinez-Carrera, I. Holker, A. Heister, A. Verrips, S.M. Hosseini-Barkooie, et al., Mutations in BICD2, which encodes a golgin and important motor adaptor, cause congenital autosomal-dominant spinal muscular atrophy, Am. J. Hum. Genet. 92 (6) (2013) 946–954.

[116] E.C. Oates, A.M. Rossor, M. Hafezparast, M. Gonzalez, F. Speziani, D.G. MacArthur, et al., Mutations in BICD2 cause dominant congenital spinal muscular atrophy and hereditary spastic paraplegia, Am. J. Hum. Genet. 92 (6) (2013) 965–973.

[117] A.M. Rossor, E.C. Oates, H.K. Salter, Y. Liu, S.M. Murphy, R. Schule, et al., Phenotypic and molecular insights into spinal muscular atrophy due to mutations in BICD2, Brain 138 (Pt 2) (2015) 293–310.

[118] M. Synofzik, L.A. Martinez-Carrera, T. Lindig, L. Schols, B. Wirth, Dominant spinal muscular atrophy due to BICD2: a novel mutation refines the phenotype, J. Neurol. Neurosurg. Psychiatry 85 (5) (2014) 590–592.

[119] B. Bansagi, H. Griffin, V. Ramesh, J. Duff, A. Pyle, P.F. Chinnery, et al., The p.Ser107Leu in BICD2 is a mutation 'hot spot' causing distal spinal muscular atrophy, Brain 138 (Pt 11) (2015) e391.

[120] C. Fiorillo, F. Moro, G. Brisca, A. Accogli, F. Trucco, R. Trovato, et al., Beyond spinal muscular atrophy with lower extremity dominance: cerebellar hypoplasia associated with a novel mutation in BICD2, Eur. J. Neurol. 23 (4) (2016) e19–21.

[121] K. Verhoeven, P. De Jonghe, K. Coen, N. Verpoorten, M. Auer-Grumbach, J.M. Kwon, et al., Mutations in the small GTP-ase late endosomal protein RAB7 cause Charcot-Marie-Tooth type 2B neuropathy, Am. J. Hum. Genet. 72 (3) (2003) 722–727.

[122] E. Perlson, S. Maday, M.M. Fu, A.J. Moughamian, E.L. Holzbaur, Retrograde axonal transport: pathways to cell death? Trends Neurosci. 33 (7) (2010) 335–344.

[123] E. Perlson, G.B. Jeong, J.L. Ross, R. Dixit, K.E. Wallace, R.G. Kalb, et al., A switch in retrograde signaling from survival to stress in rapid-onset neurodegeneration, J. Neurosci. 29 (31) (2009) 9903–9917.

[124] L.A. Ligon, B.H. LaMonte, K.E. Wallace, N. Weber, R.G. Kalb, E.L. Holzbaur, Mutant superoxide dismutase disrupts cytoplasmic dynein in motor neurons, Neuroreport 16 (6) (2005) 533–536.

[125] L.G. Bilsland, E. Sahai, G. Kelly, M. Golding, L. Greensmith, G. Schiavo, Deficits in axonal transport precede ALS symptoms in vivo, Proc. Natl. Acad. Sci. USA 107 (47) (2010) 20523–20528.

[126] P. Marinkovic, M.S. Reuter, M.S. Brill, L. Godinho, M. Kerschensteiner, T. Misgeld, Axonal transport deficits and degeneration can evolve independently in mouse models of amyotrophic lateral sclerosis, Proc. Natl. Acad. Sci. USA 109 (11) (2012) 4296–4301.

[127] J. Magrane, C. Cortez, W.B. Gan, G. Manfredi, Abnormal mitochondrial transport and morphology are common pathological denominators in SOD1 and TDP43 ALS mouse models, Hum. Mol. Genet. 23 (6) (2014) 1413–1424.

[128] E. Klinman, E.L. Holzbaur, Stress-induced CDK5 activation disrupts axonal transport via Lis1/Ndel1/dynein, Cell Rep. 12 (3) (2015) 462–473.

[129] K.R. Baldwin, V.K. Godena, V.L. Hewitt, A.J. Whitworth, Axonal transport defects are a common phenotype in *Drosophila* models of ALS, Hum. Mol. Genet. (2016).

[130] M. Niethammer, D.S. Smith, R. Ayala, J. Peng, J. Ko, M.S. Lee, et al., NUDEL is a novel Cdk5 substrate that associates with LIS1 and cytoplasmic dynein, Neuron 28 (3) (2000) 697–711.

[131] R.J. McKenney, M. Vershinin, A. Kunwar, R.B. Vallee, S.P. Gross, LIS1 and NudE induce a persistent dynein force-producing state, Cell 141 (2) (2010) 304–314.

[132] J. Huang, A.J. Roberts, A.E. Leschziner, S.L. Reck-Peterson, Lis1 acts as a "clutch" between the ATPase and microtubule-binding domains of the dynein motor, Cell 150 (5) (2012) 975–986.

[133] E. Perlson, S. Hanz, K. Ben-Yaakov, Y. Segal-Ruder, R. Seger, M. Fainzilber, Vimentin-dependent spatial translocation of an activated MAP kinase in injured nerve, Neuron 45 (5) (2005) 715–726.

[134] F. Saudou, S. Humbert, The biology of huntingtin, Neuron 89 (5) (2016) 910–926.

[135] J.P. Caviston, J.L. Ross, S.M. Antony, M. Tokito, E.L. Holzbaur, Huntingtin facilitates dynein/dynactin-mediated vesicle transport, Proc. Natl. Acad. Sci. USA 104 (24) (2007) 10045–10050.

[136] S. Engelender, A.H. Sharp, V. Colomer, M.K. Tokito, A. Lanahan, P. Worley, et al., Huntingtin-associated protein 1 (HAP1) interacts with the p150Glued subunit of dynactin, Hum. Mol. Genet. 6 (13) (1997) 2205–2212.

[137] S.H. Li, C.A. Gutekunst, S.M. Hersch, X.J. Li, Interaction of huntingtin-associated protein with dynactin P150Glued, J. Neurosci. 18 (4) (1998) 1261–1269.

[138] J.R. McGuire, J. Rong, S.H. Li, X.J. Li, Interaction of Huntingtin-associated protein-1 with kinesin light chain: implications in intracellular trafficking in neurons, J. Biol. Chem. 281 (6) (2006) 3552–3559.

[139] A.E. Twelvetrees, E.Y. Yuen, I.L. Arancibia-Carcamo, A.F. MacAskill, P. Rostaing, M.J. Lumb, et al., Delivery of GABAARs to synapses is mediated by HAP1-KIF5 and disrupted by mutant huntingtin, Neuron 65 (1) (2010) 53–65.

[140] J.P. Caviston, E.L. Holzbaur, Huntingtin as an essential integrator of intracellular vesicular trafficking, Trends Cell Biol. 19 (4) (2009) 147–155.

[141] M.M. Fu, E.L. Holzbaur, Integrated regulation of motor-driven organelle transport by scaffolding proteins, Trends Cell Biol. 24 (10) (2014) 564–574.

[142] E. Trushina, R.B. Dyer, J.D. Badger 2nd, D. Ure, L. Eide, D.D. Tran, et al., Mutant huntingtin impairs axonal trafficking in mammalian neurons in vivo and in vitro, Mol. Cell. Biol. 24 (18) (2004) 8195–8209.

[143] L.R. Gauthier, B.C. Charrin, M. Borrell-Pages, J.P. Dompierre, H. Rangone, F.P. Cordelieres, et al., Huntingtin controls neurotrophic support and survival of neurons by enhancing BDNF vesicular transport along microtubules, Cell 118 (1) (2004) 127–138.

[144] Y.C. Wong, E.L. Holzbaur, The regulation of autophagosome dynamics by huntingtin and HAP1 is disrupted by expression of mutant huntingtin, leading to defective cargo degradation, J. Neurosci. 34 (4) (2014) 1293–1305.

[145] X. Liu, C.E. Wang, Y. Hong, T. Zhao, G. Wang, M.A. Gaertig, et al., N-terminal huntingtin knock-in mice: implications of removing the N-terminal region of huntingtin for therapy, PLoS Genet. 12 (5) (2016) e1006083.

[146] S. Gunawardena, L.S. Her, R.G. Brusch, R.A. Laymon, I.R. Niesman, B. Gordesky-Gold, et al., Disruption of axonal transport by loss of huntingtin or expression of pathogenic polyQ proteins in *Drosophila*, Neuron 40 (1) (2003) 25–40.

[147] L.S. Her, L.S. Goldstein, Enhanced sensitivity of striatal neurons to axonal transport defects induced by mutant huntingtin, J. Neurosci. 28 (50) (2008) 13662–13672.

[148] Y.C. Wong, E.L. Holzbaur, Autophagosome dynamics in neurodegeneration at a glance, J. Cell Sci. 128 (7) (2015) 1259–1267.

[149] C. Wider, Z.K. Wszolek, Rapidly progressive familial parkinsonism with central hypoventilation, depression and weight loss (Perry syndrome)–a literature review, Parkinsonism Relat. Disord. 14 (1) (2008) 1–7.

[150] Z. Gan-Or, P.A. Dion, G.A. Rouleau, Genetic perspective on the role of the autophagy-lysosome pathway in Parkinson disease, Autophagy 11 (9) (2015) 1443–1457.

[151] Y. Tong, H. Yamaguchi, E. Giaime, S. Boyle, R. Kopan, R.J. Kelleher 3rd, et al., Loss of leucine-rich repeat kinase 2 causes impairment of protein degradation pathways, accumulation of alpha-synuclein, and apoptotic cell death in aged mice, Proc. Natl. Acad. Sci. USA 107 (21) (2010) 9879–9884.

[152] M.W. Dodson, T. Zhang, C. Jiang, S. Chen, M. Guo, Roles of the *Drosophila* LRRK2 homolog in Rab7-dependent lysosomal positioning, Hum. Mol. Genet. 21 (6) (2012) 1350–1363.

[153] C.H. Lin, H. Li, Y.N. Lee, Y.J. Cheng, R.M. Wu, C.T. Chien, Lrrk regulates the dynamic profile of dendritic Golgi outposts through the golgin Lava lamp, J. Cell Biol. 210 (3) (2015) 471–483.

In this chapter

Dynein dysfunction as a cause of primary ciliary dyskinesia and other ciliopathies

Niki T. Loges, Heymut Omran
University Hospital Muenster, Muenster, Germany

14.1 Introduction

Cilia are highly conserved organelles extending from almost every cell type of the human body. Their core structure—the axoneme—consists of nine doublet microtubules composed of α- and β-tubulin heterodimers. The average length and diameter comprise $6\,\mu m$ and 0.2–$0.3\,\mu m$, respectively [1]. These complex organelles consist of hundreds of different proteins, which have been characterized in various proteomics studies. In the human system axonemes display either a $9+2$ or the $9+0$ pattern. In $9+2$ cilia the nine outer doublets surround two single microtubules, whereas the single central microtubules are missing in $9+0$ cilia. Other organisms display also other axonemal structures, e.g., in rabbits $9+4$ axonemes have been described. Most nonmotile cilia have $9+0$ axonemes.

Cilia can be divided into two classes—motile or immotile. Motile cilia have large motor protein complexes (dynein arm complexes) attached to their outer doublets, which generate the force to produce cilia bending. Primary ciliary dyskinesia (PCD) is a rare hereditary disorder affecting approximately 1 in 20,000 individuals and is characterized by abnormal motility of cilia or flagella. The clinical phenotype is very complex, because defective cilia/flagella motility in various cell types such as respiratory epithelial cells, ependymal cells lining the brain ventricles, embryonic nodal cells, fallopian tube cells, and sperm cells can contribute to distinct disease manifestations. In addition in some PCD variants molecular defects do not only affect motile but also nonmotile cilia types resulting in even more complex phenotypes. However, all PCD patients suffer from recurrent respiratory infections of the upper and lower airways due to defective mucociliary airway clearance. Dysfunction of motile cilia at the embryonic node causes *situs inversus* in approximately half of the PCD individuals and has been referred to as Kartagener's syndrome.

Dyneins. https://doi.org/10.1016/B978-0-12-809470-9.00014-X

14.2 Ultrastructure of motile cilia

In *Chlamydomonas reinhardtii* the core structure of motile cilia is the "9+2" axoneme, which consists of multiple protein complexes (Fig. 14.1A–C). It contains a central pair (CP) of two single microtubules surrounded by nine outer microtubule doublets. The microtubular doublets are composed of heterodimers of α- and β-tubulin assembled into 13 and 11 protofilaments (in

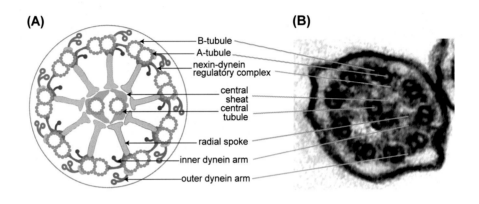

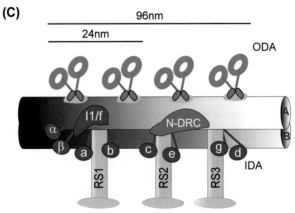

Figure 14.1 Ultrastructure of a motile 9+2 cilium. (A) Schematic diagram of a cross section of a respiratory cilium indicating the different axonemal components. Nine outer microtubule doublets (microtubule A+B) surround two single central microtubules. Attached to the outer microtubule A are the outer dynein arms and inner dynein arms. The radial spokes link the central pair and the outer microtubule doublets. (B) Transmission electron microscopy photograph of a respiratory cilium depicting the axonemal structure in a healthy control. (C) Schematic diagram of the axoneme in length based on findings obtained in *Chlamydomonas*. *Modified from K.H. Bui, T. Yagi, R. Yamamoto, R. Kamiya, T. Ishikawa, Polarity and asymmetry in the arrangement of dynein and related structures in the* Chlamydomonas *axoneme, J. Cell Biol. 198 (2012) 913–925.*

microtubules A and B, respectively). The outer microtubule doublets are connected to each other by the nexin–dynein regulatory complex (N-DRC) [3] and with the CP by radial spokes (RS). Major structures attached to each outer microtubule A are the outer and inner dynein arms (ODAs and IDAs) composed of dynein heavy chains (HC; 400–500 kDa), intermediate chains (IC; 45–140 kDa), and light chains (LC; 8–28 kDa). The conserved central and C-terminal segment of the HCs form the globular head (motor domain) whereas the N-terminal region interacts with ICs and LCs [4,5]. Axonemal dyneins are arranged in two rows along the axoneme in a 96-nm characteristic repeat (Fig. 14.1C). The 96 nm is defined by a molecular ruler that consists of two proteins: FAP59 and FAP172 in *Chlamydomonas* [6] (the human orthologues are CCDC39 and CCDC40, respectively). This molecular ruler provides the anchoring sites for the IDAs and the N-DRC. In *C. reinhardtii* the 96 nm repeat consists of four ODAs, one dimeric IDA, and six monomeric IDAs [7]. Additionally, the calmodulin- and spoke-associated complex (CSC) and the N-DRC link the RS to the outer doublets and the dynein motors [8,9]. The dynein arms are AAA (ATPase associated with various cellular activities)-type motor protein complexes, which convert the chemical energy of ATP binding and hydrolysis into mechanical force [10]. Due to ATP-dependent conformational changes and transient binding to neighboring doublets the dynein motors generate the force for ciliary bending based on the sliding between outer microtubule doublets, whereas the ODAs regulate the beat frequency and the IDAs regulate the waveform of flagellar beating [11]. The CP-RS system together with the N-DRC attached to the outer microtubule A next to the RSs and the IDAs are supposed to govern the motile machinery of the nine outer doublets by acting as a mechanochemical transducer [12–14]. Intracellular second messengers such as cAMP and cGMP, as well as intracellular calcium concentration have been shown to be involved in ciliary beat regulation [1]. cAMP regulates the ciliary beat through activation of protein kinase A [15,16]. Schmid et al. could demonstrate that a CO_2/HCO_3^- sensitive soluble adenylyl cyclase (sAC) expressed in human epithelial axonemes can regulate ciliary beat frequency by the levels of cAMP produced [17].

14.3 Outer dynein arms

Most of the knowledge about ODA composition originates from studies in the green alga, *C. reinhardtii* [18]. A wild-type strain was mutagenized by UV light and 35 strains of *Chlamydomonas* mutants missing the entire ODA were isolated by screening slow-swimming phenotypes. They comprised 10 independent genetic loci (*oda1-10*) including those of previously isolated mutants *oda38* [19] and *pf28* [20]. The 10 loci were distinct from *pf13* and *pf22*, loci for nonmotile mutants missing the outer arm [21]. These results indicated that at least 12 genes are responsible for the assembly of the ODAs [22]. In the alga *Chlamydomonas*, only one distinct type of ODA complex is known. The complex contains three HCs (α-, β-, and γ-HCs), two ICs, nine LCs, three docking complex proteins, and at least two other associated proteins [23]. Based on homology searches using the BLAST algorithms [24], Pazour et al. have identified five human ODA HC orthologues of *Chlamydomonas* β- and γ-HCs [23]. Orthologues of the β-HC are DNAH11 (chromosome 7p21), DNAH17 (chromosome 17q25), and DNAH9 (chromosome 17p12). Orthologues of the γ-HC are DNAH5 (chromosome 5p15) and DNAH8 (chromosome 6p21). No human orthologue of the *Chlamydomonas* α-HC has been identified. These findings are consistent with the concept that, like other vertebrates, human ODA complexes contain only two ODA HCs. The expression pattern of some of the four genes encoding ODA HCs has been analyzed. Northern blot analyses revealed specific expression of *DNAH5* in the lung, brain, and testis [25]. Using the in situ hybridization technique, Kispert et al. found that the cellular expression of the murine orthologue *Mdnah5* (alias *Dnahc5*) is confined in the lung to ciliated respiratory cells of the upper and lower airways [26]. In addition, the ependymal cells lining the brain ventricles, which are also covered by multiple motile cilia, express *Mdnah5* [27]. Cells of the embryonic node, which carry one specialized motile monocilium, also express *Mdnah5* [25]. Reed et al. studied in detail the expression of *DNAH9* [28], also referred to as *DNEL1*. They demonstrated specific expression in the lung, using Northern blot analysis. Reverse transcriptase-polymerase chain reaction (RT-PCR) analyses revealed RNA messages in cultured respiratory cells of tracheal and nasal origin. In addition, Reed et al. analyzed protein expression, using specific monoclonal antibodies directed against DNAH9 [28]. Immunohistochemistry and Western blot analyses showed that the ODA HC DNAH9 is present in respiratory cilia and sperm flagella [28]. Immunoelectron microscopy demonstrated that DNAH9 localizes to one subfiber of the doublet microtubules surrounding the two singlet (central complex) tubules in respiratory ciliary axonemes. RT-PCR identified *DNAH11* (*Dnahc11*) expression in trachea, testis, and lung [29]. The mouse orthologue *Lrd* (*left–right dynein*) is specifically expressed at the embryonic node [30]. Samant et al. [31] showed that the expression of both *Dnahc8* isoforms, the mouse orthologue of human *DNAH8*, is testis-specific in the adult mouse, unlike most other previously characterized mammalian axonemal HCs. However, detailed expression analyses of human *DNAH8* have not yet been performed. To investigate the molecular mechanisms involved in human ODA generation and function, specific antibodies were used to analyze the subcellular localization of the axonemal ODA HCs DNAH5, DNAH9 and DNAH11 in respiratory epithelial cells [32,33]. Confocal immunofluorescence (IF) imaging revealed characteristic patterns of subcellular localization of the analyzed ODA components. In wild-type respiratory cells, the ODA HC DNAH5 is present throughout the entire length of the ciliary axoneme, whereas the ODA HC DNAH9 localizes only to the distal ciliary compartment (Fig 14.2A). Interestingly, the ODA HC DNAH11 localizes only to the proximal part of the ciliary axonemes (Fig. 14.2B). Thus, human ODA complexes vary in their composition along the respiratory ciliary

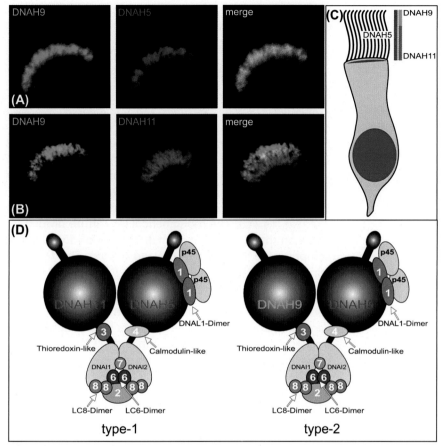

Figure 14.2 Distinct sublocalization of the outer dynein arm heavy chains in the ciliary axoneme of respiratory epithelial cells. (A) Double immunostaining using a monoclonal anti-DNAH9 and a polyclonal anti-DNAH5 antibody. DNAH9 (green) localizes only to the distal part of the cilium, while DNAH5 (red) localizes throughout the axoneme. Unspecific red signal at the ciliary base is caused by polyclonal rabbit antibodies [34]. Colocalization of both proteins is shown in yellow. Nuclei are stained in blue. (B) Double immunostaining using a monoclonal anti-DNAH11 and a polyclonal anti-DNAH9 antibody. DNAH11 (pink) localization is restricted to the proximal part of the cilium. (C) Schematic summarizing these findings. (D) Model of human respiratory outer dynein arm type-1 and type-2.

axonemes. These data indicate that in human respiratory cilia at least two ODA types are present: type-1 (DNAH11 and DNAH5 positive, proximal) and type-2 (DNAH9 and DNAH5 positive, distal) (Fig. 14.2C and D). Interestingly, in contrast to respiratory cilia, in nodal cilia DNAH11 is localized to the entire length of the axoneme [32]. The spatial diversity of ODA HCs along the axonemes probably contributes to the typical beating characteristics of cilia and sperm flagella and likewise to the various beat modes of other motile cilia types (e.g., ependymal and nodal cilia).

14.4 Inner dynein arms

Using a *C. reinhardtii* mutant with missing ODAs, dynein was extracted from axonemes of the mutant *oda1* and fractionated into seven distinct subspecies containing different HCs—named a-g. Seven IDA isoforms—one heterodimeric (two-headed) isoform (f or I1; Fig. 14.1C) located proximal of the RS1 [35] and six monomeric isoforms (a-e and g or I2 and I3; Fig. 14.1C) repeating every 96 nm along the entire axoneme—have been identified. The IDA subgroup I2 is composed of different one-headed IDAs and is located between RSs 1 and 2. Subgroup I3—also composed of different monomeric IDAs—is located distal of RS2 (Fig. 14.1C). The monomeric IDAs can be divided in three dimeric dyads based on their position in relation to the RSs [36]. Dyad1 is located to RS1 and comprises the IDA subspecies a and b, while Dyad2 is located to RS2 and contains IDA subspecies c and e. Dyad3 is localized distally of RS2 and includes IDA subspecies g and d [2].

Each IDA plays a distinct role in the generation and control of motility. Mutant flagella of *C. reinhardtii* missing distinct IDA components beat almost at normal frequency but with a reduced amplitude [37]. Furthermore, Kamiya et al. identified that almost all IDA subspecies (except subspecies f) are able to translocate microtubules in vitro and almost all subspecies (except b) generate microtubule rotation at the same time. Therefore, the IDA system appears to be involved in the initiation of a flagellar bend and the amplitude of the flagellar wave [11,38]. The IDAs consist of several subunits of HCs, ICs, and LCs. The two-headed isoform (I1) is composed of two dynein HCs (1a and 1b), three ICs (97, 138, and 140 kDa), and three LCs (8, 11, and 14 kDa) [39–42]. The β-HC of the IDA I1 is Dhc10, which is essential for the assembly and activity of the IDA I1 [43]. Dhc10, whose human orthologue is DNAH2, contains the typical axonemal HC sequences: coiled-coil domains and multiple P-loops (phosphate-binding loops). Another component of the IDA I1 is IC138, which is thought to transmit signals between the RSs and the dynein HC of I1 [44], whereas the attachment to the outer microtubule doublet is mediated by IC140. The single-headed isoforms (I2) are composed of single dynein HCs and interact with actin and the axonemal LC p28 (for the dyneins a, c, and d) or actin and the LC centrin (for the dyneins b, e, and g) [45,46]. In *IDA4 Chlamydomonas* mutants, p28 is completely missing, resulting in the absence of a subset of dynein HCs in the subgroup I2 [45]; this indicates that p28—whose human orthologue is DNALI1—is necessary for the assembly of a subset of IDAs or for their binding to the outer microtubule A. Furthermore, it is proposed that actin, p28, and centrin are associated with the stem of dyneins and function in targeting dyneins to specific sites on the microtubule A [46]. In other cytoskeletal structures actin and centrin are involved in Ca^{2+}-dependent regulatory cellular processes [47,48].

To date, little is known about the composition of the IDA complexes in mammalian cells, although in mammalian cells they seem to be as heterogeneous as in *Chlamydomonas*. Neesen et al. analyzed the first-described mammalian dynein HC—Mdhc7 (the human orthologue is DNAH1)—in mice by deleting the P1-loop (the motif responsible for ATP hydrolysis). Loss of *Mdhc7* function resulted in reduced beat amplitude of sperm flagella, leading to male infertility [49]. By comparing healthy respiratory cells with cells from PCD patients showing absence of IDAs in electron microscopy analyses, *DNAH7*—a dynein HC of the IDAs localizing along the whole axoneme—was identified [50]. It shares all known features of axonemal HCs including four P-loops and two coiled-coil domains. By BLAST searches, Pazour et al. identified further human orthologous IDA HC genes—*DNAH14, DNAH1, DNAH12, DNAH10, DNAH2,* and *DNAH6* [23]. So far, molecular defect responsible for isolated IDA defects has been reported only for DNAH1. Interestingly, mutations in *DNAH1* are reported to cause only male infertility due to morphological abnormalities of sperm flagella [51–53] consistent with findings in the *Mdhc7* knockout mice [49]. However, several genetic defects characterized by combined defects of either ODAs and IDAs or N-DRC and IDAs have been identified. All of these defects result in abnormalities of I2 IDA types because they are characterized by the absence of DNALI1 (orthologue of p28) from the ciliary axonemes. These defects will be discussed in more detail below.

14.5 Ciliopathies

Cilia assemble on most cell types of the human body to perform diverse biological functions. Defects in motile or immotile cilia are associated with various human diseases, collectively referred to as ciliopathies [54]. The function of nonmotile 9+0 cilia includes mechano- and chemosensation, as well as photoreception and olfaction. These 9+0 cilia are often also called primary cilia. Furthermore, it is known that primary cilia play an essential role in several signal transduction pathways such as the Hedgehog pathway. Diseases linked with primary cilia dysfunction are very diverse. Cystic kidney disease disorders such as nephronophthisis (NPHP) (OMIM 256100) are supposed to originate from aberrant mechanosensory function and/or cellular signaling processes caused by defective renal monocilia. Other disease manifestations include *situs inversus*, pancreatic and hepatic fibrosis, and retinal degeneration, all linked to cilia dysfunction. Until now, mutations in 20 genes linked to NPHP [55–58] have been identified and all are associated with cilia dysfunction.

Bardet–Biedl syndrome (BBS) (OMIM 209900) is a genetically heterogeneous disorder including symptoms such as retinal degeneration, cognitive impairment, obesity, diabetes, polydactyly, situs inversus, and renal abnormalities. To date, 19 BBS genes have been identified [59], of which at least seven assemble into a core complex—the BBSome—that is proposed to regulate the RAB8 (a small GTPase)-dependent vesicular trafficking of membrane proteins from the

Golgi into the ciliary membrane [60]. By preventing Rab8 (GTP) production in zebrafish, ciliation is blocked and the classical BBS phenotype results, indicating that the BBSome may regulate the entry of proteins to the cilium or on intraflagellar transport (IFT) particles.

Oral-facial-digital type-1 (OFD1) (OMIM 311200) syndrome is a heterogeneous developmental disorder characterized by polydactyly, central nervous system disorders, and 15% of cases in cystic kidney disease [61]. The OFD1 protein localizes to the basal body of the cilium and is involved in the complex mechanism of ciliary generation, because *Ofd1*-deficient mice lack node cilia and display left–right (L-R) body asymmetry defects [62]. Patients with Almström syndrome (ALMS) (OMIM 203800) display similar phenotypes to BBS patients; in addition, these patients show cardiomyopathy and liver and kidney dysfunction but do not have polydactyly. The ALMS1 protein also localizes to the basal body and the centrosome. ALMS1 appears to be important for ciliogenesis and the inactivation of *ALMS1* results in the prevention of Ca^{2+} influx into the cytosol [63]. Interestingly, there is increasing evidence that, in patients with ciliopathies such as NPHP, BBS, and OFD1, dysfunction of respiratory cilia can also be observed, suggesting that the genetic defects not only affect nonmotile but also motile cilia.

Very recently, a new ciliopathy of motile cilia, named reduced generation of multiple motile cilia (RGMC), was described [64,65]. Whereas in PCD and cystic fibrosis either dysfunction of cilia beating and abnormal mucus viscosity causes defective mucociliary clearance, in individuals with RGMC the mucociliary clearance is impaired due to reduced number of multiple motile cilia. Mutations *CCNO* and *MCIDAS* have been described to result in disturbed numbers of centrioles and defective docking of centrioles to the apical cell region. TEM analyses showed either a complete absence or a severely decreased number of cilia. The apical cell regions showed normal microvilli composition but a severe decrease of basal bodies in both *CCNO* and *MCIDAS* mutant cells. In *CCNO* mutant cells, occasionally basal bodies and attached rootlets mislocalized in the cytoplasm can be found, indicating a basal body migration defect.

The reduced number of cilia on the respiratory epithelia leads to the classical signs of a mucociliary clearance disorder with chronic infections of the upper airways (rhinitis, otitis media, nasal polyps, and sinusitis) and lower airways (recurrent pneumonia, bronchiectasis, chronic obstructive airway disease). Due to the severe lung disease, lung transplant had to be performed at the age of 34 years in two individuals, two additional individuals died due to respiratory failure at the age of 19 years and 27 years. Postnatal respiratory distress was present in all individuals. All individuals had normal situs composition. Furthermore, nonrespiratory symptoms of motile cilia disease such as hydrocephalus and female infertility were reported in three and two individuals, respectively [64,65]. An additional population-based study in Israel has shown an unexpected high prevalence of *CCNO* mutations in patients with severe mucociliary clearance disorders (6%) and suggests a more severe clinical course with rapid decline of lung function [66].

14.6 Nodal cilia

Patterning of the L-R axis is a complex process that involves establishment of a midline and orientation of the L-R asymmetry with respect to the anterior–posterior (A/P) and dorsoventral (D/V) axes. Defects in one or more of these processes result in randomization or loss of asymmetry and in changes in numbers and placement of organs (i.e., bilateral spleens). In organisms such as fishes and frogs, the D/V and A/P axes are determined at fertilization by the entry position of the sperm and distribution of the yolk [67]. In mammals, such as humans and mice, the cylindrically symmetrical embryo implants itself into the wall of the uterus. The D/V axis is specified as the proximal–distal axis from the implantation site. Subsequently, the A/P axis is randomly determined in the plane at right angles to the D/V axis [68]. Once the D/V and A/P axes are determined, the L-R axis is established. A link between motile cilia dysfunction and defects in establishing L-R body asymmetry originated from observations that half of individuals with PCD have their organs in reversed orientation, a condition called *situs inversus totalis* (also referred to as Kartagener's syndrome), which is consistent with randomization of L-R asymmetry [69]. The first morphological asymmetry apparent is the heart loop [70], but L-R asymmetric gene expression precedes morphological changes. Many studies have suggested the node as a morphological structure important for L-R determination (Fig. 14.3A; [71–73]). The node of mouse embryos appears as a roughly triangular depression with the apex point toward the anterior. This nodal pit is covered by Reichert's membrane and the cavity is filled with extraembryonic fluid. The ventral embryonic surface of the nodal pit consists of a few hundred monociliated cells (Fig. 14.3A). Monocilia from nodal pit cells, so called nodal cilia, appear as rodlike protrusions approximately 5 mm in length and 0.3 mm in diameter [71]. Nonaka et al. demonstrated that 9+0 nodal cilia contain motor protein complexes such as dynein arms and are motile. In addition they found that dysfunction of motile monocilia at the embryonic node is associated with randomization of L-R body asymmetry [73]. A similar phenotype was observed in the *inversus viscerum* (*iv*) mouse carrying recessive mutations in the axonemal dynein HC gene *Lrd*, the orthologue of the human axonemal dynein HC gene *DNAH11* [30]. *Lrd* is exclusively expressed by nodal cells at 7.5 days postcoitum [30]. In embryos heterozygous for *iv*, nodal cilia rotated as rapidly as those in wild-type embryos [74]. This rapid movement produces a leftward flow of extraembryonic fluid in the ventral node, referred to as nodal flow [73]. In contrast, the nodal cilia in *iv* homozygous embryos appeared rigid and rarely moved, resulting in *situs inversus* [74]. Interestingly, laterality defects were

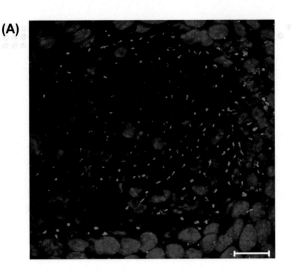

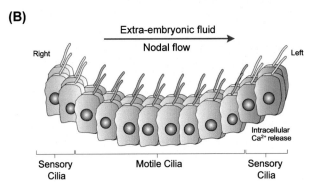

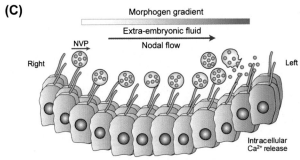

Figure 14.3 Mouse node and current models for establishing left–right (L-R) asymmetry. (A) High-resolution immunofluorescence picture of the mouse node at E8.0. Cilia are stained with an antibody directed against acetylated a-tubulin (green); nuclei are stained with Hoechst 33,342 (blue). Scale bar represents 20 mm. (B and C) The two current models explaining the mechanisms involved in establishing L-R body asymmetry. In the "two-cilia model" (B), motile cilia in the center of the node create a leftward nodal flow that is mechanically sensed through passive bending of nonmotile sensory cilia at the border of the node. Bending of the cilia on the left side leads to a left-sided intracellular release of Ca^{2+} that initiates asymmetric expression of signaling molecules (morphogen gradient). The nodal vesicular parcel (NVP) model (C) predicts that vesicles filled with morphogens (such as sonic hedgehog and retinoic acid) are secreted from the right side and transported to the left side of the embryonic node by nodal flow, where they fragment. The resulting initiation of left-sided intracellular Ca^{2+} release induces downstream signaling events (such as expression of Lefty and Nodal) that break bilaterality. In this model, the flow of extraembryonic fluid is not detected by cilia-based mechanosensation.

always accompanied by an abnormality in nodal flow, indicating that lack of nodal flow results in randomization of laterality. Indeed, artificially generated rightward flow that was sufficiently rapid to reverse the intrinsic leftward nodal flow resulted in reversal of laterality in wild-type embryos. The artificial flow was also able to direct the situs of mutant mouse embryos with immotile cilia [75]. These results provided evidence for the role of mechanical fluid flow in L-R patterning. Similar laterality-breaking mechanisms have also been proposed for zebrafish (involving Kupffer's vesicle), birds (Hensen's node), and amphibians (Spemann's organizer) [76]. In humans, mutations in several genes coding for axonemal dyneins (DNAI1, DNAI2, DNAH5, DNAH11) resulting in randomization of L-R body asymmetry have been described [25,77–83]. Two models have been proposed for how nodal flow might contribute to L-R asymmetry. The "two-cilia" model [72] predicts that there are two distinct types of cilia and that nodal flow is generated by motile cilia (characterized by Lrd expression) and sensed by nonmotile, mechanosensory cilia (characterized by polycystin-2 expression) at the periphery of the node. The "morphogen gradient" model predicts that the nodal flow results in a leftward gradient of a hypothetical morphogen (Fig. 14.3B). Tanaka et al. identified nodal vesicular parcels (NVPs) filled with sonic hedgehog and retinoic acid molecules. The secretion and release of NVPs is triggered by fibroblast growth factor signaling. NVPs are transported by the nodal flow, where they eventually fragment close to the left lateral wall of the ventral node, releasing their contents (Fig. 14.3C; [84]). Both models described an asymmetric Ca^{2+} release, which is probably involved in the subsequent proceedings of L-R determination, which in turn is based on asymmetric expression of signaling molecules, such as nodal and lefty, and transcription factors, such as Pitx2 (paired-like homeodomain transcription factor-2).

14.7 Ependymal cilia and hydrocephalus

In the brain ventricles, the synchronized beating of the ependymal 9+2 cilia creates a laminar flow of cerebrospinal fluid above the ependymal cell surface and through the cerebral aqueduct, which is termed ependymal flow [27]. A link between ependymal cilia dysfunction and enlargement of the brain ventricles (hydrocephalus) was shown by analysis of several mouse models with altered function of motile cilia. Mice lacking the axonemal dynein HC Dnahc5 (Mdnah5); the axonemal ruler component Lnks/Ccdc40; the axonemal protein SPAG6; the CP protein Hydin; or proteins involved in ciliogenesis, such as IFT88 (Tg737) or FOXJ1 (forkhead box J1; also known as HFH-4), which regulates generation of motile cilia, develops hydrocephalus [85–90]. Unlike in mice, in humans, ependymal ciliary dysmotility is not sufficient to cause hydrocephalus but increases the risk for aqueduct closure and hydrocephalus formation. Furthermore, analysis of neuron migration in the brain of Tg737[orpk] mutant mice found that the lack of ependymal flow resulted in disturbed directional migration of neuroblasts in the subventricular zone, suggesting that ependymal flow is required for the formation of a concentration gradient of guidance molecules [91]. Alternatively, other functional roles of IFT proteins might be responsible for this complex phenotype.

14.8 Sperm flagella and male infertility

The ultrastructure of the sperm flagellum and the motile 9+2 cilium are very similar but not identical, which might explain why sperm flagella dysfunction is often, but not necessarily, associated with PCD and vice versa [92]. Interestingly, male infertility of PCD patients might not necessarily be caused by a defect of sperm motility, but may also result from defective transportation of sperm cells through the ductuli efferentes. Parts of the ductuli efferentes epithelium are lined by motile cilia, which beat in the direction of the epididymis. They probably assist in the transportation of spermatozoa out of the testes, since sperms at this stage are still not activated and therefore are nonmotile [93].

14.9 Fallopian tubes and female infertility

The fallopian tubes are tubular seromucosal organs connecting the ovaries to the uterus. They play an essential role in gamete transport, fertilization, and the early development of the embryo. It consists of two major cell types—ciliated and secretory cells. It is supposed that ciliary action plays an important role in the transfer of gametes and embryos. Indeed, women with PCD (or Kartagener's syndrome) have been reported to suffer from subfertility [94–96]. Interestingly, fallopian tube cilia contain the ODA HCs DNAH5 and DNAH9, the ODA ICs DNAI1 and DNAI2, as well as the IDA LC DNALI1 and components of the N-DRC (LRRC48) and the 96-nm axonemal ruler (CCDC39), similar to respiratory cilia [97]. Although the exact role of the cilia in the fallopian tube is not clear, these cilia produce a fast and synchronized movement, which creates a laminar flow [97]. However, spontaneous conception can occur in females with severely impaired respiratory mucociliary clearance caused by absence of motor proteins in the axoneme of motile cilia. This indicates that the transport of the ovum toward the uterus is not completely dependent on cilia-generated transport [97].

14.10 Primary cilia dyskinesia

PCD is a rare hereditary disorder affecting approximately 1 in 20,000 (live births) and results from abnormalities in the structure and/or function of motile cilia and their motility. The disease phenotype is characterized by chronic upper and lower airway infections, which can cause substantial damage to the lungs.

Consolidation, atelectasis, and bronchiectasis are constant findings in adults, but may also be present already in infancy, making early PCD diagnosis substantial. In most PCD variants, the genetic defects not only affect function of motile 9+2 cilia but also result in defective 9+0 motile cilia at the embryonic node. In these PCD variants the genetic defect regularly causes randomization of L-R asymmetry. Thus, about half of PCD patients display a *situs inversus* (referred to as Kartagener's syndrome). Interestingly, *situs ambiguous* (heterotaxy) is present in ~6% of PCD patients, demonstrating the important role of nodal flow in preventing the complex cardiac defects often observed in heterotaxy patients. Björn Afzelius and others were the first to show that respiratory cilia from patients with Kartagener's syndrome often lack dynein arms on the outer doublet microtubules, resulting in an immotile cilia syndrome [98,99]. Since then, many defects in the ultrastructure and/or function of motile cilia have been identified. Because defects of cilia beating were recognized in many patients, the original name of the disease (immotile cilia syndrome) was changed to PCD.

Unfortunately, PCD is underdiagnosed or diagnosed too late. Due to the variety of defects, PCD diagnostic is difficult and complex involving several methods, such as measurement of the nasal nitric oxide production rate, high-speed video-microscopy analyses (HVMA) of the ciliary beating pattern and frequency, assessment of the ciliary composition by transmission electron microscopy (TEM) and high-resolution IF microscopy analyses, and genetic analyses [100]. Although several methods are used to diagnose PCD, all methods have limitations [100] and are not available in all PCD centers resulting in undiagnosed PCD cases.

14.11 Molecular defects affecting outer dynein arm components and docking

The most commonly encountered ultrastructural defects in PCD are characterized by defects of the ODAs. Mutations in genes encoding different components of ODA complexes cause distinct structural and/or functional defects of the ODAs. So far, mutations in five genes (*DNAI1, DNAH5, DNAH11, DNAI2,* and *TXNDC3*) have been identified and will be discussed in detail.

14.11.1 Mutations in genes encoding for ODA components

14.11.1.1 DNAI1

DNAI1 was screened in PCD patients based on a candidate approach, because *Chlamydomonas oda9* mutants deficient in the axonemal intermediate ODA chain IC78 (now known as IC1) display an axonemal defect resembling those frequently observed in PCD patients with ODA defects. It was the first gene identified to be responsible for PCD [101]. It is located on chromosome

9p21-p13 and encodes the human intermediate chain, IC1. Analysis of human *DNAI1* mutant respiratory cells with anti-DNAH5 and anti-DNAI2 antibodies revealed that DNAI1 is probably essential for the assembly of ODA type-2 as the ODA HC DNAH5 does not localize along the whole axoneme but can still be assembled within the proximal axoneme [54].

14.11.1.2 DNAH5

To identify additional genes responsible for PCD, Omran et al. [102] used a homozygosity mapping strategy and detected on chromosome 5p15 novel PCD locus. Within the critical genetic interval they localized the potential candidate gene *DNAH5* (with 80 exons). In the *C. reinhardtii* mutant *oda2*, mutations in the γ-HC (the orthologue of *DNAH5*) result in slow-swimming algae with ultrastructural ODA defects [103]. Patients who carry *DNAH5* mutations consistently display axonemal ODA defects. Other clinical manifestations include randomization of L-R body asymmetry [25,33,79]. Mutations in *DNAH5* are detected in 30% of PCD cases—the rate is even higher (~50%) when PCD cases are preselected for ODA defects, highlighting the important role of DNAH5 for the etiology of PCD. DNAH5 is an HC of the ODAs and localizes along the entire axoneme. Consistently in patients with *DNAH5* mutations, the DNAH5 protein is regularly absent from the whole ciliary axoneme. Thus, DNAH5 is present in both human ODA types—in type-1 ODAs (positive for the dynein HCs DNAH5 and DNAH11) and in type-2 ODAs (containing the dynein HCs DNAH5 and DNAH9).

14.11.1.3 DNAH11

DNAH11 is one of the two human orthologues of the *Chlamydomonas* β-HC. The gene comprises 82 exons and has been associated with Kartagener's syndrome with normal axonemal ultrastructure [32,77,80,82,104]. As a patient with homozygous nonsense mutations in *DNAH11* also suffered from cystic fibrosis, it was first uncertain whether mutations in *DNAH11* solely cause defects of laterality or also result in PCD [77]. However, several patients with PCD/Kartagener's syndrome (cystic fibrosis was excluded) have been identified with *DNAH11* mutations, thus confirming the role of DNAH11 in the pathogenesis of PCD [32,80,82,104]. Interestingly, the phenotype of respiratory cilia is characterized by an abnormal beating pattern with reduced amplitudes and recovery strokes and increased beating frequency but normal ultrastructure [82], because DNAH11 localizes only to the proximal part of the ciliary axonemes (Fig. 14.2B) [32]. Consistently, respiratory cilia of patients with *DNAH11* mutations display normal localization for the ODA components DNAH5, DNAH9, and DNAI2.

14.11.1.4 DNAI2

DNAI2, the human orthologue of the *Chlamydomonas* intermediate ODA chain IC69 (now known as IC2), was identified to be essential for ODA assembly along the whole axoneme [81]. All patients identified with loss-of-function mutations had no DNAI2 expression in respiratory cells and suffered from chronic lung disease, and four out of six patients displayed *situs inversus*. Further analysis revealed that DNAI2 is a component of the ODA type-1 and type-2. In patients with *DNAI2* mutations, DNAH5 as well as DNAH9 were absent from the ciliary axoneme.

14.11.1.5 TXNDC3

TXNDC3 is a thioredoxin nucleoside diphosphate kinase (NDK) and is orthologous to the ODA component IC1 in sea urchin. *TXNDC3* in humans has been shown by Duriez et al. [105] to be expressed in nasal cells, trachea, and testis. *Chlamydomonas* has two TXNDC3 orthologues (LC3 and LC5), which have been shown to interact with the HCs [106]. So far a single patient with mutations in *TXNDC3* has been reported. TEM revealed abnormal ODAs in some cross sections whereas other appeared normal.

14.11.1.6 DNAL1

DNAL1 is the human orthologue of *Chlamydomonas LC1* and encodes a LC of the ODA. The protein structure is highly conserved between humans and *Chlamydomonas*. Interestingly, in *Chlamydomonas*, protein activity of the γ-HC is regulated through interaction with LC1 [107]. It was also that DNAL1 and DNAH5 interact with each other, indicating that DNAL1 is important for proper DNAH5 function in humans [108]. Consistent with these findings, mutations in *DNAL1* result in absent or markedly shortened ODAs [109].

14.11.2 Mutations in genes encoding for components of the ODA docking complex

ODAs require a docking complex (ODA-DC) enabling their attachment onto microtubules. In *C. reinhardtii*, the ODA-DC is composed of three subunits, DC1, DC2 (coiled-coil proteins), and DC3 (Ca^{2+}-binding protein), which is preassembled in the cytoplasm and transported into flagella independently of ODAs [110,111]. It is hypothesized that DC1 and DC2 are important for the assembly of the ODA to the outer microtubule doublet A and DC3 for the regulation of flagellar beat or assembly [112,113]. In the last few years, several human genes encoding for ODA-DC components have been identified and mutations in these genes cause PCD with ODA defects.

14.11.2.1 CCDC114

CCDC114 (coiled-coil domain-containing protein 114) was the first ODA-DC gene characterized in humans [114,115]. *CCDC114* is the vertebrate orthologue of the *Chlamydomonas* ODA docking component *DC2*. Loss-of-function mutations in *CCDC114* result in a complete loss of ODAs resulting in ciliary immotility. Sequencing analyses of 135 PCD individuals with ODA defects documented by TEM and/or IF analysis identified homozygous *CCDC114* mutations in five PCD individuals. Interestingly, the same splice site mutation c.742G>A in *CCDC114* was identified in three of these unrelated individuals, indicating that this mutation is probably a founder mutation, which is important for further genetic testing in PCD individuals.

14.11.2.2 ARMC4

Mutations in *ARMC4*, which encodes for the armadillo repeat–containing protein 4, were identified using combined high-resolution copy-number variant and mutation analyses. Homozygous as well as compound heterozygous *ARMC4* mutations were identified in 12 PCD individuals from 10 families that segregated with the disease status in an autosomal recessive inheritance pattern [116]. Almost all mutations (except one) predicted premature termination of translation. All affected individuals displayed a clinical phenotype consistent with PCD including recurrent upper and lower airway disease as well as bronchiectasis. Additionally, 8 of the 12 affected individuals displayed *situs inversus totalis* and 4 had *situs solitus*. Thus, *ARMC4* deficiency causes PCD and randomization of L-R body asymmetry. Morpholino knockdown of *armc4* in zebrafish and analysis of the mouse mutant *Aotea* carrying a missense variation in *Armc4* confirmed that ARMC4 is critical for L-R patterning [116]. TEM and IF analyses of *ARMC4* mutant cilia revealed absence of ODAs. In all PCD variants with ODA defects, the ODA HC DNAH9, which normally localizes to the distal ODA type-2 complexes, is absent from the ciliary axonemes. Interestingly, in *ARMC4* mutant cilia, DNAH9 is present in the axoneme but mislocalized throughout the whole ciliary compartment. This specific mislocalization of DNAH9 has not been described in other PCD types so far but was consistently observed in all *ARMC4* mutant individuals, whether with *ARMC4* truncating or missense mutations [116].

14.11.2.3 CCDC151

CCDC151 is the vertebrate orthologue of *C. reinhardtii ODA10*, which was shown to be required for ODA assembly in these ciliated algae [117]. Therefore *CCDC151* was considered a reasonable PCD candidate gene, and screened for mutations by Sanger sequencing in 150 affected individuals with ODA defects documented either by TEM or by high-resolution IF analyses. In parallel, suspecting that CCDC151 might be related to ODA docking, patients showing absence of CCDC114 from the axonemes were also screened for mutations in *CCDC151* after exclusion of *CCDC114* mutations. By combined high-throughput mapping and sequencing, loss-of-function mutations in *CCDC151* were identified in five affected individuals from three independent families whose cilia showed a complete loss of ODAs and severely impaired ciliary beating [118]. Segregation analysis confirmed that the identified mutations segregated with the disease status. Interestingly, CCDC151 coimmunoprecipitates CCDC114 [118] and thus appears to be a highly evolutionarily conserved ODA-DC-related protein involved in assembly of both ODA types and their axonemal docking machinery onto ciliary microtubules.

14.11.2.4 TTC25

TTC25 encodes for a tetratricopeptide repeat (TPR) domain–containing protein with eight TPR domains. Mutations in *TTC25* were identified in three PCD individuals from two unrelated families using a whole-exome sequencing approach [119]. All three PCD-affected individuals displayed typical clinical PCD phenotypes, including recurrent upper (chronic rhinitis, chronic otitis, chronic sinusitis, and nasal polyps) and lower (chronic productive cough, recurrent pneumonia, and bronchiectasis) airway disease and one affected presented respiratory distress after birth. Two individuals displayed laterality defects, one presented with *situs inversus totalis* and one presented *situs ambiguous*. TEM cross sections of *TTC25* mutant respiratory cilia revealed absence of ODAs. Interestingly, comparison of TEM cross sections of *TTC25* mutant cilia with the TEM cross sections of *DNAH5* mutant cilia revealed that the axonemes of *TTC25* mutant cilia lack also the small projections on the doublet microtubules described to represent the ODA-DC [112]. These findings were confirmed by high-resolution IF using antibodies directed against the ODA HCs DNAH5 and DNAH9. Both ODA HCs were absent from *TTC25* mutant ciliary axonemes. This indicates that *TTC25* mutations lead to a defect of the ODA and the ODA-DC. Consistently, the localization of CCDC114 and ARMC4 are also affected by loss-of-function of *TTC25* [119]. However, TTC25 still localizes to the axonemes of cells deficient in ODA HC DNAH5, as well as to the cells of individuals with mutations in genes encoding CCDC114, CCDC151, and ARMC4. Additionally, it was shown that TTC25 coimmunoprecipitates CCDC114 [119]. Thus, TTC25 was identified as a new member involved in the ODA-DC-associated machinery. Interestingly, *TTC25* is not present in the *Chlamydomonas* genome, indicating that it evolved in higher eukaryotes.

14.11.2.5 CCDC103

CCDC103 encodes the coiled-coil domain–containing protein 103 and is mutated in the zebrafish cilia paralysis mutant *schmalhans* (*smh*[tn222]) [120]. CCDC103 is a 29 kDa protein that consists of a central RPAP3_C domain flanked by N- and C-terminal coiled-coils [121]. The *smh*[-/-] have a curved body axis, randomized L-R asymmetry, and pronephric kidney cysts, phenotypes characteristic for ciliary motility defects [122]. Mutational screening of PCD families identified individuals in six families with reduced ODAs who carried mutations in *CCDC103*. All six affected individuals carrying recessive truncating *CCDC103* mutations had typical clinical findings for PCD comprising recurrent upper and lower airway infections and bronchiectasis; two affected individuals had dextrocardia and one had *situs inversus totalis*. A homozygous amino acid exchange (p.His154Pro) was identified in five additional individuals with PCD: two of these individuals had *situs inversus totalis*, one had *situs inversus abdominalis*, one had dextrocardia and one had normal organ situs. These findings indicate that *CCDC103* mutations cause PCD and randomization of L-R body asymmetry, resembling findings in *smh* mutant zebrafish. High-resolution IF analyses of *CCDC103* mutant cilia showed a partial loss of ODA complexes along the ciliary axonemes, with some proximally localized type-1 ODA complex components, namely DNAH5 and DNAI2, still present in the ciliary axonemes. In contrast, DNAH5 and DNAI2 were absent from the distal ciliary axonemes (type-2 ODA complex). Absence of the ODA HC DNAH9 confirmed the absence of distally localized type-2 ODA complexes from the ciliary axonemes in mutant cilia. Rescue experiments in *smh* mutant zebrafish using wild-type and mutant human *CCDC103* and experiments performed in *Chlamydomonas* gave additional support that CCDC103 acts as a dynein arm attachment factor and causes PCD when mutated [120].

14.12 Molecular defects affecting cytoplasmic preassembly of dynein arms

Mechanisms responsible for cytoplasmic preassembly of dynein components, loading of those complexes onto the IFT machinery and delivery are still only poorly understood. In the last 10 years, several genetic defects responsible for combined defects of ODAs and IDAs were reported. So far nine genes encoding for dynein axonemal assembly factors (DNAAFs) involved in the cytoplasmic preassembly process of dynein arms have been identified. Distinct mechanisms are probably involved in the assembly of these complexes, because there are different types of ODA and IDA complexes (Fig. 14.2) and it is very likely that additional genetic defects will be identified. Interestingly, loss of DNAAFs such as KTU/DNAAF2 and DYX1C1/DNAAF4 results in assembly defects of ODA complex type-2 and also IDA complexes severely reducing motility of respiratory cilia but indicating that at least partial cytoplasmic preassembly of type-1 dynein arm complexes still occurs. In contrast, deficiency of other DNAAFs (i.e., LRRC50/DNAAF1) results in loss of both ODA complex types (1 and 2) and DNALI1-containing IDA complexes resulting in complete ciliary immotility and indicating that the preassembly process of dynein arm complexes is more severely disrupted.

14.13 Preassembly defects of ODA complex type-2 and DNALI1-associated IDA complexes

The first gene reported to encode a protein involved in cytoplasmic preassembly of dynein arms was *kintoun* (*ktu*/*DNAAF2*). This gene was first identified in a medaka mutant and was found to be mutated in the *pf13* mutant of *Chlamydomonas*. In humans, mutations in the *DNAAF2* gene result in PCD and Kartagener's syndrome. Dysfunction of DNAAF2 leads to the loss of ODAs type-2 (containing DNAH5 and DNAH9) and absence of IDAs containing DNALI1 from the ciliary axoneme and results in loss of ciliary motility [123]. Biochemical studies in *Chlamydomonas* and mice showed that DNAAF2/PF13 interacts with dynein ICs and HCs [123], indicating that KTU is involved in the cytoplasmic preassembly of dynein arm complexes in the cytoplasm before they are loaded onto the IFT machinery for delivery to the ciliary compartment [110].

The second preassembly factor involved in the cytoplasmic preassembly of ODA type-2 and DNALI1-containing IDAs is DNAAF4/DYX1C1. *DYX1C1* was initially described as a candidate dyslexia gene owing to a single-balanced translocation t(2; 15) (q11; q21) coincidentally segregating with dyslexia in a family [124] and subsequent SNP association studies. However, follow-up studies provided both positive [125–127] and negative [128–130] support for association with dyslexia. Interestingly, proteomic and gene expression studies indicated a possible role of *DYX1C1* in cilia [131,132]. Knockout of *Dyx1c1* results in embryonic lethality. However, homozygous mutants that survived after birth developed severe hydrocephalus by postnatal day (P) 16 and died by P21 [133], consistent with dysfunction of ependymal cilia motility, a hallmark phenotype also observed in other PCD mouse models [88,134,135]. Additionally, postnatal homozygous mutant mice showed *situs inversus totalis* or *situs ambiguous* with complex heart defects, frequent findings encountered in PCD mouse models with dysfunction of nodal cilia during early embryogenesis [136,137]. Whole-mount in situ hybridization demonstrated that *Dyx1c1* expression in the early embryo is restricted to pit cells of the embryonic node [133], consistent with a function of DYX1C1 in nodal cilia. Immotile/dyskinetic airway cilia were also observed in *Dyx1c1*-deficient mice. Further analyses of ependymal and respiratory cilia of the *Dyx1c1*-mutant mice revealed absence of ODA complexes type-1 and type-2 and DNALI1-associated IDA complexes by TEM and IF. Morpholino knockdown of zebrafish *dyx1c1* resulted in hydrocephalus, cystic kidneys, body axis curvature and L-R patterning defects consistent with motile cilia dysfunction in zebrafish [133]. Interestingly, in humans, recessive loss-of-function mutations in *DYX1C1* cause PCD with reduced fertility and disruption of the L-R body asymmetry with heart defects. In strong contrast to the literature, none of these patients exhibited dyslexia. Consistent with the findings in *Dyx1c1*-mutant mouse, ODA and IDA complexes were disrupted, and a partial loss of ODAs (type-2) and a complete loss of DNALI1-associated IDA complexes in *DYX1C1*-mutant cilia were observed. High-resolution IF microscopy of mouse respiratory cells and Western blot analyses using human lysates described a cytoplasmic localization for DYX1C1, similar to KTU/DNAAF2. Additionally, interaction between DYX1C1 and KTU/DNAAF2 demonstrated by coimmunoprecipitation and yeast-to-hybrid experiments was described [133]. These findings indicate that DYX1C1 is also an evolutionarily conserved dynein axonemal assembly factor necessary for motile cilia function.

14.14 Preassembly defects of ODA type-1 and type-2 and DNALI1-associated IDA complexes

The second gene identified encodes a protein involved in preassembly of dynein arms *DNAAF1/LRRC50* (*leucine-rich repeat containing protein 50*). *DNAAF1* was considered as a strong candidate for PCD because the reported ultrastructural phenotype of cilia and flagella in zebrafish and *Chlamydomonas* algae carrying mutations in orthologous genes showed defects of ODA and IDA structures [110,138–140]. Large genomic deletions as well as point mutations in the human *DNAAF1* gene result in a PCD variant with combined ODA and IDA defects. Functional analyses showed that *DNAAF1* deficiency disrupts assembly of ODA complexes containing DNAH9 (distally) and DNAH5 and DNAI2 (distally and proximally), as well as DNALI1-containing IDA complexes, resulting in absence of these dynein arms from the ciliary axoneme, in turn leading to immotile cilia [141]. These findings indicate that DNAAF1 plays a role in preassembly of distinct dynein arm complexes. Indeed, in parallel, Duquesnoy et al. [142] demonstrated that in *Chlamydomonas* and *Trypanosoma brucei* ODA7 is located primarily in the cell body and RNA interference (RNAi)-induced silencing of *ODA7* in *T. brucei* indicated that some dynein arm components are assembled but remain in the cytoplasm. The absence of ODA7 in both organisms results in flagellar hypomotility with selective defects of ODAs and IDAs.

C19ORF51/DNAAF3 was previously identified as the human orthologue of *Chlamydomonas pf22*. PF22 localizes to the cytoplasm, and a PF22-null mutant cannot assemble ODA complexes and only some IDA complexes. *C19ORF51/ DNAAF3* was identified in expression studies as a potential cilia-related gene [143]. Homozygous mutations in *C19ORF51/DNAAF3* result in absence of the ODA components, DNAH5, DNAH9, and DNAI2, and IDA LC DNALI1 from the ciliary axonemes of affected individuals. These findings indicate that preassembly of the ODA complexes type-1 and type-2 and DNALI1-associated IDA complexes are defective [144]. Knockdown of *dnaaf3* in zebrafish likewise disrupts dynein arm assembly and ciliary motility, causing PCD phenotypes that include hydrocephalus and laterality defects. Further biochemical studies showed altered abundance of dynein subunits in *Chlamydomonas* mutant cytoplasm, demonstrating that C19ORF51/DNAAF3 (PF22) acts at a similar stage as other DNAAF proteins, for example, KTU/DNAAF2 (PF13) [123]) and LRRC50/DNAAF1 (ODA7) [141,142], in the cytoplasmic dynein arm preassembly process. These results support the existence of a conserved process for cytoplasmic preassembly of axonemal dynein arm complexes [110,144]. Two other DNAAFs were identified using zebrafish morpholino knockdown of PCD candidate genes as an in vivo screening platform. Knockdown of two candidate genes, *c21orf59* and *c15orf26*, in zebrafish and planaria blocked preassembly of ODA complexes showing their important role for ciliary motility [145]. Similar to findings in zebrafish and planaria, mutations in *C21orf59* caused loss of ODA complexes type-1 and type-2 and DNALI1-associated IDA complexes in humans. Further DNAAFs were identified using whole-exome sequencing and high-throughput mutational analyses. Loss-of-function mutations in *ZMYND10* and *SPAG1* lead to absence of the ODA HC DNAH5 and the IDA LC DNALI1 from mutant ciliary axonemes shown by high-resolution IF. Thus, *ZMYND10* and *SPAG1* deficiency cause a loss of the ODA complexes type-1 and type-2 and DNALI1-associated IDA complexes [146,147]. Animal models support the association between *ZMYND10* and *SPAG1* and human PCD, given that *zmynd10* and *spag1* knockdown in zebrafish caused ciliary immotility leading to hydrocephalus and cystic kidneys.

LRRC6 encodes the leucine-rich repeat containing protein 6 and was first identified by means of homozygosity mapping of 15 individuals from nine consanguineous families [147–149]. Furthermore, *LRRC6* mutations were identified in 13 unrelated PCD-affected families [147]. It was considered a good candidate gene based on expression in *Chlamydomonas* flagella and human cilia [150]. Additionally, *Lrrc6* was shown to be expressed in ciliated tissues in mice [151,152]. All affected individuals had a PCD diagnosis with typical clinical features, including chronic sinusitis, bronchiectasis, recurrent otitis media, reduced nasal nitric oxide production rate, and laterality defects. Five individuals were included because of male sterility with immotile sperm flagella caused by absence of dynein arms. TEM analyses of respiratory cilia and sperm flagella of affected individuals showed absence of ODAs and IDAs. High-resolution IF microscopy revealed absence of the ODA HC DNAH5, the ODA ICs DNAI1 and DNAI2, the IDA HC DNAH7, and the IDA LC DNALI1 and confirmed the TEM findings. Additionally, it was shown that LRRC6 interacts with the dynein arm preassembly factor ZMYND10 [147]. Altogether, these findings support the role of LRRC6 as a cytoplasmic dynein arm assembly factor.

HEATR2 is a member of 10 uncharacterized HEAT-repeat containing proteins in humans, and preliminary analyses showed that *HEATR2* is a highly conserved gene in organism with motile cilia and flagella [153]. Mutational analyses performed in six Amish subfamilies and one UK-Pakistani family identified mutations in *HEATR2* in six and three affected individuals, respectively [154,155]. All affected individuals have typical PCD symptoms such as chronic sinopulmonary disease, persistent neonatal hypoxemia, infertility, reduced nasal nitric oxide production rate, and laterality defects. TEM analyses of respiratory cilia showed absence of ODAs. ShRNA silencing of the *htr2*, the *Chlamydomonas HEATR2* orthologue, resulted in absence of ODAs and altered composition of IDAs in the flagella [155]. Interestingly, IF analyses revealed absence of the ODA HC DNAH5, IC DNAI1, and the IDA LC DNALI1 but normal distribution of the IDA HC DNAH7 indicating that HEATR2 is important for the assembly of ODAs but only a subset of IDAs [154,155].

Recently, the first X-linked nonsyndromic PCD variant was described [156]. *PIH1D3* (also known as *CXorf41*) is located on chromosome Xq22.3, comprises of eight exons, and encompasses approximately 38 kb of genomic DNA.

PIH1D3 is one of four members of a protein family interacting with Hsp90 (PIH1). Studies in *Chlamydomonas* indicate that PIH1 not only functions in the preassembly of axonemal dyneins but may also function in preribosomal RNA processing [157,158]. Hemizygous mutations in *PIH1D3* were identified in two families. In one family, the mutation was confirmed in the uncle, two nephews, and one allele of the mother, consistent with an X-linked mode of inheritance [156]. Interestingly, in the second family the mutation was only identified in the affected individual, but not in the two brothers and not in the mother, indicating a de novo mutation. Both affected individuals show classical PCD symptoms such as chronic sinusitis, chronic otitis media, and chronic lower respiratory tract infections, as well as bronchiectasis in the middle lobe and mucus plugging. Additionally, both have *situs inversus totalis* and neonatal respiratory distress syndrome. HVMA analyses of respiratory cilia and sperm flagella showed complete immotility [156]. TEM analyses of ciliary cross sections showed absence of ODAs. Using antibodies directed against ODA and IDA components high-resolution IF showed complete axonemal absence of the ODA HCs, DNAH5 and DNAH9 in mutant respiratory cilia, and complete loss of the ODA ICs, DNAI1 and DNAI2 in mutant respiratory cilia [156], confirming the ODA defect previously observed by TEM. Thus, *PIH1D3* deficiency results in abnormal assembly of type-1 as well as type-2 ODA complexes. Additionally, DNAI1 and DNAI2 were both absent from mutant sperm flagella, indicating that assembly of both ODA types is also disturbed in sperm cells [156]. Interestingly, the IDA LC DNALI1 was greatly reduced or absent in the respiratory ciliary axonemes consistent with a defective cytoplasmic preassembly of ODAs and IDAs [156]. Furthermore, it was shown that PIH1D3 directly interacts and coprecipitates with the cytoplasmic dynein arm assembly factors DNAAF2 and DNAAF4. These results have clinical and genetic counseling implications for genetically unsolved male cases with a classic PCD phenotype that lack additional phenotypes such as intellectual disability or retinitis pigmentosa.

14.15 Molecular defects affecting the 96-nm axonemal ruler

One group of PCD variants was characterized by the absence of GAS8 (a component of the N-DRC) and the IDA component DNALI1 from the ciliary axonemes. TEM found characteristic changes involving reduced number of IDAs, eccentric CPs, abnormal alignment of outer doublets, and occasional displacement of outer doublets, and a beating pattern characterized by severely reduced amplitude with rigid axonemes. The clinical phenotype of this PCD variant is characterized by chronic airway disease, randomization of L-R asymmetry, and immotile sperm tails. Based on candidate approaches in dog, fish, and mice, the genes *CCDC39* and *CCDC40* were identified as possible candidate genes [85,159]. Mutational analyses of PCD patients with the previously described characteristics found changes in *CCDC39* and *CCDC40* in 13% and

60% of affected loss-of-function mutations, respectively [85,159]. Both genes encode for coiled-coil domain–containing proteins. *FAP59*, the orthologue of *CCDC39* in *Chlamydomonas*, was predicted to be essential for ciliary motility as orthologues do not exist in nonciliated organisms [160]. Interestingly, studies in *Chlamydomonas* show that FAP59 pulled down FAP172 (orthologue of CCDC40) and vice versa [6] indicating that these proteins form a complex. Additionally, the absence of the FAP59/FAP172 complex results in mislocalization of IDAs and the N-DRC. Furthermore, changes in the FAP59/FAP172 complex (e.g., elongation) resulted in aberrant numbers (e.g., duplication) of single-headed IDAs, the N-DRC, and RS2. These observations indicate that the FAP59/FAP172 complex serves as attachment site for IDAS and the N-DRC and presents a molecular ruler that determines the 96-nm repeat length and arrangements along the cilia and flagella. Consistent with findings in the *Chlamydomonas*, CCDC39 and CCDC40 absence in humans results in loss of DNALI1-containing IDA complexes and the N-DRC from the ciliary axonemes, thereby causing axonemal disorganization and abnormal ciliary beating [85,159].

14.16 Molecular defects affecting the nexin–dynein regulatory complex

The N-DRC together with the RSs and the CP is responsible for modulation of ciliary movement and regulation of dynein arms. Inhibition of the sliding forces by the N-DRC converts the sliding forces into bending of the axoneme [161,162]. So far, mutations in three genes encoding for N-DRC components have been identified: *CCDC164, CCDC65,* and *GAS8* [145,163–166]. Mutations in these genes result in isolated defects of the N-DRC and PCD.

DRC1, the *Chlamydomonas* orthologue of *CCDC164* in humans, is mutated in the *Chlamydomonas pf3* mutant, and biochemical and structural analyses have shown that the *pf3* mutant is defective in assembly of the N-DRC and several inner arm structures [3,12,13,167]. *CCDC65* was identified by morpholino knockdown of PCD candidate genes in zebrafish as an in vivo screening platform. Knockdown of *ccdc65* altered the ciliary beating pattern and resulted in strong ciliopathy phenotypes, including those involving pronephric cysts, axis curvature, L-R asymmetry defects, and hydrocephalus. Furthermore, the *Chlamydomonas ida6* mutant identifies *DRC2*, orthologous to human *CCDC65*, as an essential component of the N-DRC. The functional role of the N-DRC subunit DRC4, orthologous to human GAS8, has been studied in several organisms, including the green unicellular alga (*Chlamydomonas*), the African protozoan (*Trypanosoma*), and the zebrafish (*Danio rerio*) [168–171]. Genetic analyses of the *Chlamydomonas pf2* mutant revealed that the *PF2* encodes for DRC4 [171]. *pf2* mutant alga display reduced swimming velocities and aberrant waveforms with reduced beating amplitudes of the flagella [11,171] and fail to assemble different components (DRC3–DRC7) of the N-DRC multiprotein complex [167,172,173]. Additionally, knockdown of *gas8* in zebrafish resulted in hydrocephalus, neural tube cell death, L-R axis defects, and impaired otolith biogenesis [168]. Based on these findings, the human orthologues of *DRC1*, *DRC2*, and *DRC4* were considered excellent candidate genes for PCD.

Mutational analyses of PCD-affected individuals revealed loss-of-function mutations in *CCDC164, CCDC65,* and *GAS8* [145,163–166]. Examination of the axonemal structure of these mutant cilia by TEM found that most of sections showed no pronounced alterations in the arrangement of the outer microtubule doublets. Subtle ultrastructural changes such as substitution of peripheral doublets by single tubules or the presence of supernumerary single tubules were only observed in a few sections. Although the N-DRC is difficult to discern in routine TEM cross sections, in some cross sections (i.e., of *CCDC164* mutant cilia), the N-DRCs were missing. These findings were consistent with structural analyses of *pf3* flagella by averaging of TEM sections and cryo–electron tomography [3,12]. It is important to note that the identified TEM alterations are so subtle that they cannot be identified regularly by TEM analysis. In addition, the alterations of the beating pattern are also difficult to discern. Thus, it is expected that this PCD variant probably is frequently missed by current diagnostic approaches. To characterize the ciliary ultrastructural defect in these mutant axonemes in more detail, respiratory epithelial cells were analyzed by high-resolution IF using antibodies directed against components of the IDA (DNALI1), the 96-nm axonemal ruler (CCDC39), and the N-DRC (GAS8/DRC4 and LRRC48/DRC3). In contrast to the results obtained in *CCDC39* and *CCDC40* mutant cilia, in *CCDC164, CCDC65, and GAS8* mutant cilia only the N-DRC components GAS8/DRC4 and LRRC48/DRC3 were absent from the ciliary axonemes while DNAH5, DNALI1, and CCDC39 showed a normal localization, indicating that isolated defects of the N-DRC does not affect axonemal organization and attachment of the IDAs [145,163–166]. Thus, high-resolution IF analysis can identify isolated N-DRC defects in PCD, which underlines the high potential of this diagnostic tool. All PCD individuals with isolated N-DRC defects had typical symptoms of PCD such as neonatal respiratory distress, recurrent infections of the upper and lower respiratory airways since birth or infancy, chronic otitis media, and bronchiectasis in the middle and lingual lung lobes. Notably, none of the affected individuals had *situs inversus totalis*.

In summary, *CCDC164, CCDC65,* and *GAS8* mutant respiratory cilia show no obvious ultrastructural defects and exhibit only subtle abnormalities of ciliary beating. These PCD variants are hardly detectable by routine diagnostic tests such as high-speed video-microscopy or TEM but can be efficiently detected by IF analysis and genetic testing.

14.17 Molecular defects affecting ciliary beat regulation

14.17.1 Molecular defects affecting radial spokes

In *Chlamydomonas*, the RS is a conserved macromolecular complex containing at least 23 proteins that form a T-shaped complex, including a stalk, neck, and head compartment [174] and is assembled in the cytoplasm. The complex repeats in pairs every 96 nm along the axoneme. It is anchored to the outer microtubule doublet close to the IDAs and the N-DRC and extends to the CP [175]. Thus, its position along the axoneme is perfect for relaying signals from the CP to the dynein arms. Despite the fact that in 1979 Sturgess et al. had already described patients with missing RS complexes resulting in immotile respiratory cilia and sperm flagella [176], little is known about the composition and the exact function of RSs in humans. At least 12 RS proteins have apparent human orthologues, including RSPH9, RSPH4A, RSPH1, RSPH23, and RSPH3, underlining the high conservation of RSs throughout evolution [175]. Cryo–electron tomography studies performed in *Chlamydomonas* flagella and human respiratory cilia revealed that there are three types of RS complexes named RS1, RS2, and RS3 [174,177]. Recently, it was shown that mutations in *RSPH1*, encoding for the RS head component RSPH1, affect the RS1 and RS2 but not the RS3 complex, indicating that the composition of the RS3 complex differs from the composition of RS1 and RS2 complexes [177]. So far, mutations in five genes have been described that affect the function and composition of RS complexes in humans: *RSPH1, RSPH3, RSPH4A, RSPH9,* and *DNAJB13* [177–185]. All of the PCD-affected individuals with RS defects exhibited classical PCD symptoms, including recurrent respiratory infections, chronic cough, wheeze, bronchitis, and bronchiectasis. Interestingly, none showed laterality defects. TEM analyses of RS defective respiratory cilia revealed in some cross sections ciliary transposition defects with 9+0 or 8+1 microtubule configurations. It is important to note that these defects in the tubule composition are only present in small areas along the ciliary axonemes. This explains why electron microscopy analyses can be apparently normal in patients with RS defects [177–185]. Additionally, ciliary beating abnormalities are very subtle and only detectable by examiners experienced in HVMA [186]. Recently, it was shown that mutations in *RSPH9, RSPH4A,* and *RSPH1* can cause distinct defects of the RS head in ciliary axonemes detectable by high-resolution IF. Absence of RSPH4A due to mutations in *RSPH4A* results in deficient axonemal assembly of the RS head components RSPH1 and RSPH9. *RSPH1* mutant cilia lacking RSPH1 fail to assemble RSPH9, while *RSPH9* mutations result in axonemal absence of RSPH9 but do not affect the assembly of the other head proteins RSPH1 and RSPH4A [182]. Interestingly, the identified loss-of-function mutations, missense variants and amino acid deletions in *RSPH4A* and *RSPH9* resulted in absence of the respective proteins [182]. These findings indicate that high-resolution IF can be used to diagnose PCD with RS defects, tailor genetic diagnostics, and help to determine pathogenicity of DNA variants detected by mutational analysis.

14.17.2 Molecular defects affecting the central pair complex

The first identified molecular defect of the CP complex is caused by mutations in *HYDIN*. *HYDIN* was considered a promising candidate gene for PCD variant without *situs inversus totalis* based on findings in *hy3* mice. In *hy3* mice carrying homozygous-recessive *Hydin* mutations, there is lethality in the first weeks of life because of hydrocephalus caused by abnormal ependymal ciliary motility [88,187,188]. Interestingly, in contrast to other PCD mouse models, such as *Ccdc40-* and *Mdnah5*-deficient mice that also develop hydrocephalus [27,85], *hy3* mice do not exhibit randomization of L-R body asymmetry. By homozygosity mapping, a PCD-associated locus, chromosomal region 16q21-q23, which contains *HYDIN,* was identified. However, a nearly identical 360 kb paralogous segment (*HYDIN2*) in chromosomal region 1q21.1 complicated mutational analysis [189,190]. Homozygous mutations that affect an evolutionary conserved splice acceptor site (c.3985G>T) and that subsequently cause aberrantly spliced transcripts predicting premature protein termination in respiratory cells were identified in three PCD-affected individuals. Parallel whole-exome sequencing identified a homozygous nonsense *HYDIN* mutation (c.922A>T, p.Lys307*) in six individuals from three Faroe Island PCD-affected families that all carried an 8.8 Mb shared haplotype across *HYDIN*, indicating an ancestral founder mutation in this isolated population. All affected individuals presented classical PCD symptoms such as respiratory distress syndrome as neonates, upper airway disease including recurrent otitis media and chronic rhinosinusitis, chronic wet cough, chronic bronchitis, chronic rhinosinusitis, and bronchiectasis. Interestingly, consistent with findings in *hy3* mice, all nine PCD affected had normal body composition. However, in contrast to *hy3* mice, none of the affected individuals exhibited hydrocephalus [190]. Analyses of *HYDIN* mutant cilia by TEM exhibited a normal 9+2 axonemal composition and no alteration of ODAs and IDAs [190]. Consistent with these findings, high-resolution IF using antibodies directed against DNAH5, DNALI1, GAS8, and CCDC39 showed no alterations in structure and composition of the ODAs, IDAs, the N-DRC, and the 96-nm axonemal ruler [190]. HVMA analyses showed only very subtle respiratory cilia beating defect with reduced amplitude and a stiff beat pattern. Electron microscopy tomography (EMT) analyses of *HYDIN* mutant cilia revealed absence of one of the CP projections [190].

Unfortunately, diagnosis of those patients is often delayed, because mutations in this gene and estimated in all genes encoding parts of the CP complex do not lead to a *situs inversus totalis* or *situs ambiguous* and that structural defects of the CP complex are not regularly detectable with routine TEM.

References

[1] M. Salathe, Regulation of mammalian ciliary beating, Annu. Rev. Physiol. 69 (2007) 401–422.

[2] K.H. Bui, T. Yagi, R. Yamamoto, R. Kamiya, T. Ishikawa, Polarity and asymmetry in the arrangement of dynein and related structures in the *Chlamydomonas* axoneme, J. Cell Biol. 198 (2012) 913–925.

[3] T. Heuser, M. Raytchev, J. Krell, M.E. Porter, D. Nicastro, The dynein regulatory complex is the nexin link and a major regulatory node in cilia and flagella, J. Cell Biol. 187 (6) (2009) 921–933.

[4] S.H. Myster, J.A. Knott, K.M. Wysocki, E. O'Toole, M.E. Porter, Domains in the 1 alpha dynein heavy chain required for inner arm assembly and flagellar motility in *Chlamydomonas*, J. Cell Biol. 146 (4) (1999) 801–818.

[5] H. Sakakibara, S. Takada, S.M. King, G.B. Witman, R. Kamiya, A *Chlamydomonas* outer arm dynein mutant with a truncated beta heavy chain, J. Cell Biol. 122 (3) (1993) 653–661.

[6] T. Oda, H. Yanagisawa, R. Kamiya, M. Kikkawa, A molecular ruler determines the repeat length in eukaryotic cilia and flagella, Science 346 (2014) 857–860.

[7] S.M. King, Axonemal dyneins winch the cilium, Nat. Struct. Mol. Biol. 17 (2010) 673–674.

[8] E.E. Dymek, T. Heuser, D. Nicastro, E.F. Smith, The CSC is required for complete radial spoke assembly and wild-type ciliary motility, Mol. Biol. Cell. 22 (2011) 2520–2531.

[9] T. Heuser, E.E. Dymek, J. Lin, E.F. Smith, D. Nicastro, The CSC connects three major axonemal complexes involved in dynein regulation, Mol. Biol. Cell. 23 (2012) 3143–3155.

[10] S.A. Burgess, M.L. Walker, H. Sakakibara, P.J. Knight, K. Oiwa, Dynein structure and power stroke, Nature 13 (2003) 715–718.

[11] C.J. Brokaw, R. Kamiya, Bending patterns of *Chlamydomonas* flagella: IV. Mutants with defects in inner and outer dynein arms indicate differences in dynein arm function, Cell Motil. Cytoskelet. 8 (1) (1987) 68–75.

[12] L.C. Gardner, E. O'Toole, C.A. Perrone, T. Giddings, M.E. Porter, Components of a "dynein regulatory complex" are located at the junction between the radial spokes and the dynein arms in *Chlamydomonas* flagella, J. Cell Biol. 127 (5) (1994) 1311–1325.

[13] G. Piperno, K. Mead, W. Shestak, The inner dynein arms I2 interact with a "dynein regulatory complex" in *Chlamydomonas* flagella, J. Cell Biol. 118 (6) (1992) 1455–1463.

[14] E.F. Smith, P. Yang, The radial spokes and central apparatus: mechano-chemical transducers that regulate flagellar motility, Cell Motil. Cytoskelet. 57 (1) (2004) 8–17.

[15] M. Salathe, M.M. Pratt, A. Wanner, Cyclic AMP-dependent phosphorylation of a 26 kD axonemal protein in ovine cilia isolated from small tissue pieces, Am. J. Respir. Cell Mol. Biol. 9 (1993) 306–314.

[16] A. Schmid, G. Bai, N. Schmid, M. Zaccolo, L.E. Ostrowski, G.E. Conner, N. Fregien, M. Salathe, Real-time analysis of cAMP-mediated regulation of ciliary motility in single primary human airway epithelial cells, J. Cell Sci. 15 (2006) 4176–4186.

[17] A. Schmid, Z. Sutto, M.C. Nlend, G. Horvath, N. Schmid, J. Buck, L.R. Levin, G.E. Conner, N. Fregien, M. Salathe, Soluble adenylyl cyclase is localized to cilia and contributes to ciliary beat frequency regulation via production of cAMP, J. Gen. Physiol. 130 (2007) 99–109.

[18] L.M. DiBella, S.M. King, Dynein motor of the *Chlamydomonas* flagellum, Int. Rev. Cytol. 210 (2001) 227–268.

[19] R. Kamiya, M. Okamoto, A mutant of *Chlamydomonas reinhardtii* that lacks the flagellar outer dynein arm but can swim, J. Cell Sci. 74 (1985) 181–191.

[20] D.R. Mitchell, J.L. Rosenbaum, A motile *Chlamydomonas* flagellar mutant that lacks outer dynein arms, J. Cell Biol. 100 (1985) 1228–1234.

[21] B. Huang, G. Piperno, D.J.L. Luck, Paralyzed flagella mutants of *Chlamydomonas reinhardtii* defective for axonemal doublet microtubule arms, J. Biol. Chem. 254 (1979) 3091–3099.

[22] R. Kamiya, Mutations at twelve independent loci result in absence of outer dynein arms in *Chlamydomonas reinhardtii*, J. Cell Biol. 107 (1988) 2253–2258.

[23] G.J. Pazour, N. Agrin, B.L. Walker, G.B. Witman, Identification of predicted human outer dynein arm genes: candidates for primary ciliary dyskinesia genes, J. Med. Genet. 43 (2006) 62–73.

[24] S.F. Altschul, W. Gish, W. Miller, E.W. Myers, D.J. Lipman, Basic local alignment search tool, J. Mol. Biol. 215 (1990) 403–410.

[25] H. Olbrich, K. Haffner, A. Kispert, A. Volkel, A. Volz, G. Sasmaz, R. Reinhardt, S. Hennig, H. Lehrach, N. Konietzko, M. Zariwala, P.G. Noone, M. Knowles, H.M. Mitchison, M. Meeks, E.M. Chung, F. Hildebrandt, R. Sudbrak, H. Omran, Mutations in DNAH5 cause primary ciliary dyskinesia and randomization of left-right asymmetry, Nat. Genet. 30 (2002) 143–144.

[26] A. Kispert, M. Petry, H. Olbrich, A. Volz, U.P. Ketelsen, J. Horvath, R. Melkaoui, H. Omran, M. Zariwala, P.G. Noone, M. Knowles, Genotypeepephenotype correlations in PCD patients carrying DNAH5 mutations, Thorax 58 (2003) 552–554.

[27] I. Ibanez-Tallon, A. Pagenstecher, M. Fliegauf, H. Olbrich, A. Kispert, U.P. Ketelsen, A. North, N. Heintz, H. Omran, Dysfunction of axonemal dynein heavy chain Mdnah5 inhibits ependymal flow and reveals a novel mechanism for hydrocephalus formation, Hum. Mol. Genet. 13 (2004) 2133–2141.

[28] W. Reed, J.L. Carson, B.M. Moats-Staats, T. Lucier, P. Hu, L. Brighton, T.M. Gambling, C.H. Huang, M.W. Leigh, A.M. Collier, Characterization of an axonemal dynein heavy chain expressed early in airway epithelial ciliogenesis, Am. J. Respir. Cell Mol. Biol. 23 (2000) 734–741.

[29] C. Chapelin, B. Duriez, F. Magnino, M. Goossens, E. Escudier, S. Amselem, Isolation of several human axonemal dynein heavy chain genes: genomic structure of the catalytic site, phylogenetic analysis and chromosomal assignment, FEBS Lett. 412 (1997) 325–330.

[30] D.M. Supp, D.P. Witte, S.S. Potter, M. Brueckner, Mutation of an axonemal dynein affects left-right asymmetry in inversus viscerum mice, Nature 389 (1997) 963–996.

[31] S.A. Samant, O. Ogunkua, L. Hui, J. Fossella, S.H. Pilder, The T complex distorter 2 candidate gene, Dnahc8, encodes at least two testis-specific axonemal dynein heavy chains that differ extensively at their amino and carboxyl termini, Dev. Biol. 250 (2002) 24–43.

[32] G.W. Dougherty, N.T. Loges, J.A. Klinkenbusch, H. Olbrich, P. Pennekamp, T. Menchen, J. Raidt, J. Wallmeier, C. Werner, C. Westermann, C. Ruckert, V. Mirra, R. Hjeij, Y. Memari, R. Durbin, A. Kolb-Kokocinski, K. Praveen, M.A. Kashef, S. Kashef, F. Eghtedari, K. Häffner, P. Valmari, G. Baktai, M. Aviram, L. Bentur, I. Amirav, E.E. Davis, N. Katsanis, M. Brueckner, A. Shaposhnykov, G. Pigino, B. Dworniczak, H. Omran, DNAH11 localization in the proximal region of respiratory cilia defines distinct outer dynein arm complexes, Am. J. Respir. Cell Mol. Biol. 55 (2016) 213–224.

[33] M. Fliegauf, H. Olbrich, J. Horvath, J.H. Wildhaber, M.A. Zariwala, M. Kennedy, M.R. Knowles, H. Omran, Mislocalization of DNAH5 and DNAH9 in respiratory cells from patients with primary ciliary dyskinesia, Am. J. Respir. Crit. Care Med. 171 (2005) 1343–1349.

[34] H. Omran, N.T. Loges, Immunofluorescence staining of ciliated respiratory epithelial cells, in: Methods in Cell Biology, vol. 91, Academic Press, 2009.

[35] G. Piperno, Isolation of radial spoke heads from *Chlamydomonas* axonemes, Method. Cell. Biol. 47 (1995) 381–383.

[36] K.H. Bui, H. Sakakibara, T. Movassagh, K. Oiwa, T. Ishikawa, Molecular architecture of inner dynein arms in situ in *Chlamydomonas reinhardtii* flagella, J. Cell Biol. 183 (2008) 923–932.

[37] R. Kamiya, Exploring the function of inner and outer dynein arms with *Chlamydomonas* mutants, Cell Motil. Cytoskelet. 32 (1995) 98–102.

[38] R. Kamiya, E. Kurimoto, H. Sakakibara, T. Okagaki, A genetic approach to the function of inner and outer arm dynein, in: in: F.D. Warner, P. Satir, I.R. Gibbons (Eds.), Cell Movement, vol. 1, Alan R. Liss, Inc, New York, 1989, pp. 49–60.

[39] A. Harrison, P. Olds-Clarke, S.M. King, Identification of the t complex-encoded cytoplasmic dynein light chain tctex1 in inner arm I1 supports the involvement of flagellar dyneins in meiotic drive, J. Cell Biol. 140 (5) (1998) 1137–1147.

[40] C.A. Perrone, P. Yang, E. O'Toole, W.S. Sale, M.E. Porter, The *Chlamydomonas* IDA7 locus encodes a 140-kDa dynein intermediate chain required to assemble the I1 inner arm complex, Mol. Biol. Cell 9 (1998) 3351–3365.

[41] G. Piperno, Z. Ramanis, The proximal portion of *Chlamydomonas* flagella contains a distinct set of inner dynein arms, J. Cell Biol. 112 (4) (1991) 701–709.

[42] M.E. Porter, J. Power, S.K. Dutcher, Extragenic suppressors of paralyzed flagellar mutations in *Chlamydomonas reinhardtii* identify loci that alter the inner dynein arms, J. Cell Biol. 118 (5) (1992) 1163–1176.

[43] C.A. Perrone, S.H. Myster, R. Bower, E.T. O'Toole, M.E. Porter, Insights into the structural organization of the I1 inner arm dynein from a domain analysis of the 1beta dynein heavy chain, Mol. Biol. Cell 11 (7) (2000) 2297–2313.

[44] S.J. King, S.K. Dutcher, Phosphoregulation of an inner dynein arm complex in *Chlamydomonas reinhardtii* is altered in phototactic mutant strains, J. Cell Biol. 136 (1) (1997) 177–191.

[45] M. LeDizet, G. Piperno, The light chain p28 associates with a subset of inner dynein arm heavy chains in *Chlamydomonas* axonemes, Mol. Biol. Cell 6 (1995) 697e711 [47] K.F. Lechtreck, P. Delmotte, M.L. Robinson, M.J. Sanderson, G.B. Witman, Mutations in Hydin impair ciliary motility in mice, J. Cell Biol. 180(2008) 633–643.

[46] H.A. Yanagisawa, R. Kamiya, Association between actin and light chains in *Chlamydomonas* flagellar inner-arm dyneins, Biochem. Biophys. Res. Commun. 26 (2001) 443–447.

[47] T.D. Pollard, J.A. Cooper, Actin and actin-binding proteins. A critical evaluation of mechanisms and functions, Annu. Rev. Biochem. 55 (1986) 987–1035.

[48] M.A. Sanders, J.L. Salisbury, Centrin-mediated microtubule severing during flagellar excision in *Chlamydomonas reinhardtii*, J. Cell Biol. 108 (5) (1989) 1751–1760.

[49] J. Neesen, R. Kirschner, M. Ochs, A. Schmiedl, B. Habermann, C. Mueller, A.F. Holstein, T. Nuesslein, I. Adham, W. Engel, Disruption of an inner arm dynein heavy chain gene results in asthenozoospermia and reduced ciliary beat frequency, Hum. Mol. Genet. 10 (11) (2001) 1117–1128.

[50] Y.J. Zhang, W.K. O'Neal, S.H. Randell, K. Blackburn, M.B. Moyer, R.C. Boucher, L.E. Ostrowski, Identification of dynein heavy chain 7 as an inner arm component of human cilia that is synthesized but not assembled in a case of primary ciliary dyskinesia, J. Biol. Chem. 17 (2002) 17906–17915.

[51] A. Amiri-Yekta, C. Coutton, Z.E. Kherraf, T. Karaouzène, P. Le Tanno, M.H. Sanati, M. Sabbaghian, N. Almadani, M.A. Sadighi Gilani, S.H. Hosseini, S. Bahrami, A. Daneshipour, M. Bini, C. Arnoult, R. Colombo, H. Gourabi, P.F. Ray, Whole-exome sequencing of familial cases of multiple morphological abnormalities of the sperm flagella (MMAF) reveals new DNAH1 mutations, Hum. Reprod. 31 (2016) 2872–2880.

[52] M. Ben Khelifa, C. Coutton, R. Zouari, T. Karaouzène, J. Rendu, M. Bidart, S. Yassine, V. Pierre, J. Delaroche, S. Hennebicq, D. Grunwald, D. Escalier, K. Pernet-Gallay, P.S. Jouk, N. Thierry-Mieg, A. Touré, C. Arnoult, P.F. Ray, Mutations in DNAH1, which encodes an inner arm heavy chain dynein, lead to male infertility from multiple morphological abnormalities of the sperm flagella, Am. J. Hum. Genet. 94 (2014) 95–104.

[53] X. Wang, H. Jin, F. Han, Y. Cui, J. Chen, C. Yang, P. Zhu, W. Wang, G. Jiao, W. Wang, C. Hao, Z. Gao, Homozygous DNAH1 frameshift mutation causes multiple morphological anomalies of the sperm flagella in Chinese, Clin. Genet. (2016). http://dx.doi.org/10.1111/cge.12857.

[54] M. Fliegauf, T. Benzing, H. Omran, When cilia go bad: cilia defects and ciliopathies, Nat. Rev. Mol. Cell Biol. 8 (11) (2007) 880–893.

[55] F. Hildebrandt, E. Otto, C. Rensing, H.G. Nothwang, M. Vollmer, J. Adolphs, H. Hanusch, M. Brandis, A novel gene encoding an SH3 domain protein is mutated in Nephronophthisis type 1, Nat. Genet. 17 (2) (1997) 149–153.

[56] H. Olbrich, M. Fliegauf, J. Hoefele, A. Kispert, E. Otto, A. Volz, M.T. Wolf, G. Sasmaz, U. Trauer, R. Reinhardt, R. Sudbrak, C. Antignac, N. Gretz, G. Walz, B. Schermer, T. Benzing, F. Hildebrandt, H. Omran, Mutations in a novel gene, NPHP3, cause adolescent nephronophthisis, tapeto-retinal degeneration and hepatic fibrosis, Nat. Genet. 34 (4) (2003) 455–459.

[57] R.J. Simms, L. Eley, J.A. Sayer, Nephronophthisis, Eur. J. Hum. Genet. 17 (4) (2008) 406–416.

[58] M.T. Wolf, Nephronophthisis and related syndromes, Curr. Opin. Pediatr. 27 (2015) 201–211.

[59] R. Novas, M. Cardenas-Rodriguez, F. Irigoín, J.L. Badano, Bardet-Biedl syndrome: is it only cilia dysfunction n? FEBS Lett. 589 (2015) 3479–3491.

[60] M.V. Nachury, A.V. Loktev, Q. Zhang, C.J. Westlake, J. Pera¨nen, A. Merdes, D.C. Slusarski, R.H. Scheller, J.F. Bazan, V.C. Sheffield, P.K. Jackson, A core complex of BBS proteins cooperates with the GTPase Rab8 to promote ciliary membrane biogenesis, Cell 129 (6) (2007) 1201–1213.

[61] E. Coll, R. Torra, J. Pascual, A. Botey, J. Ara, L. Pe´rez, F. Ballesta, A. Darnell, Sporadic orofaciodigital syndrome type I presenting as end-stage renal disease, Nephrol. Dial. Transpl. 12 (5) (1997) 1040–1042.

[62] M.I. Ferrante, A. Zullo, A. Barra, S. Bimonte, N. Messaddeq, M. Studer, P. Dolle´, B. Franco, Oral-facial-digital type I protein is required for primary cilia formation and left-right axis specification, Nat. Genet. 38 (1) (2006) 112–117.

[63] G. Li, R. Vega, K. Nelms, N. Gekakis, C. Goodnow, P. McNamara, H. Wu, N.A. Hong, R. Glynne, A role for Alström syndrome protein, alms1, in kidney ciliogenesis and cellular quiescence, PLoS Genet. 3 (1) (2007) e8.

[64] M. Boon, J. Wallmeier, L. Ma, N.T. Loges, M. Jaspers, H. Olbrich, G.W. Dougherty, J. Raidt, C. Werner, I. Amirav, A. Hevroni, R. Abitbul, A. Avital, R. Soferman, M. Wessels, C. O'Callaghan, E.M. Chung, A. Rutman, R.A. Hirst, E. Moya, H.M. Mitchison, S. Van Daele, K. De Boeck, M. Jorissen, C. Kintner, H. Cuppens, H. Omran, MCIDAS mutations result in a mucociliary clearance disorder with reduced generation of multiple motile cilia, Nat. Commun. 5 (2014) 4418.

[65] J. Wallmeier, D.A. Al-Mutairi, C.T. Chen, N.T. Loges, P. Pennekamp, T. Menchen, L. Ma, H.E. Shamseldin, H. Olbrich, G.W. Dougherty, C. Werner, B.H. Alsabah, G. Kohler, M. Jaspers, M. Boon, M. Griese, S. Schmitt-Grohe, T. Zimmermann, C. Koerner-Rettberg, E. Horak, C. Kintner, F.S. Alkuraya, H. Omran, Mutations in CCNO result in congenital mucociliary clearance disorder with reduced generation of multiple motile cilia, Nat. Genet. 46 (2014) 646–651.

[66] I. Amirav, J. Wallmeier, N.T. Loges, T. Menchen, P. Pennekamp, H. Mussaffi, R. Abitbul, A. Avital, L. Bentur, G.W. Dougherty, E. Nael, M. Lavie, H. Olbrich, C. Werner, C. Kintner, H. Omran, Israeli PCD Consortium Investigators, Systematic analysis of CCNO variants in a defined population: implications for clinical phenotype and differential diagnosis, Hum. Mutat. 37 (2016) 396–405.

[67] S.F. Gilbert, Developmental Biology, seventh ed., Sinauer Associates Inc, Sunderland, MA, 2003.

[68] R.S. Beddington, E.J. Robertson, Axis development and early asymmetry in mammals, Cell 96 (1999) 195–209.

[69] L. El Zein, H. Omran, P. Bouvagnet, Lateralization defects and ciliary dyskinesia: lessons from algae, Trends Genet. 19 (2003) 162–167.

[70] M.H. Kaufmann, The Atlas of Mouse Development, first ed., Elsevier Academic Press, San Diego, CA, 1995.

[71] N. Hirokawa, Y. Tanaka, Y. Okada, Left-right determination: involvement of molecular motor KIF3, cilia, and nodal flow, Cold Spring Harb. Perspect. Biol. 1 (1) (2009) a000802.

[72] J. McGrath, M. Brueckner, Cilia are at the heart of vertebrate left-right asymmetry, Curr. Opin. Genet. Dev. 13 (2003) 385–392.

[73] S. Nonaka, Y. Tanaka, Y. Okada, S. Takeda, A. Harada, Y. Kanai, M. Kido, N. Hirokawa, Randomization of left-right asymmetry due to loss of nodal cilia generating leftward flow of extraembryonic fluid in mice lacking KIF3B motor protein, Cell 95 (1998) 829–837.

[74] Y. Okada, S. Nonaka, Y. Tanaka, Y. Saijoh, H. Hamada, N. Hirokawa, Abnormal nodal flow precedes situs inversus in iv and inv mice, Mol. Cell 4 (1999) 459–468.

[75] S. Nonaka, H. Shiratori, Y. Saijoh, H. Hamada, Determination of left-right patterning of the mouse embryo by artificial nodal flow, Nature 418 (2002) 96–99.

[76] J.J. Essner, K.J. Vogan, M.K. Wagner, C.J. Tabin, H.J. Yost, M. Brueckner, Conserved function for embryonic nodal cilia, Nature 418 (2002) 37–38.

[77] L. Bartoloni, J.L. Blouin, Y. Pan, C. Gehrig, A.K. Maiti, N. Scamuffa, C. Rossier, M. Jorissen, M. Armengot, M. Meeks, H.M. Mitchison, E.M. Chung, C.D. Delozier-Blanchet, W.J. Craigen, S.E. Antonarakis, Mutations in the DNAH11 (axonemal heavy chain dynein type 11) gene cause one form of situs inversus totalis and most likely primary ciliary dyskinesia, Proc. Natl. Acad. Sci. U.S.A. 99 (2002) 10282–10286.

[78] C. Guichard, M.C. Harricane, J.J. Lafitte, P. Godard, M. Zaegel, V. Tack, G. Lalau, P. Bouvagnet, Axonemal dynein intermediate-chain gene (DNAI1) mutations result in situs inversus and primary ciliary dyskinesia (Kartagener syndrome), Am. J. Hum. Genet. 68 (2001) 1030–1035.

[79] N. Hornef, H. Olbrich, J. Horvath, M.A. Zariwala, M. Fliegauf, N.T. Loges, J. Wildhaber, P.G. Noone, M. Kennedy, S.E. Antonarakis, J.L. Blouin, L. Bartoloni, T. Nüsslein, P. Ahrens, M. Griese, H. Kuhl, R. Sudbrak, M.R. Knowles, R. Reinhardt, H. Omran, DNAH5 mutations are a common cause of primary ciliary dyskinesia with outer dynein arm defects, Am. J. Respir. Crit. Care Med. 174 (2006) 120–126.

[80] M.R. Knowles, M.W. Leigh, J.L. Carson, S.D. Davis, S.D. Dell, T.W. Ferkol, K.N. Olivier, S.D. Sagel, M. Rosenfeld, K.A. Burns, S.L. Minnix, M.C. Armstrong, A. Lori, M.J. Hazucha, N.T. Loges, H. Olbrich, A. Becker-Heck, M. Schmidts, C. Werner, H. Omran, M.A. Zariwala, Genetic Disorders of Mucociliary Clearance Consortium, Mutations of DNAH11 in patients with primary ciliary dyskinesia with normal ciliary ultrastructure, Thorax 67 (2012) 433–441.

[81] N.T. Loges, H. Olbrich, L. Fenske, H. Mussaffi, J. Horvath, M. Fliegauf, H. Kuhl, G. Baktai, E. Peterffy, R. Chodhari, E.M. Chung, A. Rutman, C. O'Callaghan, H. Blau, L. Tiszlavicz, K. Voelkel, M. Witt, E. Zietkiewicz, J. Neesen, R. Reinhardt, H.M. Mitchison, H. Omran, DNAI2 mutations cause primary ciliary dyskinesia with defects in the outer dynein arm, Am. J. Hum. Genet. 83 (2008) 547–558.

[82] G.C. Schwabe, K. Hoffmann, N.T. Loges, D. Birker, C. Rossier, M.M. de Santi, H. Olbrich, M. Fliegauf, M. Failly, U. Liebers, M. Collura, G. Gaedicke, S. Mundlos, U. Wahn, J.L. Blouin, B. Niggemann, H. Omran, S.E. Antonarakis, L. Bartoloni, Primary ciliary dyskinesia associated with normal axoneme ultrastructure is caused by DNAH11 mutations, Hum. Mutat. 29 (2008) 289–298.

[83] M.A. Zariwala, M.W. Leigh, F. Ceppa, M.P. Kennedy, P.G. Noone, J.L. Carson, M.J. Hazucha, A. Lori, J. Horvath, H. Olbrich, N.T. Loges, A.M. Bridoux, G. Pennarun, B. Duriez, E. Escudier, H.M. Mitchison, R. Chodhari, E.M. Chung, L.C. Morgan, R.U. de Iongh, J. Rutland, U. Pradal, H. Omran, S. Amselem, M.R. Knowles, Mutations of DNAI1 in primary ciliary dyskinesia: evidence of founder effect in a common mutation, Am. J. Respir. Crit. Care Med. 174 (2006) 858–866.

[84] Y. Tanaka, Y. Okada, N. Hirokawa, FGF-induced vesicular release of sonic hedgehog and retinoic acid in leftward nodal flow is critical for left-right determination, Nature 435 (2005) 172–177.

[85] A. Becker-Heck, I.E. Zohn, N. Okabe, A. Pollock, K.B. Lenhart, J. Sullivan-Brown, J. McSheene, N.T. Loges, H. Olbrich, K. Haeffner, M. Fliegauf, J. Horvath, R. Reinhardt, K.G. Nielsen, J.K. Marthin, G. Baktai, K.V. Anderson, R. Geisler, L. Niswander, H. Omran, R.D. Burdine, The coiled-coil domain containing protein CCDC40 is essential for motile cilia function and left-right axis formation, Nat. Genet. 43 (2010) 79–84.

[86] J. Chen, H.J. Knowles, J.L. Hebert, B.P. Hackett, Mutation of the mouse hepatocyte nuclear factor/forkhead homologue 4 gene results in an absence of cilia and random left-right asymmetry, J. Clin. Invest. 102 (1998) 1077–1082.

[87] I. Ibanez-Tallon, N. Heintz, H. Omran, To beat or not to beat: roles of cilia in development and disease, Hum. Mol. Genet. 12 (2003) 27–35.

[88] K.F. Lechtreck, P. Delmotte, M.L. Robinson, M.J. Sanderson, G.B. Witman, Mutations in Hydin impair ciliary motility in mice, J. Cell Biol. 180 (2008) 633–643.

[89] R. Sapiro, I. Kostetskii, P. Olds-Clarke, G.L. Gerton, G.L. Radice, J.F. Strauss III, Male infertility, impaired sperm motility, and hydrocephalus in mice deficient in spermassociated antigen 6, Mol. Cell Biol. 22 (2002) 6298–6305.

[90] P.D. Taulman, C.J. Haycraft, D.F. Balkovetz, B.K. Yoder, Polaris, a protein involved in left-right axis patterning, localizes to basal bodies and cilia, Mol. Biol. Cell 12 (2001) 589–599.

[91] K. Sawamoto, H. Wichterle, O. Gonzalez-Perez, J.A. Cholfin, M. Yamada, N. Spassky, N.S. Murcia, J.M. Garcia-Verdugo, O. Marin, J.L. Rubenstein, M. Tessier-Lavigne, H. Okano, A. Alvarez-Buylla, New neurons follow the flow of cerebrospinal fluid in the adult brain, Science 311 (2006) 629–632.

[92] N.C. Munro, D.C. Currie, K.S. Lindsay, T.A. Ryder, A. Rutman, A. Dewar, M.A. Greenstone, W.F. Hendry, P.J. Cole, Fertility in men with primary ciliary dyskinesia presenting with respiratory infection, Thorax 49 (1994) 684–687.

[93] B.A. Afzelius, Cilia-related diseases, J. Pathol. 204 (4) (2004) 470–477.

[94] B.A. Afzelius, P. Camner, B. Mossberg, On the function of cilia in the female reproductive-tract, Fertil. Steril. 29 (1) (1978) 72–74.

[95] R.A. Lyons, E. Saridogan, O. Djahanbakhch, The reproductive significance of human fallopian tube cilia, Hum. Reprod. Update 12 (4) (2006) 363–372.

[96] H. Pedersen, Absence of dynein arms in endometrial cilia: cause of infertility? Acta Obstet. Gynecol. Scand. 62 (6) (1983) 625–627.

[97] J. Raidt, C. Werner, T. Menchen, G.W. Dougherty, H. Olbrich, N.T. Loges, R. Schmitz, P. Pennekamp, H. Omran, Ciliary function and motor protein composition of human fallopian tubes, Hum. Reprod. 30 (2015) 2871–2880.

[98] B.A. Afzelius, R. Eliasson, O. Johnsen, C. Lindholmer, Lack of dynein arms in immotile human spermatozoa, J. Cell Biol. 66 (2) (1975) 225–232.

[99] H. Pedersen, H. Rebbe, Absence of arms in the axoneme of immobile human spermatozoa, Biol. Reprod. 12 (5) (1975) 541–544.

[100] C. Werner, J.G. Onnebrink, H. Omran, Diagnosis and management of primary ciliary dyskinesia, Cilia 22 (2015) 2.

[101] G. Pennarun, E. Escudier, C. Chapelin, A.M. Bridoux, V. Cacheux, G. Roger, A. Clément, M. Goossens, S. Amselem, B. Duriez, Loss-of-function mutations in a human gene related to *Chlamydomonas reinhardtii* dynein IC78 result in primary ciliary dyskinesia, Am. J. Hum. Genet. 65 (6) (1999) 1508–1519.

[102] H. Omran, K. Haffner, A. Volkel, J. Kuehr, U.-P. Ketelsen, U.H. Ross, N. Konietzko, T. Wienker, M. Brandis, F. Hildebrandt, Homozygosity mapping of a gene locus for primary ciliary dyskinesia on chromosome 5p and identification of the heavy dynein chain DNAH5 as a candidate gene, Am. J. Respir. Cell Mol. Biol. 23 (2000) 696–702.

[103] C.G. Wilkerson, S.M. King, G.B. Witman, Molecular analysis of the gamma heavy chain of *Chlamydomonas* flagellar outer-arm dynein, J. Cell Sci. 107 (1994) 497–506.

[104] M. Pifferi, A. Michelucci, M.E. Conidi, A.M. Cangiotti, P. Simi, P. Macchia, A.L. Boner, New DNAH11 mutations in primary ciliary dyskinesia with normal axonemal ultrastructure, Eur. Respir. J. 35 (2010) 1413–1416.

[105] B. Duriez, P. Duquesnoy, E. Escudier, A.M. Bridoux, D. Escalier, I. Rayet, E. Marcos, A.M. Vojtek, J.F. Bercher, S. Amselem, A common variant in combination with a nonsense mutation in a member of the thioredoxin family causes primary ciliary dyskinesia, Proc. Natl. Acad. Sci. U.S.A. 104 (9) (2007) 3336–3341.

[106] A. Harrison, M. Sakato, H.W. Tedford, S.E. Benashski, R.S. Patel-King, S.M. King, Redoxbased control of the gamma heavy chain ATPase from *Chlamydomonas* outer arm dynein, Cell Motil. Cytoskelet. 52 (3) (2002) 131–143.

[107] S.E. Benashski, R.S. Patel-King, S.M. King, Light chain 1 from the *Chlamydomonas* outer dynein arm is a leucine-rich repeat protein associated with the motor domain of the gamma heavy chain, Biochemistry 38 (1999) 7253–7264.

[108] J. Horváth, M. Fliegauf, H. Olbrich, A. Kispert, S.M. King, H. Mitchison, M.A. Zariwala, M.R. Knowles, R. Sudbrak, G. Fekete, J. Neesen, R. Reinhardt, H. Omran, Identification and analysis of axonemal dynein light chain 1 in primary ciliary dyskinesia patients, Am. J. Respir. Cell Mol. Biol. 33 (2005) 41–47.

[109] M. Mazor, S. Alkrinawi, V. Chalifa-Caspi, E. Manor, V.C. Sheffield, M. Aviram, R. Parvari, Primary ciliary dyskinesia caused by homozygous mutation in DNAL1, encoding dynein light chain 1, Am. J. Hum. Genet. 88 (2011) 599–607.

[110] M.E. Fowkes, D.R. Mitchell, The role of preassembled cytoplasmic complexes in assembly of flagellar dynein subunits, Mol. Biol. Cell 9 (1998) 2337–2347.

[111] K. Wakabayashi, S. Takada, G.B. Witman, R. Kamiya, Transport and arrangement of the outer-dynein-arm docking complex in the flagella of *Chlamydomonas* mutants that lack outer dynein arms, Cell Motil. Cytoskelet. 48 (2001) 277–286.

[112] D.M. Casey, K. Inaba, G.J. Pazour, S. Takada, K. Wakabayashi, C.G. Wilkerson, R. Kamiya, G.B. Witman, DC3, the 21-kDa subunit of the outer dynein arm-docking complex (ODA-DC), is a novel EF-hand protein important for assembly of both the outer arm and the ODA-DC, Mol. Biol. Cell 14 (2003) 3650–3663.

[113] R. Kamiya, Functional diversity of axonemal dyneins as studied in *Chlamydomonas* mutants, Int. Rev. Cytol. 219 (2002) 115–155.

[114] M.R. Knowles, M.W. Leigh, L.E. Ostrowski, L. Huang, J.L. Carson, M.J. Hazucha, W. Yin, J.S. Berg, S.D. Davis, S.D. Dell, T.W. Ferkol, M. Rosenfeld, S.D. Sagel, C.E. Milla, K.N. Olivier, E.H. Turner, A.P. Lewis, M.J. Bamshad, D.A. Nickerson, J. Shendure, M.A. Zariwala, Genetic Disorders of Mucociliary Clearance Consortium, Exome sequencing identifies mutations in CCDC114 as a cause of primary ciliary dyskinesia, Am. J. Hum. Genet. 92 (2013) 99–106.

[115] A. Onoufriadis, T. Paff, D. Antony, A. Shoemark, D. Micha, B. Kuyt, M. Schmidts, S. Petridi, J.E. Dankert-Roelse, E.G. Haarman, J.M. Daniels, R.D. Emes, R. Wilson, C. Hogg, P.J. Scambler, E.M. Chung, UK10K, G. Pals, H.M. Mitchison, Splice-site mutations in the axonemal outer dynein arm docking complex gene CCDC114 cause primary ciliary dyskinesia, Am. J. Hum. Genet. 92 (2013) 88–98.

[116] R. Hjeij, A. Lindstrand, R. Francis, M.A. Zariwala, X. Liu, Y. Li, R. Damerla, G.W. Dougherty, M. Abouhamed, H. Olbrich, N.T. Loges, P. Pennekamp, E.E. Davis, C.M. Carvalho, D. Pehlivan, C. Werner, J. Raidt, G. Köhler, K. Häffner, M. Reyes-Mugica, J.R. Lupski, M.W. Leigh, M. Rosenfeld, L.C. Morgan, M.R. Knowles, C.W. Lo, N. Katsanis, H. Omran, ARMC4 mutations cause primary ciliary dyskinesia with randomization of left/right body asymmetry, Am. J. Hum. Genet. 93 (2013) 357–367.

[117] A.B. Dean, D.R. Mitchell, *Chlamydomonas* ODA10 is a conserved axonemal protein that plays a unique role in Outer Dynein Arm Assembly, Mol. Biol. Cell. 24 (2013) 3689–3696.

[118] R. Hjeij, A. Onoufriadis, C.M. Watson, C.E. Slagle, N.T. Klena, G.W. Dougherty, M. Kurkowiak, N.T. Loges, C.P. Diggle, N.F. Morante, G.C. Gabriel, K.L. Lemke, Y. Li, P. Pennekamp, T. Menchen, F. Konert, J.K. Marthin, D.A. Mans, S.J. Letteboer, C. Werner, T. Burgoyne, C. Westermann, A. Rutman, I.M. Carr, C. O'Callaghan, E. Moya, E.M. Chung, UK10K Consortium, E. Sheridan, K.G. Nielsen, R. Roepman, K. Bartscherer, R.D. Burdine, C.W. Lo, H. Omran, H.M. Mitchison, CCDC151 mutations cause primary ciliary dyskinesia by disruption of the outer dynein arm docking complex formation, Am. J. Hum. Genet. 95 (2014) 257–274.

[119] J. Wallmeier, H. Shiratori, G.W. Dougherty, C. Edelbusch, R. Hjeij, N.T. Loges, T. Menchen, H. Olbrich, P. Pennekamp, J. Raidt, C. Werner, K. Minegishi, K. Shinohara, Y. Asai, K. Takaoka, C. Lee, M. Griese, Y. Memari, R. Durbin, A. Kolb-Kokocinski, S. Sauer,

J.B. Wallingford, H. Hamada, H. Omran, TTC25 deficiency results in defects of the outer dynein arm docking machinery and primary ciliary dyskinesia with left-right body asymmetry randomization, Am. J. Hum. Genet. 99 (2016) 460–469.

[120] J.R. Panizzi, A. Becker-Heck, V.H. Castleman, D.A. Al-Mutairi, Y. Liu, N.T. Loges, N. Pathak, C. Austin-Tse, E. Sheridan, M. Schmidts, H. Olbrich, C. Werner, K. Häffner, N. Hellman, R. Chodhari, A. Gupta, A. Kramer-Zucker, F. Olale, R.D. Burdine, A.F. Schier, C. O'Callaghan, E.M. Chung, R. Reinhardt, H.M. Mitchison, S.M. King, H. Omran, I.A. Drummond, CCDC103 mutations cause primary ciliary dyskinesia by disrupting assembly of ciliary dynein arms, Nat. Genet. 44 (2012) 714–719.

[121] S.M. King, R.S. Patel-King, The oligomeric outer dynein arm assembly factor CCDC103 is tightly integrated within the ciliary axoneme and exhibits periodic binding to microtubules, J. Biol. Chem. 290 (2015) 7388–7401.

[122] M. Brand, C.P. Heisenberg, Y.J. Jiang, D. Beuchle, K. Lun, M. Furutani-Seiki, M. Granato, P. Haffter, M. Hammerschmidt, D.A. Kane, R.N. Kelsh, M.C. Mullins, J. Odenthal, F.J. van Eeden, C. Nüsslein-Volhard, Mutations in zebrafish genes affecting the formation of the boundary between midbrain and hindbrain, Development 123 (1996) 179–190.

[123] H. Omran, D. Kobayashi, H. Olbrich, T. Tsukahara, N.T. Loges, H. Hagiwara, Q. Zhang, G. Leblond, E. O'Toole, C. Hara, H. Mizuno, H. Kawano, M. Fliegauf, T. Yagi, S. Koshida, A. Miyawaki, H. Zentgraf, H. Seithe, R. Reinhardt, Y. Watanabe, R. Kamiya, D.R. Mitchell, H. Takeda, Ktu/PF13 is required for cytoplasmic pre-assembly of axonemal dyneins, Nature 456 (7222) (2008) 611–616.

[124] M. Taipale, N. Kaminen, J. Nopola-Hemmi, T. Haltia, B. Myllyluoma, H. Lyytinen, K. Muller, M. Kaaranen, P.J. Lindsberg, K. Hannula-Jouppi, J. Kere, A candidate gene for developmental dyslexia encodes a nuclear tetratricopeptide repeat domain protein dynamically regulated in brain, Proc. Natl. Acad. Sci. U.S.A. 100 (2003) 11553–11558.

[125] T.C. Bates, P.A. Lind, M. Luciano, G.W. Montgomery, N.G. Martin, M.J. Wright, Dyslexia and DYX1C1: deficits in reading and spelling associated with a missense mutation, Mol. Psychiatry 15 (2010) 1190–1196.

[126] C. Marino, A. Citterio, R. Giorda, A. Facoetti, G. Menozzi, L. Vanzin, M.L. Lorusso, M. Nobile, M. Molteni, Association of short-term memory with a variant within DYX1C1 in developmental dyslexia, Genes Brain Behav. 6 (2007) 640–646.

[127] K.G. Wigg, J.M. Couto, Y. Feng, B. Anderson, T.D. Cate-Carter, F. Macciardi, R. Tannock, M.W. Lovett, T.W. Humphries, C.L. Barr, Support for EKN1 as the susceptibility locus for dyslexia on 15q21, Mol. Psychiatry 9 (2004) 1111–1121.

[128] C. Marino, R. Giorda, M.L. Lorusso, L. Vanzin, N. Salandi, M. Nobile, A. Citterio, S. Beri, V. Crespi, M. Battaglia, M. Molteni, A family-based association study does not support DYX1C1 on 15q21.3 as a candidate gene in developmental dyslexia, Eur. J. Hum. Genet. 13 (2005) 491–499.

[129] H. Meng, K. Hager, M. Held, G.P. Page, R.K. Olson, B.F. Pennington, J.C. DeFries, S.D. Smith, J.R. Gruen, TDT-association analysis of EKN1 and dyslexia in a Colorado twin cohort, Hum. Genet. 118 (2005) 87–90.

[130] T.S. Scerri, S.E. Fisher, C. Francks, I.L. MacPhie, S. Paracchini, A.J. Richardson, J.F. Stein, A.P. Monaco, Putative functional alleles of DYX1C1 are not associated with dyslexia susceptibility in a large sample of sibling pairs from the UK, J. Med. Genet. 41 (2004) 853–857.

[131] R.A. Hoh, T.R. Stowe, E. Turk, T. Stearns, Transcriptional program of ciliated epithelial cells reveals new cilium and centrosome components and links to human disease, PLoS One 7 (2012) e52166.

[132] A.E. Ivliev, P.A. 't Hoen, W.M. van Roon-Mom, D.J. Peters, M.G. Sergeeva, Exploring the transcriptome of ciliated cells using in silico dissection of human tissues, PLoS One 7 (2012) e35618.

[133] A. Tarkar, N.T. Loges, C.E. Slagle, R. Francis, G.W. Dougherty, J.V. Tamayo, B. Shook, M. Cantino, D. Schwartz, C. Jahnke, H. Olbrich, C. Werner, J. Raidt, P. Pennekamp, M. Abouhamed, R. Hjeij, G. Köhler, M. Griese, Y. Li, K. Lemke, N. Klena, X. Liu, G. Gabriel, K. Tobita, M. Jaspers, L.C. Morgan, A.J. Shapiro, S.J. Letteboer, D.A. Mans, J.L. Carson, M.W. Leigh, W.E. Wolf, S. Chen, J.S. Lucas, A. Onoufriadis, V. Plagnol, M. Schmidts, K. Boldt, UK10K, R. Roepman, M.A. Zariwala, C.W. Lo, H.M. Mitchison, M.R. Knowles, R.D. Burdine, J.J. Loturco, H. Omran, DYX1C1 is required for axonemal dynein assembly and ciliary motility, Nat. Genet. 45 (2013) 995–1003.

[134] L.E. Ostrowski, W. Yin, T.D. Rogers, K.B. Busalacchi, M. Chua, W.K. O'Neal, B.R. Grubb, Conditional deletion of dnaic1 in a murine model of primary ciliary dyskinesia causes chronic rhinosinusitis, Am. J. Respir. Cell Mol. Biol. 43 (2010) 55–63.

[135] A. Sironen, N. Kotaja, H. Mulhern, T.A. Wyatt, J.H. Sisson, J.A. Pavlik, M. Miiluniemi, M.D. Fleming, L. Lee, Loss of SPEF2 function in mice results in spermatogenesis defects and primary ciliary dyskinesia, Biol. Reprod. 85 (2011) 690–701.

[136] R.J. Francis, A. Christopher, W.A. Devine, L. Ostrowski, C. Lo, Congenital heart disease and the specification of left-right asymmetry, Am. J. Physiol. Heart Circ. Physiol. 302 (2012) 2102–2111.

[137] S.Y. Tan, J. Rosenthal, X.Q. Zhao, R.J. Francis, B. Chatterjee, S.L. Sabol, K.L. Linask, L. Bracero, P.S. Connelly, M.P. Daniels, Q. Yu, H. Omran, L. Leatherbury, C.W. Lo, Heterotaxy and complex structural heart defects in a mutant mouse model of primary ciliary dyskinesia, J. Clin. Invest 117 (2007) 3742–3752.

[138] J. Freshour, R. Yokoyama, D.R. Mitchell, *Chlamydomonas* flagellar outer row dynein assembly protein ODA7 interacts with both outer row and I1 inner row dyneins, J. Biol. Chem. 282 (8) (2006) 5404–5412.

[139] J. Sullivan-Brown, J. Schottenfeld, N. Okabe, C.L. Hostetter, F.C. Serluca, S.Y. Thiberge, R.D. Burdine, Zebrafish mutations affecting cilia motility share similar cystic phenotypes and suggest a mechanism of cyst formation that differs from pkd2 morphants, Dev. Biol. 314 (2) (2007) 261e275.

[140] E. van Rooijen, R.H. Giles, E.E. Voest, C. van Rooijen, S. Schulte-Merker, F.J. van Eeden, LRRC50, a conserved ciliary protein implicated in polycystic kidney disease, J. Am. Soc. Nephrol. 19 (6) (2008) 1128–1138.

[141] N.T. Loges, H. Olbrich, A. Becker-Heck, K. Häffner, A. Heer, C. Reinhard, M. Schmidts, A. Kispert, M.A. Zariwala, M.W. Leigh, M.R. Knowles, H. Zentgraf, H. Seithe, G. Nürnberg, P. Nürnberg, R. Reinhardt, H. Omran, Deletions and point mutations of LRRC50 cause primary ciliary dyskinesia due to dynein arm defects, Am. J. Hum. Genet. 85 (6) (2009) 883–889.

[142] P. Duquesnoy, E. Escudier, L. Vincensini, J. Freshour, A.M. Bridoux, A. Coste, A. Deschildre, J. de Blic, M. Legendre, G. Montantin, H. Tenreiro, A.M. Vojtek, C. Loussert, A. Clément, D. Escalier, P. Bastin, D.R. Mitchell, S. Amselem, Loss-of-function mutations in the human ortholog of *Chlamydomonas reinhardtii* ODA7 disrupt dynein arm assembly and cause primary ciliary dyskinesia, Am. J. Hum. Genet. 85 (6) (2009) 890–896.

[143] M. Geremek, M. Bruinenberg, E. Ziętkiewicz, A. Pogorzelski, M. Witt, C. Wijmenga, Gene expression studies in cells from primary ciliary dyskinesia patients identify 208 potential ciliary genes, Hum. Genet. 129 (2011) 283–293.

[144] H.M. Mitchison, M. Schmidts, N.T. Loges, J. Freshour, A. Dritsoula, R.A. Hirst, C. O'Callaghan, H. Blau, M. Al Dabbagh, H. Olbrich, P.L. Beales, T. Yagi, H. Mussaffi, E.M. Chung, H. Omran, D.R. Mitchell, Mutations in axonemal dynein assembly factor DNAAF3 cause primary ciliary dyskinesia, Nat. Genet. 44 (2012) 381–389 S1-S2.

[145] C. Austin-Tse, J. Halbritter, M.A. Zariwala, R.M. Gilberti, H.Y. Gee, N. Hellman, N. Pathak, Y. Liu, J.R. Panizzi, R.S. Patel-King, D. Tritschler, R. Bower, E. O'Toole, J.D. Porath, T.W. Hurd, M. Chaki, K.A. Diaz, S. Kohl, S. Lovric, D.Y. Hwang, D.A. Braun, M.M. Schueler, R. Airik, E.A. Otto, M.W. Leigh, P.G. Noone, J.L. Carson, S.D. Davis, J.E. Pittman, T.W. Ferkol,

J.J. Atkinson, K.N. Olivier, S.D. Sagel, S.D. Dell, M. Rosenfeld, C.E. Milla, N.T. Loges, H. Omran, M.E. Porter, S.M. King, M.R. Knowles, I.A. Drummond, F. Hildebrandt, Zebrafish ciliopathy screen plus human mutational analysis identifies C21orf59 and CCDC65 defects as causing primary ciliary dyskinesia, Am. J. Hum. Genet. 93 (2013) 672–686.

[146] M.R. Knowles, L.E. Ostrowski, N.T. Loges, T. Hurd, M.W. Leigh, L. Huang, W.E. Wolf, J.L. Carson, M.J. Hazucha, W. Yin, S.D. Davis, S.D. Dell, T.W. Ferkol, S.D. Sagel, K.N. Olivier, C. Jahnke, H. Olbrich, C. Werner, J. Raidt, J. Wallmeier, P. Pennekamp, G.W. Dougherty, R. Hjeij, H.Y. Gee, E.A. Otto, J. Halbritter, M. Chaki, K.A. Diaz, D.A. Braun, J.D. Porath, M. Schueler, G. Baktai, M. Griese, E.H. Turner, A.P. Lewis, M.J. Bamshad, D.A. Nickerson, F. Hildebrandt, J. Shendure, H. Omran, M.A. Zariwala, Mutations in SPAG1 cause primary ciliary dyskinesia associated with defective outer and inner dynein arms, Am. J. Hum. Genet. 93 (2013) 711–720.

[147] M.A. Zariwala, H.Y. Gee, M. Kurkowiak, D.A. Al-Mutairi, M.W. Leigh, T.W. Hurd, R. Hjeij, S.D. Dell, M. Chaki, G.W. Dougherty, M. Adan, P.C. Spear, J. Esteve-Rudd, N.T. Loges, M. Rosenfeld, K.A. Diaz, H. Olbrich, W.E. Wolf, E. Sheridan, T.F. Batten, J. Halbritter, J.D. Porath, S. Kohl, S. Lovric, D.Y. Hwang, J.E. Pittman, K.A. Burns, T.W. Ferkol, S.D. Sagel, K.N. Olivier, L.C. Morgan, C. Werner, J. Raidt, P. Pennekamp, Z. Sun, W. Zhou, R. Airik, S. Natarajan, S.J. Allen, I. Amirav, D. Wieczorek, K. Landwehr, K. Nielsen, N. Schwerk, J. Sertic, G. Köhler, J. Washburn, S. Levy, S. Fan, C. Koerner-Rettberg, S. Amselem, D.S. Williams, B.J. Mitchell, I.A. Drummond, E.A. Otto, H. Omran, M.R. Knowles, F. Hildebrandt, ZMYND10 is mutated in primary ciliary dyskinesia and interacts with LRRC6, Am. J. Hum. Genet. 93 (2013) 336–345.

[148] A. Horani, T.W. Ferkol, D. Shoseyov, M.G. Wasserman, Y.S. Oren, B. Kerem, I. Amirav, M. Cohen-Cymberknoh, S.K. Dutcher, S.L. Brody, O. Elpeleg, E. Kerem, LRRC6 mutation causes primary ciliary dyskinesia with dynein arm defects, PLoS One 8 (2013) e59436.

[149] E. Kott, P. Duquesnoy, B. Copin, M. Legendre, F. Dastot-Le Moal, G. Montantin, L. Jeanson, A. Tamalet, J.F. Papon, J.P. Siffroi, N. Rives, V. Mitchell, J. de Blic, A. Coste, A. Clement, D. Escalier, A. Touré, E. Escudier, S. Amselem, Loss-of-function mutations in LRRC6, a gene essential for proper axonemal assembly of inner and outer dynein arms, cause primary ciliary dyskinesia, Am. J. Hum. Genet. 91 (2012) 958–964.

[150] J.B. Li, J.M. Gerdes, C.J. Haycraft, Y. Fan, T.M. Teslovich, H. May-Simera, H. Li, O.E. Blacque, L. Li, C.C. Leitch, R.A. Lewis, J.S. Green, P.S. Parfrey, M.R. Leroux, W.S. Davidson, P.L. Beales, L.M. Guay-Woodford, B.K. Yoder, G.D. Stormo, N. Katsanis, S.K. Dutcher, Comparative genomics identifies a flagellar and basal body proteome that includes the BBS5 human disease gene, Cell 117 (2004) 541–552.

[151] T.S. McClintock, C.E. Glasser, S.C. Bose, D.A. Bergman, Tissue expression patterns identify mouse cilia genes, Physiol. Genomics 32 (2008) 198–206.

[152] J.C. Xue, E. Goldberg, Identification of a novel testis-specific leucine-rich protein in humans and mice, Biol. Reprod. 62 (2000) 1278–1284.

[153] S. Powell, D. Szklarczyk, K. Trachana, A. Roth, M. Kuhn, J. Muller, R. Arnold, T. Rattei, I. Letunic, T. Doerks, L.J. Jensen, C. von Mering, P. Bork, eggNOG v3.0: orthologous groups covering 1133 organisms at 41 different taxonomic ranges, Nucleic Acids Res. 40 (2012) D284–D289 (Database issue).

[154] C.P. Diggle, D.J. Moore, G. Mali, P. zur Lage, A. Ait-Lounis, M. Schmidts, A. Shoemark, A. Garcia Munoz, M.R. Halachev, P. Gautier, P.L. Yeyati, D.T. Bonthron, I.M. Carr, B. Hayward, A.F. Markham, J.E. Hope, A. von Kriegsheim, H.M. Mitchison, I.J. Jackson, B. Durand, W. Reith, E. Sheridan, A.P. Jarman, P. Mill, HEATR2 plays a conserved role in assembly of the ciliary motile apparatus, PLoS Genet. 10 (2014) e1004577.

[155] A. Horani, T.E. Druley, M.A. Zariwala, A.C. Patel, B.T. Levinson, L.G. Van Arendonk, K.C. Thornton, J.C. Giacalone, A.J. Albee, K.S. Wilson, E.H. Turner, D.A. Nickerson, J. Shendure, P.V. Bayly, M.W. Leigh, M.R. Knowles, S.L. Brody, S.K. Dutcher, T.W. Ferkol, Whole-exome capture and sequencing identifies HEATR2 mutation as a cause of primary ciliary dyskinesia, Am. J. Hum. Genet. 91 (2012) 685–693.

[156] T. Paff, N.T. Loges, I. Aprea, K. Wu, Z. Bakey, E.C. Haarman, J.M.A. Daniels, E.A. Sistermans, N. Bogunovic, G.W. Dougherty, I.M. Höben, J. Große-Onnebrink, A. Matter, H. Olbrich, C. Werner, G. Pals, M. Schmidts, H. Omran, D. Micha, Mutations in PIH1D3 cause X-linked primary ciliary dyskinesia with outer and inner dynein arm defects, Am. J. Hum. Genet. (2017). http://dx.doi.org/10.1016/j.ajhg.2016.11.019.

[157] F. Dong, K. Shinohara, Y. Botilde, R. Nabeshima, Y. Asai, A. Fukumoto, T. Hasegawa, M. Matsuo, H. Takeda, H. Shiratori, T. Nakamura, H. Hamada, Pih1d3 is required for cytoplasmic preassembly of axonemal dynein in mouse sperm, J. Cell Biol. 20 (2014) 203–213.

[158] R. Yamamoto, M. Hirono, R. Kamiya, Discrete PIH proteins function in the cytoplasmic preassembly of different subsets of axonemal dyneins, J. Cell Biol. 190 (2010) 65–71.

[159] A.C. Merveille, E.E. Davis, A. Becker-Heck, M. Legendre, I. Amirav, G. Bataille, J. Belmont, N. Beydon, F. Billen, A. Cle´ment, C. Clercx, A. Coste, R. Crosbie, J. de Blic, S. Deleuze, P. Duquesnoy, D. Escalier, E. Escudier, M. Fliegauf, J. Horvath, K. Hill, M. Jorissen, J. Just, A. Kispert, M. Lathrop, N.T. Loges, J.K. Marthin, Y. Momozawa, G. Montantin, K.G. Nielsen, H. Olbrich, J.F. Papon, I. Rayet, G. Roger, M. Schmidts, H. Tenreiro, J.A. Towbin, D. Zelenika, H. Zentgraf, M. Georges, A.S. Lequarre´, N. Katsanis, H. Omran, S. Amselem, CCDC39 is required for assembly of inner dynein arms and the dynein regulatory complex and for normal ciliary motility in humans and dogs, Nat. Genet. 43 (2010) 72–78.

[160] S.S. Merchant, S.E. Prochnik, O. Vallon, E.H. Harris, S.J. Karpowicz, G.B. Witman, A. Terry, A. Salamov, L.K. Fritz-Laylin, L. Maréchal-Drouard, W.F. Marshall, L.H. Qu, D.R. Nelson, A.A. Sanderfoot, M.H. Spalding, V.V. Kapitonov, Q. Ren, P. Ferris, E. Lindquist, H. Shapiro, S.M. Lucas, J. Grimwood, J. Schmutz, P. Cardol, H. Cerutti, G. Chanfreau, C.L. Chen, V. Cognat, M.T. Croft, R. Dent, S. Dutcher, E. Fernández, H. Fukuzawa, D. González-Ballester, D. González-Halphen, A. Hallmann, M. Hanikenne, M. Hippler, W. Inwood, K. Jabbari, M. Kalanon, R. Kuras, P.A. Lefebvre, S.D. Lemaire, A.V. Lobanov, M. Lohr, A. Manuell, I. Meier, L. Mets, M. Mittag, T. Mittelmeier, J.V. Moroney, J. Moseley, C. Napoli, A.M. Nedelcu, K. Niyogi, S.V. Novoselov, I.T. Paulsen, G. Pazour, S. Purton, J.P. Ral, D.M. Riaño-Pachón, W. Riekhof, L. Rymarquis, M. Schroda, D. Stern, J. Umen, R. Willows, N. Wilson, S.L. Zimmer, J. Allmer, J. Balk, K. Bisova, C.J. Chen, M. Elias, K. Gendler, C. Hauser, M.R. Lamb, H. Ledford, J.C. Long, J. Minagawa, M.D. Page, J. Pan, W. Pootakham, S. Roje, A. Rose, E. Stahlberg, A.M. Terauchi, P. Yang, S. Ball, C. Bowler, C.L. Dieckmann, V.N. Gladyshev, P. Green, R. Jorgensen, S. Mayfield, B. Mueller-Roeber, S. Rajamani, R.T. Sayre, P. Brokstein, I. Dubchak, D. Goodstein, L. Hornick, Y.W. Huang, J. Jhaveri, Y. Luo, D. Martínez, W.C. Ngau, B. Otillar, A. Poliakov, A. Porter, L. Szajkowski, G. Werner, K. Zhou, I.V. Grigoriev, D.S. Rokhsar, A.R. Grossman, The *Chlamydomonas* genome reveals the evolution of key animal and plant functions, Science 318 (5848) (2007) 245–250.

[161] P. Satir, Studies on cilia. 3. Further studies on the cilium tip and a "sliding filament" model of ciliary motility, J. Cell Biol. 39 (1968) 77–94.

[162] K.E. Summers, I.R. Gibbons, Adenosine triphosphate-induced sliding of tubules in trypsin-treated flagella of sea-urchin sperm, Proc. Natl. Acad. Sci. U.S.A. 68 (1971) 3092–3096.

[163] A. Horani, S.L. Brody, T.W. Ferkol, D. Shoseyov, M.G. Wasserman, A. Ta-shma, K.S. Wilson, P.V. Bayly, I. Amirav, M. Cohen-Cymberknoh, S.K. Dutcher, O. Elpeleg, E. Kerem, CCDC65 mutation causes primary ciliary dyskinesia with normal ultrastructure and hyperkinetic cilia, PLoS One 8 (2013) e72299.

[164] L. Jeanson, L. Thomas, B. Copin, A. Coste, I. Sermet-Gaudelus, F. Dastot-Le Moal, P. Duquesnoy, G. Montantin, N. Collot, S. Tissier, J.F. Papon, A. Clement, B. Louis, E. Escudier, S. Amselem, M. Legendre, Mutations in GAS8, a gene encoding a nexin-dynein regulatory complex subunit, cause primary ciliary dyskinesia with axonemal disorganization, Hum. Mutat. 8 (2016) 776–785.

[165] H. Olbrich, C. Cremers, N.T. Loges, C. Werner, K.G. Nielsen, J.K. Marthin, M. Philipsen, J. Wallmeier, P. Pennekamp, T. Menchen, C. Edelbusch, G.W. Dougherty, O. Schwartz, H. Thiele, J. Altmüller, F. Rommelmann, H. Omran, Loss-of-Function GAS8 mutations cause primary ciliary dyskinesia and disrupt the nexin-dynein regulatory complex, Am. J. Hum. Genet. 97 (2015) 546–554.

[166] M. Wirschell, H. Olbrich, C. Werner, D. Tritschler, R. Bower, W.S. Sale, N.T. Loges, P. Pennekamp, S. Lindberg, U. Stenram, B. Carlén, E. Horak, G. Köhler, P. Nürnberg, G. Nürnberg, M.E. Porter, H. Omran, The nexin-dynein regulatory complex subunit DRC1 is essential for motile cilia function in algae and humans, Nat. Genet. 45 (2013) 262–268.

[167] G. Piperno, K. Mead, M. LeDizet, A. Moscatelli, Mutations in the "dynein regulatory complex" alter the ATP-insensitive binding sites for inner arm dyneins in *Chlamydomonas* axonemes, J. Cell Biol. 125 (5) (1994) 1109–1117.

[168] J.R. Colantonio, J. Vermot, D. Wu, A.D. Langenbacher, S. Fraser, J.N. Chen, K.L. Hill, The dynein regulatory complex is required for ciliary motility and otolith biogenesis in the inner ear, Nature 457 (2009) 205–209.

[169] N.R. Hutchings, J.E. Donelson, K.L. Hill, Trypanin is a cytoskeletal linker protein and is required for cell motility in African trypanosomes, J. Cell Biol. 156 (2002) 867–877.

[170] Z.P. Kabututu, M. Thayer, J.H. Melehani, K.L. Hill, CMF70 is a subunit of the dynein regulatory complex, J. Cell Sci. 123 (2010) 3587–3595.

[171] G. Rupp, M.E. Porter, A subunit of the dynein regulatory complex in *Chlamydomonas* is a homologue of a growth arrest-specific gene product, J. Cell Biol. 162 (2003) 47–57.

[172] R. Bower, D. Tritschler, K. Vanderwaal, C.A. Perrone, J. Mueller, L. Fox, W.S. Sale, M.E. Porter, The N-DRC forms a conserved biochemical complex that maintains outer doublet alignment and limits microtubule sliding in motile axonemes, Mol. Biol. Cell 24 (2013) 1134–1152.

[173] J. Lin, D. Tritschler, K. Song, C.F. Barber, J.S. Cobb, M.E. Porter, D. Nicastro, Building blocks of the nexin-dynein regulatory complex in *Chlamydomonas* flagella, J. Biol. Chem. 286 (2011) 29175–29191.

[174] G. Pigino, T. Ishikawa, Axonemal radial spokes: 3D structure, function and assembly, Bioarchitecture 2 (2012) 50–58.

[175] P. Yang, D.R. Diener, C. Yang, T. Kohno, G.J. Pazour, J.M. Dienes, N.S. Agrin, S.M. King, W.S. Sale, R. Kamiya, J.L. Rosenbaum, G.B. Witman, Radial spoke proteins of *Chlamydomonas* flagella, J. Cell Sci. 15 (2006) 1165–1174.

[176] J.M. Sturgess, J. Chao, J. Wong, N. Aspin, J.A. Turner, Cilia with defective radial spokes: a cause of human respiratory disease, N. Engl. J. Med. 11 (1979) 53–56.

[177] J. Lin, W. Yin, M.C. Smith, K. Song, M.W. Leigh, M.A. Zariwala, M.R. Knowles, L.E. Ostrowski, D. Nicastro, Cryo-electron tomography reveals ciliary defects underlying human RSPH1 primary ciliary dyskinesia, Nat. Commun. 5 (2014) 5727.

[178] L. Jeanson, B. Copin, J.F. Papon, F. Dastot-Le Moal, P. Duquesnoy, J. Cadranel, H. Corvol, A. Coste, J. Désir, A. Souayah, E. Kott, N. Collot, S. Tissier, B. Louis, A. Tamalet, J. de Blic, A. Clement, E. Escudier, S. Amselem, M. Legendre, RSPH3 mutations cause primary ciliary dyskinesia with central-complex defects and a near absence of radial spokes, Am. J. Hum. Genet. 97 (2015) 153–162.

[179] V.H. Castleman, L. Romio, R. Chodhari, R.A. Hirst, S.C. de Castro, K.A. Parker, P. Ybot-Gonzalez, R.D. Emes, S.W. Wilson, C. Wallis, C.A. Johnson, R.J. Herrera, A. Rutman, M. Dixon, A. Shoemark, A. Bush, C. Hogg, R.M. Gardiner, O. Reish, N.D. Greene, C. O'Callaghan, S. Purton, E.M. Chung, H.M. Mitchison, Mutations in radial spoke head protein genes RSPH9 and RSPH4A cause primary ciliary dyskinesia with central-microtubular-pair abnormalities, Am. J. Hum. Genet. 84 (2) (2009) 197–209.

[180] M.L. Daniels, M.W. Leigh, S.D. Davis, M.C. Armstrong, J.L. Carson, M. Hazucha, S.D. Dell, M. Eriksson, F.S. Collins, M.R. Knowles, M.A. Zariwala, Founder mutation in RSPH4A identified in patients of Hispanic descent with primary ciliary dyskinesia, Hum. Mutat. 34 (2013) 1352–1356.

[181] E. El Khouri, L. Thomas, L. Jeanson, E. Bequignon, B. Vallette, P. Duquesnoy, G. Montantin, B. Copin, F. Dastot-Le Moal, S. Blanchon, J.F. Papon, P. Lorès, L. Yuan, N. Collot, S. Tissier, C. Faucon, G. Gacon, C. Patrat, J.P. Wolf, E. Dulioust, B. Crestani, E. Escudier, A. Coste, M. Legendre, A. Touré, S. Amselem, Mutations in DNAJB13, encoding an HSP40 family member, cause primary ciliary dyskinesia and male infertility, Am. J. Hum. Genet. 99 (2016) 489–500.

[182] A. Frommer, R. Hjeij, N.T. Loges, C. Edelbusch, C. Jahnke, J. Raidt, C. Werner, J. Wallmeier, J. Große-Onnebrink, H. Olbrich, S. Cindrić, M. Jaspers, M. Boon, Y. Memari, R. Durbin, A. Kolb-Kokocinski, S. Sauer, J.K. Marthin, K.G. Nielsen, I. Amirav, N. Elias, E. Kerem, D. Shoseyov, K. Haeffner, H. Omran, Immunofluorescence analysis and diagnosis of primary ciliary dyskinesia with radial spoke defects, Am. J. Respir. Cell Mol. Biol. 53 (2015) 563–573.

[183] E. Kott, M. Legendre, B. Copin, J.F. Papon, F. Dastot-Le Moal, G. Montantin, P. Duquesnoy, W. Piterboth, D. Amram, L. Bassinet, J. Beucher, N. Beydon, E. Deneuville, V. Houdouin, H. Journel, J. Just, N. Nathan, A. Tamalet, N. Collot, L. Jeanson, M. Le Gouez, B. Vallette, A.M. Vojtek, R. Epaud, A. Coste, A. Clement, B. Housset, B. Louis, E. Escudier, S. Amselem, Loss-of-function mutations in RSPH1 cause primary ciliary dyskinesia with central-complex and radial-spoke defects, Am. J. Hum. Genet. 93 (2013) 561–570.

[184] A. Onoufriadis, A. Shoemark, M. Schmidts, M. Patel, G. Jimenez, H. Liu, B. Thomas, M. Dixon, R.A. Hirst, A. Rutman, T. Burgoyne, C. Williams, J. Scully, F. Bolard, J.J. Lafitte, P.L. Beales, C. Hogg, P. Yang, E.M. Chung, R.D. Emes, C. O'Callaghan, UK10K, P. Bouvagnet, H.M. Mitchison, Targeted NGS gene panel identifies mutations in RSPH1 causing primary ciliary dyskinesia and a common mechanism for ciliary central pair agenesis due to radial spoke defects, Hum. Mol. Genet. 23 (2014) 3362–3374.

[185] E. Ziętkiewicz, Z. Bukowy-Bieryłło, K. Voelkel, B. Klimek, H. Dmeńska, A. Pogorzelski, A. Sulikowska-Rowińska, E. Rutkiewicz, M. Witt, Mutations in radial spoke head genes and ultrastructural cilia defects in East-European cohort of primary ciliary dyskinesia patients, PLoS One 7 (2012) e33667.

[186] J. Raidt, J. Wallmeier, R. Hjeij, J.G. Onnebrink, P. Pennekamp, N.T. Loges, H. Olbrich, K. Häffner, G.W. Dougherty, H. Omran, C. Werner, Ciliary beat pattern and frequency in genetic variants of primary ciliary dyskinesia, Eur. Respir. J. 44 (2014) 1579–1588.

[187] B.E. Davy, M.L. Robinson, Congenital hydrocephalus in hy3 mice is caused by a frameshift mutation in Hydin, a large novel gene, Hum. Mol. Genet. 12 (2003) 1163–1170.

[188] H.R. Dawe, M.K. Shaw, H. Farr, K. Gull, The hydrocephalus inducing gene product, Hydin, positions axonemal central pair microtubules, BMC Biol. 5 (2007) 33.

[189] N.A. Doggett, G. Xie, L.J. Meincke, R.D. Sutherland, M.O. Mundt, N.S. Berbari, B.E. Davy, M.L. Robinson, M.K. Rudd, J.L. Weber, R.L. Stallings, C. Han, A 360-kb interchromosomal duplication of the human HYDIN locus, Genomics 88 (2006) 762–771.

[190] H. Olbrich, M. Schmidts, C. Werner, A. Onoufriadis, N.T. Loges, J. Raidt, N.F. Banki, A. Shoemark, T. Burgoyne, S. Al Turki, M.E. Hurles, UK10K Consortium, G. Köhler, J. Schroeder, G. Nürnberg, P. Nürnberg, E.M. Chung, R. Reinhardt, J.K. Marthin, K.G. Nielsen, H.M. Mitchison, H. Omran, Recessive HYDIN mutations cause primary ciliary dyskinesia without randomization of left-right body asymmetry, Am. J. Hum. Genet. 91 (2012) 672–684.

In this chapter

Severe skeletal abnormalities caused by defects in retrograde intraflagellar transport dyneins

Miriam Schmidts[1,2], Hannah M. Mitchison[3]
[1]Radboud University Medical Center, Nijmegen, The Netherlands; [2]University Hospital Freiburg, Freiburg, Germany; [3]University College London, UCL Great Ormond Street Institute of Child Health, London, United Kingdom

15.1 Introduction

Over the past two decades, detailed characterization of the evolutionarily ancient dynein transport mechanisms intrinsic to primary cilia has paralleled the discovery that they are important centers for cell signaling. Primary, nonmotile cilia (monocilia) are now well established as key sensory organelles in the body, found on the surface of most cells in eukaryotes, with essential roles in mammalian development. The mammalian primary cilium has evolved into a highly specialized cellular compartment sequestering multiple signaling proteins and coordinating their activities in response to a varied set of extracellular signaling cues specific to different tissues of the body [1]. The ciliary membrane is an organizing center extending out from the cell to interact with its extracellular surroundings; it is loaded with membrane-bound signaling receptors that detect external ligands and sensory input, triggering shifts within the cilium of downstream effector proteins that can activate intracellular gene expression responses for coordination of the cellular events required during normal embryonic development, and for tissue growth, repair, and homeostasis [2,3]. Cilia thereby regulate the cell's interaction with its external environment as well as intracellular processes including cell maintenance and cell division. Disruption of these essential processes through inherited mutations in ciliary genes causes a wide spectrum of multisystem human developmental syndromes called "ciliopathies" [4–7].

Since cilia lack an internal protein synthesis machinery, their varied functions—including the regulation of ciliary signaling—rely on intraflagellar transport (IFT) of cellular cargos between the cell and cilium, including microtubular proteins,

357

membrane receptors, and associated signaling molecules [8–11]. Primary cilia are supported by an internal axonemal cytoskeleton comprising nine microtubule doublets. IFT cargo moves in large trafficking complexes ("trains") along this axoneme, powered in opposite directions on different tracks of the doublets. On each "B-tubule," kinesin-2 motors traffic IFT-B complexes and cargos from the delivery site at the basal body in the anterograde direction (cilia base to tip), while on each "A-tubule" cytoplasmic IFT dynein-2 motors traffic IFT-A complexes plus cargo back out of the cilia in the retrograde direction [12,13]. In this way, large IFT "trains" composed of the motor, IFT complexes, cargo, and also the BBSome move around the cilium. The IFT-A and IFT-B complexes are adapters mediating motor–cargo interactions [12] and the BBSome was recently discovered to act as a cargo tether important in assembling and recycling IFT particles [14]. IFT is fully self-propagated within the axoneme and there is much evidence to show that retrograde IFT involving cytoplasmic IFT dynein-2 motors actively participates in signaling, by maintaining the ciliary composition but more specifically also by powering the ciliary export of signaling effectors for the relay of sensory information into the cell interior to activate signaling pathways [11,15].

Among the best characterized signaling modules in primary cilia are hedgehog (Hh) and noncanonical Wingless/Int1 (Wnt) signal transduction (or planar cell polarity, PCP), both embryo developmental pathways governed through the IFT-regulated entry and exit trafficking of their core components to and from the cilium [2,4]. The exact involvement of primary cilia in regulation of noncanonical Wnt signaling has been part of ongoing debate within the field [16]. Multiple other ciliary signal transduction pathways that are also proven or likely influenced by cytoplasmic IFT dynein-2 powered trafficking of their components include flow-responsive mechanosensory calcium signaling that has been proposed to function via cilia membrane–bound polycysteins [17], which may specify PCP in the brain [18]; trafficking of G protein–coupled receptors (GPCRs) [19]; and of receptor tyrosine kinase (RTK), Notch, PDGF, mammalian target of rapamycin complex 1 (mTORC1), transforming growth factor β (TGF-β), and other receptors [15,20,21,22], as well as signaling modules connected to light and odorant detection by photoreceptor and olfactory neuron cilia [1,23].

Therefore, defective IFT due to a variety of mutational causes not only results in disturbed basic cilia growth and maintenance but also severe cell signaling defects [11,15], with retrograde cytoplasmic IFT dynein-2 powered IFT likely having a bigger impact on signaling molecule trafficking than anterograde IFT. In hedgehog signaling, this may be because the Hh activator Smoothened (Smo) probably moves into cilia by diffusion rather than active transport [11,24].

The requirement for primary cilia in coordinating a broad range of signaling pathways responsible for cell proliferation and differentiation during normal embryonic development and for postnatal growth and maintenance of healthy tissues is clear, since defects of primary cilia cause a continually expanding group of ciliopathy diseases [5,25]. Ciliopathies are generally single gene disorders with diverse developmental and degenerative effects that include many skeletal dysplasias, nephronophthisis (NPHP), and other syndromes (Senior–Løken, Bardet–Biedl (BBS), Joubert (JBTS), Meckel–Gruber (MKS), orofaciodigital syndromes (OFD)). Ciliary signaling appears to be highly sensitive to even subtle ciliary changes that cause reduced cilia responses to Hh and other signaling ligands [3,26]. The importance and ubiquity in the body of primary cilia creates a wide ranging impact in medical terms, with organ and tissue involvement including skeletal dysplasia, cystic kidneys, retinal dystrophy, obesity, laterality defects, polydactyly, liver and pancreas disease, anosmia, neural tube closure, and patterning defects as well as central neurological and cognitive defects [27].

Ciliopathies tend to be rare conditions but collectively they form a disease category of major medical and economic impact that is markedly severe-lethal and involves patient care from multiple medical disciplines [28]. Here, we describe in more detail the group of primary ciliopathies that are caused by inherited mutations affecting cytoplasmic IFT dynein-2 components involved in the retrograde IFT system [11] as well as phenotypically overlapping ciliary conditions with skeletal involvement [29,30]. This devastating disease grouping generally manifests with a high perinatal and infant mortality rate and a significant extraskeletal disease burden including renal, retinal, and liver disease in surviving individuals. However, a proportion of patients surviving past adolescence can find some skeletal symptoms improve, for example, respiratory constriction can lessen [31,32].

Within the past 5 years, next-generation sequencing programs in patients combined with knowledge gained from model ciliated organisms, such as *Chlamydomonas,* zebrafish, and mice, as well as the human system, have led to step change progress in understanding the molecular basis of skeletal ciliopathy diseases; however, we describe how their variable presentation remains to be fully understood.

15.2 Role of cilia in skeletal development

The human skeleton continues to develop after birth throughout childhood. Cartilage and bones are produced from mesenchymal stem cell progenitors by ossification (osteogenesis) that continues past puberty and into the second decade of life. Two essential processes govern development of the skeletal system: (1) intramembranous ossification by which flat bones such as the skull, collarbone, and mandible develop directly from embryonic mesenchyme tissue and (2) endochondral ossification during which the other bones are formed onto previously formed cartilage "models" (anlagen) created by chondrocytes [33]. The temporary cartilage models originate from mesenchymal cells that differentiate into chondrocytes, which secrete cartilage extracellular matrix and finally, die to become replaced by bone after the invasion of blood vessels, bone marrow cells, osteoblasts, and osteoclasts [34].

Primary cilia were first described on chondrocytes during the 1960s–70s with a growing view over subsequent years that these complex organelles were unlikely to be simply vestigial structures [35]. Cilia have now been reported on mesenchymal stem cells and bone cells including osteoblasts and osteocytes [36–38]. Chondrocyte cilia show specific orientations in the bones, probably influenced by the surrounding extracellular matrix, and it has been suggested that this order can influence skeletal patterning and growth since it is disturbed in conditions such as osteoarthritis and tumors [38–40]. The fundamental and conserved role of cilia in multiple mechanosensation and chemosensation pathways required for skeletogenesis has become more apparent during recent years, since realization that skeletal dysplasias that reduced bone growth and caused lethality phenotypes in both animal models and latterly in ciliopathy patients, can arise from cilia defects.

There has been a great expansion of information coming from human studies as a result of the lowering costs and improved tractability of next-generation sequencing analysis whereby a large targeted gene panel, an entire exome or genome can be screened for mutations in multiple or all ciliary genes. Large-scale human genetic screens have revealed multiple pathogenic alleles in a number of components of the cytoplasmic IFT dynein-2 motor and associated IFT components (Table 15.1) [30,32,41–44]. Focusing on mutations of the cytoplasmic IFT dynein-2 motor causing human syndromes, we describe in this chapter the affected proteins and the clinical consequences of their dysfunction.

15.3 Clinical features of skeletal ciliopathies

15.3.1 General aspects

Skeletal ciliopathies are characterized by a spectrum of skeletal and extraskeletal findings, presenting with a wide range of symptom combinations that form a group of overlapping phenotypes, which can be difficult to distinguish, especially in fetal cases.

15.3.1.1 Skeletal findings

Common clinical features of skeletal ciliopathies include shortened ribs and a small or narrow thorax that gives rise to respiratory disturbances of variable

TABLE 18.1 Genes and phenotypes associated with skeletal ciliopathies

Condition	Inheritance	Skeletal phenotype	Renal phenotype	Retinopathy	Liver phenotype	Obesity	Developmental delay	Situs inversus	Other	Gene
Short-rib polydactyly syndromes (SRPS)	AR	Most often polydactyly, short ribs, shortened long bones, brachydactyly, abnormal pelvis configuration, sometimes oro-facial clefts	Often, NPHP-like or polycystic	Not evident in utero/at birth	Cysts and/or fibrosis may occur	–	NA (early lethality)	Rarely	Always lethal perinatal due to cardiorespiratory insufficiency resulting from severe thoracic constriction. Heart defects, gastrointestinal, cardiac defects occur.	**DYNC2H1, WDR34, WDR60, DYNC2LI1,** NEK1, WDR35, **DYNC2L1,** KIAA0586, INTU, WDPCP, C2CD3, ICK, IFT52, IFT81
Jeune asphyxiating thoracic dystrophy (JATD)	AR	Short ribs and extremities, sometimes polydactyly, abnormal pelvis configuration, scoliosis	Overall approximately 30% but mainly observed with IFT-gene mutations, rarely reported for IFT dynein-2 gene mutations	Similar to renal disease except C21orf2 mutations causing isolated retinopathy but no renal disease	Frequently elevated liver enzymes but rarely progression into liver failure	Single cases	Joubert-phenotype with CSPP1 and KIAA0586 mutations only, otherwise normal development	Not described	Often severe cardiorespiratory distress with ~30% lethality especially with IFT dynein-2 complex gene mutations; IFT gene mutations usually cause a mild rib phenotype but frequent renal and retinal disease. CEP120 gene mutations associated with OFD-like findings. Incomplete penetrance reported for TCTEX1D2 mutations.	**DYNC2H1, WDR34, WDR60, DYNC2LI1, TCTEX1D2,** IFT80, IFT172, IFT140, IFT144 (WDR19), IFT139 (TTC21 B), CEP120, CSPP1, KIAA0586, C21orf2
Mainzer–Saldino syndrome (MZSDS)	AR	Mildly shortened ribs, cone-shaped epiphyses	Always, mainly NPHP-like, rarely cystic	Always	Sometimes cholestasis and hepatic fibrosis	-	Single cases	Not described	Mild thorax phenotype, usually no cardiorespiratory lethality.	IFT140, IFT172
Axial spondylometaphyseal dysplasia (SMD)	AR	Short ribs, thorax deformities, hip dysplasia, scoliosis	–	Always	–	–	–	–		C21orf2
Endocrine–cerebro–osteo-dysplasia syndrome (ECO)	AR	Short limbs, poly(syn)dactyly, brachydactyly, clefts	Cystic	–	–	–	Possible	–	Brain malformations, facial dysmorphism.	ICK

Continued

Table 15.1 Genes and phenotypes associated with skeletal ciliopathies—cont'd

Condition	Inheritance	Skeletal phenotype	Renal phenotype	Retinopathy	Liver phenotype	Obesity	Developmental delay	Situs inversus	Other	Gene
Orofaciodigital syndrome (OFD)	AR or X-chromosome	Variable rib shortening often brachydactyly/ polydactyly, micromelia, orofacial clefts, disproportional small and oval tibia	Cystic dysplastic kidneys	–	Rarely cysts/ fibrosis	–	Brain malformations observed	–	Lobulated tongue. Coloboma, ambiguous genitalia, anal atresia, deafness have been observed.	TCTN3, OFD1, C5ORF42, TMEM231, TMEM216, TMEM107, TBC1D32, SCLT1, C2CD3, WDPCP, DDX59
Sensenbrenner syndrome (CED)	AR	Mildly shortened ribs, brachydactyly, craniosynostosis leading to dolichocephalus	Very often, mainly NPHP-like	Sometimes	Inconsistent hepatic cysts and fibrosis/ hepatic ductal plate malformation	–	Sometimes	Usually not	Mild thorax phenotype. Facial dysmorphism, thin/ sparse hair, nail and teeth dysplasia, heart defects.	IFT122, IFT144 (WDR19), IFT43, WDR35
Ellis–van Creveld syndrome (EVC)	AR	Short ribs and long bones, abnormal pelvis configuration, polydactyly of the hands	–	–	–	–	–	–	Hypoplastic nails, teeth abnormalities, heart defects.	EVC1, EVC2
Weyers acrofacial dysostosis	AD	Short stature, short extremities, polydactyly of the hands	–	–	–	–	–	–	Hypoplastic nails, teeth abnormalities.	EVC1, EVC2
Hydrolethalus syndrome	AR	Polydactyly, clefts, micrognathia	–	–	–	–	NA (lethal)	–	Midline defects, complex brain malformations.	KIF7
Acrocallosal syndrome	AR	Polydactyly	–	–	–	–	yes	–	Brain malformations, often absent corpus callosum, midline defects.	KIF7

AD, autosomal dominant; AR, autosomal recessive; NA, not applicable; X-chr., X-chromosomal. Cytoplasmic IFT dynein-2 genes highlighted in bold.

severity postnatally, short extremities, short stature, brachydactyly, and polydactyly. Craniofacial clefts can also be found in some short-rib polydactyly syndromes (SRPSs) and OFD cases.

15.3.1.2 Extraskeletal findings

The main clinical symptoms outside of the skeletal system consist of retinal
degeneration and NPHP-like [45] renal disease, although polycystic kidneys can
sometimes be observed in fetal cases. Liver disease is also sometimes reported,
predominantly cystic forms that subsequently results in fibrotic changes as well
as ductal plate malformations in fetal and newborn cases. Brain malformations
are very common in Hydrolethalus syndrome (HLS), Acrocallosal syndrome
(ACLS), and overlapping Joubert–Jeune syndrome cases while congenital heart
defects are mainly observed in Ellis–van Creveld syndrome (EVC). Ectodermal
defects such as hair, nail, and teeth abnormalities are usually only observed in
EVC and chondroectodermal dysplasia (CED) (Sensenbrenner syndrome).

Some specific radiological skeletal signs and extraskeletal features also discussed in more detail below can help to distinguish different forms of skeletal
ciliopathies. However, as a consequence of the phenotypic overlap between
some forms this can be a diagnostic challenge especially in fetal cases [46].
Very broadly, two main subgroups can be differentiated: inevitably lethal forms
versus phenotypes (potentially) compatible with life. It is widely accepted that
SRPS-labeled phenotypes are always lethal [29] as well as HLS and probably also
endocrine–cerebro–osteodysplasia (ECO) syndrome. In contrast, Jeune asphyxiating thoracic dystrophy (JATD), Mainzer–Saldino syndrome (MZSDS), CED,
OFD, EVC, and ACLS are generally compatible with life, although some individuals still do not survive. Lethality is mainly caused by cardiorespiratory failure due to a very narrow thorax and hypoplastic lungs in SRPS and JATD, heart
defects in EVC, and severe brain malformations in HLS and OFD. MZSDS and
CED, generally, have a milder thorax phenotype and therefore lethality rates are
very low in developed countries; instead the main disease burden is renal failure, retinal degeneration, and sometimes liver disease [43,47–49].

Table 15.1 has a comprehensive comparison of the overlapping clinical features of the skeletal chondrodysplasia disease group highlighting the most
important phenotypic features of the different skeletal ciliopathies. More
detailed descriptions of radiological findings are available in the clinical literature [29,30].

While a large and still increasing number of genes have been found to cause
skeletal ciliopathies when mutated, mutation in IFT dynein-2 genes seem to
specifically result in SRPS and JATD phenotypes. We will therefore focus on
these two conditions and their underlying molecular mechanisms in this chapter and also give a brief overview of phenotypically closely related and allelic
disorders.

15.3.2 Short-rib polydactyly spectrum syndrome

The SRPS spectrum includes five subphenotypes, SRPS I–V, which are all inevitably lethal in the perinatal or newborn period. These are SRPS-I (Saldino–Noonan type; OMIM 613091) [50]; SRPS-II (Majewski type; OMIM 263520) [51]; SRPS-III (Verma–Naumoff type; OMIM 613091) [52]; SRPS-IV (Beemer–Langer type; OMIM 269860) [53]; and SRPS-V (OMIM 614091) [54]. Their principle features are described below.

15.3.2.1 SRPS type I (Saldino–Noonan) and type III (Verma–Naumoff)

These subtypes are often classified together due to their phenotypic resemblance, including trident acetabulum with spurs, somewhat similar to the pattern found in JATD and EVC. SRPS-I affected individuals display extremely short extremities (labeled "flipper-like") and polydactyly is frequently observed together with extremely severe rib shortening. Additional findings include reduced ossification of the hand carpal bones, lateral bony spurs on the long bones ("banana peeling appearance") in SRPS-III or rounded metaphyses and urogenital malformations in SRPS-I. Malformations of the central nervous system (CNS) as well as the epiglottis and larynx may also occur [50,55–60].

15.3.2.2 SRPS type II (Majewski)

In contrast to SRPSs I, III, and IV, the pelvis in SRPS-II appears normal. In addition to severely shortened ribs, affected individuals suffer from an extremely shortened "ovoid" tibia that is shorter than the fibula and from premature ossification of the carpal bones. The dysplastic tibia is also a feature of OFD-IV (Mohr–Majewski syndrome, Baraitser–Burn Syndrome), which may result in problems to distinguish between fetal cases of these two conditions. Extraskeletal findings include urogenital malformations, hypoplasia of the epiglottis and larynx, as well as orofacial clefts [51,56,57,61].

15.3.2.3 SRPS type IV (Beemer–Langer)

Polydactyly is less common in SRPS-IV while other symptoms such as very short ribs, protuberant abdomen, and a hydropic fetal appearance resemble that of SRPS-II cases. Bowing of the long bones is also observed, especially affecting the radius and ulna of the forearm. The disproportionally short tibia found in SRPS-II is not a feature of SRPS-IV and this can help in distinguishing between them. The pelvis displays a hypoplastic ileum and trident appearance of the acetabulum. Extraskeletal findings include renal abnormalities such as cysts and renal hypoplasia, pancreatic cysts, and CNS malformations [53,57,62–64].

15.3.2.4 SRPS type V

This is an extremely rare SRPS phenotype including poly(syn)dactyly, cleft palate, and laterality defects such as polysplenia and intestinal malrotation, in addition to severe rib shortening and renal malformations. Campomelia and striking acromelic hypomineralization of the bones distinguishes this subtype from other forms of SRPS [54,65].

15.3.3 Jeune asphyxiating thoracic dystrophy (Jeune syndrome)

Asphyxiating thoracic dystrophy (ATD, Jeune syndrome; OMIM 208500) is classed within the SRPS spectrum but with a reduced lethality, since usually the thorax narrowing due to short ribs is milder than in the classic forms of SRPS described above. Polydactyly is also infrequent, in contrast to SRPS where it is a rather consistent feature. Some radiological features such as a specific pelvis appearance (trident acetabulum with spurs) are, however, shared between some subtypes of "proper" SRPS and JATD, as well as the general phenotype of shortened ribs, so that JATD can be regarded as at the mild end of SRPS spectrum. Similar to the milder skeletal phenotype, extraskeletal organ involvement including malformations of the brain, heart, kidneys, liver, and pancreas that are frequently observed in SRPS in prenatal ultrasound examinations, are less common in JATD.

JATD is characterized by shortened ribs resulting in a smaller than usual, sometimes bell-shaped thorax and hypoplastic lungs leading to sometimes severe respiratory distress, mainly during the first 2 years of life. Estimates of the frequency of lethality in JATD range from 20% to 60% and this is often linked to acute airway infections [32,41,66]. In addition to shortened ribs and extremities, there are pelvis abnormalities similar to those observed in SRPS-I, SRPS-III, and EVC, with trident acetabulum with spurs being present as well as so-called "handlebar clavicles" and cone-shaped epiphyses of the fingers and toes on radiographs. Polydactyly sometimes occurs as well as NPHP-like (and rarely cystic) kidney involvement causing renal failure, which is observed in about 20%–30% of JATD-affected individuals.

From human genetic studies it has become clear that these extraskeletal features are associated with certain patient genotypes such as mutations in *IFT140*, *IFT144*, and *IFT172* while being rare occurrences in cases carrying other genetic defects, including cytoplasmic IFT dynein-2 gene mutations. In contrast, JATD with isolated retinal degeneration occurs in cases with *C21orf2* mutations and this is therefore allelic to axial spondylometaphyseal dysplasia (SMD) [79]. Pancreatic lesions [67], elevated liver enzymes [47,68], and brain malformations [47,69,70] have also been described in JATD. The latter are rare in JATD, but common in another ciliopathy, JBTS, manifesting as a so-called molar tooth sign in the brain. This hallmark feature of JBTS is an abnormality detectable in brain-imaging studies such as magnetic resonance imaging (MRI), which results from the abnormal development (hypoplasia) of the cerebellar vermis and brainstem regions. It is therefore not surprising that cases of JATD with JBTS-like brain malformations both can carry mutations in *CSPP1*, also causing JBTS itself [70,71]. JATD with OFD-like features such as cleft palate has also been

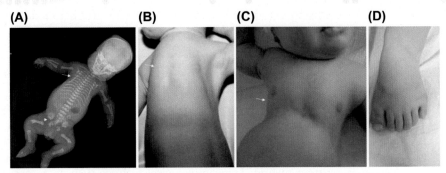

Figure 15.1 Clinical hallmarks of cytoplasmic dynein-2–associated skeletal ciliopathies. (A) Short-rib polydactyly phenotype with short horizontal ribs causing severe thoracic narrowing (*arrows*), shortened long bones and pelvis dysplasia with acetabular spurs (*arrow head*). (B, C) Examples of narrow dysplastic thorax configuration (*arrows*) in Jeune syndrome. (D) Polydactyly affecting the toes in a case of Jeune syndrome. *(A) Reproduced from McInerney-Leo et al. (2013); (B) reproduced from Schmidts et al. (2013); (C) reproduced from Halbritter et al. (2013); (D) reproduced from Schmidts et al. (2015).*

described to arise as a consequence *CEP120* mutations, which can also cause OFD, JBTS, and MKS syndrome phenotypes [72,73].

Fig. 15.1 shows the clinical hallmarks of skeletal ciliopathies that are caused by inherited mutations affecting the cytoplasmic IFT dynein-2 (SRPS and JATD).

15.3.4 Mainzer–Saldino syndrome

MZSDS (OMIM 266920) is defined as the combination of narrow thorax with impaired renal function due to NPHP-like or (poly) cystic disease as well as retinal disease. It resembles JATD very closely phenotypically and they are genetically allelic, although not for cytoplasmic IFT dynein-2 gene mutations but rather for mutations in IFT-A and IFT-B complex genes, *IFT140* and *IFT172* namely [43,47,48]. Both conditions are also called "cono-renal syndrome," due to cone-shaped epiphyses of the fingers and toes in X-rays. However, the MZSDS rib phenotype is usually mild and mechanical ventilation is usually not required [57,74–76]. Liver disease, sometimes with cholestasis as well as liver fibrosis, has also been described for some MZSDS cases [43,47]. Some of the originally reported MZSDS cases were also found to display ataxia [74,77].

15.3.5 Axial spondylometaphyseal dysplasia

Axial SMD (OMIM 602271) is allelic to JATD and is therefore mentioned in this chapter despite not being caused by mutations in a cytoplasmic IFT dynein-2 gene, but in the *C21orf2* gene encoding a protein localizing to the ciliary base [78]. The phenotype is of a ciliopathy closely resembling JATD with the retina and skeleton affected, including the typical small thorax, short extremities, and short stature. However, unlike in cases of JATD caused by *C21orf2* mutations [79], the anterior ends of the ribs appear cupped in SMD and there are irregular proximal femur epiphyses as well as malformed vertebrae (platyspondyly) that are not observed in JATD. Furthermore, although the pelvis configuration appears dysplastic, this is in contrast to JATD where a trident configuration with acetabular spurs is very common, while in SMD the iliac crest appears lacy. SMD-affected individuals develop retinal degeneration (retinitis pigmentosa, cone-rod dystrophy) similar to that observed for some JATD cases. Interestingly, renal or hepatic involvement has not been observed in either JATD or SMD cases resulting from *C21orf2* mutations, unlike in JATD cases caused by IFT-mutations (e.g. *IFT140* or *IFT172*). Additional SMD symptoms include frontal bossing and craniofacial dysmorphism [80,81].

15.3.6 Endocrine–cerebro–osteodysplasia syndrome

ECO Syndrome (OMIM 612651) was first described in affected individuals from the Amish community [82] and more recently was found to be allelic to SRPS due to mutations in *ICK* encoding intestinal cell kinase [83,84]. The main clinical features are endocrine disturbances, cerebral malformations consisting mainly not only hydrocephalus but also holoprosencephaly, absent corpus callosum, and cortical malformations, as well as skeletal abnormalities including facial clefting, short extremities, bowed forearms, ulnar deviation, poly(syn)dactyly, brachydactyly. Facial dysmorphism includes braid nasal bridge, dysplastic ears, and fused eyelids. Renal cysts are among the extraskeletal findings.

15.3.7 Orofaciodigital syndrome

The OFD spectrum of phenotypes is comparatively wide, with at least 14 subtypes to date characterized by clinical and/or genetic means. Some are lethal and especially OFD-IV (Mohr–Majewski syndrome, Baraitser–Burn Syndrome; OMIM 258860) phenotypically resembles SRPS since fetuses show a dysplastic tibia similar to SRPS-II. Thoracic narrowing of variable degree occurs as well as pre- and postaxial polydactyly, cleft lip/palate, lobulated tongue, cystic dysplastic kidneys and liver, and brain involvement [85]. Severe bilateral deafness [86] and eye coloboma [87] have also been described.

Other OFD subtypes display less clinical overlap with cytoplasmic IFT dynein-2-related phenotypes and will therefore not be described in detail here. However, shortened ribs, although not a common feature in OFD except OFD-IV, have been described in some fetal cases as well as limb shortening, brachydactyly, orofacial clefts, fibular agenesis (OFD 10), and polydactyly [88]. Apart from *OFD1* mutation–carrying males and some individuals with *C2CD3* mutations, OFD is usually compatible with life [88–90].

15.3.8 Sensenbrenner syndrome (chondroectodermal dysplasia)

Sensenbrenner syndrome (CED; OMIM 218330) is not caused by cytoplasmic IFT dynein-2 mutations itself but is another example of a ciliopathy allelic to

JATD, since in both conditions mutations in the IFT-A complex gene *IFT144/WDR19* have been identified [91]. CED is compatible with life and is fairly easily distinguished from JATD and MZSDS by multiple characteristic facial dysmorphic features such as dolichocephalus that result from craniosynostosis of the sutura sagittalis, downward-pointing palpebral fissures, small mouth, low set ears, and especially notably by ectodermal defects including nail, hair, and teeth abnormalities in addition to skin laxity causing associated with hernias [92,93].

The skeletal phenotype, however, can also be difficult to distinguish from JATD and MZSDS, due to similar short extremities including brachydactyly and variable narrowing of the ribcage; however, the thorax phenotype is notably rather mild, usually not requiring hospitalization and not causing lethality [49,94]. Extraskeletal organ involvement is fairly similar to JATD again, including an NPHP-like renal phenotype [95,96] as well as liver cysts, fibrosis, and ductal plate malformation [97–99] and less frequently congenital heart defects or retinal degeneration [49].

15.3.9 Ellis–van Creveld syndrome and Weyers acrofacial dysostosis

EVC (OMIM 225500) does not represent a cytoplasmic IFT dynein-2 mutation-caused disease entity but is another ciliopathy with phenotypes that can be strikingly similar to JATD. Very rare elsewhere in the world, this condition is more common among the Amish population in Pennsylvania. EVC is mainly characterized by shortened ribs, a dysplastic pelvis with trident appearance with spurs similar to that observed in JATD and some SRPS forms, polydactyly of the hands but not feet, dysplastic fingers and toenails, teeth abnormalities (prenatal eruption of teeth, hypodontia, and malformed teeth), and cardiac defects in many cases, primarily atrial septation defects [100–102]. Cardiac defects may give a clue to the specific diagnosis as well as ectodermal defects that are not a feature of JATD. Also, JATD patients can present in older life with additional renal, liver, and retinal disease, but this is not usually observed in EVC [103].

Weyers acrofacial dysostosis, also called Curry–Hall syndrome, is allelic to EVC, but is inherited in an autosomal-dominant pattern [104]. Features include polydactyly of the hands, teeth abnormalities, dystrophic nails, as well as short stature with short extremities, but not narrow thorax or heart defects [104–107].

15.3.10 Hydrolethalus and acrocallosal syndromes

HLS (OMIM 236680) is allelic to ACLS (OMIM 2009900) as well as JBTS caused by *KIF7* mutations and very recently, a Hydrolethalus phenotype has been classified as allelic to SRPS as well, due to newly identified defects in *KIAA0586* [108–111]. The main hallmarks of this group are polydactyly, sometimes with hallux duplication in combination with very severe brain malformations such as anencephaly, hydrocephalus, and cerebellar vermis hypoplasia that are incompatible with postnatal life. Cleft palate and micrognathia have also been described [108,112].

Although genetically overlapping, ACLS in contrast to HLS is compatible with life. Affected individuals show craniofacial dysmorphism with prominent forehead and hypertelorism as well as polydactyly/hallux duplication. The main extraskeletal features include developmental delay and brain malformations such as severe hypoplasia or agenesis of the corpus callosum with or without Dandy–Walker malformation [112,113].

15.4 Cell biological basis of skeletal ciliopathies due to IFT defects

Disturbance of IFT with malfunction of Hh and other signaling networks and the consequent devastating effects of impaired skeleton development and tumors has been well reviewed in mouse models and human skeletal ciliopathy diseases [11,38,40]. Much of the earliest hints of cilia being of importance for skeletal development and Hh signaling came from EVC mouse models that display polydactyly, short ribs, short long bones, and exhibit changes in chondrocyte maturation and differentiation affecting cartilage development [114]. Similar findings are established for IFT mutants and a number of recent reviews are available summarizing the effect of individual mutations in IFT and the cytoplasmic IFT dynein-2 motor in mouse models and human disease [11,40]. For example, while mutations in cytoplasmic IFT dynein-2 and IFT-B complex proteins cause defects in Shh signaling, mouse crosses suggest that depletion of IFT-B may suppress the Shh and cilia defects caused by the motor mutations, indicating that this system is finely balanced [115]. It remains unknown why human skeletal ciliopathies are primarily associated with retrograde, dynein IFT defects rather than anterograde, kinesin IFT defects. This could reflect that anterograde defects are less compatible with survival past early human gestation [116], or that retrograde trafficking is more important for the signaling processes required for skeletal development; it is suggested that IFT-A proteins in ciliogenesis appear to have more complex functions than those of IFT-B proteins [26].

A simplified mechanistic model of how human ciliary defects affect skeletal development is shown in Fig. 15.2. Dynein-driven cycling of proteins from the ciliary tip back to the base by the retrograde motor is in part a dynamic kinesin–dynein balance process that determines ciliary growth and functioning, where IFT of signaling molecules plays a central role as a driver most notably for skeletal ciliopathies in transduction of Hh signaling. The cilia-regulated movements of the Hh Smo receptor and its downstream Gli effector molecules depend on functioning IFT for ciliary targeting and removal [117]. In the absence of IFT,

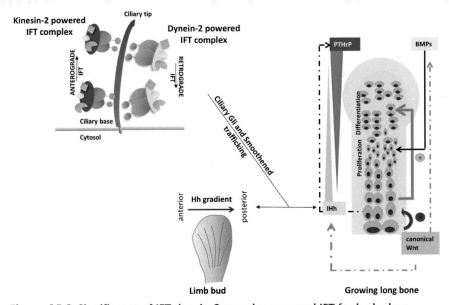

Figure 15.2 Significance of IFT dynein-2 complex powered IFT for hedgehog signaling, limb development, and bone growth. Cytoplasmic IFT dynein-2 is the motor complex for retrograde IFT, shown here as a gray-colored IFT "train" moving from the ciliary tip back to the base. The IFT dynein-2 motor moves hedgehog signaling pathway components in/out of the cilium including Smoothened and Gli molecules. A hedgehog signaling gradient is established along the anterior–posterior limb axis during development that specifies the phalanges. Hedgehog signaling is also required for long bone development, stimulating chondrocyte proliferation within the growth plate, regulated by a feedback loop connecting chondrocyte proliferation and differentiation pathways (see main text for details) and interconnected with other cell signaling pathways playing a prominent role that include canonical Wnt signaling and BMP signaling [33,190].

Smo accumulation is insufficient to activate Hh [118] and this is at least partially responsible for chondrocyte maturation defects in bone growth plates that underlie skeletal ciliopathy phenotypes [119]. Hh signal transduction is activated by binding of Shh ligand to its ciliary membrane-bound receptor Patched 1 (Ptc1), which inhibits Ptc1 cilia localization. The translocation of Ptc1 out of the cilium causes the cilia membrane GPCR protein Smo to enter, which activates the pathway through two central effector switches, the activation of ciliary GLI transcription factors and the suppression of the negative regulator SUFU. Together this switches on a cascade of downstream nuclear gene expression [120–123].

As outlined below in more detail, genetic mutations affecting components of the IFT system, most often the retrograde cytoplasmic IFT dynein-2 motor, are currently understood to be a primary cause of skeletal ciliopathy disorders. Hh signaling coordinates many patterning processes during early embryonic development, with both limb patterning and endochondral ossification, particularly, notable IFT- and Hh-dependent processes. Sonic (Shh) and Indian hedgehog (Ihh) provide positional information and trigger cell differentiation programs underlying the formation of cartilage and bone. These signaling components can signal between cells and create long-distance gradients that regulate skeletal patterning during bone growth [124]. During mammalian limb development, a hedgehog signaling gradient is established along the anterior–posterior limb axis that determines finger and toe development. Defective IFT and Hh signaling can therefore result in an incorrect number of phalanges developing (mainly polydactyly) [121,121a] (Fig. 15.2).

In addition to its role in polydactyly, Hh signaling is also thought to play a crucial role in long bone development: within the growth plate area, chondrocyte proliferation is stimulated by Ihh signaling and low levels of parathyroid hormone-related protein (PTHrP) while high levels of PTHrP and low levels of Ihh signaling promote chondrocyte differentiation. In this way, a feedback loop connects the chondrocyte proliferation and differentiation pathways (Fig. 15.2). IFT dynein-2 mutation-related Hh defects likely cause premature chondrocyte differentiation and reduced proliferation at the bone growth plates. Other cell signaling pathways are also involved, for example, PDGF signaling influences mesenchymal and bone formation [22]. Canonical Wnt signaling and BMP signaling also play a prominent role [33,190].

Null mouse IFT mutants do not generally survive past midgestation, but conditional knockouts (for example, for *Kif3a* or *Ift88*) show digit patterning abnormalities due to aberrant Shh and disrupted endochondral bone formation due to disrupted Ihh signaling [125,126]. Similarly, a presumed hypomorphic *Ift80* gene trap allele was shown to result in polydactyly as well as shortened long bones and ribs in mice [119] while *IFT80* disruption in osteoblast precursors affects their differentiation capacity and subsequently bone density [127].

15.5 Genetic basis of dynein-based skeletal ciliopathies associated with IFT defects

Noticeable phenotypic overlap exists between human ciliary chondrodysplasias across the short-rib polydactyly spectrum as well as other ciliopathies such as CED, OFD, JBTS, which parallel a significant genetic overlap as shown in Fig. 15.3 and Table 15.1. Patients affected by these different syndromes can carry mutations in the same genes, encoding IFT proteins as well as proteins localizing to the ciliary base, e.g., *CSPP1*, *CEP120*, and *KIAA0536*. The overlapping functional aspects of proteins affected by skeletal ciliopathy mutations are shown in Fig. 15.4. Precisely how mutations affecting many

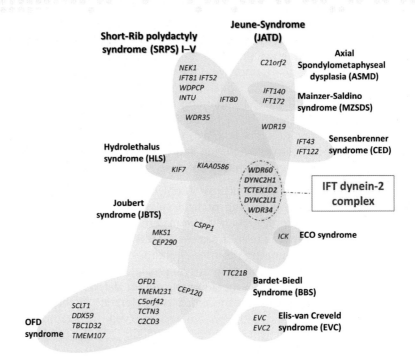

Figure 15.3 Overlapping genetic causes of the spectrum of skeletal ciliopathies. Cytoplasmic IFT dynein-2 complex genes (in red) cause both short-rib polydactyly syndromes and Jeune Syndrome. Some genes cause multiple syndromes such as *C2CD3* in cases of OFD, SRPS-like, and JBTS [89,191] and *CEP120* mutations in JBTS, JATD, and OFD [72,73]. JBTS and BBS can be caused by a large number of other different genes [70,121,192] not displayed here for space constraint reasons, as these two conditions are not classical skeletal ciliopathies.

distinct ciliary genes result in the same phenotypes is not yet clear, though their encoded proteins can have overlapping functions. Conversely, mutations in a shared gene can cause different syndromes. Examples of this include JATD and SRPS both resulting from IFT dynein-2 mutations; JATD, MZSDS, and CED from IFT complex mutations; JATD and JBTS from *CSPP1* and *CEP120* mutations; JATD and SMD from *C21orf2* mutations; SRPS-like, JBTS and OFD from *C2CD3* mutations; and SRPS and HLS from *KIAA0586* mutations. Potentially, modifier alleles or the genetic background on which such mutations occur may play a significant role in this phenotypic variability; epigenetic and stochastic factors may also contribute.

Despite the complexity of allelic disorders and phenotypically similar conditions that contribute to the spectrum of skeletal ciliopathies, some genetic and clinical distinctions can also be made. Mutations in genes encoding IFT dynein-2 subunits such as DYNC2H1, WDR34, WDR60, DYNC2LI1, and TCTEX1D2 seem

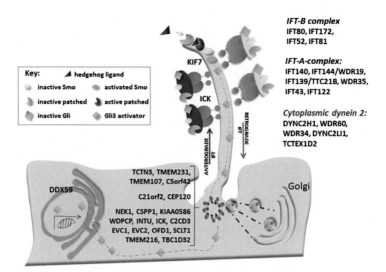

Figure 15.4 Ciliary localization and function of proteins mutated in skeletal ciliopathies. The main causes of skeletal ciliopathies are mutations in genes encoding proteins of the intraflagellar transport (IFT) machinery that are usually found along the ciliary axoneme as well as at the ciliary base. These comprise proteins of IFT complex B (IFT172, IFT80, IFT52, IFT81), IFT complex A (IFT140, IFT144/WDR19, IFT122, IFT43, WDR35, IFT139/TTC21B), the retrograde motor complex cytoplasmic IFT dynein-2 (DYNC2H1, WDR60, WDR34, DYNC2LI1, TCTEX1D2), and proteins found at the ciliary base (C21orf2, CEP120) [30,79]. Mutations result in JATD, SRPS, and CED. ICK, the causal gene for ECO syndrome and SRPS, localizes to the ciliary axoneme and base [83,84,193]. Mutations affecting a number of proteins localizing to the base of the cilium (basal body, peribasal body region, or pericentriolar matrix) also cause skeletal ciliopathies: NEK1 [131,156], CSPP1 [70], WDPCP, INTU [194,195], C2CD3 [89,191], TBC1D32 [196,197], EVC/EVC2 [198,199], KIAA0586 [108,110,111,200], and TMEM216 [201]. TCTN3, C5orf42, TMEM231, and TMEM107 mutated in OFD localize to the ciliary transition zone [202–205] while DDX59, also causing OFD when defective, is found in the cytoplasm and nucleus [206]. Many of these proteins seem to be essential for proper ciliary localization and further processing of the hedgehog signaling pathway components Patched, Smoothened (Smo), and Gli. Pathway activation requires binding of the hedgehog ligand to the pathway inhibitor of Smo, Patched, at the ciliary tip, which results in release of Smo. Activated Smo triggers Gli3 activation and Gli3 translocates to the ciliary base using IFT, from where it travels further into the nucleus to influence transcription. The binding of Patched to hedgehog ligand enhances its translocation to the cell body and Smo translocation to the cilium.

to specifically result in a JATD or SRPS specific phenotype. Genotype–phenotype correlation studies in SRPS, JATD, MZSDS, and CED also seem to suggest that mutations in IFT complex A components as well as the IFT complex B component IFT172 cause a milder rib phenotype in comparison to IFT dynein-2 mutations. Conversely, gene mutations in IFT-A and IFT-B more often than mutations in IFT dynein-2 give rise to clinically relevant extraskeletal symptoms including renal, hepatic, and retinal disease [43,47–49].

Skeletal ciliopathies, like most other ciliopathies, are usually inherited in an autosomal-recessive fashion with some exceptions including Weyers acrodental dysostosis, which is autosomal-dominant and some X-chromosomal OFD subtypes (OFD-I, OFD-VII, and OFD-VIII) [88]. Within the context of autosomal-recessive inheritance, the influence of additional alleles in ciliopathies has also been presented, firstly for BBS where Katsanis et al. proposed a "triallelic inheritance" pattern [42,128], followed by similar findings for retinal conditions and polycystic kidney disease [129,130]. While it is not in doubt that the genetic background influences skeletal ciliopathy phenotypes in humans with similar findings shown for many conditions in mice, proof that a certain allele has a specific effect has not yet been proven. Such proof is difficult to achieve since large-scale modifier studies would be needed and in contrast to investigations in mice, these cannot easily be conducted for rare human diseases.

Triallelic inheritance has not yet been described in the literature for skeletal ciliopathies, though digenic inheritance of heterozygous mutations in two different genes causing a phenotype seems possible but probably is a rare event. To date this was only described in a single case of SRPS-II with a heterozygous mutation in *NEK1* as well as *DYNC2H1* [131]. Likewise, it seems possible that protective modifier alleles may exist, one example at least being in the case of potentially partially redundant IFT dynein-2 light chain mutations identified in TCTEX1D2. *TCTEX1D2* null alleles can cause JATD in families exhibiting incomplete penetrance [132]. There are multiple examples of severe and milder skeletal ciliopathy phenotypes within the same family in affected siblings carrying the same biallelic primary mutations that range from lethal to compatible with life, contributing different phenotypic aspects across the severity spectrum, which might be explained by inheritance of different modifiers [42,79].

15.6 Human mutations in cytoplasmic IFT dynein-2 genes

In retrograde IFT, the cytoplasmic IFT dynein-2 motor exclusively transports cargos along the ciliary axoneme and—with rather confusingly terminology—is not required for transport along cytoplasmic microtubules, a function that is performed by the cytoplasmic dynein-1 motor. The term "cytoplasmic" has been assigned to distinguish the IFT dynein-2 proteins from axonemal dyneins that are required for ciliary movement, such as DNAH5 and DNAH11. While the composition of cytoplasmic dynein-1 is now quite precisely resolved, the composition of IFT dynein-2 is far less well understood.

Cytoplasmic dynein-1 in vertebrates consists of a heavy chain (DYNC1H1) dimer [133] as well as dimers of two light–intermediate chains (DYNC1LI1/DYNC1LI2)

[134,135]; dimers of intermediate chains (DYNC1I1/DYNC1I2) [136]; and dimers of three different light chains (DYNLL1/DYNLL2 (LC8) [137], DYNLT1 (TCTEX1) [138,139], and DYNLRB1/DYNLRB2 (LC7) [140–142]).

In contrast, cytoplasmic IFT dynein-2 of the axoneme was initially described to contain only a heavy chain dimer (DYNC2H1) [143,144], light–intermediate chain dimer (DYNC2LI1) [145], and one type of light chain (DYNLL1, called LC8 in *Chlamydomonas*) [141]. Subsequent work then suggested additional components, an intermediate chain represented by FAP133 in *Chlamydomonas* [146] and WDR34 in zebrafish [147]. Only recently, vertebrate IFT dynein-2 was confirmed as containing two intermediate chains WDR34 [148,149] and WDR60 [150]. The light chain composition has remained inconclusive even longer: the dynein-1 light chain TCTEX1 was the first light chain subtype found to play a role for vertebrate cilia maintenance [151,152] and very recent proteomics studies instead suggest four different subtypes of light chains: TCETX1/DYNLT1, TCTEX1D2, DYNLL1, and DYNLRB1. Potentially, also DYNLT3 and DYNLRB2 could be part of cytoplasmic IFT dynein-2 [132,153]. It has not been established if only homo- or also heterodimers regarding the intermediate chain and light chain composition occur and what the unique functions of the different chain subtypes are. Figs. 15.5 and 15.6 show IFT dynein-2 protein composition and a schematic of the protein structures.

Human mutations resulting in skeletal ciliopathies (SRPS or JATD/Jeune syndrome only) are most frequently found in the heavy chain, DYNC2H1, due to its central importance and enormous gene size. This was the first IFT dynein-2 gene coupled to a human phenotype but subsequently, with the rise of next-generation sequencing, human mutations in the intermediate chains WDR60 and WDR34 were identified and lastly in the light–intermediate chain DYNC2LI1 and the light chain TCTEX1D2. The following section will give an overview of these genes and their role in SRPS and JATD.

15.6.1 Intraflagellar transport dynein-2 heavy chain (DYNC2H1)

DYNC2H1 was the second gene found to be mutated in JATD cases after *IFT80* [154,155]. Subsequently, *DYNC2H1* mutations were also identified to cause SRPS-II [131,156], SRPS-III [155], and SRPS-I [157], proving JATD and SRPS are truly allelic with JATD at the clinically milder end of the SRPS spectrum [29].

15.6.1.1 DYNC2H1 function

DYNC2H1 is a heavy chain dynein and the DYNC2H1 dimer provides the motor function of cytoplasmic IFT dynein-2 that retrograde IFT depends upon, according to current evidence. DYNC2H1 has a predicted largest isoform of 4314 amino acids and 493 kDa mass encoded by an open reading frame of 12,945

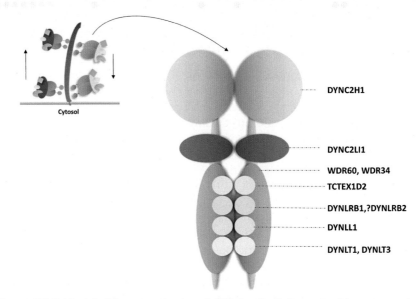

Figure 15.5 Model of human cytoplasmic IFT dynein-2. Current evidence suggests that the retrograde IFT motor consists of a homodimer of DYNC2H1 (heavy chains) and DYNCLI2 (light–intermediate chains). Both WDR60 and WDR34 have been identified as intermediate chains, if they occur as homo- or heterodimers remain to be established. Similarly, the light chain composition remains somewhat unclear with human mutations having been identified only in TCTEX1D2 to date, but DYNLRB1, DYNLL1, DYNLT1, DYNLT3, and possibly also DYNLRB2 are observed associated to the complex using protein–protein interaction studies [132,153,169].

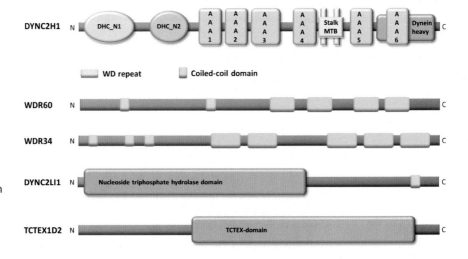

Figure 15.6 Protein structures of cytoplasmic IFT dynein-2 motor proteins. See main text for structural details (not shown to scale).

base pairs that encompasses 90 coding exons. DYNC2H1 is therefore a very large protein as shown in Fig. 15.6, containing six AAA+ domains forming the hexomeric ring-like ATP hydrolyzing motor domain, a microtubule-binding stalk domain between AAA4 and AAA5, an N-terminal tail domain (DHC_N1). AAA+ domains 1–4 are thought to bind nucleotide [158] but only AAA+ domain 1 is essential for dynein movement [159] while AAA+ domain 2, 3, and 4 are less well conserved between dyneins across species and probably are involved in regulation of the movement [160–162]. AAA+ domain 5 and 6 in contrast do not have catalytic function, mainly serving as a structural base for the stalk or buttress [163].

A linker domain (DHC_N2) is predicted to create the power stroke for movement along the microtubules by changing position depending on the nucleotide state and the protein also contains a conserved C-terminal domain arranged on top of the ATPase ring [32,164,165]. Recent work confirms the movement is generated by the linker when it physically gets into the way of the closed ring consisting of the five AAA+ domains; furthermore, the interaction surface between the stalk and motor domain is modulated by closure of the ATPase ring that provokes sliding of the helix in the stalk that in turn causes release of the microtubule-binding domains from the microtubule [166].

15.6.1.2 DYNC2H1 *human mutations and phenotype*

DYNC2H1 is the most frequently mutated gene in JATD, accounting for 30%–50% of the cases, with mutations being most frequent in Northern European descent individuals [32,41]. It is also the most frequently mutated gene causing SRPS-II [156]. Mutations have been identified all along the gene with most found to be private, except a low frequency Northern European founder mutation D3015G that affects the microtubule-binding stalk domain [32].

As for other SRPS and JATD genes, no individual harboring two full loss-of-function *DYNC2H1* null alleles (frameshift, premature stop/" nonsense" or splice site mutation) has yet been identified but instead, all human cases carry at least one presumed hypomorphic (usually missense) allele. DYNC2H1 dysfunction in human JATD and SRPS subjects seems to cause a reduced retrograde IFT rate with accumulations of IFT molecules (and presumably their cargo), as suggested by accumulation of IFT-B components IFT88 and IFT57 at the ciliary tip in fibroblasts from affected individuals, while the cilia can still be built and maintained [32]. Fig. 15.7 shows retrograde IFT disturbance in fibroblasts from *DYNC2H1* mutated individuals.

More severe (nonhuman) cases of disrupted DYNC2H1 function confirm its essential role (along with that of other IFT dynein-2 genes) during embryonic development. In *Dync2h1* knockout mice, ciliogenesis is severely impaired and incompatible with embryonic development beyond midgestation [16,115]. No

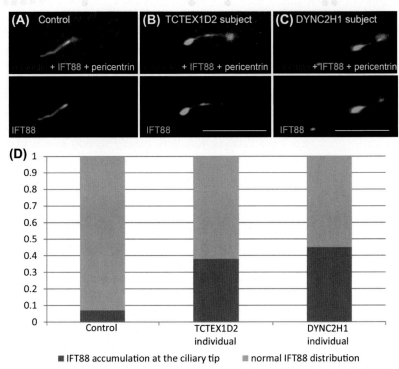

Figure 15.7 Retrograde IFT defect in JATD caused by *DYNC2H1* and *TCTEX1D2* mutations. (A–C) Compared with control cells (A), fibroblasts from JATD patients with mutations in *TCTEX1D2* (B) and *DYNC2H1* (C) display accumulation of IFT88 at the ciliary tip. The images show a single cilium in cells stained with anti-IFT88 (green), anti-acetylated alpha-tubulin marking the ciliary axoneme (red), and anti-pericentrin marking the ciliary base (white). Scale bars, 5 μm. (D) Fraction of cells (100 tested) with IFT88 accumulation at the ciliary tip. *Reproduced from Schmidts et al. (2015).*

clear phenotype–genotype correlation exists between SRPS and JATD to explain how different mutations in a common gene like *DYNC2H1* can cause their phenotypic differences, with some of the same alleles even identified in both conditions [32]. Interestingly, JATD caused by IFT dynein-2 motor mutations tends to present with a more severe rib phenotype than JATD patients with nonmotor, IFT complex gene mutations; but in contrast, affected patients usually do not seem to develop the clinically relevant extraskeletal features (NPHP-like renal disease, retinal degeneration) [32,41] seen in patients with IFT complex mutations [47,48], all except for those with *IFT80* mutations [167]. The reason for this pattern has remained elusive but may have to do with slightly different effects of IFT dynein-2 versus IFT complex mutations. For example, while the (presumably hypomorphic) IFT dynein-2 mutations probably still allow intact IFT as a process but proceeding slower/less frequently than under normal conditions, mutations in IFT complex genes may result in loss of specific cargo(s) rather than slowing down or imbalancing the overall IFT process.

15.6.2 IFT dynein-2 intermediate chains WDR60 and WDR34

In addition to a dimer of heavy chains, human cytoplasmic IFT dynein-2 also contains a dimer of intermediate chains directly binding to the heavy chain dimer. The presence of intermediate chains, although expected from analogies of cytoplasmic dynein-1 composition and IFT dynein-2 analyses in other (non-mammalian) species such as *Chlamydomonas* and *Schmidtea mediterranea* (planaria) [146,168] and later zebrafish [147]), has only been proven very recently [132,148–150,169]. The two different intermediate chains, WDR60 and WDR34, represent homologues of *Chlamydomonas* FAP163 [168] and FAP133, respectively [146].

15.6.2.1 WDR34 and WDR60 functions

WDR34 and WDR60, although of very different protein length (*WDR34* is a gene of 9 exons encoding a 536 amino acid protein while *WDR60* has 19 exons encoding a 1066 amino acid long protein), are similar in structure (Figs. 15.5 and 15.6). Both proteins contain so-called WD40 domains likely involved in protein–protein interactions including with the N-terminal part of DYNC2H1 and the dynein light chains. WDR34 has five WD40 domains toward its C-terminal and three N-terminal coiled-coil domains, while WDR60 has four C-terminal WD40 repeat domains and two N-terminal coiled-coil domains. A number of human missense mutations have been identified within these WD40 domains, presumably disturbing proper protein binding [148,149]. However, protein–protein interaction studies using patient alleles have not been performed to date, so the exact consequence of these mutations remains unclear. Whether cytoplasmic IFT dynein-2 in humans contains homodimers of WDR34 and WDR60 or heterodimers or both and how its composition may be regulated spatiotemporally has not been finally solved yet. There are indications of an asymmetric composition of the two different intermediate chains, where WDR34 may preferably associate with DYNLL1 [154] and WDR60 with TCTEX1D2 [154a]. This idea possibly allows a more diverse set of functions for the motor as suggested by Asante et al. [153]; however, in vivo data would ideally be needed to assess this question in addition to the overexpression techniques used so far.

As shown in Fig. 15.8, commercially available antibodies for both proteins reveal an immunofluorescent staining pattern around the centrioles but fail to detect any protein within the ciliary axoneme, while overexpression of GFP-labeled protein results in an axonemal protein localization [149,150,153,170].

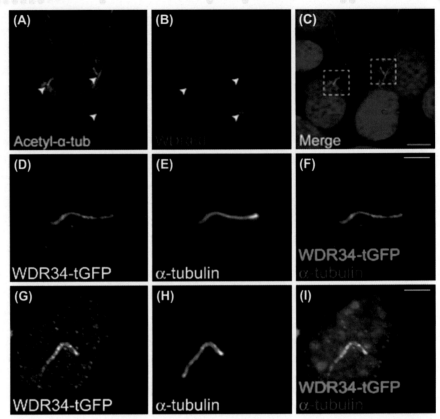

**Figure 15.8 Human IFT dynein-2 intermediate chains WDR60 and WDR34 localize
to cilia.** (A–C) Immortalized, serum-starved human chondrocytes stained with antibodies
to acetylated α-tubulin (green) and WDR60 (red), show localization of WDR60 at the
base of the cilium (*arrowheads*). Scale bars, 10 μm. (D–I) Serum-starved murine IMCD3
cells with low levels of exogenous human WDR34-tGFP expression show localization
of the protein to the axoneme. Scale bars, 5 μm. *(A–C) reproduced from McInerney-Leo et al.
(2013); (D–I) reproduced from Schmidts et al. (2013)*

The endogenous protein pool that localizes axonemally in comparison to other
compartments might be too small to be detected with these antibodies. At the
same time, the pericentriolar localization detected for endogenous proteins is
not observed with the GFP-tagged protein versions. This may be due to the
GFP tag but it is also possible that both WDR34 and WDR60 antibodies are
not fully specific and therefore it is also possible that the antibodies recognize
another pericentriolar protein, potentially an intermediate chain from cytoplasmic dynein-1, due to sequence similarities. Unfortunately, no knockout mice are
available to test this hypothesis.

The WDR34 *Chlamydomonas* homologue FAP133 has been known for a
while to represent the intermediate chain of *Chlamydomonas* IFT dynein-2
[146]. However, the WDR60 homologue FAP163 was found to represent an

intermediate chain in *Chlamydomonas* (and WD60 in planaria) much later [168], around the time when the first human mutations in JATD and SRPS were identified. siRNA knockdown of WD60 in planaria results in a severe cilia assembly defect [168], comparable to what was observed in fibroblasts of SRPS patients with biallelic WDR60 mutations [150]. Interestingly, this defect was less pronounced in a JATD patient, indicating that the degree of ciliary defects may represent the severity of the phenotype (JATD representing a milder form of SRPS) [150]. In agreement with what is observed for *Dync2h1* knockout mice, Patel-King et al. noticed that the majority of cilia in WD60 siRNA-treated planaria showed as short stubs containing an accumulation of IFT particle–like material underneath the ciliary membrane [168]. Likewise, fibroblasts of *WDR60*-mutated SRPS patients still assemble bulgy ciliary stubs that fail to extend an axoneme [150] and fibroblasts from SRPS *WDR34*–mutated patients extend shorter than normal axonemes [148]. These findings are somewhat con-trasted by reports that WDR34 depletion in mammalian cells results in a lack of cilia but mild depletion results in longer cilia [170]. However, mild siRNA-medi-ated depletion may have different effects than putative hypomorphic missense mutations that may cause complete loss of specific protein–protein interactions.

Assuming that the pericentriolar localization of WDR60 and WDR34 observed by immunostaining is real, it remains unclear what function both proteins have at this location but it can be speculated that they play a role in recruiting other proteins to the basal ciliary "building site," where there could also be a con-nection to cytoplasmic dynein-1 as previously discussed [149,153]. A putative connection to cytoplasmic dynein-1 is supported by the finding that WDR34 coprecipitates with the dynein-1 motor light chain DYNLL1, the homologue of *Chlamydomonas* LC8 [149] and likewise, FAP133 contains putative LC8-binding motifs and in addition to its flagella localization, localizes around the basal body; FAP163 also binds to LC8 in *Chlamydomonas* [168]. In *Chlamydomonas*, only one type of cytoplasmic dynein exists and it seems possible, even after the divergent development of cytoplasmic dynein-1 and IFT dynein-2 in mam-mals, that some functions and molecules still overlap between the two. Asante et al. detected binding between WDR34 with all other dynein-1 light chains (DYNLL2, DYNLT1, DYNLT3, DYNLRB1, and DYNLRB2) as well as with two reg-ulators of cytoplasmic dynein-1, NudC, and NudCD3 [153]. However, binding of the dynein-1 intermediate chain (DYNC1LI2) to dynactin, a known major interactor of cytoplasmic dynein-1, does not seem to be shared as Asante et al. failed to observe binding of WDR34 to the two main components, p150^{Glued} and p50^{dynamitin} [153].

There is a further potential role for WDR34 in regulating the NF-kappaB sig-naling pathway, which plays a role in the innate immune response. WDR34 was identified as an inhibitor of the TAK1 protein kinase, which activates the NF-kappaB pathway via the IL-1 beta and toll-like receptor pathways, several years before human WDR34 mutations causing JATD and SRPS were identified. The WD40 domains seem to be essential for the interaction [171].

WDR34 mutations probably account for roughly 10% of all JATD cases, making this the second most common disease gene after *DYNC2H1* [149]. Mutations in the larger *WDR60* gene occur much less frequently causing only 1%–2% of cases, although larger studies are needed to confirm these numbers [150]. Besides JATD, mutations in both genes can also cause SRPS [148,150]. Similarly to patients with *DYNC2H1* mutations, *WDR34* and *WDR60* JATD patients display a severe skeletal phenotype with pronounced rib shortening and often life-threatening respiratory problems after birth and occasional polydactyly, but usually no visceral involvement such as NPHP-like renal disease. This is in contrast to the rather mild thorax phenotype and usual absence of polydactyly but high rate of renal disease observed with IFT complex A and B mutations. In accordance with the likely essential function of dynein-2 in mammalian embryonic development, no patients carrying two putative null alleles for *WDR34* or *WDR60* has been identified [148–150,172,173], since such a genetic setup is probably lethal early on in pregnancy, in analogy to *DYNC2H1* mutations [174].

15.6.3 IFT dynein-2 light–intermediate chain (DYNC2LI1)

While the intermediate chain composition of IFT dynein-2 was only recently described and, regarding the light chains, protein complex composition has remained partially elusive, DYNC2LI1 (named D1LIC at the time) was identified a lot earlier as the light–intermediate chain of the motor and as a direct interaction partner of DYNC2H1 [175]. At this time Perrone et al. also found a DYNC2H1 and DYNC2LI1 complex to represent the retrograde IFT motor in *Chlamydomonas* and mammalian systems [176].

15.6.3.1 DYNC2LI1 function

DYNC2LI1 has 13 exons encoding a 351 amino acid protein with a C-terminal coiled-coil domain and a large nucleotide triphosphate hydrolase domain [177] as shown in Fig. 15.6. Like other dynein-2 proteins, DYNC2LI1 is evolutionarily very well conserved. It furthermore shows partial homology with the dynein-1 light–intermediate chain. Mammalian DYNC2LI1 has ciliary expression in kidney and localizes to the connecting cilium of retinal photoreceptors [145]. It localizes to the Golgi apparatus in COS-7 cells, suggesting possible function as an adapter between cytoplasmic IFT dynein-2 and cargo [175]. In the dynein-1 motor, cargo binding occurs via the homologous NTPase release domain of homologous DYNC1LI1 [134].

Despite the assumed ubiquitous presence of primary cilia and function of IFT dynein-2, during fetal development *DYNC2LI1* transcript expression is higher in bone tissue than other tissues as visualised by immunohistochemistry. Localization to the perichondrium, periosteum, and bone spongiosa in the area of growth plates is shown in Fig. 15.9 [177]. Analysis of fetal and adult tissue detects highest *DYNC2LI1* expression in chondrocytes, renal cells, and in the

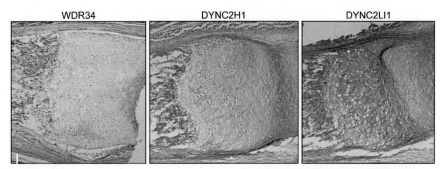

Figure 15.9 Cytoplasmic IFT dynein-2 proteins localize to the growth plate area during human fetal development. Sections from a 14-week human humerus were stained for DYNC2LI1, DYNC2H1, and WDR34. All three proteins were expressed in the perichondrium/periosteum and primary spongiosa (*closed arrows*, perichondrium/periosteum; *open arrows*, primary spongiosa). Scale bar, 50 μm. *Reproduced from S.P. Taylor, T.J. Dantas, I. Duran, S. Wu, R.S. Lachman, C. University of Washington Center for Mendelian Genomics, S.F. Nelson, D.H. Cohn, R.B. Vallee, D. Krakow, Mutations in DYNC2LI1 disrupt cilia function and cause short rib polydactyly syndrome, Nat. Commun. 6 (2015) 7092 with permission.*

brain [178], offering a potential explanation for the pronounced skeletal phenotype resulting from *DYNC2LI1* loss-of-function mutations. In human fibroblasts, DYNC2LI1 protein is detected within the transition zone of cilia, around the basal bodies and centrosomes as well as in the cytosol [178].

siRNA-mediated *DYNC2LI1* knockdown in fibroblasts results in shorter cilia with bulgy tips but not an absence of cilia [178] while knockdown in retinal epithelial (hTERT-RPE1) results in variable ciliary length defects similar to that seen after *DYNC2H1* depletion [177]. Also similar to what is observed in *DYNC2H1*-mutant cells, accumulation of IFT-B complex components such IFT88 and IFT57 is seen within bulgy cilia tips [178] suggesting defective (retrograde) trafficking; this could be rescued by reintroducing wild-type *DYNC2L1* [177]. Additional to cilia trafficking defects in a cell model, Taylor et al. observed disturbed hedgehog signaling in fibroblasts from an SRPS patient with biallelic *DYNC2LI1* mutations. Specifically the hedgehog receptor SMO, only detected in cilia from control cells after pathway stimulation, was found to enter the cilium in unstimulated *DYNC2LI1*-mutant cells, suggesting that DYNC2LI1 and therefore IFT dynein-2 in humans may play a role in regulating hedgehog signaling by controlling ciliary SMO entry in the absence of hedgehog stimulation [177]. Since human fetal tissue is rarely accessible and cartilage tissue from living patients cannot be obtained, experimental evidence such as this is sparse in the literature, supporting the hypothesis that defective Hh signaling underlies at least some of the phenotypic features observed in JATD and SRPS.

15.6.3.2 DYNC2LI1 *human mutations and phenotype*

DYNC2LI1 mutations have been described in only four JATD or SRPS families so far despite it being a very suggestive candidate gene. Similarly to patients with IFT dynein-2 heavy chain or intermediate chain mutations, all patients carried at least one putative hypomorphic allele. Phenotypic features include very small thorax, polydactyly, and shortened long bones with multitude of variable additional features including brachysyndactyly, hypoplastic nails, atrioventricular septal defect, hepatosplenomegaly, cleft lip, and asymmetric tongue. Facial dysmorphism also occured including low set ears, epicanthic fold and down-slanting palpebral fissures, broad and prominent forehead, a depressed nasal bridge, broad and up-slanting nasal tip, resembling the facial dysmorphism observed in CED. Hyperplastic gingivae and polyhydramnios were also noted during pregnancy [177,178].

15.6.4 IFT dynein-2 light chain (TCTEX1D2)

All cytoplasmic IFT dynein-2 genes have been striking candidates for skeletal ciliopathies for many years but surprisingly, mutations in a light chain were only recently identified. Since the exact light chain composition of the motor in mammals was unclear at the time (and has still not been entirely solved), *TCTEX1D2* mutations were not really expected when identified as a cause of JATD. A second surprising finding was that in contrast to mutations in other IFT dynein-2 genes causing JATD where one allele is always a presumably hypomorphic missense allele, *TCTEX1D2* mutations were all predicted to cause a complete loss of protein function. An apparently incompletely penetrant phenotype was also found in one large Arab family carrying *TCTEX1D2* mutations, with apparently unaffected (now adult) siblings and affected siblings all carrying biallelic "null" alleles [132]. A slight caveat is the chance that overlooked radiological signs, e.g., the typical pelvis configuration, can disappear or become undetectable by adulthood in affected individuals. It will be interesting to see if incomplete penetrance associated with *TCTEX1D2* mutations will be observed in more families in the future.

15.6.4.1 TCTEX1D2 function

TCTEX1D2 is a small gene of 4 exons encoding a 142 amino acid protein belonging to a protein family (TCTEX1D1–TCTEX1D4) containing a conserved similar C-terminal domain (TCTEX-domain). The function of members of this family has not been studied very well, except that TCTEX1D4 seems to represent an interaction partner protein for phosphatase 1 [179,180] and is a dynein light chain present in sperm [181]. Recent affinity proteomics studies in human cells and *Chlamydomonas* show that TCTEX1D2 (and its *Chlamydomonas* equivalent TCTEX2b) is an integral component of IFT dynein-2 since siRNA-mediated *TCTEX1D2* knockdown results in ciliogenesis defects [132,153,169]. The correct localization of TCEX1D2 to the ciliary base seems dependent on WDR60 as well as microtubules and TCTEX1D2 colocalizes with WDR60 at the microtubule organizing centers as well as at the mitotic spindle poles. Additional interaction partners include other ciliopathy proteins such as MKS3 and BBS7 as well as components of cytoplasmic dynein-1, suggesting a role within both dynein-1 and dynein-2 complexes, potentially as a cargo adapter involved in transport toward the cilium as well as within the cilium [153,169].

TCTEX1D2 depletion does not abrogate ciliogenesis in mammalian cells but reduces ciliation [169] or changes ciliary length [153] and human TCTEX1D2-deficient fibroblasts display a retrograde IFT defect, potentially less pronounced than that of DYNC2H1-deficient cells [132]. This agrees with findings in *Chlamydomonas* cells deficient for the TCTEX1D2 homologue TcTex2b where flagella growth is normal and the effect on retrograde IFT is rather modest compared to the loss of other components such as heavy chains or intermediate chains (Fig. 15.10). Similarly, morpholino-mediated antisense knockdown of zebrafish *tctex1d2* caused relatively mild ciliation defects. Anterograde trafficking seems unaffected in *Chlamydomonas* TCTEX2b mutants [132] while loss of other IFT dynein-2 components such as the DYNC2LI1 homologue D1bLIC also disturbs anterograde trafficking [182]. These rather mild effects on IFT could explain the not fully penetrant human phenotype observed. No analysis of human cytoplasmic IFT dynein-2 stability in the context of TCTEX1D2 loss has been performed but *Chlamydomonas* data suggest that the motor could be partially destabilized in this setting [132].

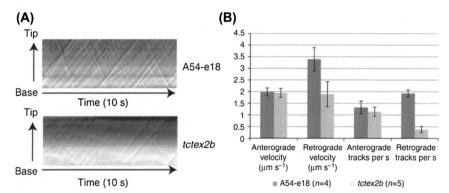

Figure 15.10 Loss of *Chlamydomonas* TCTEX1D2 impairs axonemal retrograde IFT velocity. (A) A mutant *Chlamydomonas* strain lacking the TCTEX1D2 IFT dynein-2 light chain orthologue, tctex2b, is defective in retrograde IFT recorded using DIC microscopy kymographs. Tracks with positive slopes represent IFT particles moving anterogradely, and tracks with negative slopes represent particles moving retrogradely. Compared with wild type, few retrograde tracks are visible in the *tctex2b* mutant kymograph, with a reduced slope. Retrograde particles had a larger apparent size in mutants; similar findings were reported for a temperature-sensitive *dhc1b* mutant [207]. (B) Quantitative analysis of IFT shows that the *tctex2b* mutant anterograde IFT velocity is about the same as in wild type (A54-e18 strain), while anterograde frequency is only slightly reduced, but both retrograde IFT velocity and frequency are greatly reduced. *n*, number of flagella analyzed. Error bars show standard deviation. *Reproduced from Schmidts et al. (2015).*

Although proteomics data suggest that IFT dynein-2 contains a number of additional light chains such as DYNLL1, TCTEX1, and DYNLRB1, currently without crystallography evidence its exact composition remains unclear. One confounding factor is that different light chains such as DYNLRB1 and DYNLRB2 are very similar in protein sequence and therefore, as they are very short proteins as well, they cannot be distinguished with certainty from each other during protein data analysis [132,153]. The motor's composition is probably also spatiotemporally regulated and experiments have only been performed in mammalian in vitro systems. However, given these provisos it is reasonable to hypothesize that different IFT dynein-2 light chains of the retrograde motor are partially redundant, potentially explaining why biallelic loss-of-function alleles are compatible with life, unlike for other skeletal ciliopathy genes where this situation would be incompatible with embryonic survival beyond midgestation. Furthermore, genetic redundancy and likely functional replacement by alternative light chains could also explain the rather modest effect on retrograde IFT (and lack of effect on anterograde IFT) of *TCTEX1D2* null mutations and the apparently incomplete penetrance observed in one family [132].

15.6.4.2 TCTEX1D2 *human mutations and phenotype*

To date, only three JATD families with biallelic *TCTEX1D2* mutations have been described, in which the affected individuals exhibit a purely skeletal phenotype with narrow thorax, typical JATD pelvis changes and polydactyly but no renal or retinal involvement. However, not all patients have reached adulthood so clinical signs could still develop in those organs over time. As mentioned above, biallelic putative "null" alleles distinguish TCTEX1D2 cases from patients with mutations in other cytoplasmic IFT dynein-2 genes and in addition, incomplete penetrance was suspected in one of the families [132].

15.7 Future perspective on clinical spectrum, new clinical models, and therapy

Great progress has been made within the last few years to improve understanding of the molecular composition and function of cytoplasmic IFT dynein-2, including through large-scale sequencing and proteomics programs as well as molecule-focused approaches. However, a number of open questions remain including the nature of the spatiotemporal behavior of IFT dynein-2 in terms of its intermediate and light chain composition and its putative connection to cytoplasmic dynein-1 functions. It seems likely that during different developmental phases and stages, and in different organs, dynein motor composition may be changed and adapted according to specific local IFT trafficking requirements. The recent development of versatile genetic modification tools such as CRISPR/Cas9 and TALEN technology will hopefully facilitate the elucidation of these questions in the near future.

Genetic approaches have also revealed striking genotype–phenotype correlations within human cohorts and mouse models affected by different IFT dynein–related components [3,11,30]. In ciliopathy patients, mutations in cytoplasmic IFT dynein-2 result in a severe thoracic phenotype but rarely the clinical manifestation of renal and retinal disease, compared to mutations in the associated IFT-A complex, which cause mild thoracic problems but severe renal and retinal findings. The ciliary molecular mechanisms influencing the skeletal ciliopathy mutational spectrum remain elusive and since all human patients identified so far are presumed hypomorphic for the disease, the different mutation types are not indicated at present as the biggest influence on the clinical course in different patients. It is reasonable to speculate that IFT-A mutations result in loss of specific cargo that causes distinct cell signaling deficits relevant in the kidney and eye, while IFT dynein-2 motor mutations just slow down IFT with the IFT–cargo complexes kept intact. Since cytoplasmic IFT dynein-2 is the only retrograde IFT motor identified in vertebrates it should be ubiquitously present in all primary cilia and therefore it is not highly supported to propose that tissue specific expression may be responsible for the lack of renal or retinal disease; however, compensating mechanisms in these organs cannot be excluded. More work will be required to understand the variation if any in IFT dynein-2 subunit composition across different tissues. Embryonic midgestation lethality in knockout mice and perinatal lethality in hypomorphic mouse models have hindered organ specific analysis to date, but development of conditional models may help to shed light in this research area.

IFT dynein-2 associated mutations cause significant lethality in affected individuals and while genetic diagnosis has greatly progressed due to developments in next-generation sequencing methods including the use of multiple cilia gene panels and whole-exome as well as whole-genome sequencing [183,184], unfortunately therapeutic approaches have remained restricted to symptomatic treatments such as ventilation support and/or thoracic expansion surgery in severe JATD cases. For patients such as those with IFT-A complex mutations surviving embryonic lethality, they can then become subject to life-threatening extraskeletal features such as blindness and early end-stage renal failure [47,48].

The development of targeted therapies will be the next step to take; however, there are several challenges. The main causative gene of skeletal ciliopathies, *DYNC2H1*, is too large for conventional gene replacement therapy approaches. Extensive underlying genetic and allelic heterogeneity among skeletal ciliopathies suggests that many personalized genetic medicines may need to be developed, especially since some drugs designed to target larger numbers of mutations at once—from a particular mutational class—may not be applicable. For example, due to a minority of mutations representing premature stop codon (nonsense) mutations, so-called "read-through" drugs, which can overread these variants to enhance expression of the full length protein, are not applicable to the majority of human cases [185]. CRISPR/Cas9-mediated genome editing for the correction of point mutations could be a future option

[186] or RNA-based therapeutics, for example, trans-splicing where single exons carrying mutations are corrected rather than the entire gene being replaced [187]; however, due to the mainly private mutations this would be a laborious and expensive approach.

Therefore, broader therapeutic targeting of the IFT process or of the affected downstream signaling pathways such as Hh signaling may be a more promising approach. Further basic biological studies are also required to investigate the less well-defined aspects of skeletal ciliopathies, for example, disrupted Hh signaling can explain many but not all of the phenotypes that manifest in affected individuals [116,122,123]. The rebalancing of IFT, through fine-tuning of the ratio of anterograde versus retrograde trafficking, appears to ameliorate the embryonic phenotype in mice [115] and this has potential to be helpful in humans as well. This could be achieved by genetic modification of anterograde IFT in cases of IFT dynein-2 mutations though pharmacological modifiers remain to be developed. Or, in case of IFT-B complex mutations, the DYNC2H1 inhibitor ciliobrevin [188,189] may potentially improve the phenotype. However, further work in cell and animal model systems is required before any therapeutic options can be tested in humans.

Directly restoring the IFT dynein-2 motor and its associated functions using pharmacological substances such as small molecules would be highly desirable, but at present this has not yet been achieved. Continued developments in the field of genomic and personalized medicine will continue to help to define the complete skeletal ciliopathy disease spectrum and the eventual number of molecular players involved in these disorders.

Acknowledgments

MS acknowledges funding from Radboud University (Excellence Fellowship), Radboud University Medical Center (Hypatia tenure track grant), the German Research Foundation (Deutsche Forschungsgesellschaft, DFG) (collaborative research center grant SFB1140, KIDGEM) and the European Research Council, ERC (*ERC starting grant TREATCilia,* grant agreement 716344). HMM acknowledges support from Great Ormond Street Hospital Children's Charity and the National Institute for Health Research Biomedical Research Centre at Great Ormond Street Hospital for Children NHS Foundation Trust and University College London.

References

[1] N.F. Berbari, A.K. O'Connor, C.J. Haycraft, B.K. Yoder, The primary cilium as a complex signaling center, Curr. Biol. 19 (13) (2009) R526–R535.

[2] R. Rohatgi, W.J. Snell, The ciliary membrane, Curr. Opin. Cell Biol. 22 (4) (2010) 541–546.

[3] F. Bangs, K.V. Anderson, Primary cilia and mammalian hedgehog signaling, Cold Spring Harb. Perspect. Biol. (2016).

[4] D. Huangfu, A. Liu, A.S. Rakeman, N.S. Murcia, L. Niswander, K.V. Anderson, Hedgehog signalling in the mouse requires intraflagellar transport proteins, Nature 426 (6962) (2003) 83–87.

[5] J.E. Lee, J.G. Gleeson, A systems-biology approach to understanding the ciliopathy disorders, Genome Med. 3 (9) (2011) 59.

[6] J.L. Rosenbaum, G.B. Witman, Intraflagellar transport, Nat. Rev. Mol. Cell Biol. 3 (11) (2002) 813–825.

[7] J.M. Scholey, K.V. Anderson, Intraflagellar transport and cilium-based signaling, Cell 125 (3) (2006) 439–442.

[8] T. Eguether, J.T. San Agustin, B.T. Keady, J.A. Jonassen, Y. Liang, R. Francis, K. Tobita, C.A. Johnson, Z.A. Abdelhamed, C.W. Lo, G.J. Pazour, IFT27 links the BBSome to IFT for maintenance of the ciliary signaling compartment, Dev. Cell 31 (3) (2014) 279–290.

[9] B.T. Keady, R. Samtani, K. Tobita, M. Tsuchya, J.T. San Agustin, J.A. Follit, J.A. Jonassen, R. Subramanian, C.W. Lo, G.J. Pazour, IFT25 links the signal-dependent movement of Hedgehog components to intraflagellar transport, Dev. Cell 22 (5) (2012) 940–951.

[10] Q. Wang, J. Pan, W.J. Snell, Intraflagellar transport particles participate directly in cilium-generated signaling in Chlamydomonas, Cell 125 (3) (2006) 549–562.

[11] M. He, S. Agbu, K.V. Anderson, Microtubule motors drive hedgehog signaling in primary cilia, Trends Cell Biol. 27 (2) (2017) 110–125.

[12] M. Taschner, E. Lorentzen, The intraflagellar transport machinery, Cold Spring Harb. Perspect. Biol. (10) (2016) 8.

[13] L. Stepanek, G. Pigino, Microtubule doublets are double-track railways for intraflagellar transport trains, Science 352 (6286) (2016) 721–724.

[14] Q. Wei, Y. Zhang, Y. Li, Q. Zhang, K. Ling, J. Hu, The BBSome controls IFT assembly and turnaround in cilia, Nat. Cell Biol. 14 (9) (2012) 950–957.

[15] L.B. Pedersen, J.B. Mogensen, S.T. Christensen, Endocytic control of cellular signaling at the primary cilium, Trends Biochem. Sci. 41 (9) (2016) 784–797.

[16] P.J. Ocbina, M. Tuson, K.V. Anderson, Primary cilia are not required for normal canonical Wnt signaling in the mouse embryo, PLoS One 4 (8) (2009) e6839.

[17] P.G. DeCaen, M. Delling, T.N. Vien, D.E. Clapham, Direct recording and molecular identification of the calcium channel of primary cilia, Nature 504 (7479) (2013) 315–318.

[18] S. Ohata, V. Herranz-Perez, J. Nakatani, A. Boletta, J.M. Garcia-Verdugo, A. Alvarez-Buylla, Mechanosensory genes Pkd1 and Pkd2 contribute to the planar polarization of brain ventricular epithelium, J. Neurosci. 35 (31) (2015) 11153–11168.

[19] K.I. Hilgendorf, C.T. Johnson, P.K. Jackson, The primary cilium as a cellular receiver: organizing ciliary GPCR signaling, Curr. Opin. Cell Biol. 39 (2016) 84–92.

[20] H.L. May-Simera, M.W. Kelley, Cilia, Wnt signaling, and the cytoskeleton, Cilia 1 (1) (2012) 7.

[21] C. Boehlke, F. Kotsis, V. Patel, S. Braeg, H. Voelker, S. Bredt, T. Beyer, H. Janusch, C. Hamann, M. Godel, K. Muller, M. Herbst, M. Hornung, M. Doerken, M. Kottgen, R. Nitschke, P. Igarashi, G. Walz, E.W. Kuehn, Primary cilia regulate mTORC1 activity and cell size through Lkb1, Nat. Cell Biol. 12 (11) (2010) 1115–1122.

[22] L. Schneider, C.A. Clement, S.C. Teilmann, G.J. Pazour, E.K. Hoffmann, P. Satir, S.T. Christensen, PDGFRalphaalpha signaling is regulated through the primary cilium in fibroblasts, Curr. Biol. 15 (20) (2005) 1861–1866.

[23] G.J. Pazour, R.A. Bloodgood, Targeting proteins to the ciliary membrane, Curr. Top. Dev. Biol. 85 (2008) 115–149.

[24] F. Ye, D.K. Breslow, E.F. Koslover, A.J. Spakowitz, W.J. Nelson, M.V. Nachury, Single molecule imaging reveals a major role for diffusion in the exploration of ciliary space by signaling receptors, Elife 2 (2013) e00654.

[25] H.M. Mitchison, E.M. Valente, Motile and non-motile cilia in human pathology: from function to phenotypes, J. Pathol. 241 (2) (2017) 294–309.

[26] K.F. Liem Jr., A. Ashe, M. He, P. Satir, J. Moran, D. Beier, C. Wicking, K.V. Anderson, The IFT-A complex regulates Shh signaling through cilia structure and membrane protein trafficking, J. Cell Biol. 197 (6) (2012) 789–800.

[27] F. Hildebrandt, T. Benzing, N. Katsanis, Ciliopathies, N. Engl. J. Med. 364 (16) (2011) 1533–1543.

[28] K. Baker, P.L. Beales, Making sense of cilia in disease: the human ciliopathies, Am. J. Med. Genet. C Semin. Med. Genet. 151C (4) (2009) 281–295.

[29] C. Huber, V. Cormier-Daire, Ciliary disorder of the skeleton, Am. J Med. Genet. C 160C (3) (2012) 165–174.

[30] M. Schmidts, Clinical genetics and pathobiology of ciliary chondrodysplasias, J. Pediatr. Genet. 3 (2) (2014) 46–94.

[31] J. de Vries, J.L. Yntema, C.E. van Die, N. Crama, E.A. Cornelissen, B.C. Hamel, Jeune syndrome: description of 13 cases and a proposal for follow-up protocol, Eur. J. Pediatr. 169 (1) (2010) 77–88.

[32] M. Schmidts, H.H. Arts, E.M. Bongers, Z. Yap, M.M. Oud, D. Antony, L. Duijkers, R.D. Emes, J. Stalker, J.B. Yntema, V. Plagnol, A. Hoischen, C. Gilissen, E. Forsythe, E. Lausch, J.A. Veltman, N. Roeleveld, A. Superti-Furga, A. Kutkowska-Kazmierczak, E.J. Kamsteeg, N. Elcioglu, M.C. van Maarle, L.M. Graul-Neumann, K. Devriendt, S.F. Smithson, D. Wellesley, N.E. Verbeek, R.C. Hennekam, H. Kayserili, P.J. Scambler, P.L. Beales, N.V. Knoers, R. Roepman, H.M. Mitchison, Exome sequencing identifies DYNC2H1 mutations as a common cause of asphyxiating thoracic dystrophy (Jeune syndrome) without major polydactyly, renal or retinal involvement, J. Med. Genet. 50 (5) (2013) 309–323.

[33] E.J. Mackie, L. Tatarczuch, M. Mirams, The skeleton: a multi-functional complex organ: the growth plate chondrocyte and endochondral ossification, J. Endocrinol. 211 (2) (2011) 109–121.

[34] E.J. Mackie, Y.A. Ahmed, L. Tatarczuch, K.S. Chen, M. Mirams, Endochondral ossification: how cartilage is converted into bone in the developing skeleton, Int. J. Biochem. Cell Biol. 40 (1) (2008) 46–62.

[35] N.J. Wilsman, C.E. Farnum, D.K. Reed-Aksamit, Incidence and morphology of equine and murine chondrocytic cilia, Anat. Rec. 197 (3) (1980) 355–361.

[36] P. Tummala, E.J. Arnsdorf, C.R. Jacobs, The role of primary cilia in mesenchymal stem cell differentiation: a pivotal switch in guiding lineage commitment, Cell Mol. Bioeng. 3 (3) (2010) 207–212.

[37] Z. Xiao, S. Zhang, J. Mahlios, G. Zhou, B.S. Magenheimer, D. Guo, S.L. Dallas, R. Maser, J.P. Calvet, L. Bonewald, L.D. Quarles, Cilia-like structures and polycystin-1 in osteoblasts/osteocytes and associated abnormalities in skeletogenesis and Runx2 expression, J. Biol. Chem. 281 (41) (2006) 30884–30895.

[38] X. Yuan, R.A. Serra, S. Yang, Function and regulation of primary cilia and intraflagellar transport proteins in the skeleton, Ann. NY Acad. Sci. 1335 (2015) 78–99.

[39] C.J. Haycraft, R. Serra, Cilia involvement in patterning and maintenance of the skeleton, Curr. Top. Dev. Biol. 85 (2008) 303–332.

[40] X. Yuan, S. Yang, Primary cilia and intraflagellar transport proteins in bone and cartilage, J. Dent. Res. 95 (12) (2016) 1341–1349.

[41] G. Baujat, C. Huber, J. El Hokayem, R. Caumes, C. Do Ngoc Thanh, A. David, A.L. Delezoide, A. Dieux-Coeslier, B. Estournet, C. Francannet, H. Kayirangwa, F. Lacaille, M. Le Bourgeois, J. Martinovic, R. Salomon, S. Sigaudy, V. Malan, A. Munnich, M. Le Merrer, K.H. Le Quan Sang, V. Cormier-Daire, Asphyxiating thoracic dysplasia: clinical and molecular review of 39 families, J. Med. Genet. 50 (2) (2013) 91–98.

[42] E.E. Davis, Q. Zhang, Q. Liu, B.H. Diplas, L.M. Davey, J. Hartley, C. Stoetzel, K. Szymanska, G. Ramaswami, C.V. Logan, D.M. Muzny, A.C. Young, D.A. Wheeler, P. Cruz, M. Morgan, L.R. Lewis, P. Cherukuri, B. Maskeri, N.F. Hansen, J.C. Mullikin, R.W. Blakesley, G.G. Bouffard, N.C.S. Program, G. Gyapay, S. Rieger, B. Tonshoff, I. Kern, N.A. Soliman, T.J. Neuhaus, K.J. Swoboda, H. Kayserili, T.E. Gallagher, R.A. Lewis, C. Bergmann, E.A. Otto, S. Saunier, P.J. Scambler, P.L. Beales, J.G. Gleeson, E.R. Maher, T. Attie-Bitach, H. Dollfus, C.A. Johnson, E.D. Green, R.A. Gibbs, F. Hildebrandt, E.A. Pierce, N. Katsanis, TTC21B contributes both causal and modifying alleles across the ciliopathy spectrum, Nat. Genet. 43 (3) (2011) 189–196.

[43] I. Perrault, S. Saunier, S. Hanein, E. Filhol, A.A. Bizet, F. Collins, M.A. Salih, S. Gerber, N. Delphin, K. Bigot, C. Orssaud, E. Silva, V. Baudouin, M.M. Oud, N. Shannon, M. Le Merrer, O. Roche, C. Pietrement, J. Goumid, C. Baumann, C. Bole-Feysot, P. Nitschke, M. Zahrate, P. Beales, H.H. Arts, A. Munnich, J. Kaplan, C. Antignac, V. Cormier-Daire, J.M. Rozet, Mainzer-Saldino syndrome is a ciliopathy caused by IFT140 mutations, Am. J. Hum. Genet. 90 (5) (2012) 864–870.

[44] I. Perrault, J. Halbritter, J.D. Porath, X. Gerard, D.A. Braun, H.Y. Gee, H.M. Fathy, S. Saunier, V. Cormier-Daire, S. Thomas, T. Attie-Bitach, N. Boddaert, M. Taschner, M. Schueler, E. Lorentzen, R.P. Lifton, J.A. Lawson, M. Garfa-Traore, E.A. Otto, P. Bastin, C. Caillaud, J. Kaplan, J.M. Rozet, F. Hildebrandt, IFT81, encoding an IFT-B core protein, as a very rare cause of a ciliopathy phenotype, J. Med. Genet. 52 (10) (2015) 657–665.

[45] D.A. Braun, F. Hildebrandt, Ciliopathies, Cold Spring Harb. Perspect. Biol. (3) (2017) 9.

[46] N.H. Elcioglu, C.M. Hall, Diagnostic dilemmas in the short rib-polydactyly syndrome group, Am. J. Med. Genet. 111 (4) (2002) 392–400.

[47] J. Halbritter, A.A. Bizet, M. Schmidts, J.D. Porath, D.A. Braun, H.Y. Gee, A.M. McInerney-Leo, P. Krug, E. Filhol, E.E. Davis, R. Airik, P.G. Czarnecki, A.M. Lehman, P. Trnka, P. Nitschke, C. Bole-Feysot, M. Schueler, B. Knebelmann, S. Burtey, A.J. Szabo, K. Tory, P.J. Leo, B. Gardiner, F.A. McKenzie, A. Zankl, M.A. Brown, J.L. Hartley, E.R. Maher, C. Li, M.R. Leroux, P.J. Scambler, S.H. Zhan, S.J. Jones, H. Kayserili, B. Tuysuz, K.N. Moorani, A. Constantinescu, I.D. Krantz, B.S. Kaplan, J.V. Shah, T.W. Hurd, D. Doherty, N. Katsanis, E.L. Duncan, E.A. Otto, P.L. Beales, H.M. Mitchison, S. Saunier, F. Hildebrandt, Defects in the IFT-B component IFT172 cause Jeune and Mainzer-Saldino syndromes in humans, Am. J. Hum. Genet. 93 (5) (2013) 915–925.

[48] M. Schmidts, V. Frank, T. Eisenberger, S. Al Turki, A.A. Bizet, D. Antony, S. Rix, C. Decker, N. Bachmann, M. Bald, T. Vinke, B. Toenshoff, N. Di Donato, T. Neuhann, J.L. Hartley, E.R. Maher, R. Bogdanovic, A. Peco-Antic, C. Mache, M.E. Hurles, I. Joksic, M. Guc-Scekic, J. Dobricic, M. Brankovic-Magic, H.J. Bolz, G.J. Pazour, P.L. Beales, P.J. Scambler, S. Saunier, H.M. Mitchison, C. Bergmann, Combined NGS approaches identify mutations in the intraflagellar transport gene IFT140 in skeletal ciliopathies with early progressive kidney Disease, Hum. Mutat. 34 (5) (2013) 714–724.

[49] A.E. Lin, A.Z. Traum, I. Sahai, K. Keppler-Noreuil, M.K. Kukolich, M.P. Adam, S.J. Westra, H.H. Arts, Sensenbrenner syndrome (Cranioectodermal dysplasia): clinical and molecular analyses of 39 patients including two new patients, Am. J. Med. Genet. Part A 161A (11) (2013) 2762–2776.

[50] R.M. Saldino, C.D. Noonan, Severe thoracic dystrophy with striking micromelia, abnormal osseous development, including the spine, and multiple visceral anomalies, Am. J. Roentgenol. Radium Ther. Nucl. Med. 114 (2) (1972) 257–263.

[51] F. Majewski, R.A. Pfeiffer, W. Lenz, R. Muller, G. Feil, R. Seiler, Polysyndactyly, short limbs, and genital malformations–a new syndrome? Z. fur Kinderheilkd. 111 (2) (1971) 118–138.

[52] I.C. Verma, S. Bhargava, S. Agarwal, An autosomal recessive form of lethal chondrodystrophy with severe thoracic narrowing, rhizoacromelic type of micromelia, polydacytly and genital anomalies, Birth Defects Orig. Artic. Ser. 11 (6) (1975) 167–174.

[53] F.A. Beemer, L.O. Langer Jr., J.M. Klep-de Pater, A.M. Hemmes, J.B. Bylsma, R.M. Pauli, T.L. Myers, C.C. Haws 3rd, A new short rib syndrome: report of two cases, Am. J. Med. Genet. 14 (1) (1983) 115–123.

[54] P. Mill, P.J. Lockhart, E. Fitzpatrick, H.S. Mountford, E.A. Hall, M.A. Reijns, M. Keighren, M. Bahlo, C.J. Bromhead, P. Budd, S. Aftimos, M.B. Delatycki, R. Savarirayan, I.J. Jackson, D.J. Amor, Human and mouse mutations in WDR35 cause short-rib polydactyly syndromes due to abnormal ciliogenesis, Am. J. Hum. Genet. 88 (4) (2011) 508–515.

[55] B. Le Marec, E. Passarge, P. Dellenbach, J. Kerisit, J. Signargout, B. Ferrand, J. Senecal, Lethal neonatal forms of chondroectodermal dysplasia. Apropos of 5 cases, Ann. Radiol. 16 (1) (1973) 19–26.

[56] J. Spranger, B. Grimm, M. Weller, G. Weissenbacher, J. Herrmann, E. Gilbert, R. Krepler, Short rib-polydactyly (SRP) syndromes, types Majewski and Saldino-Noonan, Z. fur Kinderheilkd. 116 (2) (1974) 73–94.

[57] J. Spranger, L.O. Langer, M.H. Weller, J. Herrmann, Short rib-polydactyly syndromes and related conditions, Birth Defects Orig. Artic. Ser. 10 (9) (1974) 117–123.

[58] R. Bernstein, J. Isdale, M. Pinto, J. Du Toit Zaaijman, T. Jenkins, Short rib-polydactyly syndrome: a single or heterogeneous entity? A re-evaluation prompted by four new cases, J. Med. Genet. 22 (1) (1985) 46–53.

[59] P. Naumoff, L.W. Young, J. Mazer, A.J. Amortegui, Short rib-polydactyly syndrome type 3, Radiology 122 (2) (1977) 443–447.

[60] S.S. Yang, L.O. Langer Jr., A. Cacciarelli, B.B. Dahms, E.R. Unger, J. Roskamp, N.D. Dinno, H. Chen, Three conditions in neonatal asphyxiating thoracic dysplasia (Jeune) and short rib-polydactyly syndrome spectrum: a clinicopathologic study, Am. J. Med. Genet. Suppl. 3 (1987) 191–207.

[61] H. Chen, S.S. Yang, E. Gonzalez, M. Fowler, A. Al Saadi, Short rib-polydactyly syndrome, Majewski type, Am. J. Med. Genet. 7 (2) (1980) 215–222.

[62] N. Kovacs, I. Sarkany, G. Mohay, K. Adamovich, T. Ertl, G. Kosztolanyi, R. Kellermayer, High incidence of short rib-polydactyly syndrome type IV in a Hungarian Roma subpopulation, Am. J. Med. Genet. A 140 (24) (2006) 2816–2818.

[63] E. Passarge, Familial occurrence of a short rib syndrome with hydrops fetalis but without polydactyly, Am. J. Med. Genet. 14 (2) (1983) 403–405.

[64] S.S. Yang, J.A. Roth, L.O. Langer Jr., Short rib syndrome Beemer-Langer type with polydactyly: a multiple congenital anomalies syndrome, Am. J. Med. Genet. 39 (3) (1991) 243–246.

[65] P. Kannu, J.H. McFarlane, R. Savarirayan, S. Aftimos, An unclassifiable short rib-polydactyly syndrome with acromesomelic hypomineralization and campomelia in siblings, Am. J. Med. Genet. Part A 143A (21) (2007) 2607–2611.

[66] F. Oberklaid, D.M. Danks, V. Mayne, P. Campbell, Asphyxiating thoracic dysplasia. Clinical, radiological, and pathological information on 10 patients, Arch. Dis. Child. 52 (10) (1977) 758–765.

[67] M.S. Hopper, J.E. Boultbee, A.R. Watson, Polyhydramnios associated with congenital pancreatic cysts and asphyxiating thoracic dysplasia. A case report, South Afr. Med. J. = Suid-Afrikaanse Tydskrif Vir Geneeskunde 56 (1) (1979) 32–33.

[68] L.M. Yerian, L. Brady, J. Hart, Hepatic manifestations of Jeune syndrome (asphyxiating thoracic dystrophy), Semin. Liver Dis. 23 (2) (2003) 195–200.

[69] A.M. Lehman, P. Eydoux, D. Doherty, I.A. Glass, D. Chitayat, B.Y. Chung, S. Langlois, S.L. Yong, R.B. Lowry, F. Hildebrandt, P. Trnka, Co-occurrence of Joubert syndrome and Jeune asphyxiating thoracic dystrophy, Am. J. Med. Genet. Part A 152A (6) (2010) 1411–1419.

[70] K. Tuz, R. Bachmann-Gagescu, D.R. O'Day, K. Hua, C.R. Isabella, I.G. Phelps, A.E. Stolarski, B.J. O'Roak, J.C. Dempsey, C. Lourenco, A. Alswaid, C.G. Bonnemann, L. Medne, S. Nampoothiri, Z. Stark, R.J. Leventer, M. Topcu, A. Cansu, S. Jagadeesh, S. Done, G.E. Ishak, I.A. Glass, J. Shendure, S.C. Neuhauss, C.R. Haldeman-Englert, D. Doherty, R.J. Ferland, Mutations in CSPP1 cause primary cilia abnormalities and Joubert syndrome with or without Jeune asphyxiating thoracic dystrophy, Am. J. Hum. Genet. 94 (1) (2014) 62–72.

[71] N. Akizu, J.L. Silhavy, R.O. Rosti, E. Scott, A.G. Fenstermaker, J. Schroth, M.S. Zaki, H. Sanchez, N. Gupta, M. Kabra, M. Kara, T. Ben-Omran, B. Rosti, A. Guemez-Gamboa, E. Spencer, R. Pan, N. Cai, M. Abdellateef, S. Gabriel, J. Halbritter, F. Hildebrandt, H. van Bokhoven, M. Gunel, J.G. Gleeson, Mutations in CSPP1 lead to classical Joubert syndrome, Am. J. Hum. Genet. 94 (1) (2014) 80–86.

[72] R. Shaheen, M. Schmidts, E. Faqeih, A. Hashem, E. Lausch, I. Holder, A. Superti-Furga, U.K. Consortium, H.M. Mitchison, A. Almoisheer, R. Alamro, T. Alshiddi, F. Alzahrani, P.L. Beales, F.S. Alkuraya, A founder CEP120 mutation in Jeune asphyxiating thoracic dystrophy expands the role of centriolar proteins in skeletal ciliopathies, Hum. Mol. Genet. 24 (5) (2015) 1410–1419.

[73] S. Roosing, M. Romani, M. Isrie, R.O. Rosti, A. Micalizzi, D. Musaev, T. Mazza, L. Al-Gazali, U. Altunoglu, E. Boltshauser, S. D'Arrigo, B. De Keersmaecker, H. Kayserili, S. Brandenberger, I. Kraoua, P.R. Mark, T. McKanna, J. Van Keirsbilck, P. Moerman, A. Poretti, R. Puri, H. Van Esch, J.G. Gleeson, E.M. Valente, Mutations in CEP120 cause Joubert syndrome as well as complex ciliopathy phenotypes, J. Med. Genet. 53 (9) (2016) 608–615.

[74] F. Mainzer, R.M. Saldino, M.B. Ozonoff, H. Minagi, Familial nephropathy associated with retinitis pigmentosa, cerebellar ataxia and skeletal abnormalities, Am. J. Med. 49 (4) (1970) 556–562.

[75] M. Popovic-Rolovic, N. Calic-Perisic, G. Bunjevacki, D. Negovanovic, Juvenile nephronophthisis associated with retinal pigmentary dystrophy, cerebellar ataxia, and skeletal abnormalities, Arch. Dis. Child. 51 (10) (1976) 801–803.

[76] D.G. Robins, T.A. French, T.M. Chakera, Juvenile nephronophthisis associated with skeletal abnormalities and hepatic fibrosis, Arch. Dis. Child. 51 (10) (1976) 799–801.

[77] A. Giedion, Phalangeal cone shaped epiphysis of the hands (PhCSEH) and chronic renal disease–the conorenal syndromes, Pediatr. Radiol. 8 (1) (1979) 32–38.

[78] Z. Wang, A. Iida, N. Miyake, K.M. Nishiguchi, K. Fujita, T. Nakazawa, A. Alswaid, M.A. Albalwi, O.H. Kim, T.J. Cho, G.Y. Lim, B. Isidor, A. David, C.F. Rustad, E. Merckoll, J. Westvik, E.L. Stattin, G. Grigelioniene, I. Kou, M. Nakajima, H. Ohashi, S. Smithson, N. Matsumoto, G. Nishimura, S. Ikegawa, Axial spondylometaphyseal dysplasia is caused by C21orf2 mutations, PLoS One 11 (3) (2016) e0150555.

[79] G. Wheway, M. Schmidts, D.A. Mans, K. Szymanska, T.M. Nguyen, H. Racher, I.G. Phelps, G. Toedt, J. Kennedy, K.A. Wunderlich, N. Sorusch, Z.A. Abdelhamed, S. Natarajan, W. Herridge, J. van Reeuwijk, N. Horn, K. Boldt, D.A. Parry, S.J. Letteboer, S. Roosing, M. Adams, S.M. Bell, J. Bond, J. Higgins, E.E. Morrison, D.C. Tomlinson, G.G. Slaats, T.J. van Dam, L. Huang, K. Kessler, A. Giessl, C.V. Logan, E.A. Boyle, J. Shendure, S. Anazi, M. Aldahmesh, S. Al Hazzaa, R.A. Hegele, C. Ober, P. Frosk, A.A. Mhanni, B.N. Chodirker, A.E. Chudley, R. Lamont, F.P. Bernier, C.L. Beaulieu, P. Gordon, R.T. Pon, C. Donahue, A.J. Barkovich, L. Wolf, C. Toomes, C.T. Thiel, K.M. Boycott, M. McKibbin, C.F. Inglehearn, U.K. Consortium, G. University of Washington Center for Mendelian, F. Stewart, H. Omran, M.A. Huynen, P.I. Sergouniotis, F.S. Alkuraya, J.S. Parboosingh, A.M. Innes, C.E. Willoughby, R.H. Giles, A.R. Webster, M. Ueffing, O. Blacque, J.G. Gleeson, U. Wolfrum, P.L. Beales, T. Gibson, D. Doherty, H.M. Mitchison, R. Roepman, C.A. Johnson, An SiRNA-based Functional Genomics Screen for the Identification of Regulators of Ciliogenesis and Ciliopathy Genes, Nat. Cell Biol. 17 (8) (2015) 1074–1087.

[80] B. Isidor, S. Baron, P. Khau van Kien, A.M. Bertrand, A. David, M. Le Merrer, Axial spondylometaphyseal dysplasia: confirmation and further delineation of a new SMD with retinal dystrophy, Am. J. Med. Genet. A 152A (6) (2010) 1550–1554.

[81] S. Ehara, O.H. Kim, S. Maisawa, S. Takasago, G. Nishimura, Axial spondylometaphyseal dysplasia, Eur. J. Pediatr. 156 (8) (1997) 627–630.

[82] P. Lahiry, J. Wang, J.F. Robinson, J.P. Turowec, D.W. Litchfield, M.B. Lanktree, G.B. Gloor, E.G. Puffenberger, K.A. Strauss, M.B. Martens, D.A. Ramsay, C.A. Rupar, V. Siu, R.A. Hegele, A multiplex human syndrome implicates a key role for intestinal cell kinase in development of central nervous, skeletal, and endocrine systems, Am. J. Hum. Genet. 84 (2) (2009) 134–147.

[83] M.M. Oud, C. Bonnard, D.A. Mans, U. Altunoglu, S. Tohari, A.Y. Ng, A. Eskin, H. Lee, C.A. Rupar, N.P. de Wagenaar, K.M. Wu, P. Lahiry, G.J. Pazour, S.F. Nelson, R.A. Hegele, R. Roepman, H. Kayserili, B. Venkatesh, V.M. Siu, B. Reversade, H.H. Arts, A novel ICK mutation causes ciliary disruption and lethal endocrine-cerebro-osteodysplasia syndrome, Cilia 5 (2016) 8.

[84] S. Paige Taylor, M. Kunova Bosakova, M. Varecha, L. Balek, T. Barta, L. Trantirek, I. Jelinkova, I. Duran, I. Vesela, K.N. Forlenza, J.H. Martin, A. Hampl, G. University of Washington Center for Mendelian, M. Bamshad, D. Nickerson, M.L. Jaworski, J. Song, H. Wan Ko, D.H. Cohn, D. Krakow, P. Krejci, An inactivating mutation in intestinal cell kinase, ICK, impairs hedgehog signalling and causes short Rib-polydactyly syndrome, Hum. Mol. Genet. (2016).

[85] M. Baraitser, The orofaciodigital (OFD) syndromes, J. Med. Genet. 23 (2) (1986) 116–119.

[86] N.C. Nevin, P.S. Thomas, Orofaciodigital syndrome type IV: report of a patient, Am. J. Med. Genet. 32 (2) (1989) 151–154.

[87] L.C. Ades, W.K. Clapton, A. Morphett, L.L. Morris, E.A. Haan, Polydactyly, campomelia, ambiguous genitalia, cystic dysplastic kidneys, and cerebral malformation in a fetus of consanguineous parents: a new multiple malformation syndrome, or a severe form of oral-facial-digital syndrome type IV? Am. J. Med. Genet. 49 (2) (1994) 211–217.

[88] B. Franco, C. Thauvin-Robinet, Update on oral-facial-digital syndromes (OFDS), Cilia 5 (2016) 12.

[89] C.R. Cortes, A.M. McInerney-Leo, I. Vogel, M.C. Rondon Galeano, P.J. Leo, J.E. Harris, L.K. Anderson, P.A. Keith, M.A. Brown, M. Ramsing, E.L. Duncan, A. Zankl, C. Wicking, Mutations in human C2CD3 cause skeletal dysplasia and provide new insights into phenotypic and cellular consequences of altered C2CD3 function, Sci. Rep. 6 (2016) 24083.

[90] A.L. Ruess, S. Pruzansky, E.F. Lis, Intellectual development and the OFD syndrome: a review, Cleft Palate J. 2 (1965) 350–356.

[91] C. Bredrup, S. Saunier, M.M. Oud, T. Fiskerstrand, A. Hoischen, D. Brackman, S.M. Leh, M. Midtbo, E. Filhol, C. Bole-Feysot, P. Nitschke, C. Gilissen, O.H. Haugen, J.S. Sanders, I. Stolte-Dijkstra, D.A. Mans, E.J. Steenbergen, B.C. Hamel, M. Matignon, R. Pfundt, C. Jeanpierre, H. Boman, E. Rodahl, J.A. Veltman, P.M. Knappskog, N.V. Knoers, R. Roepman, H.H. Arts, Ciliopathies with skeletal anomalies and renal insufficiency due to mutations in the IFT-A gene WDR19, Am. J. Hum. Genet. 89 (5) (2011) 634–643.

[92] M.J. Amar, R. Sutphen, B.G. Kousseff, Expanded phenotype of cranioectodermal dysplasia (Sensenbrenner syndrome), Am. J. Med. Genet. 70 (4) (1997) 349–352.

[93] H. Arts, N. Knoers, Cranioectodermal dysplasia, in: R.A. Pagon et al. (Ed.), GeneReviews(R), 1993. Seattle (WA).

[94] L.S. Levin, J.C. Perrin, L. Ose, J.P. Dorst, J.D. Miller, V.A. McKusick, A heritable syndrome of craniosynostosis, short thin hair, dental abnormalities, and short limbs: cranioectodermal dysplasia, J. Pediatr. 90 (1) (1977) 55–61.

[95] G.D. Lang, I.D. Young, Cranioectodermal dysplasia in sibs, J. Med. Genet. 28 (6) (1991) 424.

[96] T. Eke, G. Woodruff, I.D. Young, A new oculorenal syndrome: retinal dystrophy and tubulointerstitial nephropathy in cranioectodermal dysplasia, Br. J. Ophthalmol. 80 (5) (1996) 490–491.

[97] M. Zaffanello, F. Diomedi-Camassei, M.L. Melzi, G. Torre, F. Callea, F. Emma, Sensenbrenner syndrome: a new member of the hepatorenal fibrocystic family, Am. J. Med. Genet. Part A 140 (21) (2006) 2336–2340.

[98] J. Walczak-Sztulpa, J. Eggenschwiler, D. Osborn, D.A. Brown, F. Emma, C. Klingenberg, R.C. Hennekam, G. Torre, M. Garshasbi, A. Tzschach, M. Szczepanska, M. Krawczynski, J. Zachwieja, D. Zwolinska, P.L. Beales, H.H. Ropers, A. Latos-Bielenska, A.W. Kuss, Cranioectodermal dysplasia, Sensenbrenner syndrome, is a ciliopathy caused by mutations in the IFT122 gene, Am. J. Hum. Genet. 86 (6) (2010) 949–956.

[99] A.E. Konstantinidou, H. Fryssira, S. Sifakis, C. Karadimas, P. Kaminopetros, G. Agrogiannis, S. Velonis, P.G. Nikkels, E. Patsouris, Cranioectodermal dysplasia: a probable ciliopathy, Am. J. Med. Genet. Part A 149A (10) (2009) 2206–2211.

[100] R.W. Ellis, S. van Creveld, A syndrome characterized by ectodermal dysplasia, polydactyly, chondro-dysplasia and congenital morbus cordis: report of three cases, Arch. Dis. Child. 15 (82) (1940) 65–84.

[101] V.A. McKusick, R. Eldridge, J.A. Hostetler, J.A. Egeland, Dwarfism in the amish, Trans. Assoc. Am. Phys. 77 (1964) 151–168.

[102] M.G. Blackburn, R.E. Belliveau, Ellis-van Creveld syndrome. A report of previously undescribed anomalies in two siblings, Am. J. Dis. Child. 122 (3) (1971) 267–270.

[103] G. Baujat, M. Le Merrer, Ellis-van Creveld syndrome, Orphanet J. Rare Dis. 2 (2007) 27.

[104] H. Weyers, Ueber eine korrelierte Missbildung der Kiefer und Extremitatenakren (Dysostosis acro-facialis), Fortschr Roentgenstr. 77 (1952) 5.

[105] M. Roubicek, J. Spranger, Weyers acrodental dysostosis in a family, Clin. Genet. 26 (6) (1984) 587–590.

[106] M. Roubicek, J. Spranger, Syndrome of polydactyly, conical teeth and nail dysplasia, Am. J. Med. Genet. 20 (1) (1985) 205–207.

[107] C.J. Curry, B.D. Hall, Polydactyly, conical teeth, nail dysplasia, and short limbs: a new autosomal dominant malformation syndrome, Birth Defects Orig. Artic. Ser. 15 (5B) (1979) 253–263.

[108] C. Alby, K. Piquand, C. Huber, A. Megarbane, A. Ichkou, M. Legendre, F. Pelluard, F. Encha-Ravazi, G. Abi-Tayeh, B. Bessieres, S. El Chehadeh-Djebbar, N. Laurent, L. Faivre, L. Sztriha, M. Zombor, H. Szabo, M. Failler, M. Garfa-Traore, C. Bole, P. Nitschke, M. Nizon, N. Elkhartoufi, F. Clerget-Darpoux, A. Munnich, S. Lyonnet, M. Vekemans, S. Saunier, V. Cormier-Daire, T. Attie-Bitach, S. Thomas, Mutations in KIAA0586 cause lethal ciliopathies ranging from a hydrolethalus phenotype to short-rib polydactyly syndrome, Am. J. Hum. Genet. 97 (2) (2015) 311–318.

[109] R. Bachmann-Gagescu, I.G. Phelps, J.C. Dempsey, V.A. Sharma, G.E. Ishak, E.A. Boyle, M. Wilson, C. Marques Lourenco, M. Arslan, G. University of Washington Center for Mendelian, J. Shendure, D. Doherty, KIAA0586 is mutated in Joubert Syndrome, Hum. Mutat. 36 (9) (2015) 831–835.

[110] M.C. Malicdan, T. Vilboux, J. Stephen, D. Maglic, L. Mian, D. Konzman, J. Guo, D. Yildirimli, J. Bryant, R. Fischer, W.M. Zein, J. Snow, M. Vemulapalli, J.C. Mullikin, C. Toro, B.D. Solomon, J.E. Niederhuber, N.C.S. Program, W.A. Gahl, M. Gunay-Aygun, Mutations in human homologue of chicken talpid3 gene (KIAA0586) cause a hybrid ciliopathy with overlapping features of Jeune and Joubert syndromes, J. Med. Genet. 52 (12) (2015) 830–839.

[111] L.A. Stephen, H. Tawamie, G.M. Davis, L. Tebbe, P. Nurnberg, G. Nurnberg, H. Thiele, M. Thoenes, E. Boltshauser, S. Uebe, O. Rompel, A. Reis, A.B. Ekici, L. McTeir, A.M. Fraser, E.A. Hall, P. Mill, N. Daudet, C. Cross, U. Wolfrum, R.A. Jamra, M.G. Davey, H.J. Bolz, TALPID3 controls centrosome and cell polarity and the human ortholog KIAA0586 is mutated in Joubert syndrome (JBTS23), Elife (2015) 4.

[112] A. Putoux, S. Thomas, K.L. Coene, E.E. Davis, Y. Alanay, G. Ogur, E. Uz, D. Buzas, C. Gomes, S. Patrier, C.L. Bennett, N. Elkhartoufi, M.H. Frison, L. Rigonnot, N. Joye, S. Pruvost, G.E. Utine, K. Boduroglu, P. Nitschke, L. Fertitta, C. Thauvin-Robinet, A. Munnich, V. Cormier-Daire, R. Hennekam, E. Colin, N.A. Akarsu, C. Bole-Feysot, N. Cagnard, A. Schmitt, N. Goudin, S. Lyonnet, F. Encha-Razavi, J.P. Siffroi, M. Winey, N. Katsanis, M. Gonzales, M. Vekemans, P.L. Beales, T. Attie-Bitach, KIF7 mutations cause fetal hydrolethalus and acro-callosal syndromes, Nat. Genet. 43 (6) (2011) 601–606.

[113] A. Schinzel, W. Schmid, Hallux duplication, postaxial polydactyly, absence of the corpus callosum, severe mental retardation, and additional anomalies in two unrelated patients: a new syndrome, Am. J. Med. Genet. 6 (3) (1980) 241–249.

[114] V.L. Ruiz-Perez, H.J. Blair, M.E. Rodriguez-Andres, M.J. Blanco, A. Wilson, Y.N. Liu, C. Miles, H. Peters, J.A. Goodship, Evc is a positive mediator of Ihh-regulated bone growth that localises at the base of chondrocyte cilia, Development 134 (16) (2007) 2903–2912.

[115] P.J. Ocbina, J.T. Eggenschwiler, I. Moskowitz, K.V. Anderson, Complex interactions between genes controlling trafficking in primary cilia, Nat. Genet. 43 (6) (2011) 547–553.

[116] C.R. Cortes, V. Metzis, C. Wicking, Unmasking the ciliopathies: craniofacial defects and the primary cilium, Wiley Interdisc. Rev. Dev. Biol. 4 (6) (2015) 637–653.

[117] K.C. Corbit, P. Aanstad, V. Singla, A.R. Norman, D.Y. Stainier, J.F. Reiter, Vertebrate Smoothened functions at the primary cilium, Nature 437 (7061) (2005) 1018–1021.

[118] S.R. May, A.M. Ashique, M. Karlen, B. Wang, Y. Shen, K. Zarbalis, J. Reiter, J. Ericson, A.S. Peterson, Loss of the retrograde motor for IFT disrupts localization of Smo to cilia and prevents the expression of both activator and repressor functions of Gli, Dev. Biol. 287 (2) (2005) 378–389.

[119] S. Rix, A. Calmont, P.J. Scambler, P.L. Beales, An Ift80 mouse model of short rib polydactyly syndromes shows defects in hedgehog signalling without loss or malformation of cilia, Hum. Mol. Genet. 20 (7) (2011) 1306–1314.

[120] R. Rohatgi, L. Milenkovic, M.P. Scott, Patched1 regulates hedgehog signaling at the primary cilium, Science 317 (5836) (2007) 372–376.

[121] S.C. Goetz, P.J. Ocbina, K.V. Anderson, The primary cilium as a Hedgehog signal transduction machine, Methods Cell Biol. 94 (2009) 199–222.

[121a] P. Satir, S.T. Christensen, Overview of structure and function of mammalian cilia, Annu Rev Physiol. 69 (2007), 377–400.

[122] S.C. Goetz, K.V. Anderson, The primary cilium: a signalling centre during vertebrate development, Nat. Rev. Genet. 11 (5) (2010) 331–344.

[123] N. Sasai, J. Briscoe, Primary cilia and graded Sonic Hedgehog signaling, Wiley Interdiscip. Rev. Dev. Biol. 1 (5) (2012) 753–772.

[124] H.W. Ehlen, L.A. Buelens, A. Vortkamp, Hedgehog signaling in skeletal development, Birth Defects Res. C Embryo Today 78 (3) (2006) 267–279.

[125] C.J. Haycraft, Q. Zhang, B. Song, W.S. Jackson, P.J. Detloff, R. Serra, B.K. Yoder, Intraflagellar transport is essential for endochondral bone formation, Development 134 (2) (2007) 307–316.

[126] B. Song, C.J. Haycraft, H.S. Seo, B.K. Yoder, R. Serra, Development of the post-natal growth plate requires intraflagellar transport proteins, Dev. Biol. 305 (1) (2007) 202–216.

[127] X. Yuan, J. Cao, X. He, R. Serra, J. Qu, X. Cao, S. Yang, Ciliary IFT80 balances canonical versus non-canonical hedgehog signalling for osteoblast differentiation, Nat. Commun. 7 (2016) 11024.

[128] N. Katsanis, S.J. Ansley, J.L. Badano, E.R. Eichers, R.A. Lewis, B.E. Hoskins, P.J. Scambler, W.S. Davidson, P.L. Beales, J.R. Lupski, Triallelic inheritance in Bardet-Biedl syndrome, a Mendelian recessive disorder, Science 293 (5538) (2001) 2256–2259.

[129] H. Khanna, E.E. Davis, C.A. Murga-Zamalloa, A. Estrada-Cuzcano, I. Lopez, A.I. den Hollander, M.N. Zonneveld, M.I. Othman, N. Waseem, C.F. Chakarova, C. Maubaret, A. Diaz-Font, I. MacDonald, D.M. Muzny, D.A. Wheeler, M. Morgan, L.R. Lewis, C.V. Logan, P.L. Tan, M.A. Beer, C.F. Inglehearn, R.A. Lewis, S.G. Jacobson, C. Bergmann, P.L. Beales, T. Attie-Bitach, C.A. Johnson, E.A. Otto, S.S. Bhattacharya, F. Hildebrandt, R.A. Gibbs, R.K. Koenekoop, A. Swaroop, N. Katsanis, A common allele in RPGRIP1L is a modifier of retinal degeneration in ciliopathies, Nat. Genet. 41 (6) (2009) 739–745.

[130] C. Bergmann, J. von Bothmer, N. Ortiz Bruchle, A. Venghaus, V. Frank, H. Fehrenbach, T. Hampel, L. Pape, A. Buske, J. Jonsson, N. Sarioglu, A. Santos, J.C. Ferreira, J.U. Becker, R. Cremer, J. Hoefele, M.R. Benz, L.T. Weber, R. Buettner, K. Zerres, Mutations in multiple PKD genes may explain early and severe polycystic kidney disease, J. Am. Soc. Nephrol. 22 (11) (2011) 2047–2056.

[131] C. Thiel, K. Kessler, A. Giessl, A. Dimmler, S.A. Shalev, S. von der Haar, M. Zenker, D. Zahnleiter, H. Stoss, E. Beinder, R. Abou Jamra, A.B. Ekici, N. Schroder-Kress, T. Aigner, T. Kirchner, A. Reis, J.H. Brandstatter, A. Rauch, NEK1 mutations cause short-rib polydactyly syndrome type majewski, Am. J. Hum. Genet. 88 (1) (2011) 106–114.

[132] M. Schmidts, Y. Hou, C.R. Cortes, D.A. Mans, C. Huber, K. Boldt, M. Patel, J. van Reeuwijk, J.M. Plaza, S.E. van Beersum, Z.M. Yap, S.J. Letteboer, S.P. Taylor, W. Herridge, C.A. Johnson, P.J. Scambler, M. Ueffing, H. Kayserili, D. Krakow, S.M. King, UK10K, P.L. Beales, L. Al-Gazali, C. Wicking, V. Cormier-Daire, R. Roepman, H.M. Mitchison, G.B. Witman, TCTEX1D2 mutations underlie Jeune asphyxiating thoracic dystrophy with impaired retrograde intraflagellar transport, Nat. Commun. 6 (2015) 7074.

[133] I.R. Gibbons, D.J. Asai, W.J. Tang, B.H. Gibbons, A cytoplasmic dynein heavy chain in sea urchin embryos, Biol. Cell/Under Auspices Eur. Cell Biol. Organ. 76 (3) (1992) 303–309.

[134] S.H. Tynan, A. Purohit, S.J. Doxsey, R.B. Vallee, Light intermediate chain 1 defines a functional subfraction of cytoplasmic dynein which binds to pericentrin, J. Biol. Chem. 275 (42) (2000) 32763–32768.

[135] S.M. Hughes, K.T. Vaughan, J.S. Herskovits, R.B. Vallee, Molecular analysis of a cytoplasmic dynein light intermediate chain reveals homology to a family of ATPases, J. Cell Sci. 108 (Pt 1) (1995) 17–24.

[136] M.A. Crackower, D.S. Sinasac, J. Xia, J. Motoyama, M. Prochazka, J.M. Rommens, S.W. Scherer, L.C. Tsui, Cloning and characterization of two cytoplasmic dynein intermediate chain genes in mouse and human, Genomics 55 (3) (1999) 257–267.

[137] T. Dick, K. Ray, H.K. Salz, W. Chia, Cytoplasmic dynein (ddlc1) mutations cause morphogenetic defects and apoptotic cell death in *Drosophila melanogaster*, Mol. Cell. Biol. 16 (5) (1996) 1966–1977.

[138] L.M. DiBella, S.E. Benashski, H.W. Tedford, A. Harrison, R.S. Patel-King, S.M. King, The Tctex1/Tctex2 class of dynein light chains. Dimerization, differential expression, and interaction with the LC8 protein family, J. Biol. Chem. 276 (17) (2001) 14366–14373.

[139] N.J. Pavlos, T.S. Cheng, A. Qin, P.Y. Ng, H.T. Feng, E.S. Ang, A. Carrello, C.H. Sung, R. Jahn, M.H. Zheng, J. Xu, Tctex-1, a novel interaction partner of Rab3D, is required for osteoclastic bone resorption, Mol. Cell. Biol. 31 (7) (2011) 1551–1564.

[140] J. Jiang, L. Yu, X. Huang, X. Chen, D. Li, Y. Zhang, L. Tang, S. Zhao, Identification of two novel human dynein light chain genes, DNLC2A and DNLC2B, and their expression changes in hepatocellular carcinoma tissues from 68 Chinese patients, Gene 281 (1–2) (2001) 103–113.

[141] K.K. Pfister, P.R. Shah, H. Hummerich, A. Russ, J. Cotton, A.A. Annuar, S.M. King, E.M. Fisher, Genetic analysis of the cytoplasmic dynein subunit families, PLoS Genet. 2 (1) (2006) e1.

[142] K.K. Pfister, E.M. Fisher, I.R. Gibbons, T.S. Hays, E.L. Holzbaur, J.R. McIntosh, M.E. Porter, T.A. Schroer, K.T. Vaughan, G.B. Witman, S.M. King, R.B. Vallee, Cytoplasmic dynein nomenclature, J. Cell Biol. 171 (3) (2005) 411–413.

[143] E.A. Vaisberg, P.M. Grissom, J.R. McIntosh, Mammalian cells express three distinct dynein heavy chains that are localized to different cytoplasmic organelles, J. Cell Biol. 133 (4) (1996) 831–842.

[144] B.H. Gibbons, D.J. Asai, W.J. Tang, T.S. Hays, I.R. Gibbons, Phylogeny and expression of axonemal and cytoplasmic dynein genes in sea urchins, Mol. Biol. Cell 5 (1) (1994) 57–70.

[145] A. Mikami, S.H. Tynan, T. Hama, K. Luby-Phelps, T. Saito, J.E. Crandall, J.C. Besharse, R.B. Vallee, Molecular structure of cytoplasmic dynein 2 and its distribution in neuronal and ciliated cells, J. Cell Sci. 115 (Pt 24) (2002) 4801–4808.

[146] P. Rompolas, L.B. Pedersen, R.S. Patel-King, S.M. King, Chlamydomonas FAP133 is a dynein intermediate chain associated with the retrograde intraflagellar transport motor, J. Cell Sci. 120 (Pt 20) (2007) 3653–3665.

[147] B.L. Krock, I. Mills-Henry, B.D. Perkins, Retrograde intraflagellar transport by cytoplasmic dynein-2 is required for outer segment extension in vertebrate photoreceptors but not arrestin translocation, Invest. Ophthalmol. Vis. Sci. 50 (11) (2009) 5463–5471.

[148] C. Huber, S. Wu, A.S. Kim, S. Sigaudy, A. Sarukhanov, V. Serre, G. Baujat, K.H. Le Quan Sang, D.L. Rimoin, D.H. Cohn, A. Munnich, D. Krakow, V. Cormier-Daire, WDR34 mutations that cause short-rib polydactyly syndrome type III/severe asphyxiating thoracic dysplasia reveal a role for the NF-kappaB pathway in cilia, Am. J. Hum. Genet. 93 (5) (2013) 926–931.

[149] M. Schmidts, J. Vodopiutz, S. Christou-Savina, C.R. Cortes, A.M. McInerney-Leo, R.D. Emes, H.H. Arts, B. Tuysuz, J. D'Silva, P.J. Leo, T.C. Giles, M.M. Oud, J.A. Harris, M. Koopmans, M. Marshall, N. Elcioglu, A. Kuechler, D. Bockenhauer, A.T. Moore, L.C. Wilson, A.R. Janecke, M.E. Hurles, W. Emmet, B. Gardiner, B. Streubel, B. Dopita, A. Zankl, H. Kayserili, P.J. Scambler, M.A. Brown, P.L. Beales, C. Wicking, E.L. Duncan, H.M. Mitchison, Mutations in the gene encoding IFT dynein complex component WDR34 cause Jeune asphyxiating thoracic dystrophy, Am. J. Hum. Genet. 93 (5) (2013) 932–944.

[150] A.M. McInerney-Leo, M. Schmidts, C.R. Cortes, P.J. Leo, B. Gener, A.D. Courtney, B. Gardiner, J.A. Harris, Y. Lu, M. Marshall, P.J. Scambler, P.L. Beales, M.A. Brown, A. Zankl, H.M. Mitchison, E.L. Duncan, C. Wicking, Short-rib polydactyly and Jeune syndromes are caused by mutations in WDR60, Am. J. Hum. Genet. 93 (3) (2013) 515–523.

[151] K.J. Palmer, L. MacCarthy-Morrogh, N. Smyllie, D.J. Stephens, A role for Tctex-1 (DYNLT1) in controlling primary cilium length, Eur. J. Cell Biol. 90 (10) (2011) 865–871.

[152] A. Li, M. Saito, J.Z. Chuang, Y.Y. Tseng, C. Dedesma, K. Tomizawa, T. Kaitsuka, C.H. Sung, Ciliary transition zone activation of phosphorylated Tctex-1 controls ciliary resorption, S-phase entry and fate of neural progenitors, Nat. Cell Biol. 13 (4) (2011) 402–411.

[153] D. Asante, N.L. Stevenson, D.J. Stephens, Subunit composition of the human cytoplasmic dynein-2 complex, J. Cell Sci. 127 (Pt 21) (2014) 4774–4787.

[154] M. Schmidts, J. Vodopiutz, S. Christou-Savina, C.R. Cortés, A.M. McInerney-Leo, R.D. Emes, H.H. Arts, B. Tüysüz, J. D'Silva, P.J. Leo, T.C. Giles, M.M. Oud, J.A. Harris, M. Koopmans, M. Marshall, N. Elçioglu, A. Kuechler, D. Bockenhauer, A.T. Moore, L.C. Wilson, A.R. Janecke, M.E. Hurles, W. Emmet, B. Gardiner, B. Streubel, B. Dopita, A. Zankl, H. Kayserili, P.J. Scambler, M.A. Brown, P.L. Beales, C. Wicking; UK10K, E.L. Duncan, H.M. Mitchison, Mutations in the gene encoding IFT dynein complex component WDR34 cause Jeune asphyxiating thoracic dystrophy, Am. J. Hum. Genet. 93 (5) (November 7, 2013) 932–944. http://doi:10.1016/j.ajhg.2013.10.003. Epub October 31, 2013.

[154a] N. Dagoneau, M. Goulet, D. Genevieve, Y. Sznajer, J. Martinovic, S. Smithson, C. Huber, G. Baujat, E. Flori, L. Tecco, D. Cavalcanti, A.L. Delezoide, V. Serre, M. Le Merrer, A. Munnich, V. Cormier-Daire, DYNC2H1 mutations cause asphyxiating thoracic dystrophy and short rib-polydactyly syndrome, type III, Am. J. Hum. Genet. 84 (5) (2009) 706–711.

[155] A.E. Merrill, B. Merriman, C. Farrington-Rock, N. Camacho, E.T. Sebald, V.A. Funari, M.J. Schibler, M.H. Firestein, Z.A. Cohn, M.A. Priore, A.K. Thompson, D.L. Rimoin, S.F. Nelson, D.H. Cohn, D. Krakow, Ciliary abnormalities due to defects in the retrograde transport protein DYNC2H1 in short-rib polydactyly syndrome, Am. J. Hum. Genet. 84 (4) (2009) 542–549.

[156] J. El Hokayem, C. Huber, A. Couve, J. Aziza, G. Baujat, R. Bouvier, D.P. Cavalcanti, F.A. Collins, M.P. Cordier, A.L. Delezoide, M. Gonzales, D. Johnson, M. Le Merrer, A. Levy-Mozziconacci, P. Loget, D. Martin-Coignard, J. Martinovic, G.R. Mortier, M.J. Perez, J. Roume, G. Scarano, A. Munnich, V. Cormier-Daire, NEK1 and DYNC2H1 are both involved in short rib polydactyly Majewski type but not in Beemer Langer cases, J. Med. Genet. 49 (4) (2012) 227–233.

[157] N. Badiner, S.P. Taylor, K. Forlenza, R.S. Lachman, G. University of Washington Center for Mendelian, M. Bamshad, D. Nickerson, D.H. Cohn, D. Krakow, Mutations in DYNC2H1, the cytoplasmic dynein 2, heavy chain 1 motor protein gene, cause short-rib polydactyly type I, Saldino-Noonan type, Clin. Genet. (2016).

[158] I.R. Gibbons, B.H. Gibbons, G. Mocz, D.J. Asai, Multiple nucleotide-binding sites in the sequence of dynein beta heavy chain, Nature 352 (6336) (1991) 640–643.

[159] T. Kon, M. Nishiura, R. Ohkura, Y.Y. Toyoshima, K. Sutoh, Distinct functions of nucleotide-binding/hydrolysis sites in the four AAA modules of cytoplasmic dynein, Biochemistry 43 (35) (2004) 11266–11274.

[160] A.P. Carter, Crystal clear insights into how the dynein motor moves, J. Cell Sci. 126 (Pt 3) (2013) 705–713.

[161] A.J. Roberts, T. Kon, P.J. Knight, K. Sutoh, S.A. Burgess, Functions and mechanics of dynein motor proteins, Nat. Rev. Mol. Cell Biol. 14 (11) (2013) 713–726.

[162] E.S. Gleave, H. Schmidt, A.P. Carter, A structural analysis of the AAA+ domains in *Saccharomyces cerevisiae* cytoplasmic dynein, J. Struct. Biol. 186 (3) (2014) 367–375.

[163] H. Schmidt, A.P. Carter, Review: structure and mechanism of the dynein motor ATPase, Biopolymers 105 (8) (2016) 557–567.

[164] A.P. Carter, R.D. Vale, Communication between the AAA+ ring and microtubule-binding domain of dynein, Biochem. Cell Biol. 88 (1) (2010) 15–21.

[165] A.P. Carter, C. Cho, L. Jin, R.D. Vale, Crystal structure of the dynein motor domain, Science 331 (6021) (2011) 1159–1165.

[166] H. Schmidt, R. Zalyte, L. Urnavicius, A.P. Carter, Structure of human cytoplasmic dynein-2 primed for its power stroke, Nature 518 (7539) (2015) 435–438.

[167] P.L. Beales, E. Bland, J.L. Tobin, C. Bacchelli, B. Tuysuz, J. Hill, S. Rix, C.G. Pearson, M. Kai, J. Hartley, C. Johnson, M. Irving, N. Elcioglu, M. Winey, M. Tada, P.J. Scambler, IFT80, which encodes a conserved intraflagellar transport protein, is mutated in Jeune asphyxiating thoracic dystrophy, Nat. Genet. 39 (6) (2007) 727–729.

[168] R.S. Patel-King, R.M. Gilberti, E.F. Hom, S.M. King, WD60/FAP163 is a dynein intermediate chain required for retrograde intraflagellar transport in cilia, Mol. Biol. Cell 24 (17) (2013) 2668–2677.

[169] A.A. Gholkar, S. Senese, Y.C. Lo, J. Capri, W.J. Deardorff, H. Dharmarajan, E. Contreras, E. Hodara, J.P. Whitelegge, P.K. Jackson, J.Z. Torres, Tctex1d2 associates with short-rib polydactyly syndrome proteins and is required for ciliogenesis, Cell Cycle 14 (7) (2015) 1116–1125.

[170] D. Asante, L. Maccarthy-Morrogh, A.K. Townley, M.A. Weiss, K. Katayama, K.J. Palmer, H. Suzuki, C.J. Westlake, D.J. Stephens, A role for the Golgi matrix protein giantin in ciliogenesis through control of the localization of dynein-2, J. Cell Sci. 126 (Pt 22) (2013) 5189–5197.

[171] D. Gao, R. Wang, B. Li, Y. Yang, Z. Zhai, D.Y. Chen, WDR34 is a novel TAK1-associated suppressor of the IL-1R/TLR3/TLR4-induced NF-kappaB activation pathway, Cell Mol. Life Sci. 66 (15) (2009) 2573–2584.

[172] C. Cossu, F. Incani, M.L. Serra, A. Coiana, G. Crisponi, L. Boccone, M.C. Rosatelli, New mutations in DYNC2H1 and WDR60 genes revealed by whole-exome sequencing in two unrelated Sardinian families with Jeune asphyxiating thoracic dystrophy, Clin. Chim. Acta 455 (2016) 172–180.

[173] A.M. McInerney-Leo, J.E. Harris, P.J. Leo, M.S. Marshall, B. Gardiner, E. Kinning, H.Y. Leong, F. McKenzie, W.P. Ong, J. Vodopiutz, C. Wicking, M.A. Brown, A. Zankl, E.L. Duncan, Whole exome sequencing is an efficient, sensitive and specific method for determining the genetic cause of short-rib thoracic dystrophies, Clin. Genet. 88 (6) (2015) 550–557.

[174] Y. Qiao, J. Wen, F. Tang, S. Martell, N. Shomer, P.C. Leung, M.D. Stephenson, E. Rajcan-Separovic, Whole exome sequencing in recurrent early pregnancy loss, Mol. Hum. Reprod. 22 (5) (2016) 364–372.

[175] P.M. Grissom, E.A. Vaisberg, J.R. McIntosh, Identification of a novel light intermediate chain (D2LIC) for mammalian cytoplasmic dynein 2, Mol. Biol. Cell 13 (3) (2002) 817–829.

[176] C.A. Perrone, D. Tritschler, P. Taulman, R. Bower, B.K. Yoder, M.E. Porter, A novel dynein light intermediate chain colocalizes with the retrograde motor for intraflagellar transport at sites of axoneme assembly in Chlamydomonas and Mammalian cells, Mol. Biol. Cell 14 (5) (2003) 2041–2056.

[177] S.P. Taylor, T.J. Dantas, I. Duran, S. Wu, R.S. Lachman, C. University of Washington Center for Mendelian Genomics, S.F. Nelson, D.H. Cohn, R.B. Vallee, D. Krakow, Mutations in DYNC2LI1 disrupt cilia function and cause short rib polydactyly syndrome, Nat. Commun. 6 (2015) 7092.

[178] K. Kessler, I. Wunderlich, S. Uebe, N.S. Falk, A. Giessl, J.H. Brandstatter, B. Popp, P. Klinger, A.B. Ekici, H. Sticht, H.G. Dorr, A. Reis, R. Roepman, E. Seemanova, C.T. Thiel, DYNC2LI1 mutations broaden the clinical spectrum of dynein-2 defects, Sci. Rep. 5 (2015) 11649.

[179] L. Korrodi-Gregorio, A. Margarida Lopes, S.L. Esteves, S. Afonso, A. Lemos de Matos, A.A. Lissovsky, O.A. da Cruz e Silva, E.F. da Cruz e Silva, P.J. Esteves, M. Fardilha, An intriguing shift occurs in the novel protein phosphatase 1 binding partner, TCTEX1D4: evidence of positive selection in a pika model, PLoS One 8 (10) (2013) e77236.

[180] L. Korrodi-Gregorio, S.I. Vieira, S.L. Esteves, J.V. Silva, M.J. Freitas, A.K. Brauns, G. Luers, J. Abrantes, P.J. Esteves, E.S.O.A. da Cruz, M. Fardilha, E.S.E.F. da Cruz, TCTEX1D4, a novel protein phosphatase 1 interactor: connecting the phosphatase to the microtubule network, Biol. Open 2 (5) (2013) 453–465.

[181] M.J. Freitas, L. Korrodi-Gregorio, F. Morais-Santos, E. Cruz e Silva, M. Fardilha, TCTEX1D4 interactome in human testis: unraveling the function of dynein light chain in spermatozoa, OMICS 18 (4) (2014) 242–253.

[182] J. Reck, A.M. Schauer, K. VanderWaal Mills, R. Bower, D. Tritschler, C.A. Perrone, M.E. Porter, The role of the dynein light intermediate chain in retrograde IFT and flagellar function in Chlamydomonas, Mol. Biol. Cell 27 (15) (2016) 2404–2422.

[183] H.L. Rehm, Evolving health care through personal genomics, Nat. Rev. Genet. 18 (4) (2017) 259–267.

[184] R. Shaheen, K. Szymanska, B. Basu, N. Patel, N. Ewida, E. Faqeih, A. Al Hashem, N. Derar, H. Alsharif, M.A. Aldahmesh, A.M. Alazami, M. Hashem, N. Ibrahim, F.M. Abdulwahab, R. Sonbul, H. Alkuraya, M. Alnemer, S. Al Tala, M. Al-Husain, H. Morsy, M.Z. Seidahmed, N. Meriki, M. Al-Owain, S. AlShahwan, B. Tabarki, M.A. Salih, W. Ciliopathy, T. Faquih, M. El-Kalioby, M. Ueffing, K. Boldt, C.V. Logan, D.A. Parry, N. Al Tassan, D. Monies, A. Megarbane, M. Abouelhoda, A. Halees, C.A. Johnson, F.S. Alkuraya, Characterizing the morbid genome of ciliopathies, Genome Biol. 17 (1) (2016) 242.

[185] K.M. Keeling, X. Xue, G. Gunn, D.M. Bedwell, Therapeutics based on stop codon readthrough, Annu. Rev. Genomics Hum. Genet. 15 (2014) 371–394.

[186] C. Fellmann, B.G. Gowen, P.C. Lin, J.A. Doudna, J.E. Corn, Cornerstones of CRISPR-Cas in drug discovery and therapy, Nat. Rev. Drug Discov. (2016).

[187] A. Berger, S. Maire, M.C. Gaillard, J.A. Sahel, P. Hantraye, A.P. Bemelmans, mRNA trans-splicing in gene therapy for genetic diseases, Wiley Interdiscip. Rev. RNA 7 (4) (2016) 487–498.

[188] A.J. Firestone, J.S. Weinger, M. Maldonado, K. Barlan, L.D. Langston, M. O'Donnell, V.I. Gelfand, T.M. Kapoor, J.K. Chen, Small-molecule inhibitors of the AAA+ATPase motor cytoplasmic dynein, Nature 484 (7392) (2012) 125–129.

[189] B.D. Engel, H. Ishikawa, J.L. Feldman, C.W. Wilson, P.T. Chuang, J. Snedecor, J. Williams, Z. Sun, W.F. Marshall, A cell-based screen for inhibitors of flagella-driven motility in Chlamydomonas reveals a novel modulator of ciliary length and retrograde actin flow, Cytoskelet. Hob. 68 (3) (2011) 188–203.

[190] H.M. Kronenberg, Developmental regulation of the growth plate, Nature 423 (6937) (2003) 332–336.

[191] C. Thauvin-Robinet, J.S. Lee, E. Lopez, V. Herranz-Perez, T. Shida, B. Franco, L. Jego, F. Ye, L. Pasquier, P. Loget, N. Gigot, B. Aral, C.A. Lopes, J. St-Onge, A.L. Bruel, J. Thevenon, S. Gonzalez-Granero, C. Alby, A. Munnich, M. Vekemans, F. Huet, A.M. Fry, S. Saunier, J.B. Riviere, T. Attie-Bitach, J.M. Garcia-Verdugo, L. Faivre, A. Megarbane, M.V. Nachury, The oral-facial-digital syndrome gene C2CD3 encodes a positive regulator of centriole elongation, Nat. Genet. 46 (8) (2014) 905–911.

[192] E. Forsythe, P.L. Beales, Bardet-Biedl syndrome, Eur. J Hum. Genet. 21 (1) (2013) 8–13.

[193] T. Chaya, Y. Omori, R. Kuwahara, T. Furukawa, ICK is essential for cell type-specific ciliogenesis and the regulation of ciliary transport, EMBO J. 33 (11) (2014) 1227–1242.

[194] J. Saari, M.A. Lovell, H.C. Yu, G.A. Bellus, Compound heterozygosity for a frame shift mutation and a likely pathogenic sequence variant in the planar cell polarity-ciliogenesis gene WDPCP in a girl with polysyndactyly, coarctation of the aorta, and tongue hamartomas, Am. J. Med. Genet. A 167A (2) (2015) 421–427.

[195] A. Aguilar, Ciliopathies: CPLANE regulates intraflagellar transport, Nat. Rev. Nephrol. 12 (7) (2016) 376.

[196] N. Adly, A. Alhashem, A. Ammari, F.S. Alkuraya, Ciliary genes TBC1D32/C6orf170 and SCLT1 are mutated in patients with OFD type IX, Hum. Mutat. 35 (1) (2014) 36–40.

[197] H.W. Ko, R.X. Norman, J. Tran, K.P. Fuller, M. Fukuda, J.T. Eggenschwiler, Broad-minded links cell cycle-related kinase to cilia assembly and hedgehog signal transduction, Dev. Cell 18 (2) (2010) 237–247.

[198] V.L. Ruiz-Perez, S.E. Ide, T.M. Strom, B. Lorenz, D. Wilson, K. Woods, L. King, C. Francomano, P. Freisinger, S. Spranger, B. Marino, B. Dallapiccola, M. Wright, T. Meitinger, M.H. Polymeropoulos, J. Goodship, Mutations in a new gene in Ellis-van Creveld syndrome and Weyers acrodental dysostosis, Nat. Genet. 24 (3) (2000) 283–286.

[199] S.W. Tompson, V.L. Ruiz-Perez, H.J. Blair, S. Barton, V. Navarro, J.L. Robson, M.J. Wright, J.A. Goodship, Sequencing EVC and EVC2 identifies mutations in two-thirds of Ellis-van Creveld syndrome patients, Hum. Genet. 120 (5) (2007) 663–670.

[200] Y. Yin, F. Bangs, I.R. Paton, A. Prescott, J. James, M.G. Davey, P. Whitley, G. Genikhovich, U. Technau, D.W. Burt, C. Tickle, The Talpid3 gene (KIAA0586) encodes a centrosomal protein that is essential for primary cilia formation, Development 136 (4) (2009) 655–664.

[201] E.M. Valente, C.V. Logan, S. Mougou-Zerelli, J.H. Lee, J.L. Silhavy, F. Brancati, M. Iannicelli, L. Travaglini, S. Romani, B. Illi, M. Adams, K. Szymanska, A. Mazzotta, J.E. Lee, J.C. Tolentino, D. Swistun, C.D. Salpietro, C. Fede, S. Gabriel, C. Russ, K. Cibulskis, C. Sougnez, F. Hildebrandt, E.A. Otto, S. Held, B.H. Diplas, E.E. Davis, M. Mikula, C.M. Strom, B. Ben-Zeev, D. Lev, T.L. Sagie, M. Michelson, Y. Yaron, A. Krause, E. Boltshauser, N. Elkhartoufi, J. Roume, S. Shalev, A. Munnich, S. Saunier, C. Inglehearn, A. Saad, A. Alkindy, S. Thomas, M. Vekemans, B. Dallapiccola, N. Katsanis, C.A. Johnson, T. Attie-Bitach, J.G. Gleeson, Mutations in TMEM216 perturb ciliogenesis and cause Joubert, Meckel and related syndromes, Nat. Genet. 42 (7) (2010) 619–625.

[202] S. Thomas, M. Legendre, S. Saunier, B. Bessieres, C. Alby, M. Bonniere, A. Toutain, L. Loeuillet, K. Szymanska, F. Jossic, D. Gaillard, M.T. Yacoubi, S. Mougou-Zerelli, A. David, M.A. Barthez, Y. Ville, C. Bole-Feysot, P. Nitschke, S. Lyonnet, A. Munnich, C.A. Johnson, F. Encha-Razavi, V. Cormier-Daire, C. Thauvin-Robinet, M. Vekemans, T. Attie-Bitach, TCTN3 mutations cause Mohr-Majewski syndrome, Am. J. Hum. Genet. 91 (2) (2012) 372–378.

[203] E.C. Roberson, W.E. Dowdle, A. Ozanturk, F.R. Garcia-Gonzalo, C. Li, J. Halbritter, N. Elkhartoufi, J.D. Porath, H. Cope, A. Ashley-Koch, S. Gregory, S. Thomas, J.A. Sayer, S. Saunier, E.A. Otto, N. Katsanis, E.E. Davis, T. Attie-Bitach, F. Hildebrandt, M.R. Leroux, J.F. Reiter, TMEM231, mutated in orofaciodigital and Meckel syndromes, organizes the ciliary transition zone, J. Cell Biol. 209 (1) (2015) 129–142.

[204] E. Lopez, C. Thauvin-Robinet, B. Reversade, N.E. Khartoufi, L. Devisme, M. Holder, H. Ansart-Franquet, M. Avila, D. Lacombe, P. Kleinfinger, I. Kaori, J. Takanashi, M. Le Merrer, J. Martinovic, C. Noel, M. Shboul, L. Ho, Y. Guven, F. Razavi, L. Burglen, N. Gigot, V. Darmency-Stamboul, J. Thevenon, B. Aral, H. Kayserili, F. Huet, S. Lyonnet, C. Le Caignec, B. Franco, J.B. Riviere, L. Faivre, T. Attie-Bitach, C5orf42 is the major gene responsible for OFD syndrome type VI, Hum. Genet. 133 (3) (2014) 367–377.

[205] M. Romani, F. Mancini, A. Micalizzi, A. Poretti, E. Miccinilli, P. Accorsi, E. Avola, E. Bertini, R. Borgatti, R. Romaniello, S. Ceylaner, G. Coppola, S. D'Arrigo, L. Giordano, A.R. Janecke, M. Lituania, K. Ludwig, L. Martorell, T. Mazza, S. Odent, L. Pinelli, P. Poo, M. Santucci, S. Signorini, A. Simonati, R. Spiegel, F. Stanzial, M. Steinlin, B. Tabarki, N.I. Wolf, F. Zibordi, E. Boltshauser, E.M. Valente, Oral-facial-digital syndrome type VI: is C5orf42 really the major gene? Hum. Genet. 134 (1) (2015) 123–126.

[206] H.E. Shamseldin, A. Rajab, A. Alhashem, R. Shaheen, T. Al-Shidi, R. Alamro, S. Al Harassi, F.S. Alkuraya, Mutations in DDX59 implicate RNA helicase in the pathogenesis of orofaciodigital syndrome, Am. J. Hum. Genet. 93 (3) (2013) 555–560.

[207] B.D. Engel, H. Ishikawa, K.A. Wemmer, S. Geimer, K. Wakabayashi, M. Hirono, B. Craige, G.J. Pazour, G.B. Witman, R. Kamiya, W.F. Marshall, The role of retrograde intraflagellar transport in flagellar assembly, maintenance, and function, J. Cell Biol. 199 (1) (2012) 151–167.

In this chapter

Ciliary dynein dysfunction caused by chronic alcohol exposure

Michael Price[1], Fan Yang[2], Joseph H. Sisson[1], Maureen Wirschell[2]
[1]University of Nebraska Medical Center, Omaha, NE, United States; [2]University of Mississippi Medical Center, Jackson, MS, United States

16.1 Overview

Cilia are highly conserved across evolution at both the structural and molecular levels. Ciliary movement is driven by systematic regulation of both the outer dynein arm (ODA) and the inner dynein arm (IDA) activities. Detailed characterization of the subunit composition and assembly of the dynein motors into the axoneme is detailed in Chapter 5 (vol. 1 of this book).

Ciliopathies are a recently defined class of human diseases caused by defects in assembly or function of motile and nonmotile cilia [1–7]. Ciliary dysfunction underlies a growing list of genetic conditions that affect multiple organ systems throughout the body. Pathologies associated with ciliary dysfunction range from blindness [8,9], chronic lung infections [10], male and female reproductive defects [11,12], cyst formation in the kidney and liver [13–15], congenital heart defects [16], laterality malformations [17], skeletal malformations [18], obesity [19,20], diabetes [21,22], and more. The list of genetic mutations resulting in ciliopathies is expanding and includes a number of genes encoding ciliary proteins, or proteins involved in building ciliary organelles [10]. The detailed structure of cilia/flagella, the dynein motors that generate their movement, and the regulatory pathways that control dynein motor function are detailed in other chapters (see Chapters 5, 6, 8, 9, and 11 (vol. 1 of this book) and Chapters 1 and 2 (this volume)). Here, we focus on the ODA motors of cilia and discuss the impact of alcohol on ODA function and ciliary beat frequency (CBF). This chapter highlights our current understanding of the acquired ciliopathy called alcohol-induced ciliary dysfunction (AICD).

In the last two decades, we have begun to appreciate the impact of excessive and sustained alcohol consumption on lung function [23–25]. Individuals with prolonged alcohol use disorders (AUDs) experience a number of health-related issues including AICD. Alcohol impairs airway cilia responsiveness, which results in defective mucociliary clearance [24]. In addition, prolonged alcohol

Dyneins. https://doi.org/10.1016/B978-0-12-809470-9.00016-3

consumption causes a reduction in daily sperm production and in sperm motility [26]. The detailed mechanism underlying AICD is unclear, but evidence demonstrates a role for nitric oxide (NO) [27] and cyclic nucleotide–dependent kinases [28,29]. Furthermore, treatment with antioxidants [30] and phosphatase inhibitors [31] can prevent or reverse AICD in murine and bovine systems.

This chapter focuses on alcohol's impact on the lung and the consequences of prolonged alcohol exposure on mucociliary clearance. We summarize findings in both mammalian and protist systems used to study AICD and propose models for how alcohol changes the function of the ODAs in cilia. We conclude this chapter with a discussion of future directions for AICD studies and propose treatment strategies for AICD.

16.2 Alcohol and mucociliary function

16.2.1 Importance of mucociliary clearance

The conducting airways of the lung include the trachea and bronchi. These airways are commonly harmed by viral infections, smoking, and environmental exposures. Another important lung exposure is drinking alcohol that is known to impair airway function [32] and increase the risk of bronchitis, pneumonia, and other lung diseases in people who drink heavily [33]. Because drinking is common in the population [24], the role that alcohol exposure plays in altering proximal airways function is especially important since clearance of inhaled dust, pathogens, and debris is primarily cleared through mucociliary clearance.

Normal mucociliary clearance requires the coordinated beating of cilia and the discrete secretion of mucus. Mucociliary clearance generates a caudad to cephalad flow of mucus so that inhaled particles, pathogens, and debris are moved up and out of the lung airways to the oropharynx, where they can be swallowed or expectorated. This propulsion of particles trapped in mucus is driven by cilia. Within cilia, dyneins are the motor molecules that apply the force to generate mucus flow. From a physiology perspective, the flow of mucus out of the lung can be expressed as the rate of cilia beating (CBF) x the periciliary fluid height, which approximates the stroke volume. When expressed this way, it is clear that there is a direct relationship between CBF and mucus flow. The dependence of clearance on cilia beating is clearly demonstrated in people with genetic ciliopathies such as primary ciliary dyskinesia (PCD), which results in severe airway diseases such as bronchiectasis and pneumonia. PCD patients have no effective mucociliary clearance, leaving coughing as the only way to clear mucus from the lungs [34]. So it is clear that ciliary beat is required for effective lung clearance. More recently, we are learning that acquired ciliopathies, related to exposures such as alcohol, also play a role in the pathogenesis of lung diseases.

Considerable research has demonstrated that alcohol has profound effects on ciliary function [32]. Normally, the lung is pathogen-free below the vocal cords so inhaled pathogens that breach the vocal cords are quickly removed through mucociliary clearance. In heavy drinkers, however, ciliary function is impaired, which results in aspirated pathogens persisting in the mid to lower airways, resulting in pneumonia and bronchitis [33]. Not surprisingly, pneumonia in people with alcohol use disorders is much higher than in the general population [33]. Mechanistically the exposure of ciliary dyneins to alcohol is predictable when both the physiology of the bronchial circulation and the volatility of ethanol are considered together.

The lung is distinctive because it has two circulations: (1) the pulmonary circulation exchanges carbon dioxide for oxygen by presenting venous blood to the alveolar surface, and (2) the bronchial circulation arises from the aorta where it delivers fully oxygenated blood to nourish the conducting airways. Ciliated cells lining the conducting airways take in oxygen through diffusion from this bronchial circulation. Importantly, alcohol in the bronchial circulation off-gasses into the airway lumen based on the volatility of alcohol and the concentration gradient that is present from the bronchial circulation to the airway lumen. Once in the airway lumen, this alcohol-rich vapor cools as it moves out of the lung, so alcohol condensation occurs on the luminal surface. This alcohol condensate is then reabsorbed back into the bronchial circulation. This results in alcohol recirculation within the bronchial capillaries and high levels of alcohol cycling through ciliated cells. This alcohol off-gassing into the ciliated airways has been confirmed by the Breathalyzer test that is used to estimate blood alcohol levels by measuring the concentration of ethanol in exhaled air [35]. Indeed, quantitative studies in humans demonstrate that more than 99% of the exhaled alcohol measured by the Breathalyzer test comes exclusively from the bronchial circulation [36] from which it passes across the ciliated pseudostratified columnar epithelium of the proximal airways. For this reason, it is not surprising that the ciliated cell is exposed to high concentrations of alcohol, which might alter ciliary function during intoxication. Because the ODA motors control CBF, it is logical to hypothesize that alcohol alters ODA function (see below).

16.2.2 Linkage of CBF to the ODAs in mammalian airway cilia

Experiments in mammalian ciliated airway cells exposed to alcohol demonstrate that exposure to modest concentrations of alcohol causes brief, transient stimulation of CBF [37]. The time frame of "brief" alcohol depends on the model system. (1) In live rats fed 36% of their calories as alcohol, CBF increases by 30%–40% over control-fed rats after 1 week of drinking alcohol [38]. In this model, blood alcohol concentrations of 45 mM were routinely obtained. (2) In naïve primary ciliated bovine bronchial cells exposed to 25–100 mM alcohol in vitro, alcohol stimulated CBF after 1 h of alcohol exposure, raising CBF ~30%

over baseline [37]. This stimulation was sustained for 2–4h before returning back to baseline. (3) Using an organelle preparation of isolated, demembranated, bovine cilia extracted from naïve tracheae, alcohol (specifically ethanol) stimulated CBF in reactivated axonemes within 1–2 min of ATP reactivation [39]. CBF frequencies of 30%–40% over baseline were achieved. These isolated axonemes were exquisitely sensitive to alcohol with maximum stimulation occurring with alcohol concentrations of 1 mM [39]. To put this very low alcohol concentration in context, the legal blood alcohol concentration limit for safe driving is less than 22 mM [39]. These data indicate that CBF in multiple models and systems is rapidly and transiently responsive to alcohol exposure. Because CBF is largely dependent on activation of the ODA motors [40], there is a tight association between alcohol and regulation of ODAs. Regardless of the model system, the mechanism of this brief transient stimulation of cilia by alcohol is fairly well understood.

Mechanistic studies designed to explore the signal transduction pathway through which alcohol stimulates cilia have revealed that the first step in CBF stimulation is the activation of nitric oxide synthase (NOS; Fig. 16.1, left panel). The rapidity of alcohol-triggered NO production strongly supports the activation of a constitutive NOS, either the endothelial (eNOS) or the neuronal (nNOS) isoforms. Recent studies demonstrate that in mammalian airway cells, eNOS is triggered through the chaperone protein, heat shock protein 90 (HSP90) [41]. When HSP90 closely associates with eNOS, NO production rapidly increases, activating the canonical cGMP-dependent kinase (PKG) activation pathway. Specifically, increased NO production activates soluble guanylyl cyclase causing the production of cyclic GMP and PKG activation, which is necessary for CBF stimulation (Fig. 16.1, left panel). Concurrently, alcohol also activates a soluble adenylyl cyclase (sAC), which triggers the activation of the cAMP-dependent protein kinase A (PKA). NO-dependent alcohol stimulation of cilia requires the sequential, dual kinase–activation pathway of increased beating through more dynein activation [39,40,42]. While the link is clearly established with this sequential, dual kinase–signaling pathway, one would predict that an increase in CBF should improve clearance in the lungs. While this does happen following brief alcohol exposure, the effects of prolonged alcohol exposure are quite different than the effects of brief alcohol exposure described above.

When people, or animals, have prolonged alcohol intake, such as drinking daily for several weeks, and/or with very high doses of alcohol, cilia become resistant to β-adrenergic stimulation [29]. The initial transient increase in CBF by alcohol lasts for only a few hours in vitro [37] and for 1 week in vivo [38], which is followed by complete desensitization of this dual kinase–dependent activation pathway [40,42]. Once desensitization of airway cilia from prolonged alcohol has occurred, the ciliary apparatus is resistant to further β-adrenergic or cyclic nucleotide stimulation. Despite desensitization of β-agonist stimulation

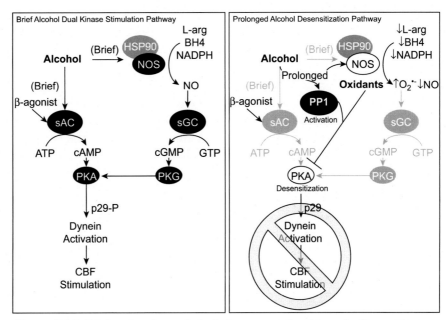

Figure 16.1 Brief and prolonged alcohol exposure signal transduction pathways. *Left,* brief alcohol results in the sequential activation of a dual kinase pathway to stimulate dynein and ciliary beat frequency (CBF). Brief alcohol stimulates the association of heat shock protein 90 (HSP90) with nitric oxide synthase (NOS) to increase canonical nitric oxide (NO) production. NO stimulates soluble guanylyl cyclase to convert GTP to cGMP, thereby stimulating cGMP-dependent kinase (PKG). Alcohol simultaneously activates soluble adenylyl cyclase (sAC) to convert ATP to cAMP, thereby stimulating cAMP-dependent kinase (PKA) and phosphorylation of a 29-kDa protein (p29). Sequential activation of PKG then PKA by alcohol maximally stimulates dynein and CBF. *Right,* prolonged alcohol exposure blunts dynein and CBF activation by β-agonists. Prolonged alcohol depletes NOS substrates L-arginine (L-arg), tetrahydrobiopterin (BH₄), and nicotinamide adenine dinucleotide phosphate (NADPH) resulting in increased oxidant production and kinase desensitization. Prolonged alcohol exposure activates PP1 resulting in desensitization of PKA to cAMP, or β-adrenergic, stimulation thereby preventing dynein activation and increased CBF.

pathways, baseline, homeostatic CBF remains the same, meaning that prolonged alcohol does not decrease baseline CBF. The consequence is that, in the prolonged alcohol exposure model, as one might expect in a person who drinks heavily, there is a failure to increase clearance of the lungs when aspiration of inhaled pathogens occurs, breaching the upper airway defenses. The result is that bacteria entering the lower airways are not cleared, leading to bronchitis and pneumonia. This desensitization of the cilia activation pathway is linked directly to desensitization of PKA, which is the critical later step in the activation of ODAs by β-agonists (β-agonist>cAMP>PKA>dynein>CBF). While the mechanism of this cilia desensitization by prolonged alcohol exposure is partially understood, the upstream mechanism of how alcohol drives desensitization of the ODAs is not clear. However, there is evolving evidence that AICD is linked to cilia-associated oxidative stress.

16.2.3 Desensitization of dynein activation by prolonged alcohol exposure: putative redox mechanisms

Prolonged alcohol ingestion alters the antioxidant/oxidant stress balance in the lung [23,43]. This redox imbalance leads to the generation of reactive oxygen and nitrogen species (ROS,RNS) and increases lung susceptibility to oxidant-mediated injury [43,44]. Evidence that AICD is an oxidant-dependent mechanism comes from in vivo studies of mice drinking alcohol (20% weight/volume) for 12 weeks with concomitant feeding of antioxidants (N-acetyl cysteine and procysteine, 0.163 mg/mL and 0.35% w/v in drinking water, respectively). The alcohol-drinking mice developed CBF desensitization to β-agonists as early as 6 weeks. However, mice that were coadministered alcohol plus antioxidants did not develop AICD at any time throughout the study. Furthermore, AICD was reversed by a 1-week period of abstinence in the alcohol-fed group [30]. These data are consistent with a reversible, redox-dependent signaling mechanism in AICD.

The most likely source of oxidants in the context of alcohol exposure in the airway is via the production of NO driven by alcohol. NO is a highly reactive molecule first characterized as the signaling molecule originally called endothelial-derived relaxing factor (EDRF) [45]. Biological NO comes primarily from the three isoforms of the homodimeric NOS enzymes: nNOS (NOS1), eNOS (NOS3), and inducible NOS (iNOS or NOS2) [46–48]. Both nNOS and eNOS have been localized to the motile cilia apparatus in mammalian cells [27,49]. The localization of key cilia motility enzymes in cilia highlights a role for short-lived reactive signaling intermediates to regulate cilia function [27,50]. The NOS enzymes share the common mechanism of catalyzing the reaction of oxygen, nicotinamide adenine dinucleotide phosphate (NADPH), tetrahydrobiopterin (BH_4), and L-arginine (L-arg) to produce the free radical NO and L-citrulline [46]. NO then acts as a signaling intermediate by autocrine or paracrine mechanisms [51]. A well-characterized "canonical" example is the covalent binding of NO to the heme of soluble guanylyl cyclase to stimulate the production of cGMP, which activates PKG (Fig. 16.1) [46,51]. In contrast, signaling properties of aberrant, or noncanonical, NO production are poorly characterized. In conditions of prolonged alcohol exposure where NADPH, BH_4, or L-arginine are depleted, the passing of electrons between the domains of the NOS enzyme becomes uncoupled, resulting in additional production of the reactive free radical anion, superoxide ($O_2^{\cdot-}$; Fig. 16.1, right panel) [30,52]. The reaction of $O_2^{\cdot-}$ with NO, one of the fastest known biological chemical reactions, results in the formation of the highly reactive anion, peroxynitrite ($ONOO^-$) [52,53]. NO and $ONOO^-$, as RNS in the presence of oxygen, can catalyze the reversible oxidation of thiol-rich proteins such as thioredoxin [54]. The *Chlamydomonas* ODA intermediate chain 1 (IC1), its sea urchin homologue, and the human homologue (DNAI1) contain a thioredoxin domain that regulates protein partner binding and is a likely target of alcohol-induced oxidative stress or signaling. Moreover, proteins with redox-sensitive domains have been localized to cilia and in close proximity to the ODA [55,56].

16.2.4 Kinase–phosphatase relationships in AICD

As described previously, key cilia-localized kinases (PKA and PKG) are necessary for maximal stimulation of CBF by alcohol (Fig. 16.1, left panel). In addition, the desensitization of PKA is common to the prolonged alcohol desensitization of CBF. These kinases likely coexist with phosphatases in a regulatory complex with ODA components. In contrast to kinase inactivation by prolonged alcohol exposure, in vitro studies using mammalian models demonstrate that phosphatase activation occurs with prolonged alcohol exposure. More specifically, a series of phosphatase inhibitor studies in isolated bovine axonemes and mouse airway epithelial cells show the specific activation of PP1 in the absence of changes in total PP1 [31]. *In vitro* inhibition of PP1 by addition of the recombinant, endogenous, and specific inhibitor, inhibitor-2, restored PKA and CBF responsiveness to cAMP and β-agonists, respectively. *In vitro* phosphorylation studies revealed that stimulation of PKA in alcohol-naïve axonemes resulted in the phosphorylation of a 29 kiloDalton (kDa) protein (p29). This phosphorylation is absent with prolonged alcohol exposure. Coincident with PP1 inhibition in the context of alcohol was the reappearance of p29 phosphorylation upon stimulation of PKA. PP1 inhibition alone did not produce phosphorylation of p29 [31]. These data suggest that axoneme-localized PP1 negatively regulates axoneme-localized PKA, and that PKA-dependent phosphorylation of p29 is a positive regulator of dynein activity and CBF (Fig. 16.1, right panel). Structure-based studies have revealed that PP1 contains putative, reactive cysteine residues (Cys 155 and Cys 158) close to the active site [57]. This suggests that PP1 is subject to direct redox control, likely linking the oxidant component of AICD to phosphatase activation. Moreover, these data suggest that AICD is a phenomenon primarily localized to the axoneme. Many of the axoneme-localized kinases and phosphatases that regulate dynein activity are highly conserved throughout evolution allowing for the use of model organisms to probe the mechanism of AICD.

16.3 Alcohol and *Chlamydomonas* flagella

16.3.1 Alcohol-induced ciliary dysfunction is conserved

Our knowledge of AICD is completely based on studies in mammalian model systems including bovine and murine airway cilia. However, these systems have distinct limitations for studies of cilia and AICD including genetic variability of bovine trachea and limited ability to purify cilia in large quantities, among others. In contrast, *Chlamydomonas* is advantageous in many aspects for the study of cilia. It is a powerful, genetic system that offers a large repertoire of well-defined motility mutants, and it is a superior biochemical system for purifying flagella for many downstream biochemical, pharmacological, proteomic, and ultrastructural studies. Analyses of swimming speed and protist behavior offer simple methods for rapid assessment of ciliary function. Indeed, many genes and their encoded proteins that contribute to ciliary assembly and function were originally discovered and characterized in *Chlamydomonas* [58,59].

Given the wide array of experimental advantages offered by this protist, model system, it is an exceptional choice for discovery of the underlying mechanisms that drive AICD and for identification of specific ciliary proteins impacted by alcohol. A recent study by Yang et al. [60] revealed that alcohol treatment results in decreased forward swimming velocities in wild-type *Chlamydomonas* cells. The reduced swimming speeds are caused by a significant reduction in CBF [60]. These results demonstrated that AICD is a highly conserved consequence of alcohol exposure.

16.3.2 Alcohol impacts the ODA

Analyses of mutant *Chlamydomonas* strains, which fail to assemble the ODA in the axoneme, demonstrate that the ODAs contribute primarily to normal CBF and power, parameters required for normal ciliary movement. Given that alcohol exposure to wild-type *Chlamydomonas* cells resulted in decreased CBF, it is likely that alcohol affects ODA function. Consistently, alcohol does not affect *Chlamydomonas* mutant cells lacking the ODA in the axoneme; *oda* mutants show no further decrease in forward swimming speed in response to alcohol exposure [60].

Taking further advantage of novel *oda* mutants only available in the *Chlamydomonas* model system, Yang et al. [60] further demonstrated that alcohol targets specific ODA motor subunits. Unique mutants lacking only the ODA–α-HC, or the motor domains of the ODA–β-HC, or γ-HC showed differential sensitivity to alcohol; the β-HC and γ-HC subunits are required for AICD, whereas, the α-HC is only partially affected by alcohol. Importantly, the β-HCs and γ-HCs are highly conserved in mammalian ODAs [61,62]. These results have provided direct evidence for alcohol targeting specific ciliary dynein motors.

16.3.3 Alcohol alters phosphorylation of specific ciliary proteins

In *Chlamydomonas*, genetic, biochemical, and pharmacological studies have revealed both calcium and phosphoregulatory pathways for regulation of the dynein arm activities (detailed in Chapters 8 and 9 (vol. 1 of this book)). The

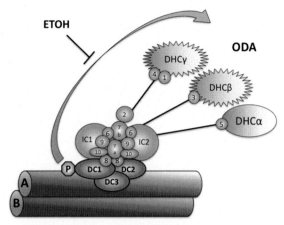

Figure 16.2 Model for AICD in *Chlamydomonas*. The outer dynein arm (ODA) is attached to the outerdoublet microtubules (A and B tubes) by the heterotrimeric ODA-DC, which is comprised of DC1, DC2, and DC3. Alcohol alters a pathway for control of ODA function that involves phosphorylation of DC1 and the activity of the β- and γ-HC ODA subunits.

pathways involve the central pair, radial spoke, N-DRC, kinases, phosphatases, and calcium-binding proteins [63–66]. As detailed above, in mammalian airway cilia, chronic alcohol exposure downregulates PKG and PKA. However, the downstream ciliary targets affected by misregulated kinase activity remain unknown due to the experimental limitations of mammalian airway models for gene discovery and protein identification. By using an anti-phosphothreonine antibody, it was determined that phosphorylation of DC1, a subunit of the ODA-DC, is altered by alcohol. Moreover, DC1 phosphorylation correlates with AICD, indicating a novel role for DC1 in a regulatory pathway that controls ODA function that is affected in AICD [60] (Fig. 16.2).

By using *Chlamydomonas* as a model system, the downstream mechanism underlying AICD has been partially revealed. To further study the signaling pathway in regulation of AICD and ODA function, it will be important to identify key phosphorylation sites in DC1 and the kinases and phosphatases that are involved in the AICD pathway.

Besides the phosphoregulatory pathway, AICD may also be mediated by altered redox state of ciliary proteins. As detailed above, considerable evidence shows that prolonged alcohol exposure increases lung oxidative stress, likely by increasing ROS–RNS production (nitro-oxidant stress) and depletion of critical antioxidants such as glutathione (GSH) that protect the lung epithelium from injury and inflammation caused by oxidizing conditions [67–71]. During oxidant stressing conditions, such as prolonged alcohol consumption, NOS may become uncoupled, diverting NO toward production of $O_2^{\cdot-}$ and $ONOO^-$ as described above [72].

Interestingly, in *Chlamydomonas*, CBF and ODA function are also modulated by changes in reducing–oxidizing conditions [73]. Specifically, oxidizing conditions reduce CBF mediated by the ODA, and result in formation of both intra- and intermolecular disulfide bonds in two ODA thioredoxin light chains (LC3 and LC5) and one ODA-DC subunit (DC3). How these changes ultimately reduce CBF is unclear. Given that the effects of alcohol on ciliary motility are conserved between mammals and algae, the use of experimentally malleable systems will continue to be a powerful approach to identify ciliary protein targets of alcohol and to reveal the molecular mechanisms responsible for AICD.

16.4 New questions and future directions

Cilia-localized phosphatases and oxidants, as ultimate regulators of dynein activation and CBF, are implicated as therapeutic targets in AICD. Key understanding of the generation of oxidants and activation of phosphatases will direct targeted therapies. Key questions that still remain include:

"What is the precise source and oxidant species generated to cause AICD?" As described above, associated with generation of oxidants is the depletion of NOS substrates (L-arginine, BH_4, and NADPH). Airway-localized cilia are amenable to unique therapy and can be directly targeted by inhaled aerosols. Supplementation of L-arginine or BH_4 in diet or directly by aerosol may improve NOS coupling, restore canonical NO generation, and mitigate oxidant stress. Indeed, there are several aerosolized varieties of antioxidants, such as ascorbate (vitamin C) and acetylcysteine (Mucomyst); however, there are currently no prescribed therapies for AICD and inhaled antioxidants for other mucociliary disorders (cystic fibrosis) have largely failed clinically [74–76]. Preclinical data suggest these failures are due to imprecision of targeting the specific oxidant species and localization. This highlights the importance of understanding the precise molecular perturbation in AICD.

"How do oxidants cause AICD?" is another important question in understanding the role of alcohol in the regulation of dynein activity and CBF. There are several oxidant-sensitive components of axonemal dynein activation, including but not limited to: ODA-docking complex subunits such as DC3, a redox-sensitive, calcium-binding protein [55]; nucleoside diphosphate kinases [56], necessary for the recycling of ADP to ATP as the fuel for dynein; and the kinases and phosphatases previously described. Which of these components is the redox-sensitive switch to cause AICD is still unknown.

"What are the kinase and phosphatase targets in AICD?" The recent, preclinical discovery that PP1 inhibition rescues AICD suggests that phosphatase inhibitors may play a key role in future therapies of AICD. Unfortunately, PP1 is a ubiquitous enzyme expressed in multiple cellular compartments and necessary for viability of cells. Studies of okadaic acid and tautomycin, polyketide inhibitors of PP1 and PP2A, produced as toxins from shellfish and sponges, have revealed the necessity of local and specific phosphatase inhibition as these compounds are fatal in microgram quantities [77,78]. Despite this, many phosphatase inhibitors have been developed as cancer therapeutics, but only recently have small molecule PP1-specific inhibitors been discovered [79]. An alternative, yet more abstract approach, would be to disrupt PP1 targeting to the axoneme, PP1–dynein interactions, or phosphorylation targets. Work in *Chlamydomonas* has revealed phosphorylation of DC1 in the AICD pathway [60]. Moreover, it is clear that PP1 inhibition restores phosphorylation of the p29 target of PKA coincident with increased dynein activity in mammalian cilia [31]. The precise relationship of PP1 in restoring PKA activity and how PKA activates dynein in the context of alcohol is unresolved. Understanding the precise role for these phosphorylation events to regulate dynein activity will guide therapeutic approaches for AICD.

In summary, alcohol has distinct effects on dynein activity and CBF. Brief alcohol exposure stimulates dynein activity and CBF through a NO-dependent pathway in mammalian cells. In contrast, prolonged alcohol exposure drives oxidant production to disrupt key phosphorylation events preventing stimulation of dynein and CBF above baseline. Finally, this alcohol-mediated dynein dysfunction seems to be evolutionarily conserved based on studies in bovine, murine, and protist models and is tightly associated with impaired mucociliary clearance and subsequent pulmonary infections in humans who drink heavily.

References

[1] A. Horani, S.L. Brody, T.W. Ferkol, Picking up speed: advances in the genetics of primary ciliary dyskinesia, Pediatr. Res. 75 (1–2) (2014) 158–164.

[2] N.A. Zaghloul, N. Katsanis, Zebrafish assays of ciliopathies, in: H.W. Detrich 3rd, M. Westerfiedl, L.I. Zon (Eds.), The Zebrafish: Disease Models and Chemical Screens, Academic Press, Oxford, UK, 2011, pp. 257–272.

[3] A.M. Waters, P.L. Beales, Ciliopathies: an expanding disease spectrum, Pediatr. Nephrol. 26 (7) (2011) 1039–1056.

[4] F. Hildebrandt, T. Benzing, N. Katsanis, Ciliopathies, N. Engl. J. Med. 364 (16) (2011) 1533–1543.

[5] S.A. Brugmann, D.R. Cordero, J.A. Helms, Craniofacial ciliopathies: a new classification for craniofacial disorders, Am. J. Med. Genet. A 152A (12) (2010) 2995–3006.

[6] J.L. Tobin, P.L. Beales, The nonmotile ciliopathies, Genet. Med. 11 (6) (2009) 386–402.

[7] A. D'Angelo, B. Franco, The dynamic cilium in human diseases, Pathogenetics 2 (1) (2009) 3.

[8] K. Nikopoulos, P. Farinelli, B. Giangreco, C. Tsika, B. Royer-Bertrand, M.K. Mbefo, et al., Mutations in CEP78 cause cone-rod dystrophy and hearing loss associated with primary-cilia defects, Am. J. Hum. Genet. 99 (3) (2016) 770–776.

[9] E.N. Suspitsin, E.N. Imyanitov, Bardet-biedl syndrome, Mol. Syndromol. 7 (2) (2016) 62–71.

[10] M.R. Knowles, M. Zariwala, M. Leigh, Primary ciliary dyskinesia, Clin. Chest Med. 37 (3) (2016) 449–461.

[11] P.F. Ceccaldi, F. Carre-Pigeon, Y. Youinou, B. Delepine, P.E. Bryckaert, G. Harika, et al., Kartagener's syndrome and infertility: observation, diagnosis and treatment, J. Gynecol. Obstet. Biol. Reprod. Paris. 33 (3) (2004) 192–194.

[12] B.A. Afzelius, R. Eliasson, Male and female infertility problems in the immotile-cilia syndrome, Eur. J. Respir. Dis. Suppl. 127 (1983) 144–147.

[13] T.V. Masyuk, A.I. Masyuk, N.F. La Russo, Therapeutic targets in polycystic liver disease, Curr. Drug Targets (2015).

[14] S. Abdul-Majeed, S.M. Nauli, Polycystic diseases in visceral organs, Obstet. Gynecol. Int. 2011 (2011) 609370.

[15] M. Gunay-Aygun, Liver and kidney disease in ciliopathies, Am. J. Med. Genet. C Semin. Med. Genet. 151C (4) (2009) 296–306.

[16] M. Brueckner, Heterotaxia, congenital heart disease, and primary ciliary dyskinesia, Circulation 115 (22) (2007) 2793–2795.

[17] M. Brueckner, Cilia propel the embryo in the right direction, Am. J. Med. Genet. 101 (4) (2001) 339–344.

[18] C. Huber, V. Cormier-Daire, Ciliary disorder of the skeleton, Am. J. Med. Genet. C Semin. Med. Genet. 160c (3) (2012) 165–174.

[19] E.C. Mariman, R.G. Vink, N.J. Roumans, F.G. Bouwman, C.T. Stumpel, E.E. Aller, et al., The cilium: a cellular antenna with an influence on obesity risk, Br. J. Nutr. 116 (4) (2016) 576–592.

[20] A. Szmigielska, G. Krzemien, M. Roszkowska-Blaim, E. Obersztyn, Polydactyly and obesity – the clinical manifestation of ciliopathy: a boy with Bardet-Biedl syndrome, Dev. Period Med. 20 (2) (2016) 105–109.

[21] H. Lee, J. Song, J.H. Jung, H.W. Ko, Primary cilia in energy balance signaling and metabolic disorder, BMB Rep. 48 (12) (2015) 647–654.

[22] E.C. Oh, S. Vasanth, N. Katsanis, Metabolic regulation and energy homeostasis through the primary Cilium, Cell Metab. 21 (1) (2015) 21–31.

[23] D.M. Boe, R.W. Vandivier, E.L. Burnham, M. Moss, Alcohol abuse and pulmonary disease, J. Leukocyte Biol. 86 (5) (2009) 1097–1104.

[24] J.H. Sisson, J.A. Stoner, D.J. Romberger, J.R. Spurzem, T.A. Wyatt, J. Owens-Ream, et al., Alcohol intake is associated with altered pulmonary function, Alcohol 36 (1) (2005) 19–30.

[25] M. Sapkota, T.A. Wyatt, Alcohol, aldehydes, adducts and airways, Biomolecules 5 (4) (2015) 2987–3008.

[26] A. Paula Franco Punhagui, H. Rodrigues Vieira, G. Eloisa Munhoz De Lion Siervo, R. da Rosa, G. Scantamburlo Alves Fernandes, Ethanol exposure during peripubertal period increases the mast cell number and impairs meiotic and spermatic parameters in adult male rats, Microsc. Res. Tech. 79 (6) (2016) 541–549.

[27] S.L. Stout, T.A. Wyatt, J.J. Adams, J.H. Sisson, Nitric oxide-dependent cilia regulatory enzyme localization in bovine bronchial epithelial cells, J. Histochem. Cytochem. 55 (5) (2007) 433–442.

[28] T.A. Wyatt, J.R. Spurzem, K. May, J.H. Sisson, Regulation of ciliary beat frequency by both PKA and PKG in bovine airway epithelial cells, Am. J. Physiol. 275 (4 Pt 1) (1998) L827–L835.

[29] T.A. Wyatt, J.H. Sisson, Chronic ethanol downregulates PKA activation and ciliary beating in bovine bronchial epithelial cells, Am. J. Physiol. Lung Cell Mol. Physiol. 281 (3) (2001) L575–L581.

[30] S.M. Simet, J.A. Pavlik, J.H. Sisson, Dietary antioxidants prevent alcohol-induced ciliary dysfunction, Alcohol 47 (8) (2013) 629–635.

[31] M.E. Price, J.A. Pavlik, J.H. Sisson, T.A. Wyatt, Inhibition of protein phosphatase 1 reverses alcohol-induced ciliary dysfunction, Am. J. Physiol. Lung Cell Mol. Physiol. 308 (6) (2015) L577–L585.

[32] J.H. Sisson, Alcohol and airways function in health and disease, Alcohol (Fayettev. NY) 41 (5) (2007) 293–307.

[33] M.D. Lebowitz, Respiratory symptoms and disease related to alcohol consumption, Am. Rev. Respir. Dis. 123 (1) (1981) 16–19.

[34] P.G. Noone, M.W. Leigh, A. Sannuti, S.L. Minnix, J.L. Carson, M. Hazucha, et al., Primary ciliary dyskinesia: diagnostic and phenotypic features, Am. J. Respir. Crit. Care Med. 169 (4) (2004) 459–467.

[35] S.C. George, A.L. Babb, M.P. Hlastala, Modeling the concentration of ethanol in the exhaled breath following pretest breathing maneuvers, Ann. Biomed. Eng. 23 (1) (1995) 48–60.

[36] M.P. Hlastala, The alcohol breath test–a review, J. Applied Physiology (Bethesda, Md 1985) 84 (2) (1998) 401–408.

[37] J.H. Sisson, Ethanol stimulates apparent nitric oxide-dependent ciliary beat frequency in bovine airway epithelial cells, Am. J. Physiol. 268 (4 Pt 1) (1995) L596–L600.

[38] T.A. Wyatt, M.J. Gentry-Nielsen, J.A. Pavlik, J.H. Sisson, Desensitization of PKA-stimulated ciliary beat frequency in an ethanol-fed rat model of cigarette smoke exposure, Alcohol Clin. Exp. Res. 28 (7) (2004) 998–1004.

[39] J.H. Sisson, J.A. Pavlik, T.A. Wyatt, Alcohol stimulates ciliary motility of isolated airway axonemes through a nitric oxide, cyclase, and cyclic nucleotide-dependent kinase mechanism, Alcohol Clin. Exp. Res. 33 (4) (2009) 610–616.

[40] T.A. Wyatt, M.A. Forget, J.M. Adams, J.H. Sisson, Both cAMP and cGMP are required for maximal ciliary beat stimulation in a cell-free model of bovine ciliary axonemes, Am. J. Physiol. Lung Cell Mol. Physiol. 288 (3) (2005) L546–L551.

[41] S.M. Simet, J.A. Pavlik, J.H. Sisson, Proteomic analysis of bovine axonemes exposed to acute alcohol: role of endothelial nitric oxide synthase and heat shock protein 90 in cilia stimulation, Alcohol Clin. Exp. Res. 37 (4) (2013) 609–615.

[42] T.A. Wyatt, M.A. Forget, J.H. Sisson, Ethanol stimulates ciliary beating by dual cyclic nucleotide kinase activation in bovine bronchial epithelial cells, Am. J. Pathol. 163 (3) (2003) 1157–1166.

[43] L. Kaphalia, W.J. Calhoun, Alcoholic lung injury: metabolic, biochemical and immunological aspects, Toxicol. Lett. 222 (2) (2013) 171–179.

[44] J.A. Polikandriotis, H.L. Rupnow, S.C. Elms, R.E. Clempus, D.J. Campbell, R.L. Sutliff, et al., Chronic ethanol ingestion increases superoxide production and NADPH oxidase expression in the lung, Am. J. Respir. Cell Mol. Biol. 34 (3) (2006) 314–319.

[45] L.J. Ignarro, G.M. Buga, K.S. Wood, R.E. Byrns, G. Chaudhuri, Endothelium-derived relaxing factor produced and released from artery and vein is nitric oxide, Proc. Natl. Acad. Sci. USA 84 (24) (1987) 9265–9269.

[46] M.A. Marletta, Nitric oxide synthase: function and mechanism, Adv. Exp. Med. Biol. 338 (1993) 281–284.

[47] M. Munakata, Pulmonary nitric oxide synthase isoform expression and their functional significance, Nihon Rinsho Jpn. J. Clin. Med. 54 (2) (1996) 358–363.

[48] C. Nathan, Q.W. Xie, Nitric oxide synthases: roles, tolls, and controls, Cell 78 (6) (1994) 915–918.

[49] C.L. Jackson, J.S. Lucas, W.T. Walker, H. Owen, I. Premadeva, P.M. Lackie, Neuronal NOS localises to human airway cilia, Nitric Oxide 44 (0) (2015) 3–7.

[50] A. Schmid, G. Bai, N. Schmid, M. Zaccolo, L.E. Ostrowski, G.E. Conner, et al., Real-time analysis of cAMP-mediated regulation of ciliary motility in single primary human airway epithelial cells, J. Cell Sci. 119 (Pt 20) (2006) 4176–4186.

[51] P. Lane, S.S. Gross, Cell signaling by nitric oxide, Semin. Nephrol. 19 (3) (1999) 215–229.

[52] N. Kuzkaya, N. Weissmann, D.G. Harrison, S. Dikalov, Interactions of peroxynitrite, tetrahydrobiopterin, ascorbic acid, and thiols: implications for uncoupling endothelial nitric-oxide synthase, J. Biol. Chem. 278 (25) (2003) 22546–22554.

[53] H. Botti, M.N. Moller, D. Steinmann, T. Nauser, W.H. Koppenol, A. Denicola, et al., Distance-dependent diffusion-controlled reaction of *NO and O2*- at chemical equilibrium with ONOO, J. Phys. Chem. B 114 (49) (2010) 16584–16593.

[54] W.H. Koppenol, NO nomenclature? Nitric Oxide 6 (1) (2002) 96–98.

[55] D.M. Casey, T. Yagi, R. Kamiya, G.B. Witman, DC3, the smallest subunit of the Chlamydomonas flagellar outer dynein arm-docking complex, is a redox-sensitive calcium-binding protein, J. Biol. Chem. 278 (43) (2003) 42652–42659.

[56] C.M. Sadek, A. Jimenez, A.E. Damdimopoulos, T. Kieselbach, M. Nord, J.A. Gustafsson, et al., Characterization of human thioredoxin-like 2. A novel microtubule-binding thioredoxin expressed predominantly in the cilia of lung airway epithelium and spermatid manchette and axoneme, J. Biol. Chem. 278 (15) (2003) 13133–13142.

[57] J.S. Fetrow, N. Siew, J. Skolnick, Structure-based functional motif identifies a potential disulfide oxidoreductase active site in the serine/threonine protein phosphatase-1 subfamily, FASEB J. 13 (13) (1999) 1866–1874.

[58] J. Pan, Cilia and ciliopathies: from Chlamydomonas and beyond, Sci. China C Life Sci. 51 (6) (2008) 479–486.

[59] L. Vincensini, T. Blisnick, P. Bastin, 1001 model organisms to study cilia and flagella, Biol. Cell 103 (3) (2011) 109–130.

[60] F. Yang, J. Pavlik, L. Fox, C. Scarbrough, W.S. Sale, J.H. Sisson, et al., Alcohol-induced ciliary dysfunction targets the outer dynein arm, Am. J. Physiol. Lung Cell Mol. Physiol. 308 (6) (2015) L569–L576.

[61] D.J. Asai, D.E. Wilkes, The dynein heavy chain family, J. Eukaryot. Microbiol. 51 (1) (2004) 23–29.

[62] D.R. Mitchell, Speculations on the evolution of 9+2 organelles and the role of central pair microtubules, Biol. Cell 96 (9) (2004) 691–696.

[63] C.A. Elam, W.S. Sale, M. Wirschell, The regulation of dynein-driven microtubule sliding in Chlamydomonas flagella by axonemal kinases and phosphatases, in: in: M.K. Stephen, J.P. Gregory (Eds.), Methods in Cell Biology, vol. 92, Academic Press, 2009, pp. 133–151.

[64] E.F. Smith, P. Yang, The radial spokes and central apparatus: mechano-chemical transducers that regulate flagellar motility, Cell Motil. Cytoskelet. 57 (1) (2004) 8–17.

[65] M. Wirschell, T. Hendrickson, W.S. Sale, Keeping an eye on I1: I1 dynein as a model for flagellar dynein assembly and regulation, Cell Motil. Cytoskelet. 64 (8) (2007) 569–579.

[66] M. Wirschell, R. Yamamoto, L. Alford, A. Gokhale, A. Gaillard, W.S. Sale, Regulation of ciliary motility: conserved protein kinases and phosphatases are targeted and anchored in the ciliary axoneme, Arch. Biochem. Biophys. 510 (2) (2011) 93–100.

[67] L.A. Brown, F.L. Harris, R. Bechara, D.M. Guidot, Effect of chronic ethanol ingestion on alveolar type II cell: glutathione and inflammatory mediator-induced apoptosis, Alcohol Clin. Exp. Res. 25 (7) (2001) 1078–1085.

[68] L.A. Brown, F.L. Harris, D.M. Guidot, Chronic ethanol ingestion potentiates TNF-alpha-mediated oxidative stress and apoptosis in rat type II cells, Am. J. Physiol. Lung Cell Mol. Physiol. 281 (2) (2001) L377–L386.

[69] F. Holguin, I. Moss, L.A. Brown, D.M. Guidot, Chronic ethanol ingestion impairs alveolar type II cell glutathione homeostasis and function and predisposes to endotoxin-mediated acute edematous lung injury in rats, J. Clin. Invest. 101 (4) (1998) 761–768.

[70] Y. Liang, S.M. Yeligar, L.A.S. Brown, Chronic-alcohol-Abuse-induced oxidative stress in the development of acute respiratory distress syndrome, Sci. World J. 2012 (2012) 9.

[71] M. Moss, D.M. Guidot, M. Wong-Lambertina, T. Ten Hoor, R.L. Perez, L.A. Brown, The effects of chronic alcohol abuse on pulmonary glutathione homeostasis, Am. J. Respir. Crit. Care Med. 161 (2 Pt 1) (2000) 414–419.

[72] M.C. Verhaar, P.E. Westerweel, A.J. van Zonneveld, T.J. Rabelink, Free radical production by dysfunctional eNOS, Heart 90 (5) (2004) 494–495.

[73] K. Wakabayashi, S.M. King, Modulation of Chlamydomonas reinhardtii flagellar motility by redox poise, J. Cell Biol. 173 (5) (2006) 743–754.

[74] A.M. Cantin, T.B. White, C.E. Cross, H.J. Forman, R.J. Sokol, D. Borowitz, Antioxidants in cystic fibrosis: conclusions from the CF antioxidant workshop, Bethesda, Maryland, November 11–12, 2003, Free Radic. Biol. Med. 42 (1) (2007) 15–31.

[75] C.E. Cross, A. van der Vliet, C.A. O'Neill, S. Louie, B. Halliwell, Oxidants, antioxidants, and respiratory tract lining fluids. Environmental health perspectives, 102 (Suppl. 10) (1994) 185–191.

[76] G.E. Hatch, Asthma, inhaled oxidants, and dietary antioxidants, Am. J. Clin. Nutr. 61 (Suppl. 3) (1995) 625S–630S.

[77] P. Cohen, C.F. Holmes, Y. Tsukitani, Okadaic acid: a new probe for the study of cellular regulation, Trends Biochem. Sci. 15 (3) (1990) 98–102.

[78] C. MacKintosh, S. Klumpp, Tautomycin from the bacterium Streptomyces verticillatus. Another potent and specific inhibitor of protein phosphatases 1 and 2A, FEBS Lett. 277 (1–2) (1990) 137–140.

[79] S. Mitsuhashi, N. Matsuura, M. Ubukata, H. Oikawa, H. Shima, K. Kikuchi, Tautomycetin is a novel and specific inhibitor of serine/threonine protein phosphatase type 1, PP1, Biochem. Biophys. Res. Commun. 287 (2) (2001) 328–331.

In this chapter

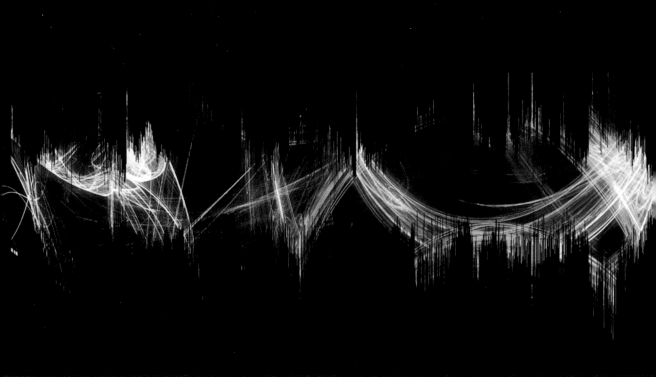

Dynein-based motility of pathogenic protozoa

Simon Imhof, Kent L. Hill
University of California, Los Angeles, CA, United States

17.1 Introduction: impact of flagellated protozoan parasites on human health and agriculture

Flagellated protozoa include several parasitic species that are pathogenic to humans, animals, and plants (Fig. 17.1). These parasites cause tremendous human suffering worldwide and limit economic development in some of the world's most impoverished regions. Examples include human pathogens responsible for malaria, trichomoniasis, epidemic diarrhea, Chagas disease Leishmaniasis, and sleeping sickness. Together, these diseases afflict more than half a billion people.

In addition to their direct importance in human infections, parasitic protozoa that depend on dynein-mediated flagellar motility contribute to the destruction of human resources through infection of livestock and agriculturally important crops. The plant pathogen *Phytopthora*, for example, uses flagellar motility to spread through the soil and infect new plant hosts [1]. *Phytopthora cinnamon* is listed as one of the 100 most invasive species, infecting a wide range of trees. Livestock pathogens that are also a threat to humans include those responsible for zoonotic infections, such as *Trypanosoma brucei*, the causative agent of Nagana in cattle and sleeping sickness in humans in Africa, and *Balantidium coli*, which mainly infects the gastrointestinal (GI) tract of pigs but it is also the only known ciliate that can infect humans [2]. Flagellated and ciliated parasites are also a problem in aquaculture. For example, the ciliate *Ichthyophthirius multifiliis* causing white spot disease can be found all over the globe where freshwater fish are cultivated [3]. Chytrid fungi cause chytridiomycosis in amphibians and employ a flagellated zoospore stage for movement in water sources [4].

Flagellar motility in these various parasites is used to migrate along or into tissues, to resist peristalsis or to find new hosts. Studies directly interrogating the role of motility in parasite transmission and pathogenesis have been done in

Dyneins. http://dx.doi.org/10.1016/B978-0-12-809470-9.00017-5

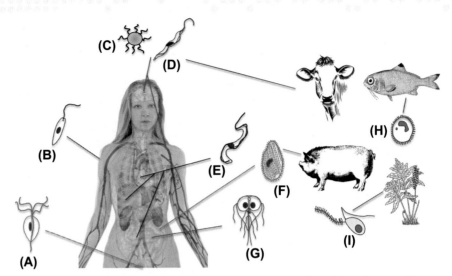

Figure 17.1 Eukaryotic pathogens relying on dynein-mediated flagellar motility cause a broad spectrum of diseases in humans and lead to economic losses in food production. (A) *Trichomonas vaginalis* causing trichomoniasis. (B) *Leishmania* spp. causing visceral and cutaneous leishmaniasis. (C) *Plasmodium* spp. causing malaria. (D) *Trypanosoma brucei* causing sleeping sickness in humans and Nagana in cattle. (E) *Trypanosoma cruzi* causing Chagas disease. (F) *Balantidium coli* causing balantidiasis in pigs and humans. (G) *Giardia intestinalis* causing giardiasis. (H) *Ichthyophthirius multifiliis* causing white spot disease in freshwater fish. (I) *Phythophtora* spp. causing a variety of diseases in plants. The major human tissues affected are indicated.

some organisms, but in general, the implied importance of motility remains to be directly tested. Unlike bacterial and viral pathogens, there are no effective vaccines against parasitic protozoa and drugs available for treatment are antiquated, often toxic and increasingly ineffective. As such, new approaches for battling parasitic diseases are desperately needed and flagellar motility presents potential as a novel target for therapeutic intervention.

Beyond their role as medically and economically important pathogens, parasites present interesting biological examples to study, often with unusual features of flagellum biology for dissection. Protozoan pathogens also encompass a great phylogenetic diversity and include representatives from some of the earliest branching organisms in the eukaryotic lineage (Fig. 17.2). These include groups that are very divergent from the most commonly studied model organisms and therefore provide divergent examples in which to study fundamentals of flagellar motility. For some of these organisms, emerging tools and systems for molecular genetics and systems biology are opening new opportunities for functional analysis [5,82]. In this review we describe flagellar motility with respect to some of the major protozoan parasites infecting humans. Due to its experimental accessibility *T. brucei* is the most extensively studied organism within this group and is thus the main focus.

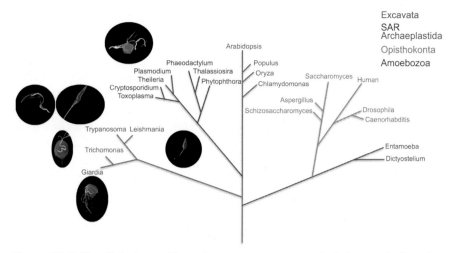

Figure 17.2 Flagellated parasitic protozoa encompass great phylogenetic diversity. Many important human pathogens with flagella are evolutionary distinct from the most common model organisms employed for cilium biology studies. *Drosophila*, zebrafish, mouse, and human, all belong to a single clade, Opisthokonta, while *Chlamydomonas* belongs to the Archaeplastida. Many of the major eukaryotic pathogens belong to the SAR and Excavata supergroups. Colorized scanning EM images are shown for select organisms, with the cell body pseudocolored in blue and the flagellum in gold. Large-scale images are available as supplemental figures. Note that the *Plasmodium* example shows in vitro–induced exaflagellation, with the parasite emerging from a red blood cell (red). *Phylogenetic tree adapted from V.L. Koumandou, C. Boehm, K.A. Horder, M.C. Field, Evidence for recycling of invariant surface transmembrane domain proteins in African trypanosomes, Eukaryot. Cell 12 (2013) 330–342.*

17.2 Biology and mechanism of flagellar motility in parasite infections

17.2.1 Kinetoplastids

Kinetoplastids are a widespread and diverse group of flagellated protists, which depend on a single flagellum for cell motility. The group is found on all continents of the globe and includes free-living species as well as obligate parasites. Parasitic species infect aquatic and terrestrial hosts, ranging from invertebrates to vertebrates and plants, thus presenting important burdens on global human health and agriculture.

Among kinetoplastids the most widely studied group is the family of Trypanosomatidae, which includes the human pathogens *T. brucei*, *Trypanosoma cruzi*, and *Leishmania* spp. All of these pathogens are dependent on a blood sucking arthropod vector for transmission from one human host to another. *Leishmania* parasites are transmitted by sandflies and cause cutaneous or visceral leishmaniasis. Cutaneous leishmaniasis is self-limiting, presenting as a large ulcer at the sand fly bite site that usually heals within a few months to 1 year and leaves a disfiguring scar. In visceral leishmaniasis the infection spreads to visceral organs, causing liver and spleen damage with the disease being potentially lethal if untreated [6]. *T. cruzi* is transmitted by triatomine bugs and causes Chagas disease. Clinical manifestations in the first weeks to months are nonspecific or even absent and symptoms often emerge only years after the initial infection, corresponding to the chronic phase. Damage to the heart and digestive system are common and untreated infection may be fatal [7]. Chagas disease is endemic in South and Central America and is becoming an increasing problem in the United States where the vector is found in several southern states [8].

T. brucei and related species, collectively known as "African trypanosomes," are transmitted by tsetse flies of the genus *Glossina*, causing sleeping sickness in humans and Nagana in domestic animals throughout sub-Saharan Africa [9]. In humans, the first clinical symptoms after infection are recurring fever, headache, and joint pain. During this stage of the disease parasites are found in the bloodstream and lymphatic system. In stage 2 of the disease, parasites penetrate the blood brain barrier to invade the central nervous system (CNS). Following CNS invasion, severe neurological symptoms appear including disruption of the sleep cycle, which leads to the name sleeping sickness. If untreated, sleeping sickness is fatal.

17.2.1.1 Role of motility in disease transmission and pathogenesis

17.2.1.1.1 Trypanosoma brucei

17.2.1.1.1.1 Transmission cycle in the tsetse fly *T. brucei* is extracellular in all life cycle stages and depends on its own flagellum for cell motility. Perhaps the most obvious need for parasite motility is during transmission through the insect vector. When a tsetse fly takes a blood meal from an infected host it takes up bloodstream-form trypanosomes that differentiate into "procyclic" forms in the tsetse midgut. As with other blood sucking arthropods, tsetse flies produce a chitinous material called the peritrophic matrix between the blood meal and the midgut epithelium [10]. To establish a persistent infection, trypanosomes must migrate through or around the peritrophic matrix to enter the ectoperitrophic space, which lies between the peritrophic matrix and the gut epithelium [11]. Once established in the ectoperitrophic space, trypanosomes then move along the GI tract, into the mouthparts, and then to the salivary glands, where they attach via their flagellum to the gland epithelium [12–15]. Attached parasites then differentiate into mammalian infective metacyclic forms that are injected into the mammalian host during a subsequent blood meal [16].

Prior to making it to the salivary gland parasites are not infectious to a mammalian host. Therefore, parasite movement to the salivary gland is considered to be critical for transmission through the fly and recent work supports this view. Knockout of the outer arm dynein subunit DNAI1 impairs parasite motility and these mutants show a decreased rate of establishing midgut infection [17]. Even in those cases when mutant parasites established a midgut infection, they were unable to reach the salivary glands and therefore unable to complete the cycle. These results support the idea that there are at least two steps in the transmission cycle where parasite motility is important; first, to reach the ecto-peritrophic space and second, to migrate into the salivary glands. A limitation to these studies is that the motility mutant used also exhibits a growth defect, which might contribute to the transmission defect. Nonetheless, this important work provides the first direct evidence supporting a role for dynein-dependent parasite motility in fly transmission. Beyond its role in driving parasite propulsion, dynein-dependent axonemal beating contributes to final stages of cell separation during cytokinesis in procyclic stages but not bloodstream stages [18–20].

17.2.1.1.1.2 Pathogenesis in the mammalian host *T. brucei* is transmitted to a mammalian host when an infected tsetse fly takes a blood meal. Parasites presumably have direct access to the blood, but infection may first proceed in the dermis and from there parasites move to the lymphatic system and then into the bloodstream [9,21]. Parasites proliferate in the bloodstream and eventually penetrate the blood vessel endothelium moving into extravascular tissue spaces, including the CNS. Parasites can also reenter the bloodstream from extravascular compartments [22,23]. It is reasonable to expect that parasite motility facilitates movement between tissue compartments, e.g., dermis to bloodstream, bloodstream to CNS, and CNS to bloodstream, though this has not yet been experimentally tested. It is less clear whether parasite motility would be needed within the bloodstream.

Studies of *T. brucei* motility during mammalian infection have been hampered by the lack of motility mutants in the infectious bloodstream life cycle stage. Early studies employed RNAi to deplete specific flagellar proteins and these were almost invariably lethal [19,24,25], leading to speculation that perturbing motility itself might be lethal in bloodstream stage *T. brucei* [24]. However, subsequent studies demonstrated this not to be the case by generating motility mutants that are fully viable. These studies further indicated that structural defects rather than defective motility underlie the lethal phenotype of RNAi lines targeting flagellum proteins [26,27].

Using systems for simultaneous, inducible gene knockdown and inducible expression, Ralston and colleagues generated a conditional mutant with impaired motility but minimal impact on cell doubling time [27]. They targeted residues in the outer arm dynein subunit LC1 thought to bind dynein heavy chain and shown in *Chlamydomonas* to be required for LC1 function [28]. Mice infected with LC1 mutants via intraperitoneal (IP) injection develop infections with blood parasitemia, infection time course and pathogenic features that are indistinguishable from infection by WT parasites [29]. The results demonstrate that *T. brucei* can withstand a motility defect and still establish an infection in the bloodstream. LC1 experiments were done using an IP infection route and an acute infection model that proceeds to lethal outcome before CNS invasion. As such, it was not possible to assess dermis/ bloodstream and bloodstream/CNS movements. Nonetheless, these studies provide an important advance by generating the first viable bloodstream stage motility mutants and providing the first direct test of the requirement for motility in trypanosome pathogenesis. The stage is now set to further explore the role of parasite motility within the mammalian host. It will be of particular interest to examine more severe motility mutants, different routes of infection, and to employ a chronic infection model, where CNS invasion can be assessed.

Trypanosomes in the bloodstream must evade the host immune system. One part of their strategy is periodic switching of Variant Surface Glycoproteins (VSGs) that form a protective surface coat [30]. Another part of their strategy is to internalize and degrade any VSG-antibody complexes that form [31]. Forward parasite motility is suggested to facilitate movement of VSG-antibody complexes to the flagellar pocket at the posterior pole of the cell where they are internalized by endocytosis [31]. While this presents interesting model, some aspects of the model remain to be tested and its contribution to infection has not yet been examined in mice.

17.2.1.1.1.3 *Social motility* In its natural environment, *T. brucei* is in constant contact with host tissue surfaces, particularly in the tsetse fly. When cultivated on surfaces, procyclic (fly midgut stage) parasites use flagellar motility to assemble into multicellular groups that sense and respond collectively to extracellular signals [32]. Once they reach a specific cell density parasites in groups begin coordinated movement outward, generating symmetric projections that radiate away from the site of inoculation [32,33]. This group behavior was termed social motility (SoMo) based on similarities to social motility in bacteria. SoMo is developmentally regulated, and its defects correlate with fly midgut infection defects, supporting a role for this group motility in vivo [33,34]. Social motility is dependent on regulation of axonemal dynein, because mutants with defects in the axonemal dynein regulatory complex (DRC) are social motility negative [32]. This result also demonstrates that group movements outward result from active parasite motility rather than simple expansion by cell doubling. Interestingly, parasites within a group can detect other parasites nearby and adjust their movements in response, demonstrating the ability of these organisms to control axonemal beating in response to environmental signals [32]. These combined studies pave the way for addressing the question of how trypanosomes react to external factors to control flagellar beating and thereby navigate through the host environment.

17.2.1.1.2 Leishmania

Leishmania parasites switch between a motile extracellular stage in the sand fly vector and an immotile intracellular stage in the mammalian host. When motile promastigote forms are injected into the host dermis during a blood meal the parasites are phagocytosed by macrophages and other phagocytes [35]. The parasite then differentiates into an immotile amastigote form, with a very short flagellum, within the phagolysosome of the host cell [36]. In vitro infection of macrophages with *Leishmania* showed that the parasite enters the host cell in a polarized manner with the flagellar tip facing the macrophage, ergo in the direction of parasite movement [36]. Flagellar beating remains very active inside of the host cell leading to reorientation of the parasite flagellum toward the macrophage periphery and oscillatory parasite movement. Thus, an actively beating flagellum is implied to play a role during the early phase of cell invasion.

When macrophages infected with *Leishmania* amastigotes are taken up by a sand fly during a blood meal, the decrease in temperature and increase in pH triggers differentiation into motile promastigote forms with a beating flagellum at the anterior pole. Promastigote forms multiply within the blood meal and after a few days migrate to the anterior part of the gut where they penetrate the peritrophic matrix and attach to the gut epithelium [37]. This migration and the crossing of the peritrophic matrix to escape the blood meal is essential to avoid expulsion during defecation. Promastigote forms in the ectoperitrophic space and then differentiate into infective metacyclic forms that can be transmitted during a subsequent blood meal. Although not yet experimentally tested, it is intuitive that dynein-mediated flagellar motility is essential for parasite movement in the gut, penetration of the peritrophic matrix, and hence, transmission to a new human host. There is some evidence to indicate that parasite movements within the vector are directed by chemotaxis and/or osmotaxis [38–41].

17.2.1.1.3 Trypanosoma cruzi

When an infected triatomine bug takes a blood meal it excretes infective flagellated trypomastigote cells in the feces. The trypanosomes then enter the bite wound or penetrate mucosal membranes and infect various cell types. Inside the host cell they differentiate into immotile amastigote forms, which multiply and eventually differentiate into motile trypomastigote forms. The host cell lyses and extracellular trypomastigotes can infect new host cells to start a new replication cycle, or infect a triatomine bug when taken up by a blood meal. In the triatomine bug, epimastigote forms multiply in the midgut and then differentiate into infective metacyclic forms in the hindgut. There are no studies undertaken so far that address the importance of motility during the *T. cruzi* life cycle, but there are several points where it is likely that motility or flagellar beating plays a role. It is reasonable, for example, to expect that active motility is required to penetrate mucosal membranes. Likewise, when trypomastigotes egress from the host cell, flagellar motility may facilitate infection of a new host cell, as well as parasite movement through tissues. In the triatomine bug, epimastigote forms might be transported to the hindgut passively, but to do so in a controlled manner via active parasite motility would ensure a higher transmission rate.

17.2.1.2 Motility mechanisms

Trypanosomatid motility occurs exclusively through dynein-dependent flagellar beating. Flagellum arrangement and parasite morphology influence flagellar beating and cell propulsion, so these topics are considered first.

Trypanosomatids are polarized, uniflagellated cells and each life cycle stage has a distinctive morphology defined by positioning of the flagellum relative to other cell structures. In all forms, the flagellum exits the cytoplasm through the flagellar pocket, which also functions as the exclusive site of endocytosis and secretion [42]. A variety of morphotypes exist, but two of the most prominent are trypomastigote and promastigote forms (Fig. 17.3). In trypomastigotes,

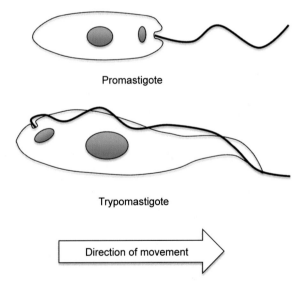

Promastigote

Trypomastigote

Direction of movement

Figure 17.3 Two major motile morphotypes of kinetoplastids. The flagellum of the promastigote form exits the cytoplasm on the anterior pole of the cell and extends free of the cell body. The flagellum of the trypomastigote form emerges from the cytoplasm at the posterior pole and is laterally attached to the cell body for most of its length, with only a small segment extending beyond the cell on the anterior side.

the flagellum remains laterally attached to the cell body after it exits the flagellar pocket, extending in a helical path along the cell body to the end of the cell [43,44]. This unique arrangement is mediated by a specialized flagellum attachment zone (FAZ) on the cell body and has implications for motility of the organism. In promastigote forms, the flagellum extends free from the cell after exit through the flagellar pocket, similar to what is observed in other eukaryotes.

Cell motility is driven by a flagellar wave that travels in the opposite direction of that observed in most eukaryotic flagella. Beating initiates at the tip of the flagellum and propagates toward the flagellar pocket, causing the cell to move in the opposite direction, with the flagellum tip leading [45,46]. In trypomastigote forms, since the flagellum is helically attached to the cell body, the flagellar wave is directly transmitted to the cell body, causing the whole cell to rotate and undulate as it moves forward (Movie 17.1). This unique motility gives the genus its name as *Trypanosoma*, comes from the Greek for "auger body" [47]. Flagellar beating is highly dynamic, three dimensional, and in many cases erratic, making it challenging to describe in terms of typical beat parameters such as frequency and amplitude [48,49]. High speed video microscopy studies have suggested that trypanosomes may alternate between left-handed and right-handed helical rotation [49], as observed for *Plasmodium* gametes [50], although subsequent work has questioned this model [48]. Biomechanical analyses indicate that the helical waveform derives from viscous drag on the cell body as it translocates forward.

Auger-like propulsion of trypanosomes is thought to facilitate tissue penetration, although this idea has not yet been directly tested. Interestingly, the direction of flagellar wave propagation can be reversed, resulting in tumbling or reverse motility and leading to "run and tumble" like movements in liquid culture [19,51,52]. Beat reversals may occur stochastically, or in response to encountering objects in the environment [48], evoking thoughts of avoidance responses observed in *Paramecia* [53]. RNAi knockdown studies have demonstrated that outer arm dyneins are required for tip-to-base flagellar beating [19,54]. Thus, outer dyneins in trypanosomes are integral contributors to type of flagellar wave generated.

As with other eukaryotes, flagellum beating is driven by a canonical 9+2 axoneme with 9 outer doublet microtubules surrounding a central pair of singlet microtubules [55,56]. Targeted disruption of individual axonemal subcomplexes has demonstrated a requirement for outer dyneins [19,27,54], inner dyneins [57], the central pair apparatus, radial spokes, and the nexin–dynein regulatory complex [18,25,27,58,59]. Given distinctive motility features in these pathogens and the contribution of motility to disease transmission and pathogenesis, axonemal subcomplexes warrant investigation as potential therapeutic targets.

Despite underlying conservation, the trypanosome flagellum displays a number of distinctive features compared with *Chlamydomonas*, a de facto model for cilium biology. These include a central pair apparatus having a fixed orientation relative to outer doublets [18,19,60], tip-to-base beating with periodic beat reversals, and a highly dynamic and irregular helical waveform that is transmitted directly to the cell body to drive cell propulsion [52]. While distinctive arrangement of flagellar structures certainly contribute to the parasite's unusual motility, it is likely that dynein motors and regulatory activities also participate. The cohort of dynein heavy-chain genes differs from that of humans and *Chlamydomonas* [20,61] and it will be important to determine distribution of these dynein heavy-chain isoforms along the axoneme through high-resolution 3D studies [62–64]. As noted above, outer dyneins dictate beat direction. Comparative genomic studies revealed lineage-specific isoforms of several conserved motility components, including subunits of the DRC [65]. It is possible these duplications reflect specialization of axonemal doublets due to attachment of the paraflagellar rod (PFR), an extra-axonemal filament that runs parallel to the axoneme and is directly connected to axonemal doublets 4–7. The PFR exerts mechanical as well as regulatory influence on axoneme beating, although its exact role is uncertain [63,66–68].

17.2.2 Giardia

Giardia intestinalis (lamblia) is the causative agent of giardiasis, which presents clinically as malabsorptive diarrhea. Giardiasis is one of the most prevalent intestinal infections in humans, accounting for several hundred million cases each year. In some regions, infection rates nearing 100% have been reported [69,70]. The parasite is transmitted through ingestion of contaminated food or water containing infective cysts, which may persist in the environment for several weeks to months. On reaching the small intestine, infective cysts undergo excystation and differentiate into flagellated trophozoites. Trophozoite stages may then attach to the gut epithelium via a specialized microtubule-based structure, the ventral disk [71,72]. Gut attachment helps the parasite avoid elimination by gut peristalsis and destroys gut microvilli, leading to disease symptoms. Trophozoites that make it to the large intestine undergo encystment, to generate infective cysts that are shed in the feces, thus completing the parasite life cycle [71].

Flagellar motility is essential for survival and pathogenesis of *Giardia*, although the exact areas of influence are poorly understood. With respect to cell propulsion, flagellar-driven motility helps the parasite move to suitable attachment sites in the gut and may also participate in detachment, although it was shown that flagellar motility is not directly required to maintain attachment [73].

Moreover, *Giardia* trophozoites divide while attached to the intestinal epithelium and flagellar beating is essential for cell division and implicated in the formation of cysts and excystation [74]. Thus, as observed for *T. brucei* [18,19] and *Tetrahymena* [75] dynein-dependent axonemal motility contributes functions beyond cell propulsion.

Giardia trophozoites are bilaterally symmetric with two nuclei and a complex microtubule cytoskeleton that includes eight flagella organized in four bilaterally symmetrical pairs: the anterior, caudal, posteriolateral, and ventral pair [74]. The eight flagellar basal bodies are located in close proximity to the organism's two nuclei, with a canonical transition zone and 9+2 axoneme extending from each basal body. *Giardia* axonemes are distinguished by having a long cytoplasmic region before exiting the cell as a membrane-encased flagellum. Such an arrangement presents interesting questions regarding the use of intraflagellar transport (IFT) for flagellum assembly. Two independent studies suggest IFT-independent and IFT-dependent mechanisms may operate in the same cell, driving assembly of cytoplasmic and membrane-enclosed portions, respectively [74,76,77].

Individual flagellar pairs show different beating patterns, combining to produce an elaborate cell propulsion mechanism (Movie 17.2). Anterior flagella show a helical beating pattern, for example, while the beating pattern of ventral flagella is sigmoidal in a plane parallel to that of the cell body. Caudal flagella exhibit a flexing movement rather than a typical flagellar beat. How beat pattern and frequency are regulated and coordinated among eight flagella to accomplish effective cell propulsion is not known. Contributing to individual flagellar waveforms are lineage-specific extra-axonemal structures associated with specific flagellar pairs, such as the caudal complex that is connected to the cytoplasmic portion of caudal axonemes [74]. Whether these structures play an active or passive role in flagellar beating remains to be determined. Another parasitic species with an elaborated beating pattern of different flagella is *Spironucleus*, as can be seen in Movie 17.5, which causes "hole in the head" disease in fish.

17.2.3 Plasmodium

Various subspecies of *Plasmodium* parasites are transmitted by *Anopheles* mosquitos and cause malaria in humans. These parasites group with ciliates in the alveolata clade. The global impact of malaria is immense. In 2014, 97 countries and territories had ongoing malaria transmission (WHO), with an estimated 3.3 billion people at risk, approximately 200 million reported cases each year and more than 500,000 deaths. *Plasmodium* is nonflagellated throughout the majority of its life cycle, but is critically dependent on flagellar motility during fertilization in the mosquito midgut. Nonflagellated male and female gametocytes are produced in the bloodstream of the human host and taken up by the *Anopheles* mosquito during a blood meal. In the mosquito midgut, the male gametocyte undergoes exflagellation to produce highly motile microgametes that use dynein-dependent axonemal motility to find and fertilize female gametes (Movie 17.1). During exflagellation, the parasite undergoes three successive rounds of DNA replication and nuclear division without undergoing cell division to yield a single cell with eight nuclei and eight basal bodies that each extend a motile axoneme within the cytoplasm. The resulting eight axonemes protrude from the cell, taking membrane and nucleus along to generate eight motile microgametes. This is a very rapid process; DNA replication, exflagellation, and fertilization of female macrogametes are completed within 30–60 min after the uptake by the mosquito.

The short-lived male microgamete is the only life cycle stage where *Plasmodium* relies on dynein-mediated motility. Each microgamete is essentially a stripped-down flagellum, having a 9+2 axoneme with minimal cytoplasmic material and no organelles other than the nucleus. As such, the entire cell is roughly the size of the axoneme. Recent work indicates that the microgamete moves using bihelical motility, in which axonemal beating generates a helical waveform that alternates between left-handed and right-handed helical turns [50]. Given the minimalist nature of the *Plasmodium* flagellum, these organisms offer advantages for studying fundamental mechanics of axoneme-dependent cell motility, including deciphering the minimal requirements and mechanisms necessary to produce chiral beating in the axoneme [50].

17.2.4 Trichomonas vaginalis

Trichomoniasis is the most common curable sexually transmitted disease, with 3.7 million people infected in the United States. Transmission may be facilitated by the fact only about 30% of people having the infection show symptoms [78]. *Trichomonas* is found in the urogenital tract of infected humans, where the parasite can switch between a free-swimming trophozoite stage and an amoeboid-adherent stage [79]. The free-swimming trophozoite possesses four flagella at the anterior pole of the cell and one flagellum that is attached to the cell body. The role of the attached flagellum is not known while the four anterior flagella produce a flagellar beat that drives locomotion. On contact with vaginal epithelial cells, the parasite transforms into the amoeboid stage within 1 minute, although the flagellar apparatus is not degraded. So the cells can rapidly switch between these two stages and swim to different locations within the urogenital tract. The role of dynein-mediated motility has not been studied in *Trichomonas* and while it is likely to be important for the parasite to spread through the host, no experimental data are yet available.

17.3 Final remarks

Despite the enormous impact of protozoan parasites capable of dynein-mediated motility on human health and food production, the regulation of flagellar motility and importance for pathogenesis remain vastly understudied. Many protozoan parasites are believed to be among the earliest branches of the eukaryotic tree, with a long, distinct evolutionary history compared to the most commonly used model organisms like *Chlamydomonas* or mammals. Studying the flagellum and dynein-mediated motility in these parasites, therefore, has a great potential for new discovery, ranging from parasite-specific functions and potential drug targets to insight into conserved functions and fundamental aspects of flagellar motility to biomechanical principles of microbial cell propulsion. The "brain-eating amoeba," *Naegleria fowleri*, for example, is an opportunistic parasite that has the unusual ability to switch between an amoeboid stage using actin-mediated motility and a nonfeeding flagellated stage, which employs axonemal motility for propulsion. During differentiation into the flagellated stage, two new basal bodies are generated de novo. So this organism represents an important a model for studying de novo biogenesis of basal bodies and motile axonemes. Meanwhile *T. brucei* is emerging as a very accessible model organism and a member of the early branching Excavata clade in the eukaryotic evolutionary tree. In summary, protozoan parasites represent a largely understudied group of organisms holding a great potential for future discoveries important for human health and fundamental processes of dynein-mediated motility.

Supplementary data

Supplementary data related to this chapter can be found online at doi:10.1016/B978-0-12-809470-9.00017-5

Acknowledgments

We are grateful to colleagues who provided unpublished SEM images for Fig. 17.2: *Trichomonas* (Marlene Benchimal, Universidade Federal do Rio de Janeiro), *T. cruzi* (David Engman, Cedars-Sinai Hospital, Los Angeles), *Leishmania*, and *Giardia* (Andrew Hemphill, University of Bern); and movies; and unpublished movies: Movies 17.2 and 17.5 (Scott Dawson, UC Davis). Movie 17.1 (Laurence Wilson, University of York; Lucy Carter and Sarah Reece, U. Edinburgh), the image of *Plasmodium*, is adapted from Ref. [80]. Funding: NIH (AI052348); Swiss National Foundation.

References

[1] S. Kamoun, O. Furzer, J.D.G. Jones, H.S. Judelson, G.S. Ali, R.J.D. Dalio, S.G. Roy, L. Schena, A. Zambounis, F. Panabieres, et al., The Top 10 oomycete pathogens in molecular plant pathology, Mol. Plant Pathol. 16 (2015) 413–434.

[2] F.L. Schuster, L. Ramirez-Avila, Current world status of *Balantidium coli*, Clin. Microbiol. Rev. 21 (2008) 626–638.

[3] H. Dickerson, T. Clark, *Ichthyophthirius multifiliis*: a model of cutaneous infection and immunity in fishes, Immunol. Rev. 166 (1998) 377–384.

[4] P.L. Pereira, A.M. Torres, D.F. Soares, M. Hijosa-Valsero, E. Bécares, Chytridiomycosis: a global threat to amphibians, Rev. Sci. Tech. (2013).

[5] B. Morga, P. Bastin, Getting to the heart of intraflagellar transport using Trypanosoma and Chlamydomonas models: the strength is in their differences, Cillia (2013).

[6] I. Okwor, J. Uzonna, Social and economic burden of human leishmaniasis, Am. J. Trop. Med. Hyg. 94 (2016) 489–493.

[7] M. Traina, S. Meymandi, J.S. Bradfield, Heart failure secondary to Chagas disease: an emerging problem in non-endemic areas, Curr. Heart Fail Rep. (2016).

[8] J.H. Theis, M. Tibayrenc, S.K. Ault, D.T. Mason, Agent of Chagas' disease from Honduran vector capable of developing in California insects: implications for cardiologists, Am. Heart J. (1985).

[9] D. Malvy, F. Chappuis, Sleeping sickness, Clin. Microbiol. Infect. 17 (2011) 986–995.

[10] D. Hegedus, M. Erlandson, C. Gillott, U. Toprak, New insights into peritrophic matrix synthesis, architecture, and function, Annu. Rev. Entomol. 54 (2009) 285–302.

[11] W. Gibson, M. Bailey, The development of *Trypanosoma brucei* within the tsetse fly midgut observed using green fluorescent trypanosomes, Kinetoplastid Biol. Dis. (2003).

[12] N.A. Dyer, C. Rose, N.O. Ejeh, A. Acosta-Serrano, Flying tryps: survival and maturation of trypanosomes in tsetse flies, Trends Parasitol. 29 (2013) 188–196.

[13] R. Sharma, E. Gluenz, L. Peacock, W. Gibson, K. Gull, M. Carrington, The heart of darkness: growth and form of Trypanosoma brucei in the tsetse fly, Trends Parasitol. 25 (2009) 517–524.

[14] L. Tetley, K. Vickerman, Differentiation in *Trypanosoma brucei*: host-parasite cell junctions and their persistence during acquisition of the variable antigen coat, J. Cell Sci. (1985).

[15] J. Van Den Abbeele, Y. Claes, D. van Bockstaele, D. Le Ray, M. Coosemans, *Trypanosoma brucei* spp. development in the tsetse fly: characterization of the post-mesocyclic stages in the foregut and proboscis, Parasitology 118 (Pt 5) (1999) 469–478.

[16] B. Rotureau, I. Subota, J. Buisson, P. Bastin, A new asymmetric division contributes to the continuous production of infective trypanosomes in the tsetse fly, Development 139 (2012) 1842–1850.

[17] B. Rotureau, C.P. Ooi, D. Huet, S. Perrot, P. Bastin, Forward motility is essential for trypanosome infection in the tsetse fly, Cell Microbiol. (2013).

[18] K.S. Ralston, A.G. Lerner, D.R. Diener, K.L. Hill, Flagellar motility contributes to cytokinesis in *Trypanosoma brucei* and is modulated by an evolutionarily conserved dynein regulatory system, Eukaryot. Cell 5 (2006) 696–711.

[19] C. Branche, L. Kohl, G. Toutirais, J. Buisson, J. Cosson, P. Bastin, Conserved and specific functions of axoneme components in trypanosome motility, J. Cell Sci. 119 (2006) 3443–3455.

[20] K.S. Ralston, K.L. Hill, The flagellum of *Trypanosoma brucei*: new tricks from an old dog, Int. J. Parasitol. 38 (2008) 869–884.

[21] G. Caljon, N. Van Reet, C. De Trez, M. Vermeersch, D. Perez-Morga, J. Van den Abbeele, The dermis as a delivery site of trypanosoma brucei for tsetse flies, PLos Pathog. 12 (2016).

[22] S. Trindade, F. Rijo-Ferreira, T. Carvalho, D. Pinto-Neves, F. Guegan, F. Aresta-Branco, F. Bento, S.A. Young, A. Pinto, J. Van den Abbeele, et al., Trypanosoma brucei parasites occupy and functionally adapt to the adipose tissue in mice, Cell Host Microbe 19 (2016) 837–848.

[23] F.W. Jennings, D.D. Whitelaw, P.H. Holmes, H.G. Chizyuka, G.M. Urquhart, The brain as a source of relapsing *Trypanosoma brucei* infection in mice after chemotherapy, Int. J. Parasitol. (1979).

[24] R. Broadhead, H.R. Dawe, H. Farr, S. Griffiths, S.R. Hart, N. Portman, M.K. Shaw, M.L. Ginger, S.J. Gaskell, P.G. McKean, et al., Flagellar motility is required for the viability of the bloodstream trypanosome, Nature 440 (2006) 224–227.

[25] K.S. Ralston, K.L. Hill, Trypanin, a component of the flagellar dynein regulatory complex, is essential in bloodstream form African trypanosomes, PLos Pathog. 2 (2006) 873–882.

[26] K.S. Ralston, Z.P. Kabututu, J.H. Melehani, M. Oberholzer, K.L. Hill, The *Trypanosoma brucei* flagellum: moving parasites in new directions, Annu. Rev. Microbiol. 63 (2009) 335–362.

[27] K.S. Ralston, N.K. Kisalu, K.L. Hill, Structure-function analysis of dynein light chain 1 identifies viable motility mutants in bloodstream-form trypanosoma brucei, Eukaryot. Cell 10 (2011) 884–894.

[28] R.S. Patel-King, S.M. King, An outer arm dynein light chain acts in a conformational switch for flagellar motility, J. Cell Biol. 186 (2009) 283–295.

[29] N.K. Kisalu, G. Langousis, L.A. Bentolila, K.S. Ralston, K.L. Hill, Mouse infection and pathogenesis by *Trypanosoma brucei* motility mutants, Cell Microbiol. 16 (2014) 912–924.

[30] A. Schwede, M. Carrington, Bloodstream form Trypanosome plasma membrane proteins: antigenic variation and invariant antigens, Parasitology 137 (2010) 2029–2039.

[31] M. Engstler, T. Pfohl, S. Herminghaus, M. Boshart, G. Wiegertjes, N. Heddergott, P. Overath, Hydrodynamic flow-mediated protein sorting on the cell surface of trypanosomes, Cell 131 (2007) 505–515.

[32] M. Oberholzer, M.A. Lopez, B.T. McLelland, K.L. Hill, Social motility in african trypanosomes, PLoS Pathog. 6 (2010) e1000739.

[33] S. Imhof, S. Knusel, K. Gunasekera, X.L. Vu, I. Roditi, Social motility of African trypanosomes is a property of a distinct life-cycle stage that occurs early in tsetse fly transmission, 2014.

[34] S. Imhof, X.L. Vu, P. Butikofer, I. Roditi, A glycosylation mutant of trypanosoma brucei links social motility defects in vitro to impaired colonization of tsetse flies in vivo, Eukaryot. Cell 14 (2015) 588–592.

[35] J.P. de Menezes, E.M. Saraiva, B. da Rocha-Azevedo, The site of the bite: *Leishmania* interaction with macrophages, neutrophils and the extracellular matrix in the dermis, Parasites Vectors (2016).

[36] C.L. Forestier, C. Machu, C. Loussert, P. Pescher, G.F. Spath, Imaging host cell-*Leishmania* interaction dynamics implicates parasite motility, lysosome recruitment, and host cell wounding in the infection process, Cell Host Microbe 9 (2011) 319–330.

[37] A. Dostalova, P. Volf, *Leishmania* development in sand flies: parasite-vector interactions overview, Parasites Vectors 5 (2012).

[38] J.S. Oliveira, M.N. Melo, N.F. Gontijo, A sensitive method for assaying chemotaxic responses of *Leishmania* promastigotes, Exp. Parasitol. 96 (2000) 187–189.

[39] V.C. Barros, J.S. Oliveira, M.N. Melo, N.F. Gontijo, *Leishmania* amazonensis: chemotaxic and osmotaxic responses in promastigotes and their probable role in development in the phlebotomine gut, Exp. Parasitol. 112 (2006) 152–157.

[40] E. Diaz, A.K. Zacarias, S. Perez, O. Vanegas, L. Kohidai, M. Padron-Nieves, A. Ponte-Sucre, Effect of aliphatic, monocarboxylic, dicarboxylic, heterocyclic and sulphur-containing amino acids on *Leishmania* spp. chemotaxis, Parasitology 142 (2015) 1621–1630.

[41] P.A. Bates, *Leishmania* sand fly interaction: progress and challenges, Curr. Opin. Microbiol. 11 (2008) 340–344.

[42] M.C. Field, M. Carrington, The trypanosome flagellar pocket, Nat. Rev. Microbiol. 7 (2009) 775–786.

[43] S. Lacomble, S. Vaughan, C. Gadelha, M.K. Morphew, M.K. Shaw, J.R. McIntosh, K. Gull, Three-dimensional cellular architecture of the flagellar pocket and associated cytoskeleton in trypanosomes revealed by electron microscope tomography, J. Cell Sci. 122 (2009) 1081–1090.

[44] S. Vaughan, L. Kohl, I. Ngai, R.J. Wheeler, K. Gull, A repetitive protein essential for the flagellum attachment zone filament structure and function in *Trypanosoma brucei*, Protist 159 (2008) 127–136.

[45] C. Gadelha, B. Wickstead, K. Gull, Flagellar and ciliary beating in trypanosome motility, Cell Motil. Cytoskelet. 64 (2007) 629–643.

[46] G. Ballesteros-Rodea, M. Santillan, S. Martinez-Calvillo, R. Manning-Cela, Flagellar motility of *Trypanosoma cruzi* epimastigotes, 2012.

[47] M. Gruby, Recherches et observations sur une nouvelle espèce d'hématozoaire, *Trypanosoma sanguinis*, vol. 17, Académie des Sciences, Paris, 1843, pp. 1134–1136.

[48] N. Heddergott, T. Kruger, S.B. Babu, A. Wei, E. Stellamanns, S. Uppaluri, T. Pfohl, H. Stark, M. Engstler, Trypanosome motion represents an adaptation to the crowded environment of the vertebrate bloodstream, PLoS Pathog. 8 (2012) e1003023.

[49] J.A. Rodriguez, M.A. Lopez, M.C. Thayer, Y.Z. Zhao, M. Oberholzer, D.D. Chang, N.K. Kisalu, M.L. Penichet, G. Helguera, R. Bruinsma, et al., Propulsion of African trypanosomes is driven by bihelical waves with alternating chirality separated by kinks, Proc. Natl. Acad. Sci. U.S.A. 106 (2009) 19322–19327.

[50] L.G. Wilson, L.M. Carter, S.E. Reece, High-speed holographic microscopy of malaria parasites reveals ambidextrous flagellar waveforms, Proc. Natl. Acad. Sci. U.S.A. 110 (2013) 18769–18774.

[51] K.L. Hill, Biology and mechanism of trypanosome cell motility, Eukaryot. Cell 2 (2003) 200–208.

[52] T.L. Jahn, E.C. Bovee, Infectious Blood Diseases of Man and Animals, vol. 1, Academic Press, 1968, pp. 393–436.

[53] A. Hamel, C. Fisch, L. Combettes, P. Dupuis-Williams, C.N. Baroud, Transitions between three swimming gaits in Paramecium escape, Proc. Natl. Acad. Sci. U.S.A. 108 (2011) 7290–7295.

[54] D.M. Baron, Z.P. Kabututu, K.L. Hill, Stuck in reverse: loss of LC1 in Trypanosoma brucei disrupts outer dynein arms and leads to reverse flagellar beat and backward movement, J. Cell Sci. 120 (2007) 1513–1520.

[55] G. Langousis, K.L. Hill, Motility and more: the flagellum of Trypanosoma brucei, Nat. Rev. Microbiol. 12 (2014) 505–518.

[56] A. Schneider, T. Sherwin, R. Sasse, D.G. Russell, K. Gull, T. Seebeck, Subpellicular and flagellar microtubules of *Trypanosoma brucei* brucei contain the same alpha-tubulin isoforms, J. Cell Biol. (1987).

[57] A.L. Springer, D.F. Bruhn, K.W. Kinzel, N.F. Rosenthal, R. Zukas, M.M. Klingbeil, Silencing of a putative inner arm dynein heavy chain results in flagellar immotility in Trypanosoma brucei, Mol. Biochem. Parasitol. 175 (2011) 68–75.

[58] Z.P. Kabututu, M. Thayer, J.H. Melehani, K.L. Hill, CMF70 is a subunit of the dynein regulatory complex, J. Cell Sci. 123 (2010) 3587–3595.

[59] H.T. Nguyen, J. Sandhu, G. Langousis, K.L. Hill, CMF22 is a broadly conserved axonemal protein and is required for propulsive motility in trypanosoma brucei, Eukaryot. Cell 12 (2013) 1202–1213.

[60] C. Gadelha, B. Wickstead, P.G. McKean, K. Gull, Basal body and flagellum mutants reveal a rotational constraint of the central pair microtubules in the axonemes of trypanosomes, J. Cell Sci. 119 (2006) 2405–2413.

[61] B. Wickstead, K. Gull, Dyneins across eukaryotes: a comparative genomic analysis, Traffic 8 (2007) 1708–1721.

[62] L.C. Hughes, K.S. Ralston, K.L. Hill, Z.H. Zhou, Three-dimensional structure of the trypanosome flagellum suggests that the paraflagellar rod functions as a biomechanical spring, PLoS One 7 (2012).

[63] A.Y. Koyfman, M.F. Schmid, L. Gheiratmand, C.J. Fu, H.A. Khant, D. Huang, C.Y. He, W. Chiu, Structure of *Trypanosoma brucei* flagellum accounts for its bihelical motion, Proc. Natl. Acad. Sci. U.S.A. 108 (2011) 11105–11108.

[64] J.L. Hoog, C. Bouchet-Marquis, J.R. McIntosh, A. Hoenger, K. Gull, Cryo-electron tomography and 3-D analysis of the intact flagellum in *Trypanosoma brucei*, J. Struct. Biol. 178 (2012) 189–198.

[65] D.M. Baron, K.S. Ralston, Z.P. Kabututu, K.L. Hill, Functional genomics in *Trypanosoma brucei* identifies evolutionarily conserved components of motile flagella, J. Cell Sci. 120 (2007) 478–491.

[66] N. Portman, S. Lacomble, B. Thomas, P.G. McKean, K. Gull, Combining RNA interference mutants and comparative proteomics to identify protein components and dependences in a eukaryotic flagellum, J. Biol. Chem. 284 (2009) 5610–5619.

[67] M. Oberholzer, G. Marti, M. Baresic, S. Kunz, A. Hemphill, T. Seebeck, The *Trypanosoma brucei* cAMP phosphodiesterases TbrPDEB1 and TbrPDEB2: flagellar enzymes that are essential for parasite virulence, FASEB J. 21 (2007) 720–731.

[68] C. Santrich, L. Moore, T. Sherwin, P. Bastin, C. Brokaw, K. Gull, J.H. LeBowitz, A motility function for the paraflagellar rod of *Leishmania* parasites revealed by PFR-2 gene knockouts, Mol. Biochem. Parasitol. 90 (1997) 95–109.

[69] P.A. Flanagan, *Giardia* – diagnosis, clinical course and epidemiology – a review, Epidemiol. Infect. 109 (1992) 1–22.

[70] E. Einarsson, S. Ma'ayeh, S.G. Svard, An up-date on *Giardia* and giardiasis, Curr. Opin. Microbiol. 34 (2016) 47–52.

[71] J. Ankarklev, J. Jerlstrom-Hultqvist, E. Ringqvist, K. Troell, S.G. Svard, Behind the smile: cell biology and disease mechanisms of *Giardia* species, Nat. Rev. Microbiol. 8 (2010) 413–422.

[72] H.G. Elmendorf, S.C. Dawson, M. McCaffery, The cytoskeleton of *Giardia lamblia*, Int. J. Parasitol. 33 (2003) 3–28.

[73] S.A. House, K. Richter Dj, J.K. Pham, S.C. Dawson, *Giardia* flagellar motility is not directly required to maintain attachment to surfaces, PLoS Pathog. (2011).

[74] S.C. Dawson, S.A. House, Life with eight flagella: flagellar assembly and division in *Giardia*, Curr. Opin. Microbiol. 13 (2010) 480–490.

[75] J.M. Brown, C. Hardin, J. Gaertig, Rotokinesis, a novel phenomenon of cell locomotion-assisted cytokinesis in the ciliate *Tetrahymena thermophila*, Cell Biol. Int. 23 (1999) 841–848.

[76] J.C. Hoeng, S.C. Dawson, S.A. House, M.S. Sagolla, J.K. Pham, J.J. Mancuso, J. Lowe, W.Z. Cande, High-resolution crystal structure and in vivo function of a kinesin-2 homologue in *Giardia intestinalis*, Mol. Biol. Cell 19 (2008) 3124–3137.

[77] M.L. Carpenter, W.Z. Cande, Using Morpholinos for gene knockdown in *Giardia intestinalis*, Eukaryot. Cell 8 (2009) 916–919.

[78] D. Leitsch, Recent advances in the *Trichomonas vaginalis* field, LID, 2016, http://dx.doi.org/10.12688/f1000research.7594.1. LID – F1000 Faculty Rev-162 (pii). F1000Res.

[79] G. Kusdian, C. Woehle, W.F. Martin, S.B. Gould, The actin-based machinery of *Trichomonas vaginalis* mediates flagellate-amoeboid transition and migration across host tissue, Cell Microbiol. 15 (2013) 1707–1721.

[80] O. Billker, S. Dechamps, R. Tewari, G. Wenig, B. Franke-Fayard, V. Brinkmann, Calcium and a calcium-dependent protein kinase regulate gamete formation and mosquito transmission in a malaria parasite, Cell 117 (2004) 503–514.

[81] V.L. Koumandou, C. Boehm, K.A. Horder, M.C. Field, Evidence for recycling of invariant surface transmembrane domain proteins in African trypanosomes, Eukaryot. Cell 12 (2013) 330–342.

[82] M. Oberholzer, G. Langousis, H.T. Nguyen, E.A. Saada, M.M. Shimogawa, Z.O. Jonsson, S.M. Nguyen, J.A. Wohlschlegel, K.L. Hill, Independent analysis of the flagellum surface and matrix proteomes provides insight into flagellum signaling in mammalian-infectious *Trypanosoma brucei*, Mol. Cell Proteomics 10 (2011).

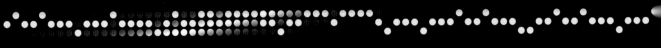

In this chapter

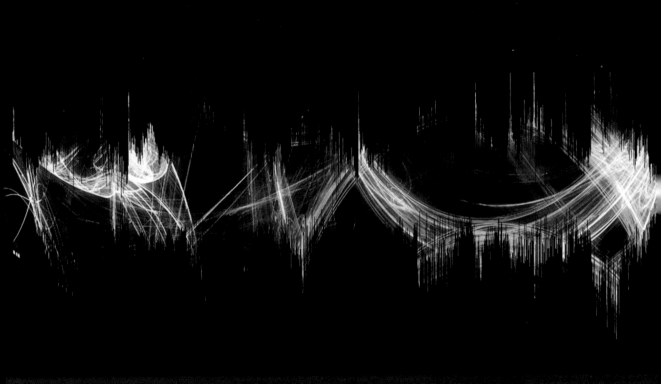

Dynein axonemal light chain 4: involvement in congenital mirror movement disorder

John B. Vincent[1,2]

[1]Campbell Family Mental Health Research Institute, Toronto, ON, Canada; [2]University of Toronto, Toronto, ON, Canada

18.1 Introduction

Dynein Axonemal Light Chain 4 (DNAL4) protein is understood to be part of the axonemal (or ciliary and flagellar) complex of dynein molecules (including dynein heavy and light chains) (GO:0005858) and a component of the microtubule-based dynein motor complex [13]. In humans, two known axonemal light chain proteins, DNAL1 and DNAL4, sit at the outer edge of the complex. DNAL1 is a 190 residue protein (estimated 21.5 kDa), encoded by an eight-exon gene on chromosome 14q24.3. DNAL4 is a 105-residue protein (estimated 12 kDa), encoded by a four-exon gene on chromosome 22q13.1. DNAL1 and 4 show very limited sequence homology (Fig. 18.1).

18.2 Axonemal dynein components: biology and disease

Axonemal dynein components are believed to be important for polarity. For example, in medaka (Japanese rice fish, *Oryzias latipes*), axonemal intermediate chain 2 gene mutants show a defect in the left–right polarity of organs [18]. In humans, mutations in numerous dynein axonemal components, and regulatory or assembly genes, including genes encoding dynein axonemal heavy chains, intermediate chains, assembly factors, regulatory complex subunits result in primary ciliary dyskinesia (PCD) with or without *situs inversus*, typically with autosomal recessive inheritance (reviewed in Ref. [20]). Mutations in *DNAL1* gene have also been linked to autosomal recessive PCD. PCD is characterized by recurrent respiratory tract infections due to defects in the cilia lining the respiratory tracts. *Situs inversus* is a nonpathological

Dyneins. https://doi.org/10.1016/B978-0-12-809470-9.00018-7
437

```
DNAL1    MAKATTIKEALARWEEKTGQRPSEAKEIKLYAQIPPIEKMD-------ASLSMLANCEK
DNAL4    ------------MGETEGKKD--EADYKRLQTFPLVRHSDMPEEMRVETMELCVTACEK

DNAL1    LSLSTNCIEKIANLNGLKNLRILSLGRNNIKNLNGLEAVGDTLEELWISYNFIEKLKG--
DNAL4    FSNNNESAAKMIK--------------------ETMDKKFGSSWHVV--IGEGFGFE

DNAL1    -IHIMKKLKILYMSNNL-VKDWAEFVKLAELPCLEDLVFVGNPLEEKHSAENNWIEEATK
DNAL4    ITHEVKNLLYLYFGGTLAVCVWK---------CS------------------------

DNAL1    RVPKLKKLDGTPVIKGDEEEDN
DNAL4    ---------------------
```

Figure 18.1 Clustal Omega (www.ebi.ac.uk) multiple sequence alignment of human dynein axonemal light chain proteins, DNAL1 (Q4LDG9) and DNAL4 (O96015). *Yellow highlights* indicate sequence identity.

condition in which polarity of organs may be reversed, e.g., a right-hand side located heart. Mutations in the dynein axonemal heavy chain 1 gene, *DNAH1*, have been shown to cause male infertility through abnormalities in the sperm flagella [3].

18.2.1 DNAL4: biology and disease

18.2.1.1 DNAL4 expression and localization

We recently reported the mapping and identification of a splice mutation leading to a 28 amino acid deletion in DNAL4, in a large consanguineous Pakistani family with an apparent autosomal recessive form of congenital mirror movements (MRMVs) (MRMV3; MIM [1]). A summary of this study is outlined below. Studies of dynein components in Chlamydomonas have shown that knockout of LC10, which is the orthologue of DNAL4, show only modest disruption of cilial/flagellar beat [27]. Immunostaining of murine tissues showed strong expression of Dnal4 both in apical cilia in the bronchial epithelium and in sperm flagella. However, in the MRMV family reported by us with a splice site mutation in *DNAL4* [1], there is no evidence of either *situs inversus* or PCD. Although sperm motility was not checked specifically, there is clearly no apparent fertility issue in this family. This suggests that DNAL4 may also only play a minor ciliary/flagellar role. Data from the St. Jude Brain Gene Expression Map (BGEM: http://www.stjudebgem.org) in situ hybridization database [15] suggest, somewhat surprisingly, that the *DNAL4* mouse orthologue, *Dnal4* or *Dnalc4*, is expressed at much higher levels than other dynein components during embryogenesis and is prominently expressed in hippocampus, suggesting that DNAL4 may have additional functions unrelated to a role in the axoneme [27]. Analysis of mouse in situ hybridization data for *Dnal4* gene expression from the Allen Brain Atlas [14] shows strong expression in granular regions of the cerebellum, olfactory bulb, and pyramidal layers and dentate gyrus of the hippocampus (Fig. 18.2). Expression in the medial habenula and habenular and posterior commissures was also noted. Despite anticipated links between

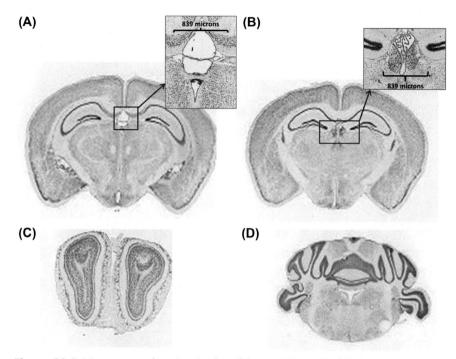

Figure 18.2 Mouse coronal section (male, wild type, P56, C57BL/6J), in situ hybridization using RNA antisense to *Dnal4*: http://mouse.brain-map.org/experiment/show/74047778. (A) Signal on CA1 to 3, pyrimidal layer, and dentate gyrus of hippocampus. Zoom on habenular and posterior commissures. (B) Signal on CA1 to 3, pyrimidal layer, and dentate gyrus of hippocampus. Zoom on third ventricle and medial habenula. (C) Main olfactory bulb, signal on granular and mitral layers. (D) Cerebellum, granular layers. *Image credit: Allen Institute (Allen Institute for Brain Science, Allen Human Brain Atlas, 2010. Available from:* human.brain-map.org).

the corpus callosum and MRMV, *DNAL4* mRNA expression in the corpus callosum was notably weak or absent. Intriguingly, we also note that the orthologues of MRMV1 and MRMV2 genes, deleted in colorectal carcinoma (*DCC*) and *RAD51*, as well as the ligand for *DCC*, *NTN1*, all appear to be expressed in similar regions, including the medial habenula (Fig. 18.3). The medial habenula is a region of the diencephalon, sitting just below and adjacent to the corpus callosum, has not previously been implicated in MRMV. The habenula, an ancient and somewhat mysterious brain structure, is believed to have a number of diverse functions, but particularly relating to motor suppression (reviewed in Ref. [12]), and thus there may be an important link to the motor suppression required for unimanual movements. It should, however, be noted that these are from in situ RNA hybridization experiments, and DNAL4 protein localization may differ significantly. That being said, protein expression analysis of DNAL4 through immunohistochemistry (IHC) supports high expression in human hippocampus and cerebellum (The Human Protein Atlas; http://www.proteinatlas.

(A) Rad51

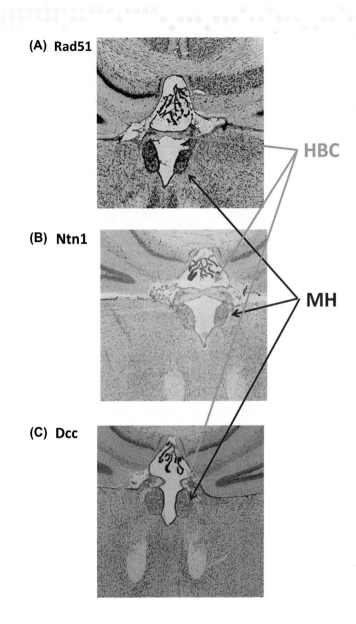

HBC

(B) Ntn1

MH

(C) Dcc

org), but a more specific brain dissection using IHC would be needed to determine precise locations of the protein. Both protein and RNA expression studies indicate DNAL4 also shows relatively high expression in lung, testis, fallopian tube, and placenta. Data from the BioGPS database also show highest expression for mouse *Dnalc4* (array: Affymetrix MOE430, probe: 1416870_at) in testis, and then nucleus accumbens, and in humans, *DNAL4* is expressed highest in

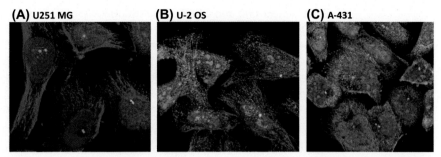

(A) U251 MG   **(B)** U-2 OS   **(C)** A-431

Figure 18.4 Immunohistochemical analysis for DNAL4. Cell lines stained with Rabbit anti-Dnal4 polyclonal antibody HPA003647 in green (Atlas Antibodies, Sigma–Aldrich), nucleus (DAPI; blue), and microtubules (antitubulin antibody; red): (A) U-251 MG: malignant glioblastoma cell line showing mainly centrosomal staining. (B) U-2 OS: human osteosarcoma cell line showing nuclear, nucleolar, cytoplasmic, and centrosomal staining. (C) A-431: human squamous carcinoma cells showing nuclear, nucleolar, cytoplasmic, and centrosomal staining. *Images from* http://www.proteinatlas.org; *M. Uhlén, P. Oksvold, L. Fagerberg, E. Lundberg, K. Jonasson, M. Forsberg, M. Zwahlen, C. Kampf, K. Wester, S. Hober, H. Wernerus, L. Björling, F. Ponten, Towards a knowledge-based human protein atlas, Nat. Biotechnol. 28 (2010) 1248–1250; M. Uhlén, L. Fagerberg, B.M. Hallström, C. Lindskog, P. Oksvold, A. Mardinoglu, Å. Sivertsson, C. Kampf, E. Sjöstedt, A. Asplund, I. Olsson, K. Edlund, E. Lundberg, S. Navani, C.A. Szigyarto, J. Odeberg, D. Djureinovic, J.O. Takanen, S. Hober, T. Alm, P.H. Edqvist, H. Berling, H. Tegel, J. Mulder, J. Rockberg, P. Nilsson, J.M. Schwenk, M. Hamsten, K. von Feilitzen, M. Forsberg, L. Persson, F. Johansson, M. Zwahlen, G. von Heijne, J. Nielsen, F. Pontén, Proteomics. Tissue-based map of the human proteome, Science 347 (2015) 1260419 shared under a Creative Commons Attribution License.*

testis and thyroid, and then in whole brain (array: Affymetrix U133A, probe: 204008_at). It should also be noted that, according to the available Allen Brain Atlas mouse data [14], *Dnal4, Dcc, Rad51,* and *Ntn1* are all expressed in the hippocampus, CA1 and CA3 pyrimidal layers and dentate gyrus, also in the granular layer in cerebellum. The dentate gyrus is important for forming memories but also for spatial behavior Page 442 [30].

IHC studies in cells reported by Protein Atlas (http://www.proteinatlas.org) indicate that DNAL4 protein localizes to cytoplasmic regions, the nucleus, nucleolus, and centrosome (Fig. 18.4). The centrosomal location, particularly prominent in the U-251 cells, may suggest a role in the dynein component of astral microtubules during cell division.

18.2.1.2 DNAL4 protein interactions

Protein interaction analysis using the BioGRID data repository (Biological General Repository for Interaction Datasets; http://thebiogrid.org/) version 3.2.108 indicates no reported interactions between DNAL4 and other axonemal dynein components, yet several yeast-two-hybrid screening show interaction with an 8 kDa cytoplasmic light chain LC8-type 2 dynein component, DYNLL2 [24,25]. Protein interactions are also indicated between DNAL4 and FHL5, SCTR, CAMK2D,

GNA14, TRIM54, and GPBP1. Interestingly, BioGRID lists a number of interactions for DYNLL2, including a various components of postsynaptic scaffolding such as DLG4 (known more commonly as PSD-95), DLG2, DLGAP1, SHANK2, HOMER3, and GPRIN2. Both DYNLL2 and DYNLL1 are also known interactors of gephyrin, a postsynaptic scaffolding protein that clusters GABA and glycine neurotransmitter receptors at inhibitory synapses [9]. Interaction between DYNLL2 (Dlc2) and DLGAP1 (GKAP) organizes postsynaptic scaffold complex to enhance N-methyl-D-aspartate (NMDA) receptor signaling [17].

Additionally, there appears to be no obvious direct link through protein interaction database analysis between DNAL4 (or via DYNLL2) and RAD51, DCC or its ligand netrin-1 (NTN1). It has recently been shown that DCC links NTN1 signaling to microtubules via the neuronal β-tubulin isoform TUBB3 [21]. TUBB3 is known to play a crucial role in axonal guidance. A study in Drosophila indicated that an 8 kDa cytoplasmic dynein light chain protein is necessary for axonal guidance, and suggested that disruption of dynein function could affect microtubule bundling, and thereby disrupt directional axonal growth [19]. Given its size and the neuronal role implicated by its protein interactions, DYNLL2 would be a strong candidate for this. While DNAL4 is unlikely to bind directly to the microtubule assembly, it is very likely that the disruption of DNAL4 caused by the deletion of 28 amino acids in the MRMV3 family would disrupt either homodimerization, or interactions with the dynein complex, possibly via DYNLL2, or interaction between the dynein complex and cargo molecules, which could ultimately result in disrupted axonal growth.

18.2.1.3 Gene ontology for DNAL4

Gene ontology, also using BioGRID analysis through AmiGO2, also suggests a role for DNAL4 in the neurotrophin TRK receptor signaling pathway (GO:0048011) and in the related neurotrophin signaling pathway (GO:0038179). The protein tyrosine phosphatase, nonreceptor type 11 (PTPN11), is also present in these pathways (GO:0048011 and GO:0038179) and is also included in the axogenesis pathway (GO:0007409) along with NTN1 and DCC. So, PTPN11 may represent a common link between DCC/netrin and DNAL4.

18.2.1.4 Identification of DNAL4 in mirror movement disorder

18.2.1.4.1 Mirror movement disorder
Performing unilateral limb movements requires communication between the brain hemispheres— chiefly through the corpus callosum—to restrict output to the primary motor cortex (M1) contralateral to the intended limb movement (reviewed in Ref. [2]). Modulation of transcallosal interhemispheric inhibition is required to prevent mirroring, and a network of cortical and subcortical areas required for performing unilateral movements is believed to include the supplementary motor area, dorsal premotor cortex, the ipsilateral motor cortex (M1), and the basal ganglia (reviewed in Ref. [2]).

First reports of MRMVs were from Erlenmeyer in 1879, but the term was coined by Cohen et al. [6] and described involuntary contralateral movements mirroring the intended limb movement. MRMV may occasionally be present in young children, gradually disappearing within the first decade of life, and it is presumed that MRMV is a necessary stage in the maturating motor network [4]. However if it persists beyond the first 10 years it is referred to as congenital MRMV. MRMV may be present in all limbs but is most common in the fingers and hands, with intensity increasing with the complexity of the voluntary movement. MRMV impair tasks requiring coordination of both hands such as tying shoe-laces, buttoning shirts, etc., and it is associated with upper limb pain during lengthy manual activities. A severe form of congenital MRMV affecting both hands and forearms was identified in a 15-year-old girl, possibly related to structural abnormalities of the motor network, and/or altered decussation of the corticospinal tract [5,10,23].

Congenital MRMV is rare, with a strong genetic component, and mainly inherited in an autosomal-dominant fashion, with the possibility that some sporadic cases may be caused by recessive forms [22]. There is known genetic heterogeneity; heterozygous mutations in *DCC* (MIM 120470) have been reported in families with autosomal-dominant congenital MRMV (MRMV1 [MIM 157600]). The *DCC* gene encodes for the receptor for NTN1 (MIM 601614). Disruption of the DCC/NTN1 signaling pathway, which promotes axon guidance toward the midline, results in abnormal ipsilateral connections [8,26].

A heterozygous truncating mutation in the gene *RAD51* [MIM 179617] was reported by Depienne et al. [7] in a large congenital MRMV family with incomplete penetrance (MRMV2 [MIM 614508]). The mechanism linking *RAD51* haploinsufficiency to MRMV remains unclear. Insufficient *RAD51*-related DNA repairs during corticogenesis may lead to increased apoptosis and altered central nervous system (CNS) development [7]. Although a role for *RAD51* in axonal guidance is currently not known, recent studies show that MRMV2 patients to have abnormal decussation of the corticospinal tract, abnormal interhemispheric inhibition, and bilateral activation of primary motor areas during intended unimanual movements, also abnormal supplementary motor area activity while performing unimanual or bimanual movements [11].

18.2.1.5 MRMV3 is linked to mutation in DNAL4 [1]

A large family from Sindh province, Pakistan, segregating congenital MRMV disorder was identified and ascertained for genetic studies. Although complex, from analysis of the pedigree it would appear that autosomal recessive inheritance is the most likely model of transmission (Fig. 18.5). MRMV was observed in fingers and hands, but proximal upper limbs were not involved, and MRMVs were absent in the lower limbs. No other neurological or behavioral comorbidity was present. Pain and/or cramping during sustained manual activity were not

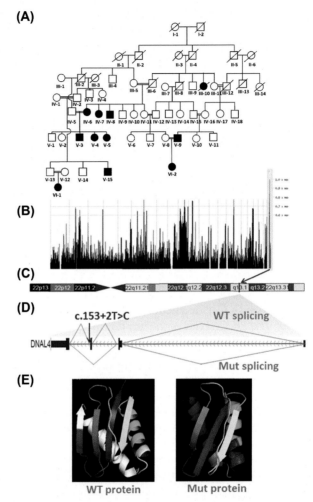

Figure 18.5 (A) Pedigree of Pakistani MRMV3/*DNAL4* family. Affected individuals are indicated by *dark-shaded symbols*. Affected individuals IV:7, IV:8, V:3, V:15, and VI:2, and unaffected individual V:14 were included in the microarray analysis. (B) Homozygosity Mapper output plot for all autosomes (numbered 1–22, left to right), showing maximum homozygous-by-descent (HBD) (across all affecteds, but not the unaffected individual) only on chromosome 22 (indicated by *yellow highlight*). The *y*-axis indicates the fraction of the maximum homozygosity score (1.0). (C) Ideogram of chromosome 22 indicating region of HBD. (D) Ideogram indicating location of the NM_005740.2: c.153+2T>C *DNAL4* variant at the intron 3 donor splice site (*red arrow*) mutant (Mut), and the wild-type (WT) RNA splicing with light blue lines above the horizontal line, and mutant splicing below the line. (E) Predicted 3D structures for wild-type DNAL4, and the mutant DNAL4 with the 28 amino acid stretch encoded by exon 3 missing. The *pink bars* in the WT protein represent the sequence flanking the 28 residue deletion section in the wild-type model and the mutant model.

reported by affected individuals. *Situs inversus* was not present, and there was no history of infections that might be indicative of ciliary dyskinesia. Onset of symptoms was noted at around 2–3 years of age; however, in some cases symptoms were reported anecdotally to be present from 2–3 months. Known dominant genes for MRMV (*RAD51* and *DCC*) were excluded by Sanger sequencing. Microarray genotype analysis indicated a ~3.3 Mb homozygous-by-descent (HBD) region on chromosome 22q13.1 (Chr22:36605976–39904648; hg19; Fig. 18.1B) that is shared by all affected family members, but not in unaffected members. Within this region, whole-exome sequencing identified a mutation two nucleotides from the exon 3/intron 3 donor splice site in the *DNAL4* gene, *DNAL4*: chr22:39176929A>G; NM_005740.2: c.153+2T>C. Coding variation in all other genes within this HBD region were excluded either by whole exome sequencing or, where sequence coverage was insufficient, by Sanger sequencing. This variant has not been found in any large scale control datasets (e.g., http://gnomad.broadinstitute.org). Linkage analysis using Simwalk2 gave a highly significant maximum location score [directly comparable to a multipoint LOD (logarithm of odds) score] of 6.197 for markers D22S280 and *DNAL4* (NM_005740): c.153+2T>C. RT-PCR and sequencing of patient mRNA showed that the result of this splice mutation is skipping of exon 3 (Fig. 18.2B and C), but maintaining the same reading frame. The removal of exon 3 results in a deletion of 28 amino acids (residues 24 to 51 inclusive) from the 105-residue protein. See Fig. 18.5E for protein 3D structure predictions for wild-type and mutant DNAL4.

Mutation screening in 13 idiopathic MRMV cases from Canada and a further 17 cases from France showed no mutations in *DNAL4* [1,16], indicating that DNAL4 is likely to be a rare cause of MRMV.

Evolutionary comparative protein sequence analysis for DNAL4 using Clustal Omega alignment (http://www.ebi.ac.uk) shows it to be highly conserved across the kingdom Animalia (or Metazoa). Conservation is at greater than 98% identity across all mammalian species. Conservation appears to be particularly strong across the stretch of the DNAL4 protein that is deleted in the MRMV3 family (residues 24–51), with 54% of residues conserved across all species, compared to 27% for the rest of the protein (see Fig. 18.6).

It is also of interest that preweaning lethality has been noted in mice with homozygous knockout of *Dnalc4* (Dnal4[tm1b(EUCOMM)Wtsi]; http://www.mouse-phenotype.org). This might suggest that the mutation reported for MRMV3 is likely to be hypomorphic.

18.3 Conclusions

While there is a clear genetic link between the genetic mutation in *DNAL4* identified and MRMV3, the role of the DNAL4 protein, either in normal biology or in disease, has yet to be established. Given their very different protein

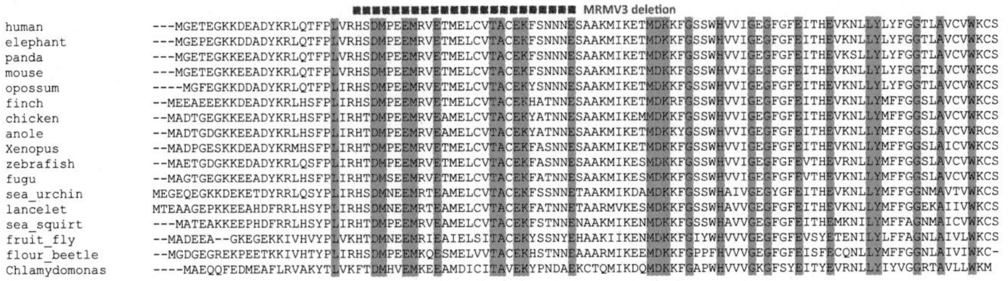

Figure 18.6 Analysis of evolutionary conservation of DNAL4 protein using Clustal Omega (www.ebi.ac.uk) multiple sequence alignment. The red bar above the alignment indicates the position of the 28 deleted amino acids. *Blue shading* indicates amino acids that are conserved 100% across the species included in the alignment. Protein sequences used: human: NP_005,731.1; elephant: XP_003419805.1; giant panda: XP_002914599.1; mouse: NP_059498.2; opossum: XP_001367599.1; chicken: NP_001006242.1; finch: XP_005421907.1; anole: XP_003221024.1; *Xenopus*: NP_001087464.1; zebrafish: NP_001003603.1; fugu: XP_003976859.1; lancelet: XP_002595410.1; sea squirt: XP_002126866.1; sea urchin: XP_794465.1; flour beetle: XP_967210.1; fruit fly: NP_610734.1; *Chlamydomonas*: ACC68802.1.

interacting pathways, expression profiles, and associated diseases, the functional role of DNAL4 is likely to differ significantly from that of the other dynein axonemal light chain, DNAL1, and DNAL4's main function may be outside the axoneme. We hypothesize that the *DNAL4* mutation identified in the Pakistani MRMV family may lead to inability of DNAL4 either to homodimerize or to bind to DYNLL2 and that this subsequent functional disruption of either the paired DNAL4 dimer or of DYNLL2, and thus the overall dynein complex, impairs the retrograde transport required for NTN1-induced axon outgrowth and pathfinding during development, either within commissural callosal regions, or within the nonmirroring network. The involvement of the medial habenula, implicated by mouse in situ hybridization of not only *Dnal4* but also of other MRMV genes such as *Rad51*, may suggest a role for this region of the brain and for MRMV genes in the motor suppression that is essential for unilateral limb movements.

References

[1] I. Ahmed, K. Mittal, T.I. Sheikh, M.A. Rafiq, A. Mikhailov, N. Vasli, M. Ohadi, H. Mahmood, G.A. Rouleau, A. Bhatti, M. Ayub, M. Srour, P. John, J.B. Vincent, The dynein axonemal light chain 4 gene on 22q13.1 is associated with an autosomal recessive form of congenital mirror movement disorder, Hum. Genet. 133 (2014) 1419–1429.

[2] V. Beaulé, S. Tremblay, H. Theoret, Interhemispheric control of unilateral movement, Neural Plast. (2012) 627816.

[3] M. Ben Khelifa, C. Coutton, R. Zouari, T. Karaouzène, J. Rendu, M. Bidart, S. Yassine, V. Pierre, J. Delaroche, S. Hennebicq, D. Grunwald, D. Escalier, K. Pernet-Gallay, P.S. Jouk, N. Thierry-Mieg, A. Touré, C. Arnoult, P.F. Ray, Mutations in DNAH1, which encodes an inner arm heavy chain dynein, lead to male infertility from multiple morphological abnormalities of the sperm flagella, Am. J. Hum. Genet. 94 (2014) 95–104.

[4] C. Bonnet, A. Roubertie, D. Doummar, N. Bahi-Buisson, V. Cochen de Cock, E. Roze, Developmental and benign movement disorders in childhood, Mov. Disord. 25 (2010) 1317–1334.

[5] M. Cincotta, A. Borgheresi, L. Balzini, L. Vannucchi, G. Zeloni, A. Ragazzoni, F. Benvenuti, G. Zaccara, G. Arnetoli, U. Ziemann, Separate ipsilateral and contralateral corticospinal projections in congenital mirror movements: neurophysiological evidence and significance for motor rehabilitation, Mov. Disord. 18 (2003) 1294–1300.

[6] L.G. Cohen, J. Meer, I. Tarkka, S. Bierner, D.B. Leiderman, R.M. Dubinsky, J.N. Sanes, B. Jabbari, B. Branscum, M. Hallett, Congenital Mirror Movements. Abnormal organization of motor pathways in two patients, Brain 114 (1991) 381–403.

[7] C. Depienne, D. Bouteiller, A. Meneret, S. Billot, S. Groppa, S. Klebe, F. Charbonnier-Beaupel, J.C. Corvol, J.P. Saraiva, N. Brueggemann, K. Bhatia, M. Cincotta, V. Brochard, C. Flamand-Roze, W. Carpentier, S. Meunier, Y. Marie, M. Gaussen, G. Stevanin, R. Wehrle, M. Vidailhet, C. Klein, I. Dusart, A. Brice, E. Roze, RAD51 haploinsufficiency causes congenital mirror movements in humans, Am. J. Hum. Genet. 90 (2012) 301–307, http://dx.doi.org/10.1016/j.ajhg.2011.12.002.

[8] C. Depienne, M. Cincotta, S. Billot, D. Bouteiller, S. Groppa, V. Brochard, C. Flamand, C. Hubsch, S. Meunier, F. Giovannelli, S. Klebe, J.C. Corvol, M. Vidailhet, A. Brice, E. Roze, A novel DCC mutation and genetic heterogeneity in congenital mirror movements, Neurology 76 (2011) 260–264.

[9] J.C. Fuhrmann, S. Kins, P. Rostaing, O. El Far, J. Kirsch, M. Sheng, A. Triller, H. Betz, M. Kneussel, Gephyrin interacts with Dynein light chains 1 and 2, components of motor protein complexes, J. Neurosci. 22 (2002) 5393–5402.

[10] C. Galléa, T. Popa, S. Billot, A. Méneret, C. Depienne, E. Roze, Congenital mirror movements: a clue to understanding bimanual motor control, J. Neurol. 258 (2011) 1911–1919.

[11] C. Gallea, T. Popa, C. Hubsch, R. Valabregue, V. Brochard, P. Kundu, B. Schmitt, E. Bardinet, E. Bertasi, C. Flamand-Roze, N. Alexandre, C. Delmaire, A. Méneret, C. Depienne, C. Poupon, L. Hertz-Pannier, M. Cincotta, M. Vidailhet, S. Lehericy, S. Meunier, E. Roze, RAD51 deficiency disrupts the corticospinal lateralization of motor control, Brain 136 (2013) 3333–3346.

[12] O. Hikosaka, The habenula: from stress evasion to value-based decision making, Nat. Rev. Neurosci. 11 (2010) 503–513.

[13] M. Iwasaki, T. Kuwata, Y. Yamazaki, N.A. Jenkins, N.G. Copeland, M. Osato, Y. Ito, E. Kroon, G. Sauvageau, T. Nakamura, Identification of cooperative genes for NUP98-HOXA9 in myeloid leukemogenesis using a mouse model, Blood 105 (2005) 784–793.

[14] E.S. Lein, M.J. Hawrylycz, N. Ao, M. Ayres, A. Bensinger, A. Bernard, A.F. Boe, M.S. Boguski, K.S. Brockway, E.J. Byrnes, L. Chen, L. Chen, T.M. Chen, M.C. Chin, J. Chong, B.E. Crook, A. Czaplinska, C.N. Dang, S. Datta, N.R. Dee, A.L. Desaki, T. Desta, E. Diep, T.A. Dolbeare, M.J. Donelan, H.W. Dong, J.G. Dougherty, B.J. Duncan, A.J. Ebbert, G. Eichele, L.K. Estin, C. Faber, B.A. Facer, R. Fields, S.R. Fischer, T.P. Fliss, C. Frensley, S.N. Gates, K.J. Glattfelder, K.R. Halverson, M.R. Hart, J.G. Hohmann, M.P. Howell, D.P. Jeung, R.A. Johnson, P.T. Karr, R. Kawal, J.M. Kidney, R.H. Knapik, C.L. Kuan, J.H. Lake, A.R. Laramee, K.D. Larsen, C. Lau, T.A. Lemon, A.J. Liang, Y. Liu, L.T. Luong, J. Michaels, J.J. Morgan, R.J. Morgan, M.T. Mortrud, N.F. Mosqueda, L.L. Ng, R. Ng, G.J. Orta, C.C. Overly, T.H. Pak, S.E. Parry, S.D. Pathak, O.C. Pearson, R.B. Puchalski, Z.L. Riley, H.R. Rockett, S.A. Rowland, J.J. Royall, M.J. Ruiz, N.R. Sarno, K. Schaffnit, N.V. Shapovalova, T. Sivisay, C.R. Slaughterbeck, S.C. Smith, K.A. Smith, B.I. Smith, A.J. Sodt, N.N. Stewart, K.R. Stumpf, S.M. Sunkin, M. Sutram, A. Tam, C.D. Teemer, C. Thaller, C.L. Thompson, L.R. Varnam, A. Visel, R.M. Whitlock, P.E. Wohnoutka, C.K. Wolkey, V.Y. Wong, M. Wood, M.B. Yaylaoglu, R.C. Young, B.L. Youngstrom, X.F. Yuan, B. Zhang, T.A. Zwingman, A.R. Jones, Genome-wide atlas of gene expression in the adult mouse brain, Nature 445 (2007) 168–176.

[15] S. Magdaleno, P. Jensen, C.L. Brumwell, A. Seal, K. Lehman, A. Asbury, T. Cheung, T. Cornelius, D.M. Batten, C. Eden, S.M. Norland, D.S. Rice, N. Dosooye, S. Shakya, P. Mehta, T. Curran, BGEM: an in situ hybridization database of gene expression in the embryonic and adult mouse nervous system, PLoS Biol. 4 (2006) e86.

[16] A. Méneret, O. Trouillard, M. Vidailhet, C. Depienne, E. Roze, Congenital mirror movements: no mutation in DNAL4 in 17 index cases, J. Neurol. 261 (2014) 2030–2031.

[17] E. Moutin, F. Raynaud, L. Fagni, J. Perroy, GKAP-DLC2 interaction organizes the postsynaptic scaffold complex to enhance synaptic NMDA receptor activity, J. Cell Sci. 125 (2012) 2030–2040.

[18] Y. Nagao, J. Cheng, K. Kamura, R. Seki, A. Maeda, D. Nihei, S. Koshida, Y. Wakamatsu, T. Fujimoto, M. Hibi, H. Hashimoto, Dynein axonemal intermediate chain 2 is required for formation of the left-right body axis and kidney in medaka, Dev. Biol. 347 (2010) 53–61.

[19] R. Phillis, D. Statton, P. Caruccio, R.K. Murphey, Mutations in the 8 kDa dynein light chain gene disrupt sensory axon projections in the Drosophila imaginal CNS, Development 122 (1996) 2955–2963.

[20] R. Popatia, K. Haver, A. Casey, Primary ciliary dyskinesia: an update on new diagnostic modalities and review of the literature, Pediatr. Allergy Immunol. Pulmonol. 27 (2014) 51–59.

[21] C. Qu, T. Dwyer, Q. Shao, T. Yang, H. Huang, G. Liu, Direct binding of TUBB3 with DCC couples netrin-1 signaling to intracellular microtubule dynamics in axon outgrowth and guidance, J. Cell Sci. 126 (2013) 3070–3081.

[22] P. Rasmussen, Persistent mirror movement. A clinical study of 17 children, adolescents and young adults, Dev. Med. Child. Neurol. 35 (1993) 699–707.

[23] F. Regli, G. Filippa, M. Wiesendanger, Hereditary mirror movements, Arch. Neurol. 16 (1967) 620–623.

[24] T. Rolland, M. Taşan, B. Charloteaux, S.J. Pevzner, Q. Zhong, N. Sahni, S. Yi, I. Lemmens, C. Fontanillo, R. Mosca, A. Kamburov, S.D. Ghiassian, X. Yang, L. Ghamsari, D. Balcha, B.E. Begg, P. Braun, M. Brehme, M.P. Broly, A.R. Carvunis, D. Convery-Zupan, R. Corominas, J. Coulombe-Huntington, E. Dann, M. Dreze, A. Dricot, C. Fan, E. Franzosa, F. Gebreab, B.J. Gutierrez, M.F. Hardy, M. Jin, S. Kang, R. Kiros, G.N. Lin, K. Luck, A. MacWilliams, J. Menche, R.R. Murray, M. Palagi, M.M. Poulin, X. Rambout, J. Rasla, P. Reichert, V. Romero, E. Ruyssinck, J.M. Sahalie, A. Scholz, A.A. Shah, A. Sharma, Y. Shen, K. Spirohn, S. Tam, A.O. Tejeda, S.A. Trigg, J.C. Twizere, K. Vega, J. Walsh, M.E. Cusick, Y. Xia, A.L. Barabási, L.M. Iakoucheva, P. Aloy, J. De Las Rivas, J. Tavernier, M.A. Calderwood, D.E. Hill, T. Hao, F.P. Roth, M. Vidal, A proteome-scale map of the human interactome network, Cell 159 (2014) 1212–1226.

[25] J.F. Rual, K. Venkatesan, T. Hao, T. Hirozane-Kishikawa, A. Dricot, N. Li, G.F. Berriz, F.D. Gibbons, M. Dreze, N. Ayivi-Guedehoussou, N. Klitgord, C. Simon, M. Boxem, S. Milstein, J. Rosenberg, D.S. Goldberg, L.V. Zhang, S.L. Wong, G. Franklin, S. Li, J.S. Albala, J. Lim, C. Fraughton, E. Llamosas, S. Cevik, C. Bex, P. Lamesch, R.S. Sikorski, J. Vandenhaute, H.Y. Zoghbi, A. Smolyar, S. Bosak, R. Sequerra, L. Doucette-Stamm, M.E. Cusick, D.E. Hill, F.P. Roth, M. Vidal, Towards a proteome-scale map of the human protein-protein interaction network, Nature 437 (2005) 1173–1178.

[26] M. Srour, J.B. Riviére, J.M. Pham, M.P. Dubé, S. Girard, S. Morin, P.A. Dion, G. Asselin, D. Rochefort, P. Hince, S. Diab, N. Sharafaddinzadeh, S. Chouinard, H. Théoret, F. Charron, G.A. Rouleau, Mutations in DCC cause congenital mirror movements, Science 328 (2010) 592.

[27] C.A. Tanner, P. Rompolas, R.S. Patel-King, O. Gorbatyuk, K. Wakabayashi, G.J. Pazour, S.M. King, Three members of the LC8/DYNLL family are required for outer arm dynein motor function, Mol. Biol. Cell. 19 (2008) 3724–3734.

[28] M. Uhlén, P. Oksvold, L. Fagerberg, E. Lundberg, K. Jonasson, M. Forsberg, M. Zwahlen, C. Kampf, K. Wester, S. Hober, H. Wernerus, L. Björling, F. Ponten, Towards a knowledge-based human protein atlas, Nat. Biotechnol. 28 (2010) 1248–1250.

[29] M. Uhlén, L. Fagerberg, B.M. Hallström, C. Lindskog, P. Oksvold, A. Mardinoglu, Å. Sivertsson, C. Kampf, E. Sjöstedt, A. Asplund, I. Olsson, K. Edlund, E. Lundberg, S. Navani, C.A. Szigyarto, J. Odeberg, D. Djureinovic, J.O. Takanen, S. Hober, T. Alm, P.H. Edqvist, H. Berling, H. Tegel, J. Mulder, J. Rockberg, P. Nilsson, J.M. Schwenk, M. Hamsten, K. von Feilitzen, M. Forsberg, L. Persson, F. Johansson, M. Zwahlen, G. von Heijne, J. Nielsen, F. Pontén, Proteomics. Tissue-based map of the human proteome, Science 347 (2015) 1260419.

[30] G.F. Xavier, V.C. Costa, Dentate gyrus and spatial behaviour, Prog. Neuropsychopharmacol. Biol. Psychiatry 33 (2009) 762–773.

In this chapter

Does dynein influence the non-Mendelian inheritance of chromosome 17 homologues in male mice?*

Stephen H. Pilder
Temple University School of Medicine, Philadelphia, PA, United States

19.1 Prevailing models of the genetic and functional basis of phenotypes specific to male mice carrying *t*-haplotypes during the early years of the genomics era

Modern mouse *t*-haplotypes are closely related members of a family of inversion polymorphisms of the proximal one-third of chromosome 17 (the *t*-complex). These variants descended from a common ancestor ~10,000–100,000 years ago, and presently exist at a relatively low equilibrium frequency in feral populations of all subspecies of the *Mus musculus* species complex [1]. All family members are characterized by four nearly region-spanning, non-overlapping, megabase-pair inversions relative to the wild-type (+) form of the region (Fig. 19.1) [2]. Because these inversions suppress recombination between + and *t*, the progeny of a +/*t*-heterozygous parent usually receives either *t* or +en bloc.

Intriguingly, males, but not females, heterozygous for a *t*-haplotype transmit the variant *t*-homologue to >95% of their progeny, a phenotype known as transmission ratio distortion (TRD) or meiotic drive. The huge selective advantage in favor of the driving *t*-chromosome is offset by a number of forces, the two most-often cited being: (1) most *t*-haplotypes carry recessive embryonic lethal mutations, and (2) males homozygous for a nonlethal *t*-haplotype (t^X/t^X) or doubly heterozygous for two complementing lethal *t*-haplotypes (t^X/t^Y) are consistently sterile [3,4].

---

*This chapter has been reprinted from the first edition without modification.

Dyneins. https://doi.org/10.1016/B978-0-12-809470-9.00019-9

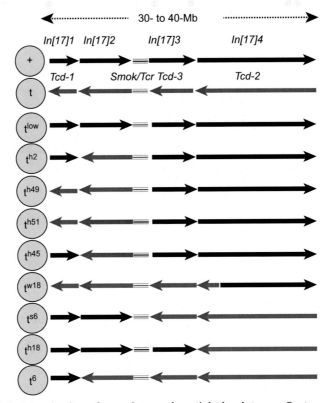

Figure 19.1 Organization of complete and partial *t*-haplotypes. Centromeric regions are denoted by *circles* on the left side of the diagram with the name of each chromosome indicated within the *circle* (+ denotes wild-type chromosome 17, *t*, a complete *t*-haplotype, and *t* followed by a superscript indicating a partial *t*-haplotype). *Black bars* indicate +-DNA, *red bars*, *t*-DNA, and *arrowheads* represent the relative orientation of inverted segments of chromatin in + versus *t*. The positions of inversions 1–4 (*In[17]1-In[17]4*) are shown above the wild-type chromosome, and the locations of distorters (*Tcd1-3*) and *Smok/Tcr* are indicated between the wild-type and complete *t*-haplotype regional representations. The uninverted *Smok* and *Tcr* loci are represented by either a *black or red triplicate of lines* without *arrows*, respectively. Inversions and the *Smok/Tcr* region are drawn approximately to scale.

The *t*-haplotypes were discovered over eighty years ago, yet for most of that time this large chromosomal expanse has hidden its most fascinating secrets—the identities of the genes that cause the non-Mendelian inheritance of *t* from +/*t*-heterozygous males and the sterility of *t*/*t*-males—in its nearly impenetrable inversions. Nonetheless, just prior to the era of genomics and proteomics, Lyon [4,5] elucidated the genetic basis of *t*-related male TRD and sterility, by following the effects on chromosomal transmission and fertility of males carrying a series of rare recombinants between +- and *t*-homologues known as partial *t*-haplotypes (Fig. 19.1). The result of Lyon's labor was a cogent hypothesis that proposed a model containing the following tenets: (1) TRD occurs only when the *t*-allele of a cis-acting factor called the *t*-complex responder (Tcr) is present heterozygously; (2) the *t*-alleles of three both cis- and trans-acting factors of varying strengths called *t*-complex distorters (Tcds) operate additively on Tcr to drive the transmission of whatever chromosome is carrying the *t*-allele of Tcr to higher levels; (3) if Tcd^{t}s are absent, the chromosome containing Tcr^t is transmitted to only 10%–20% of the progeny; (4a) homozygosity for Tcd-2^t or (4b) heterozygosity for Tcd-2^t plus homozygosity for Tcd-1^t (and thus lack of appropriate complementing wild-type function) causes male sterility (in the case of 4b, the specific combination of *t*-alleles causes profound subfertility, rather than absolute sterility as in 4a); and (5) *Tcr* maps centrally in the *t*-complex, *Tcd-1* is near its proximal end, *Tcd-2* is located distally, and *Tcd-3* maps between *Tcr* and *Tcd-2* (Fig. 19.1). Together, these findings suggested that the *Tcr* alleles are expressed during the haploid phase of spermatogenesis (spermiogenesis) and are somehow retained in the cell of origin, despite the fact that male germ cells derived from the same stem cell (spermatogonium) develop in a syncitium that allows transfer of mRNAs and proteins between daughter cells. This implies that the allelic nature of *Tcr* determines the fate of each daughter cell, since it is assumed that they are otherwise identical.

High-resolution marker mapping studies at the dawn of the genomics era [2] placed *Tcd-1* in the centromeric inversion (*In[17]1*), *Tcd-3* in the middle inversion (*In[17]3*), *Tcd-2* in the large distal inversion (*In[17]4*), and *Tcr* in a small uninverted region of the *t*-complex between *In[17]2* and *In[17]3* (Fig. 19.1). Additionally, Silver and Olds-Clarke [6] demonstrated that, unlike other known cases of male meiotic drive where the wild-type chromosome-carrying spermatid is destroyed during development, equal quantities of +-homologue carrying and *t*-homologue carrying sperm from +/*t*-males are present in the female uterus 90 min after coitus. The consensus conclusion derived from these data is that the function of the +-carrying sperm produced by a +/*t*-heterozygous male must be relatively more defective in its capacity to fertilize than the *t*-carrying sperm produced by the same male.

Following this logic, Olds-Clarke and Johnson [7] published a comprehensive study of the effects of complete *t*-haplotypes on sperm function in fertilization. Cauda epididymal (mature) sperm from +/+, t^X/+, t^Y/+, and t^X/t^Y males were incubated in either in vitro fertilization medium minus the addition of exogenous calcium (IVF-, a minimal medium that *does not* support fertilization in vitro), or a medium that *does support* fertilization in vitro (IVF, a minimal medium containing 1.7 mM added Ca^{2+}). Three key events that measure fertilizing capacity of sperm—hyperactivated motility (a type of vigorous, but nonlinear flagellar movement required for fertility), in vitro capacitation (the ability to undergo the acrosome reaction quickly when challenged by zona pellucida proteins), and the spontaneous acrosome reaction, were determined over a 4-h time course and compared between genotypes. As expected, no significant difference between genotypes for these three important calcium-dependent events was observed in IVF medium. However, in IVF medium, sperm from t^X/+, t^Y/+, and t^X/t^Y males exhibited only minor differences from sperm produced by +/+ males in their ability to capacitate and spontaneously undergo acrosomal exocytosis, yet highly significant differences in the initiation and maintenance of the hyperactivated state of motility: after 1 h in IVF medium, nearly 60% of spermatozoa from t^X/+ and t^Y/+ males exhibited hyperactivated motility, compared to a high of about 35% of sperm from +/+ males at 2 h of incubation. Quantitatively, the +/t-precocious hyperactivation phenotype was attributable to a rapid decline in linearity to ultralow t^X/t^Y levels, accompanied by a more gradual drop in curvilinear velocity (or vigorous movement) to about a mean half-way between the means of wild-type and t^X/t^Y levels by 4 h of incubation. However, the flagella of nearly all sperm from t^X/t^Y males demonstrated premature hyperactivation by 5 min of incubation, although this flagellar phenotype abruptly ended as a precipitous decline in vigor accompanying the constant, abnormally low level of linearity ensued. Additionally, the flagella of 90%–95% of the motile sperm from t^X/t^Y males became "frozen" in an aberrant curvature phenotype called "curlicue", a chronic negative bend throughout the entire flagellum (Fig. 19.2A and B). The movement of these sperm resembled little more than nonrhythmic, twitching in place, producing slow circular swimming.

Interestingly, from the earliest time point and beyond, an increasingly high percentage of the flagella of sperm from t^X/+-and t^Y/+-mice exhibited a flagellar curvature phenotype called "fishhook" (Fig. 19.2C and D) in which the middle-piece of the flagellum displayed an almost chronic, negative bend, while the principal piece displayed nearly normal curvature, with abnormal bending occurring only infrequently in that portion of the flagellum. The abnormality of this phenotype, however, had more to do with its premature occurrence rather than the occurrence itself, since wild-type sperm flagella display a "fishhook"-like phenotype when the axonemal free Ca^{2+} concentration rises to high levels [8].

Surprisingly, demembranated, reactivated flagella of sperm from +/t males had been shown to display a negative curvature phenotype significantly more rapidly than the demembranated, reactivated flagella of sperm from +/+ males [9]. Taken together, these results suggested that the altered motility of sperm produced by t-bearing males was a calcium-dependent phenomena and that the calcium hypersensitive element(s) were intrinsic to the axonemal and/or peri-axonemal structures of the mouse sperm tail.

Independent studies 14 years apart also revealed that at early time points (up to 2 h of incubation in IVF medium), the swimming speed, path shape, and flagellar beat frequency of sperm populations from +/t males were bimodally

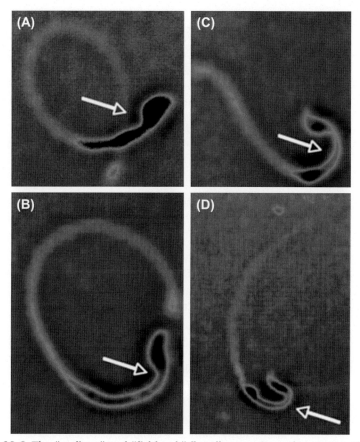

Figure 19.2 The "curlicue" and "fishhook" flagellar curvature phenotypes. High speed (Adobe Photoshop "solarized") photographs captured from videotapes recorded at 30 frames per second of motile t^X/t^Y-sperm (A, B) and $+/t$-sperm (C, D) swimming in an 80-micron deep slide chamber in complete IVF medium. The lightly minced caudae epididymides of test males were incubated at 37°C in 5% CO_2 in air for 15 min to allow sperm to swim out of the caudae. After removal of epididymal tissue, sperms were incubated for an additional 15-min in fresh media before videotaping for 5 min on slides maintained at 37°C. *Arrows* indicate that the bend of the middle piece of $+/t$-sperm is generally more acute than that of t^X/t^Y-sperm. However, it is noteworthy that both forms of curvature are always negative (direction of curvature of the flagellum is always opposite to the direction of curvature of the sperm head). The chronic nature of this one-sided curvature suggests that both phenotypes may result from defects in the ability of dyneins to promote a recovery stroke.

distributed [7,10]: one peak was closer to the unimodal peak of sperm from +/+ males, which exhibited vigorous, progressive motility, and a consistent beat pattern, while the other was closer to the unimodal peak of sperm from t^X/t^Y males, which displayed slow, nonprogressive motility, with no regular beat configuration. These data tied TRD to altered sperm motility, with half of

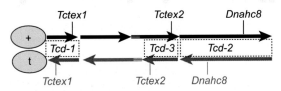

Figure 19.3 The location of three axonemal dynein subunits in the *t*-complex.
+- and *t*-chromosomes are shown as in Fig. 19.1, and the relative positions of *t-complex testis expressed-1* and *-2* (*Tctex1, Tctex2*) (inner and outer arm light chains, respectively), and *Dnahc8*, a unique, principal piece-specific outer arm gamma heavy chain subunit, are indicated.

the sperm population (presumed to be the *t*-sperm population) exhibiting relatively better flagellar function than the other half (presumed to be the +-sperm population).

Armed with convincing models of the genetic and functional bases of *t*-related male TRD and sterility, plus technological advances in physical gene mapping and analysis, investigators began a series of systematic searches for the elusive Tcr and Tcd factors. Several of these studies were published somewhat before the final determination that altered flagellar function was most likely at the heart of *t*-TRD and male sterility, primarily relying on novel differential cDNA subtraction, transcription, and sequence analyses to compare +- and *t*-alleles of genes physically linked to either the large *Tcr* or *Tcd* loci via restriction fragment length polymorphisms. Interestingly, among the first Tcd candidates isolated by these techniques were two flagellar dynein light chains (although their identities as such did not become apparent until later), aptly named *t-complex testis expressed-1* and *-2* (*Tctex1* and *Tctex2*), mapping to the *Tcd-1* and *Tcd-3* loci (Fig. 19.3), respectively [11–13]. These early investigations were followed by a string of studies that employed mice heterozygous for a *t*-haplotype and a series of interspecific recombinant chromosome 17 homologues (*S-+* homologues), which were either able or unable to rescue the flagellar phenotypes associated with *t/t*-male sterility. Therefore, mice carrying various *S-+/t* chromosome 17 genotypes were employed to map the key factors responsible for the "curlicue" phenotype within two submegabase regions of the large *Tcd-2* locus [14–18]. The first of three genes localized in this way was identified as *Dnahc8* (Fig. 19.3), a gene encoding a sperm-specific γ-type axonemal outer dynein arm heavy chain, restricted to the flagellar principal piece in mice and (putatively) in other mammals [19–22].

Below, I will summarize some of the properties of these three dynein subunits (in addition to their physical map locations) that for nearly 15 years have supported the idea that dyneins influence TRD.

19.2 Properties of *t-complex testis expressed-1*

Tctex1 (*Dynlt1*), the first member of a family of structurally homologous dynein light chains, maps to the *Tcd-1* locus in the most proximal *t*-inversion (*In[17]1*). Though it was unknown at the time of its identification, *Tctex1* is present on wild-type mouse chromosome 17 as two paralogous copies (*Dynlt1c* and *AC175035.1*) approximately 200-kb apart. These paralogues encode proteins that are identical except in their C-terminal exons, which differ in both sequence and length (Ensembl Genome Database, NCBI m37 Mouse Assembly). Though the number of *t*-haplotype copies has not yet been elucidated, the *t*-protein is homologous in sequence and length to the more proximal, longer (113 amino acids) wild-type paralogue (Fig. 19.4A). Regardless of gene number, the steady state level of $Tctex1^t$ mRNA is fourfold that of $Tctex1^+$ mRNA in the testis. In addition, *Tctex1* is expressed at a higher level in the testis than in other tissues [13].

Although first characterized as a light chain subunit of brain cytoplasmic dynein 1 [24], TCTEX1 was subsequently localized to the dimeric axonemal inner dynein arm I1 in the flagella of *Chlamydomonas reinhardtii* and also to the mouse sperm tail [25]. Significantly, recent findings have demonstrated that inner arm I1 appears to play a central regulatory role in coordinating flagellar beating and determining waveform in response to signals transduced from the central pair of axonemal microtubules through the radial spokes of the axoneme [26,27]. More recently, an additional Chlamydomonas sequence homologue of *Tctex1* was shown to be a dynein light chain (LC9) bound to the intermediate chains of the outer dynein arm [28]. The data from deletion mutants strongly suggest that its absence has a negative effect on flagellar beat frequency.

In late 1999, Herrmann and colleagues identified *Tcr* and demonstrated that it encoded a spermiogenesis-specifically expressed fusion product of sperm motility kinase 1 (*Smok1*) and another, unrelated kinase [29]. Almost 6 years later, but before any *Tcd* had been positively identified, the solution structure of TCTEX1 was solved [23], revealing that dimeric TCTEX1 is most likely bound to both the "cargo" subunit of axonemal microtubule doublets (the A subfiber) and to the dynein intermediate chain IC138 at opposite ends of the TCTEX1 dimer (Fig. 19.4B). Interestingly, a single mutation (out of three) in $TCTEX1^t$ ($Q_{41}H$) alters a highly conserved residue that is exposed close to the intermonomer groove (Fig. 19.4A). This mutation directly abuts an exposed serine in the Chlamydomonas TCTEX1 structure (a threonine in mouse TCTEX1), producing a potential new target for phosphorylation. Additionally, during the time interval between the identification of *Tcr* and solving the structure of TCTEX1, analysis of chromosomal deletions of, within, and surrounding the *Tcd-1* locus

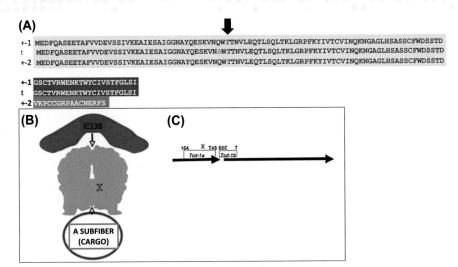

(A)

+-1 MEDFQASEETAFVVDEVSSIVKEAIESAIGGNAYQESKVNQWTTNVLEQTLSQLTKLGRPFKYIVTCVINQKNGAGLHSASSCFWDSSTD
t   MEDFQASEETAFVVDEVSSIVKEAIESAIGGNAYQESKVN*H*WTTNVLEQTLSQLTKLGRPFKYIVTCVINQKNGAGLHSASSCFWDSSTD
+-2 MEDFQASEETAFVVDEVSSIVKEAIESAIGGNAYQESKVNQWTTNVLEQTLSQLTKLGRPFKYIVTCVINQKNGAGLHSASSCFWDSSTD

+-1 GSCTVRWENKTWYCIVSTFGLSI
t   GSCTVRWENKTWYCIVSTFGLSI
+-2 VKPCCGRPAACNERFS

(B) IC138 **(C)**

X A SUBFIBER (CARGO)

164   X T48 66E   T
      Tcd-1a   Tcd-1b

Figure 19.4 *t-Complex testis expressed-1* (*Tctex1*) sequence, structure, and location in *Tcd-1a*. (A) Sequence of two wild-type paralogues of *Tctex1* is shown with alignment of the *t*-amino acid sequence sandwiched between them. The two wild-type sequences differ only in their C-terminal exons (indicated by *white letters on a red background* vs. *white letters on a blue background* for TCTEX1-1 and TCTEX1-2, respectively). The *t*-sequence closely aligns with TCTEX1-1, except for three amino acid changes, only one of which is considered likely to have an effect on function (shown as a *blue H*). The threonine whose phosphorylation status might be affected by this change is shown in all three proteins as a *red T* and is indicated by the *vertical arrow*. (B) A freehand sketch of one view of the solution structure of dimeric TCTEX1 (gray; from Ref. [23] with *arrows* above and below pointing at the location of the intermonomer groove between them. The picture indicates that the dimer binds an intermediate chain (above) and interacts directly with cargo (below) at opposite poles. The "X" marks the location of the Q→H mutation in the *t*-haplotype protein. (C) Division of *t*-complex distorter (*Tcd-1*) into *Tcd-1a* and *Tcd-1b*, with *Tctex1* (blue "X") located in the former sublocus. The approximate boundaries of these loci are: *D17Mit164* (*164*) proximally and *D17Leh48* (*T48*) distally for *Tcd-1a*, and *D17Leh66E* proximally and the locus of *Brachyury* (*T*) for *Tcd-1b*.

suggested that (1) Tcd-1 was a hypomorph or amorph [30], and (2) the locus could be divided into at least two subloci (Fig. 19.4C), *Tcd-1a* (possibly containing *Tctex1* near its distal end), and a more distal *Tcd-1b* sublocus [31,32].

19.3 Properties of *t-complex testis expressed-2*

Tctex2 (*Tcte3*) was first identified in the same cDNA subtractive screening experiment in which *Tctex1* was isolated [13]. Subsequent mapping studies placed it in the proximal end of *In[17]4* [11], but higher resolution mapping and sequencing has since positioned it distally in *In[17]3* (Ensembl Genome Database, NCBI m37 Mouse Assembly).

Initially *Tctex2* was described as a gene(s) encoding a membrane-associated protein found exclusively on the surface of the sperm tail [12], but its identity as a structural homologue of TCTEX1, and an outer arm axonemal dynein light chain (LC2) in Chlamydomonas flagella, was later demonstrated [33]. Unlike *Tctex1*, the *t*-haplotype form of *Tctex2* shows an abnormally low level of mRNA expression, and its sequence contains numerous in-frame, but potentially deleterious mutations [12]; In addition, targeted deletion of *Tctex2* in Chlamydomonas results in the loss of outer dynein arm assembly and subsequent ultraslow, jerky flagellar movement. Moreover, these deletion mutants fail to appropriately respond to photoshock, a calcium-dependent property of the outer dynein arm consisting of backward swimming in response to intense light [34]. Thus, although untested, it is possible that the multiple amino acid changes in the TCTEX2^t may have a calcium-sensitive gain-of-function (hypermorphic) effect.

Interestingly, in wild-type mouse chromosome 17, *Tctex2* is present in three nearly identical copies (Fig. 19.5), each separated from the next more-distal copy by a distance of approximately 17-kb. Homozygous deletion of the most

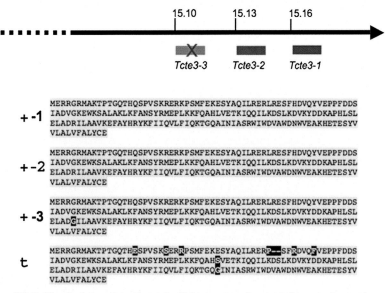

Figure 19.5 Physical map location of wild-type paralogues of *t-complex testis expressed-2* (*Tctex2*) (*Tcte3*) and comparison of the +- and *t*-sequences. *Top*, Physical map positions of three wild-type paralogues of *Tctex2* in *In[17]3*. Numbers are in megabase pairs from the centromere and all three genes are transcribed toward the centromere. The red "X" in *Tcte3-3* indicates that it has undergone homozygous deletion from the genome by targeted disruption [35] resulting in incompletely penetrant sperm motility and azoospermia (due to increased testicular apoptosis) phenotypes but no other notable problems. *Bottom*, comparison of the amino acid sequences of the wild-type (+) and *t*-haplotype (*t*) homologues of *Tctex2* demonstrating numerous potentially deleterious mutations in the *t*-homologue (*yellow letters on a black background*).

proximal copy results in several incompletely penetrant phenotypes, including sterility due to either asthenozoospermia in some cases or poor sperm motility in others [35]. These poorly motile sperms exhibit an inability to move from the uterus into the oviduct, previously demonstrated to be a deficiency of sperm from males carrying two *t*-haplotypes [36].

19.4 Properties of *Dnahc8*

Interspecific recombinant chromosome 17 homologue mapping was initially employed to determine the origin, number, and extent of the *t*-inversions [2]. Two useful byproducts of these experiments were: (1) *M. musculus* males heterozygous for an introgressed chromosome 17 homologue (*S*; from the aboriginal mouse species, *Mus spretus*) and the endogenous wild-type chromosome 17 homologue (+) were fertile (*S*/+ males), while, *M. musculus* males heterozygous for the *S*-chromosome 17 homologue and the endogenous chromosome 17 *t*-variant were, without exception, sterile (*S*/*t* males); and (2) the *S*-homologue and the +-homologue were capable of recombination at wild-type rates in the three inversions that housed the *Tcd* loci. Thus, males heterozygous for a large assortment of *S*-+ recombinant chromosome 17 homologues and a complete *t*-haplotype were employed in breeding studies to map *t*-sterility loci within large inversions to a high resolution [15–18,37–39].

Motility of sperm from sterile recombinants was assessed quantitatively by computer assisted sperm motility analysis and high-resolution video-microscopy. Physical mapping of recombinant breakpoints from informative chromosome 17 homologues eventually led to the identification of three factors that together appeared to cause male sterility, precocious hyperactivation, and "curlicue" in ~95% of motile sperm flagella. Each factor was mapped to a submegabase region of the distal one-third of *In[17]4+* (Fig. 19.6A), with the most proximal of these containing the axonemal dynein heavy chain, *Dnahc8* [20,40–42]. Interestingly, the *S*-allele of *Dnahc8* turned out to be a null allele in the *M. musculus* background, suggesting that the hemizygous (or homozygous) *Dnahc8^t* allele was a major cause of flagellar dysfunction in males homozygous for the large *Tcd-2* locus in *In[17]4*.

Sequence analysis of *Dnahc8+* demonstrated that the gene consisted of 93 exons totaling more than 14-kb, within a genomic region of over 250-kb. Phylogenetic analysis of the computer-generated translation product of this message demonstrated that DNAHC8 was most closely related to the Chlamydomonas axonemal outer dynein arm γ-heavy chain. Both +/+ and *t*/*t* mice produced two alternatively spliced transcripts, the more abundant one coding for a protein of 4732 amino acids, while the less-abundant and less-stable transcript encoded a C-terminally truncated protein of 4203 residues [21,22].

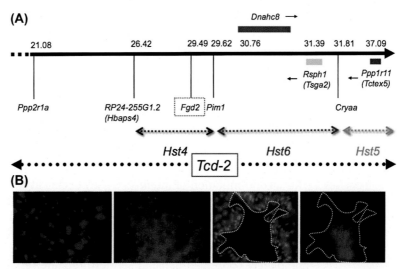

Figure 19.6 Physical map location of Dnahc8 in In[17]4 and expression of DNAHC8 in the testis. (A) *Dnahc8* is located in the proximal half of the *Hybrid Sterility 6 (Hst6)* locus between *Fgd2*, located in the *Hybrid Sterility 4 (Hst4)* locus and *Rsph1*, located more distally in the *Hst6* locus. The *Hst6* locus plus the *Hybrid Sterility 5 (Hst5)* locus have been subdivided into three subloci (*curlicueA, curlicueB1,* and *curlicueB2*; [40], based on the location of a trio of genes whose *t*-alleles are thought to be responsible for expression of the "curlicue" phenotype (*Dnahc8, Rsph1,* and *Ppp1r11*; [41]). (B) Immunohistochemistry of wild-type testis sections probed with an affinity purified antibody raised against a portion of the unique N-terminal extension of DNAHC8. In panels 1 and 3 from the left, DAPI [(4′,6-diamidino-2-phenylindole) is a fluorescent stain that binds strongly to A-T rich regions in DNA] staining has been utilized to show nuclear morphology. Immunofluorescent localizations of DNAHC8 in panels 2 and 4 (sections showing an earlier and later stage of spermatid development, respectively) demonstrate intense, punctate fluorescence over the region of round spermatids approximately where flagellar growth would originate in panel 2, and equally intense fluorescence in the center of the lumen of a seminiferous tubule section in panel 4, an area where the principal piece of sperm tails resides. The *dashed yellow borders* in panels 3 and 4 define the outer boundaries of the lumen.

Expression of the gene was testis specific, and transcription of both +- and *t*-alleles occurred exclusively in late pachytene primary and secondary spermatocytes. Translation was also specific to these cells, as mRNAs were not observed in round or elongating spermatids by in situ hybridization [21]. Both +- and *t*-polypeptides were visualized by immunohistochemistry and immunocytochemistry of the testis and cauda epididymal sperm, respectively, and were shown to localize to the cytoplasm of pachytene spermatocytes, the tail bud in round spermatids, and the principal piece of the flagellum in testicular and cauda epididymal sperm (Fig. 19.6B). These findings demonstrated that at least in mammals, the subunits of the outer dynein arm are not identical from base to tip of the tail [22].

Significantly, comparative sequence analysis of isolated *t*- and +-cDNAs indicated that the *t*-message contains 17 nonsynonymous mutations scattered throughout the protein, several coding for nonconservative changes at otherwise highly conserved residues in all annotated mammalian genomes [22]. Of particular interest were two mutations, one ($P_{3408}T$) occurring at the end of the outward α-helical branch of the antiparallel coiled-coil stalk on which sits the ATP-dependent microtubule (MT) binding site. This mutation aligns with a residue formerly reported to play a significant role in ATP-dependent MT binding in cytoplasmic dyneins, and thus, could affect microtubule translocation [22,43].

The second mutation of interest ($G_{127}R$) occurs at the N-terminus of the most highly conserved region of the unique mammalian DNAHC8 N-terminal extension [21,22]. The exchange of arginine for what is always a hydrophobic residue in all annotated wild-type mammalian DNAHC8 orthologous protein sequences creates a protein kinase C phosphorylation target of a completely conserved flanking serine/threonine residue (S_{129} in mouse). This serine resides just upstream of a highly conserved serine/arginine/lysine rich region and an IQ calmodulin-binding motif (Fig. 19.7A). While yet untested, it is possible that phosphorylation of S_{129} might alter the conformation of the region, and thus, modify the behavior of the downstream IQ motif, potentially affecting the way in which a (putative) calmodulin light chain (LC4) in Chlamydomonas binds to DNAHC8 and/or interacts with the A subfiber of the outer microtubule doublet [44,45].

The DNAHC8 N-terminal 167 amino acid sequence is completely unrelated to any region found in any other dynein, and the N-terminal 75% of the mouse sequence appears to be similar only to the orthologous sequences in mammals. Interestingly, in mouse DNAHC8, the first ~60 residues of the N-terminus are 25% glu/asp (mostly glu) and ~36% pro with no basic residues. This sequence has similarity in its proline content and position to a region of sperm- and spermatid-specific isoform of glyceraldehyde 3-phosphate dehydrogenase (GAPDS) known to anchor it to the fibrous sheath of mammalian sperm [46]. While the significance of this feature is unknown at this time, it is tempting to speculate that the unique N-terminus of DNAHC8 has a role in targeting this heavy chain and the dynein particle to which it belongs to the developing principal piece of the sperm tail where the fibrous sheath, a mammalian specific periaxonemal structure [47], resides [15,16]. More importantly, Smok/Tcr has been recently localized exclusively to the principal piece of the sperm tail by immunocytochemical approaches using both light and transmission electron microscopy, where it appears to be linked to both the longitudinal columns and circumferential ribs of the fibrous sheath and the inner surface of the outer dense fibers (Fig. 19.7B), adjacent to the outer doublet MTs of the axoneme, close to the outer dynein arms [48].

(A)

```
mouse-t    SVLSDRL-SQ-SSRRPSKFRRSMTGIPNLQETLKEKQARFREARENRK
mouse-+    SVLSDGI-SQ-SSRRPSKFRRSMTGIPNLQETLKEKQARFREARENRK
cow        SMLSDVL-SQ-SSHRSSKYRRSMSGIPNLQETLKERQARYRDARENRK
marmos.    SVISDVL-SSPSSWGSSRYRRSMSGIPNLQETLKERQARYRDARESRR
dog        RRGPSML-S-QSSRRSSRYHRSMSGIPNLQETLKERQARFRDARESRK
guin.pig   ASLAEGQ-SENSSRRPSKYLRSMSGIPNLQETLKEKQARFREARENRK
sloth      SMLSEML-SQ-SSRRSSKYRRSMSGIPNLQETLKERQARFREAREGRK
armadil.   SVISEVF-S-QSSRRSSKYRRSMSGIPNLQETLKERQARFREAREGRK
l.hdg.hog  SVISEVL-SQSSRRSSRYRRSVSGIPNFQETLKEKQARFREARETRK
hdg.hog    STVSDVL-SQ-SSRRSSKYRRSMSGIPNLQETLKEKQARFREAREGRR
gorilla    SVISEVL-SLPSSRRSSRYCRSMSGLPNLQETLKERQARFREARESRR
human      SVISEVL-SLPSSRRSSRYRRSMSGLPNLQETLKERQARFREARESRR
elephant   SVLSEML-SQSSSRRSSKYRRSMSGIPNLQETLKERQARFREAREGRK
macaque    SVISEVL-SSPSSRRSSKYRRSMSGLPTLQETLKERQARFRDARESRR
walaby     VPSSEGL-AA-QPRRSSKFRRSMSCIQAVQETMKERQARYKEAREGRR
mous.lem   SVLSEVL-SSASSRRSSKYRRSMSGIPNLQETLKEKQARFREARENRK
opossum    GKVPSIEGST-SSRRPSKFFRSLSCIQAFQETLKEKQARFRDARENRK
mbat       QSMLSAL-SQ-SSHRSSKYFRSMSGIP-LQETLKERQARFRDARESRR
pika       SVLSEVL-SQ-SSRRSSKYRRSMSGIPSFQETLKEKQARYRDARENRK
platypus   SALDSGPVA--YTRRMSKFRRSTSGVQALQETLKEKQARYREAREGRK
rabbit     SVLSEVL-SQ-SSPRSSKYRRSMSGIPNFQETMKEKQARFRDARENRK
bushbaby   SVLSEML-SL-SPRRSSKYRRSMSGIPNLQETLKEKQARFRDARENRK
chimp      SVISEVL-SLPSSRRSSRYRRSMSGLPNLQETLKERQARFREARESRR
orangat.   SVISEVL-SLPSSRRSSRYRRSMSGLPNLQETLKERQARFREARESRR
hyrax      TVLSEVL-SQSSSRRSSKYRRSMSGIPTLQETLKERQARFREARENRK
megabat    SVLSEIL-SQ-GSLRSSKYRRSMSGIPNLQETLKEKQARFREARENRK
rat        SVLSDGL-SQ-SSRRSSKFRRSMTGIPNLQETLKEKQARFREAREGRR
pig        SMLSDVL-SQ-SSGKSSKYRRSMSGIPNLQETLKERQARFRDAREGRK
treeshrew  SVLSSVF-SSQSSRRSSKYRRSMSGIPNLQETLKEKQARFREAESRK
dolphin    PVPSEVL-SQ-SSRRSSKYLRSMSGVTNLQETLKERQARFRDARENRK
alpaca     SMLSDVL-SQ-SSRSSSRYRRSMSGIPSLQETLKERQARFRDARESRK
```

(B)

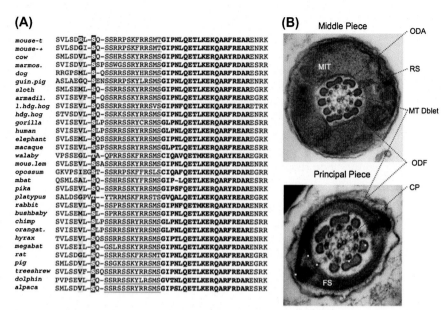

Middle Piece
ODA
MIT
RS
MT Dblet
ODF
Principal Piece
CP
FS

Figure 19.7 The unique N-terminal region of mammalian DNAHC8. (A) The first mutation in DNAHC8^t is a nonconservative change at the N-terminus of the most highly conserved region of the unique DNAHC8 N-terminal extension. This region contains a serine/arginine/lysine-rich sequence (*underlined*) upstream of a putative IQ calmodulin-binding motif (*bold lettering*). At the very N-terminus of this sequence, there is less conservation, but certain amino acids retain specific character traits in all annotated mammalian proteomes. An alignment of the *t*-haplotype sequence with the wild-type mouse sequence and 29 additional annotated mammalian sequences demonstrates that the first mutation in DNAHC8^t occurs in a residue that is always hydrophobic in wild-type orthologues (*black letter on yellow background*), and changes this residue (a glycine in the case of the +-mouse sequence) to an arginine (*yellow letter on a black background*). Bioinformatic analyses of this change with numerous programs indicate that the change would promote phosphorylation by a protein kinase C-like enzyme of a completely conserved serine/threonine (*white letter on blue background*) two amino acids (in the mouse) downstream of the $G_{127}R$ mutation. The effect of this mutation, if any, has not been ascertained, but it is tempting to speculate that a conformational change related to such a phosphorylation event could have a deleterious effect on any interaction of the downstream IQ motif with a calmodulin light chain. (B) Transmission electron micrographs of cross-sections through the middle piece (top) and principal piece (bottom) of the mammalian sperm tail. MIT corresponds to the mitochondrial sheath, a middle piece–specific periaxonemal structure, and FS indicates the fibrous sheath, a principal piece-specific periaxonemal structure. Other structures found throughout the flagellum: *CP*, central pair of singlet microtubules; *MT Dblet*, microtubule doublet; *ODA*, outer dynein arms; *ODF*, outer dense fibers; *RS*, radial spokes. Both DNAHC8 and Smok/Tcr localize specifically to structures in the principal piece, with the *yellow dots* indicating sites of Smok/Tcr localization in the FS, ODF, and between the ODF and outer doublet microtubules, putatively in the vicinity of the N-terminus of DNAHC8 at the base of the ODA.

19.5 Chromosomal deletion analysis modifies Lyon's model

Lyon predicted that homozygosity for distorters was the basis for male sterility [4]. In support of this hypothesis, Lyon later demonstrated that a large deletion (T22H) on a wild-type chromosome that eliminated the entire *Tcd-1* locus could completely recapitulate the *Tcd-1* effects on male sterility and TRD when respectively heterozygous with a complete or distal partial *t*-haplotype containing *Tcr* [30]. These data reinforced the idea that the *Tcd-1* distorter was identical to the *Tcd-1* sterility gene, since both appeared to display severe loss-of-function (hypomorphic or amorphic) phenotypes.

As previously stated, data had already suggested that the *Tcd-1* locus could be divided into at least two subloci, *Tcd-1a* and *Tcd-1b*, and sterility was shown to map exclusively to *Tcd-1a* [31]. Since *Tctex1* mapped within the *Tcd-1a* sublocus, it remained a candidate for both the distorter, Tcd-1a, and the associated sterility factor.

A more recent analysis of the *Tcd-1a* sublocus employed bacterial artificial chromosome (BAC) transgenesis in an effort to rescue male sterility and/or TRD caused by deletion of the *Tcd-1a* region [49]. A BAC containing two proximal genes from the locus, the testis-expressed genes *Synj2* and *Serac1* (whose *t*-alleles carry deleterious mutations), but not *Tctex1*, was capable of rescuing the sterility phenotype caused by deletion of the locus, but had no effect on TRD (Fig. 19.8). These results suggested that while *Tctex1ᵗ* might still be the *Tcd-1a* distorter component, it was not the cause of male sterility mapping to the *Tcd-1a* locus, thus, modifying Lyon's original concept of identity between atleast one of the *t*-distorters and *t*-sterility genes.

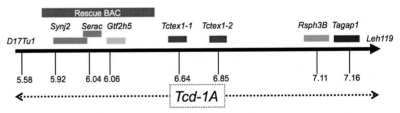

Figure 19.8 The position of the sterility genes in *Tcd-1a* relative to distorter candidates. The physical map positions of testis expressed *Synj2* and *Serac* are shown in relationship to the physical locations of *Tctex1-1*, *Tctex1-2*, *Rsph3b*, and *Tagap1* in the *Tcd-1a* locus. Numbers indicate megabase pairs from the centromere. A chromosome carrying a deletion of the entire region causes (1) male sterility when in trans to a complete *t*-haplotype and (2) elevation of the transmission level of the *t⁶* partial *t*-haplotype (see Fig. 19.1) when in trans to it. A bacterial artificial chromosome (BAC) transgene carrying only *Synj2* and *Serac* (and *Gtf2h5*, a gene not expressed in the testis) is able to rescue fertility in the first case, but has no effect on transmission ratio distortion in the second case, indicating that not all sterility genes are distorters and vice versa.

19.6 Identification of *t*-complex distorters

Prior to the discovery of any of the *Tcd* genes by Herrmann and colleagues, but after his laboratory had identified *Tcr* and described it as a member of a family of "sperm motility kinases" (*Smok*^{Tcr}, encoding a dominant negative kinase), Herrmann and colleagues reasoned that since *Tcr* appeared to be a loss-of-function kinase, *Tcd* genes would most likely code for products that act in signaling pathways upstream of Tcr to enhance Tcr activity in *t*-sperm to near wild-type Smok1 levels, while, at the same time, driving wild-type Smok1 activity in +-sperm to aberrantly high levels [29]. Thus, according to this simple and elegant model of TRD, distorters should act upstream of Smok/Tcr as either amorphic/hypomorphic negative regulators of Smok/Tcr activity, or hypermorphic positive regulators of Smok/Tcr activity, but not as downstream targets (such as dyneins) (Fig. 19.9A). In fact, in the ensuing search for distorters by Herrmann and colleagues, this description of what distorters must or must not look like became one of four key limiting criteria for identifying *Tcd* candidate genes whose function in TRD could then be tested in transgenic mice. The other three criteria were identical to those used by most investigators since Lyon first proposed her model of TRD [5], including: (1) genomic location of the candidate, (2) testis expression of the candidate, and (3) substantial differences in sequence and/or expression of the *t*-allele of the candidate relative to those properties of the wild-type allele.

Interestingly, in the decade since the identification of *Tcr*, the new strategy of limiting *Tcd* genes to those that encode upstream hypermorphic activators or hypomorphic inhibitors of Smok/Tcr action has resulted in the identification of two genes that appear to have distorter activity. Both genes encode Rho-GTPase (putatively RhoA) regulators. The first, *Tagap1*, is a *Tcd-1a* candidate (presently named *Tagap1*^{Tcd1A}), coding for a GTPase-activating protein (GAP), a negative regulator of RhoA activity [50].

A single copy on wild-type chromosome 17 is ubiquitously expressed, but shows a relatively low level of mRNA expression in testis. Four copies are present in *t*-haplotypes, each containing multiple nonsynonymous mutations. Although the *t*-isoforms constitute three distinct polymorphic classes, each has an intact RhoGAP domain. Additionally, *t*-allele mRNA expression levels are considerably higher than +- allele levels in testis.

Because chromosomal deletion analysis had already demonstrated that *Tcd-1* is amorphic or hypomorphic, it follows that deletion of *Tagap1*$^+$ should demonstrate positive TRD when the deletion is in trans to t^6, a partial *t*-haplotype lacking only the *Tcd-1*^{t} alleles (see Fig. 19.1). Furthermore, an overexpressing wild-type allele of *Tagap1* should demonstrate negative TRD when in trans from t^6. However, tests of both hypermorphic and amorphic alleles of *Tagap1*$^+$

Dyneins

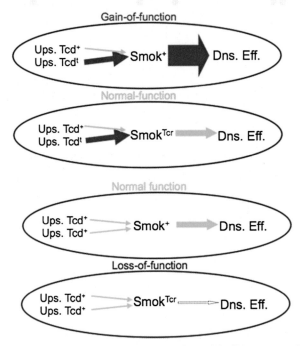

Figure 19.9 Model of transmission ratio distortion with distorters acting only upstream of Smok/Tcr. *Tcr* (*Smok^Tcr*) encodes a dominant loss-of-function kinase, a variant of wild-type Smok (*Smok^+*), and each is present only in the cell of its origin. Thus in sperm from a +/*t*-male, the activity of wild-type Smok in +-sperm is boosted to aberrantly high levels when activated by upstream gain-of-function *t*-distorters (Ups.-Tcd^t), leading to dysregulation of downstream effector (Dns. Eff.) molecules and relatively poor sperm motility (Top Oval, gain-of-function effect). However, the same upstream gain-of-function *t*-distorters are present in *t*-sperm and they restore Tcr activity to nearly wild-type levels, thus, promoting nearly wild-type motility (Second Oval, relatively normal function effect). When only +-distorters (Ups.-Tcd^+) are present, a Smok^+-containing sperm behaves just like a wild-type sperm from a +/+-male (Third Oval, normal function effect), but a Tcr-containing sperm fails to sufficiently activate downstream effector molecules (Bottom Oval, loss-of-function effect).

demonstrated that hypermorphic *Tagap1^+* caused a small but significant elevation of *t^6* transmission (88% vs. 80% for control +/*t^6* males), while targeted deletion of *Tagap1^+* caused a significantly greater depression of *t^6* transmission. Thus, for *Tagap1^t* to demonstrate positive TRD, it would have to be a hypermorph. The explanation given for this seeming contradiction was that *Tagap1* must be hypostatic to *Tcd-1b* (or synergize with it), and the resultant must encode a hypomorph/amorph of a Smok/Tcr inhibitor. However, because the steady state level and functionality of the Tagap1^t protein was not assessed, it is conceivable that *Tagap1^t*, whose multiple isoforms carry numerous potentially deleterious mutations, could demonstrate dominant negative activity and, thus,

466

be a hypomorph. A "suppressor of meiotic drive" function for $Tagap1^t$ would be consistent with such a finding. Evidence of suppressors of drive have been obtained in TRD experiments using laboratory strains where TRD is not continually monitored and selective pressure for drive has been removed [1].

Using the same criteria as those used to identify *Tagap1* as a strong candidate for *Tcd-1a*, Herrmann's laboratory identified a *Tcd-2* candidate gene, *Fgd2*, coding for a guanine nucleotide exchange factor (GEF), a promoter of the Rho active state [51]. $Fgd2^+$ is located just proximal to the *Hybrid Sterility 6* (*Hst6*) locus containing *Dnahc8* [16,20] and lies within the *Hybrid Sterility 4* (*Hst4*) locus (see Fig. 19.6A) [37]. Although the *t*-allele codes for a protein that differs from the wild-type isoform by only a single amino acid ($S_{234}G$), the +- and *t*-isoforms demonstrate different mRNA expression profiles in the testis.

Wild-type *Fgd2* mRNA is ubiquitous, and testis expression commences in early spermatocytes. Two isoforms are expressed, a nonabundant full length mRNA and an abundant shorter isoform, transcribed from a promoter within the gene. Because translation of the shorter transcript would produce a polypeptide that deletes much of the N-terminus of the GEF domain, it was suggested that a product of this abundant isoform might act as a dominant inhibitor of the larger, less abundant isoform. Alternatively, the shorter transcript might undergo nonsense-mediated decay, as appears possible in Bauer et al. [51].

As for transcription of *t*-mRNA, the wild-type transcript abundance pattern appeared to be reversed, with *t/t* producing at least threefold more of the longer testis mRNA than most wild-type inbred strains of mice but also considerably less of the shorter transcript. The enhanced expression level of the longer mRNA in *t/t* mice suggested that $Fgd2^t$ would encode a hypermorph of a positive regulator of Smok/Tcr activity. Indeed, targeted deletion of the locus on a wild-type chromosome reduced the transmission of a partial *t*-haplotype containing only *Tcd-1* and *Tcr* (t^{h49}, see Fig. 19.1) from 47% to 35% when in trans to t^{h49}, suggesting that a hypermorphic *t*-allele should raise the transmission level of t^{h49} by up to 24%. However, a test of a gain-of-function allele was not performed, thus precluding our ability to know with any sense of accuracy, exactly how high the transmission level of t^{h49} might be elevated by it. Nonetheless, the *t*-allele of *Fgd2* is an excellent candidate for a distorter and has been designated $Fgd2^{Tcd2}$.

19.7 What about dyneins?

After many years of searching for the molecular basis of *t*-TRD, have we finally come close to the end of the journey? Maybe. Maybe not. I remember Olds-Clarke's cautionary thoughts about being overly restrictive when defining the molecular basis of sperm capacitation [52], herein paraphrased with regard to meiotic drive: TRD may be, in many ways, like World War II, with several differ-ent theatres of operation, related only by the need for coordinating its activities around Tcr, and one common end: to transmit the *t*-haplotype to all prog-eny (and thus, win the Great Chromosomal War). In other words, it matters little whether a protein acts upstream or downstream of Tcr; if it increases the transmission level of the *t*-haplotype at the expense of the transmission of the +-chromosome, it must be a distorter. Thus, it may be too early to restrict all potential mutant downstream effectors (Fig. 19.10) of Smok/Tcr function from consideration as conditionally defective mediators (downstream distorters) of the Smok/Tcr signal that act to raise moderately distorted transmission levels of *t* to even higher levels.

This idea is in some ways reinforced by the evolutionary history of *t*-haplotypes. Numerous studies suggest an ancient origin for *t*-haplotypes, dating back more than two million years [2,53,54]. However, modern *t*-haplotypes are very closely related at the phylogenetic level, indicating that they must be very recent descendants of a common ancestor [55]. To reconcile an ancient ori-gin with the phylogenetic relatedness of contemporary *t*-haplotypes, Ardlie and Silver [1] suggested that if suppressors of meiotic drive were, at one time, widespread in feral populations of mice, the perseverance of *t*-haplotypes in the population would require recurrent mutation to higher distorting alleles of already existing distorters and the addition of new distorter loci subsequently locked in by selection for inversions. Given the intricate, nonlinear evolution-ary history of contemporary *t*-haplotypes, plus Olds-Clarke's caution, one could propose a model of TRD that takes into account mutant downstream effectors of Smok/Tcr activity whose punishing effects are meted out conditionally as responses to increasing levels of Smok/Tcr activity.

That being said, although it has been common knowledge for some time that mutant dynein subunits present in the mouse sperm tail axoneme (at least one of them exclusively) map to three of the *Tcd* loci, the fact is that none of them have ever been directly tested for their potential effects on TRD. Instead, natu-rally occurring or targeted deletions of wild-type isoforms have been employed, more often than not in Chlamydomonas (but occasionally in mice), to inves-tigate both qualitative and quantitative downstream effects on dynein arm assembly, dynein function, or fertility, while the potential effects of the *t*-alleles of these subunits (and other flagellar components mapping to the *t*-complex; see Fig. 19.10) on TRD have not been assessed in vivo. Some of these assess-ments may be technically difficult. However, experiments similar to those devised by Herrmann's group [50,51] to identify *Tagap1* and *Fgd2* as likely upstream distorters are feasible, and should be performed on the plethora of already identified potential downstream effector molecules mapping to *Tcd* loci, some already known to carry potentially deleterious mutations [40,41]. So stay tuned.

Dyneins

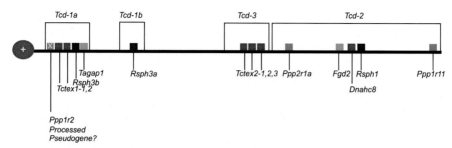

Figure 19.10 Mapped *Tcd* candidates. The relative map positions within known distorter loci on the wild-type chromosome of 14 potential or (all but) proven distorters are shown as *squares*, with dynein subunits colored red, radial spoke proteins, blue, regulators of Rho activity, green, and ser/thr protein phosphatase 1 and 2 regulatory or scaffolding subunits, gray. All are expressed in testis, and most have been shown to be sperm proteins. The candidacies of *t-complex testis expressed-1* (*Tctex1*), *t-complex testis expressed-2*, *Dnahc8*, *Fgd2*, and *Tagap1* have already been described in the text. Other candidates are described below:

1. *Rsph1* (*Tsga2*, *Meichroacidin*, *Morn 40*, *Rsp44*) encodes a radial spoke protein that in mouse is also found in abundance in both the fibrous sheath and outer dense fibers of the sperm tail [40,56]. Our (unpublished) preliminary evidence from coimmunoprecipitation studies suggest that it binds to both activity-promoting regulatory subunits and inhibitory subunits of ser/thr protein phosphatase 1 (PP1) in testis and sperm, and to IQUB, a sperm-specific calmodulin binding protein. Thus, it is potentially active as a calcium-sensitive switching mechanism in signaling pathways in the sperm tail. The *t*-allele carries numerous potentially deleterious mutations at highly conserved sites in all mammalian sequences. It could act either upstream or downstream of Smok/Tcr. It is a *Tcd-2* candidate.

2. *Rsph3a* and *Rsph3b* are paralogous radial spoke protein sequences. The former maps to the *Tcd-1b* locus and the latter, to the *Tcd-1a* locus, just proximal to *Tagap1*. In mammals, radial spoke protein 3 is an AKAP, binding to ERK1/2 [57]. It also binds IQUB in sperm. Whether the *t*-sequence of either or both is mutated or expressed differently than the allelic +-sequence is unknown.

3. *Ppp1r11* (formerly *Tctex5*; [11]) encodes an evolutionary ancient and powerful inhibitory subunit of PP1. It is present in catalytically inactive complexes with PP1γ2 in the testis and sperm [58], and genetic evidence suggests that it has a synergistic relationship with *Rsph1* in the expression of "curlicue" [40,41]. The *t*-variant carries numerous nonconservative mutations at otherwise completely conserved residues in mammalian orthologues [41]. Our (unpublished) preliminary coimmunoprecipitation data suggest that it interacts with RSPH1 in the testis and sperm tail. Because the inhibition of ser/thr phosphatase activity promotes vigorous, progressive sperm motility, *Ppp1r11* can be considered a downstream *Tcd-2* candidate.

4. *Ppp2r1a* codes for the major scaffolding subunit of ser/thr protein phosphatase 2A. It is expressed in the testis and is found in sperm and maps to the proximal end of the *Tcd-2* locus. The sequence and expression of its *t*-allele have not been studied.

5. Two alleles of what are believed to be processed pseudogene copies of *Ppp1r2*, another ubiquitous inhibitor of PP1, are located between *Serac* and *Tctex1* (*RP23-456P8.3* and *AC18309.1-201*; Ensembl Mouse Genome Database-NCBI-M37). They appear to be identical, single exon paralogues of *Ppp1r2*, with an intact coding region, and relatively few mutations, none of which encode frame shifts or premature stop codons. Our preliminary (unpublished) data suggest that one or both produce a testis-specific transcript that can code for a PPP1R2 variant with an additional 109 amino acid N-terminal extension containing interesting features (not shown as part of the Ensembl transcript), and that this transcript may, indeed, be translated. Interestingly, the entire sequence is not present in *t*-haplotypes.

References

[1] K.G. Ardlie, L.M. Silver, Low frequency of mouse *t* haplotypes in wild populations is not explained by modifiers of meiotic drive, Genetics 144 (1996) 1787–1797.

[2] M.F. Hammer, J. Schimenti, L.M. Silver, Evolution of mouse chromosome 17 and the origin of inversions associated with t haplotypes, Proc. Natl. Acad. Sci. USA 86 (1989) 3261–3265.

[3] D. Bennett, The T-locus of the mouse, Cell 6 (1975) 441–454.

[4] M.F. Lyon, Male sterility of the mouse t-complex is due to homozygosity of the distorter genes, Cell 44 (1986) 357–363.

[5] M.F. Lyon, Transmission ratio distortion in mouse t-haplotypes is due to multiple distorter genes acting on a responder locus, Cell 37 (1984) 621–628.

[6] L.M. Silver, P. Olds-Clarke, Transmission ratio distortion of mouse t-haplotypes is not a consequence of wild-type sperm degeneration, Dev. Biol. 105 (1984) 250–252.

[7] P. Olds-Clarke, L.R. Johnson, t haplotypes in the mouse compromise sperm flagellar function, Dev. Biol. 155 (1993) 14–25.

[8] D.F. Katz, E.Z. Drobnis, J.W. Overstreet, Factors regulating mammalian sperm migration through the female reproductive tract and oocyte vestments, Gamete Res. 22 (1989) 443–469.

[9] C.B. Lindemann, J.S. Goltz, K.S. Kanous, T.K. Gardner, P. Olds-Clarke, Evidence for an increased sensitivity to Ca^{2+} in the flagella of sperm from tw32/+mice, Mol. Reprod. Dev. 26 (1990) 69–77.

[10] D.F. Katz, R.P. Erickson, M. Nathanson, Beat frequency is bimodally distributed in spermatozoa from T/t12 mice, J. Exp. Zool. 210 (1979) 529–535.

[11] H. Ha, C.A. Howard, Y.I. Yeom, K. Abe, H. Uehara, K. Artzt, D. Bennett, Several testis-expressed genes in the mouse t-complex have expression differences between wild-type and t-mutant mice, Dev. Genet. 12 (1991) 318–332.

[12] L.Y. Huw, A.S. Goldsborough, K. Willison, K. Artzt, Tctex2: a sperm tail surface protein mapping to the *t*-complex, Dev. Biol. 170 (1995) 183–194.

[13] E. Lader, H.S. Ha, M. O'Neill, K. Artzt, D. Bennett, tctex-1: a candidate gene family for a mouse *t* complex sterility locus, Cell 58 (1989) 969–979.

[14] L.R. Johnson, S.H. Pilder, P. Olds-Clarke, The cellular basis for interaction of sterility factors in the mouse *t* haplotype, Genet. Res. Camb. 66 (1995) 189–193.

[15] D.M. Phillips, S.H. Pilder, P.J. Olds-Clarke, L.M. Silver, Factors that may regulate assembly of the mammalian sperm tail deduced from a mouse *t* complex mutation, Biol. Reprod. 49 (1993) 1347–1352.

[16] S.H. Pilder, P. Olds-Clarke, D.M. Phillips, L.M. Silver, Hybrid sterility-6: a mouse t complex locus controlling sperm flagellar assembly and movement, Dev. Biol. 159 (1993) 631–642.

[17] A.A. Redkar, P. Olds-Clarke, L.M. Dugan, S.H. Pilder, High resolution mapping of sperm function defects in the t complex fourth inversion, Mamm. Genome 9 (1998) 825–830.

[18] S.A. Samant, J. Fossella, L.M. Silver, S.H. Pilder, Mapping and cloning recombinant breakpoints demarcating the hybrid sterility 6-specific sperm tail assembly defect, Mamm. Genome 10 (1999) 88–94.

[19] M. Fliegauf, H. Olbrich, J. Horvath, J.H. Wildhaber, M.A. Zariwala, M. Kennedy, M.R. Knowles, H. Omran, Mis-localization of DNAH5 and DNAH9 in respiratory cells from primary ciliary dyskinesia patients, Am. J. Respir. Crit. Care Med. (2005) 1343–1349.

[20] J. Fossella, S.A. Samant, L.M. Silver, S.M. King, K.T. Vaughan, P. Olds-Clarke, K.A. Johnson, A. Mikami, R.B. Vallee, S.H. Pilder, An axonemal dynein at the Hybrid Sterility 6 locus: implications for t haplotype-specific male sterility and the evolution of species barriers, Mamm. Genome 11 (2000) 8–15.

[21] S.A. Samant, O. Ogunkua, L. Hui, J. Fossella, S.H. Pilder, The t complex distorter 2 candidate gene, Dnahc8, encodes at least two testis-specific axonemal dynein heavy chains that differ extensively at their amino and carboxyl termini, Dev. Biol. 250 (2002) 24–43.

[22] S.A. Samant, O.O. Ogunkua, L. Hui, J. Lu, Y. Han, J.M. Orth, S.H. Pilder, The mouse t complex distorter/sterility candidate, Dnahc8, expresses a c-type axonemal dynein heavy chain isoform confined to the principal piece of the sperm tail, Dev. Biol. 285 (2005) 57–69.

[23] H. Wu, M.W. Maciejewski, S. Takebe, S.M. King, Solution structure of the Tctex1 dimer reveals a mechanism for dynein-cargo interactions, Structure 13 (2005) 213–223.

[24] S.M. King, J.F. Dillman 3rd, S.E. Benashski, R.J. Lye, S. Patel-King, K.K. Pfister, The mouse t-complex-encoded protein Tctex-1 is a light chain of brain cytoplasmic dynein, J. Biol. Chem. 271 (1996) 32281–32287.

[25] A. Harrison, P. Olds-Clarke, S.M. King, Identification of the t complex-encoded cytoplasmic dynein light chain Tctex1 in inner arm I1 supports the involvement of flagellar dyneins in meiotic drive, J. Cell Biol. 140 (1998) 1137–1147.

[26] N. Kotani, H. Sakakibara, S.A. Burgess, H. Kojima, K. Oiwa, Mechanical properties of inner-arm dynein-f (dynein I1) studied with in vitro motility assays, Biophys. J. 93 (2007) 886–894.

[27] C.B. Lindemann, K.A. Lesich, Flagellar and ciliary beating: the proven and the possible, J. Cell Sci. 123 (2010) 519–528.

[28] L.M. DiBella, O. Gorbatyuk, M. Sakato, K. Wakabayashi, R.S. Patel-King, G.J. Pazour, G.B. Witman, S.M. King, Differential light chain assembly influences outer arm dynein motor function, Mol. Biol. Cell. 16 (2005) 5661–5674.

[29] B.G. Herrmann, B. Koschorz, K. Wertz, K.J. McLaughlin, A. Kispert, A protein kinase encoded by the t complex responder gene causes non-mendelian inheritance, Nature 402 (1999) 141–146.

[30] M.F. Lyon, Deletion of mouse t-complex distorter-1 produces an effect like that of the t-form of the distorter, Genet. Res. 59 (1992) 27–33.

[31] M.F. Lyon, J.C. Schimenti, E.P. Evans, Narrowing the critical regions for mouse t complex transmission ratio distortion factors by use of deletions, Genetics 155 (2000) 793–801.

[32] A. Planchart, Y. You, J.C. Schimenti, Physical mapping of male fertility and meiotic drive QTLs in the mouse t complex using chromosome deficiencies, Genetics 155 (2000) 803–812.

[33] R.S. Patel-King, S.E. Benashski, A. Harrison, S.M. King, A Chlamydomonas homologue of the putative murine t complex distorter Tctex-2 is an outer arm dynein light chain, J. Cell Biol. 137 (1997) 1081–1090.

[34] G.J. Pazour, A. Koutoulis, S.E. Benashski, B.L. Dickert, H. Sheng, R.S. Patel-King, S.M. King, G.B. Witman, LC2, the Chlamydomonas homologue of the t complex-encoded protein Tctex2, is essential for outer dynein arm assembly, Mol. Biol. Cell 10 (1999) 3507–3520.

[35] S. Rashid, P. Grzmil, J.-D. Drenckhahn, A. Meinhardt, I. Adham, W. Engel, J. Neesen, Disruption of the murine dynein light chain gene Tcte3-3 results in asthenozoospermia, Reproduction 139 (2010) 99–111.

[36] P. Olds-Clarke, Motility characteristics of sperm from the uterus and oviducts of female mice after mating to congenic males differing in sperm transport and fertility, Biol. Reprod. 34 (1986) 453–467.

[37] S.H. Pilder, M.F. Hammer, L.M. Silver, A novel mouse chromosome 17 hybrid sterility locus: implications for the origin of t haplotypes, Genetics 129 (1991) 237–246.

[38] S.H. Pilder, Identification and linkage mapping of Hst7, a new M. spretus/M. m. domesticus chromosome 17 hybrid sterility locus, Mamm. Genome 8 (1997) 290–291.

[39] S.H. Pilder, P. Olds-Clarke, J.M. Orth, W.F. Jester, L. Dugan, Hst7: a male sterility mutation perturbing sperm motility, flagellar assembly, and mitochondrial sheath differentiation, J. Androl. 18 (1997) 663–671.

[40] L. Hui, J. Lu, Y. Han, S.H. Pilder, The mouse T complex gene Tsga2, encoding polypeptides located in the sperm tail and anterior acrosome, maps to a locus associated with sperm motility and sperm-egg interaction abnormalities, Biol. Reprod. 74 (2006) 633–643.

[41] S.H. Pilder, J. Lu, Y. Han, L. Hui, S.A. Samant, O.O. Olugbemiga, K.W. Meyers, L. Cheng, S. Vijayaraghavan, The molecular basis of "curlicue": a sperm motility abnormality linked to the sterility of t haplotype homozygous male mice, Soc. Reprod. Fertil. Suppl. 63 (2007) 123–133.

[42] K.T. Vaughn, A. Mikami, B.M. Paschal, E.L.F. Holzbaur, S.M. Hughes, C.J. Echeverri, K.J. Moore, D.J. Gilbert, N.G. Copeland, N.A. Jenkins, R.B. Vallee, Multiple mouse chromosomal loci for dynein-based motility, Genomics 36 (1996) 29–38.

[43] M.P. Koonce, I. Tikhonenko, Functional elements within the dynein microtubule-binding domain, Mol. Biol. Cell 11 (2000) 523–529.

[44] S.M. King, Sensing the mechanical state of the axoneme and integration of Ca^{2+} signaling by outer arm dynein, Cytoskeleton 67 (2010) 207–213.

[45] M. Sakato, H. Sakakibara, S.M. King, Chlamydomonas outer arm dynein alters conformation in response to Ca^{2+}, Mol. Biol. Cell 18 (2007) 3620–3634.

[46] D.O. Bunch, J.E. Welch, P.L. Magyar, E.M. Eddy, D.A. O'Brien, Glyceraldehyde 3-phosphate dehydrogenase-S protein distribution during mouse spermatogenesis, Biol. Reprod. 58 (1998) 834–841.

[47] D.W. Fawcett, The mammalian spermatozoon, Dev. Biol. 44 (1975) 394–436.

[48] N. Veron, H. Bauer, A.Y. Weiße, G. Luder, M. Werber, B.G. Herrmann, Retention of gene products in syncytial spermatids promotes non-Mendelian inheritance as revealed by the t complex responder, Genes Dev. 23 (2009) 2705–2710.

[49] J.C. Schimenti, J.L. Reynolds, A. Planchart, Mutations in Serac1 or Synj2 cause proximal t haplotype-mediated male mouse sterility but not transmission ratio distortion, Proc. Natl. Acad. Sci. USA 102 (2005) 3342–3347.

[50] H. Bauer, J. Willert, B. Koschorz, B.G. Herrmann, The t complex-encoded GTPase-activating protein Tagap1 acts as a transmission ratio distorter in mice, Nat. Genet. 37 (2005) 969–973.

[51] H. Bauer, N. Veron, J. Willert, B.G. Herrmann, The t-complex-encoded guanine nucleotide exchange factor Fgd2 reveals that two opposing signaling pathways promote transmission ratio distortion in the mouse, Genes Dev. 21 (2007) 143–147.

[52] P. Olds-Clarke, Unresolved issues in mammalian fertilization, Int. Rev. Cytol. 232 (2003) 129–184.

[53] C. Delarbre, Y. Kashi, P. Boursot, J.S. Beckmann, P. Kourilsky, F. Bonhomme, G. Gachelin, Phylogenetic distribution in the genus Mus of t complex-specific DNA and protein markers: inferences on the origin of t-haplotypes, Mol. Biol. Evol. 5 (1988) 120–133.

[54] J. Klein, M. Nizetic, Z. Golubic, M. Dembic, F. Figueroa, Evolution of the H-2 genes on t chromosomes, in: B. Pernis, H.J. Vogel (Eds.), Cell Biology of the Major Histocornpatability Complex, Academic Press, New York, 1985, pp. 97–106.

[55] M.F. Hammer, L.M. Silver, Phylogenetic analysis of the alpha-globin pseudogene-4 (Hba-ps4) locus in the house mouse species complex reveals a stepwise evolution of t haplotypes, Mio. Biol. Evol. 10 (1993) 971–1001.

[56] K. Tokuhiro, M. Hirose, Y. Miyagawa, A. Tsujimura, S. Irie, A. Isotani, M. Okabe, Y. Toyama, C. Ito, K. Toshimori, K. Takeda, S. Oshio, H. Tainaka, J. Tsuchida, A. Okuyama, Y. Nishimune, H. Tanaka, Meichroacidin containing the membrane occupation and recognition nexus motif is essential for spermatozoa morphogenesis, J. Biol. Chem. 283 (2008) 19039–19048.

Dyneins

[57] A. Jivan, S. Earnest, Y.-C. Juang, M.H. Cobb, Radial spoke protein 3 is a mammalian protein kinase A-anchoring protein that binds ERK1/2, J. Biol. Chem. 284 (2009) 29437–29445.

[58] L. Cheng, S. Pilder, A.C. Nairn, S. Ramdas, S. Vijayaraghavan, Ppigamma2 and PPP1R11 are parts of a multimeric complex in developing testicular germ cells in which their steady state levels are reciprocally related, PLoS One 4 (2009) e4861.

Index

'*Note:* Page numbers followed by "f" indicate figures, "t" indicate tables and "b" indicate boxes.'

A

A/P axes. *See* Anterior–posterior axes (A/P axes)

AAA domain
AAA-type motor protein complexes, 318–319
AAA1, 23, 38, 42–43, 122
sites, 173–174
AAA1–2 gap, 28
AAA2, 42–43, 121
AAA3, 37, 43, 121–123, 122f
AAA4, 4, 37, 43, 121–123
AAA4S, 37
AAA5, 42–43, 121
AAA5S, 37, 39
large domain, 28
AAA6, 121
role in dynein motility, 121–123

AAA+. *See* ATPases associated with various cellular activities (AAA+)

AAAS. *See* α-Helical small domains (AAAS)

Acetylcysteine, 412

ACLS. *See* Acrocallosal syndrome (ACLS)

Acquired ciliopathy, 403

Acrocallosal syndrome (ACLS), 363, 368–369

Activated dynein:dynactin:BicD2 complexes, 161

Adenosine triphosphate (ATP), 8–9
binding, 119
hydrolysis, 42–43, 47–48
hydrolysis cycle, 113

Adenylyl–imidodiphosphate (AMP–PNP), 8–9

ADHD. *See* Attention deficit hyperactivity disorder (ADHD)

ADP–vanadate, 23

Advanced electron microscopy techniques, dynein molecules properties of, 15–23
axonemal dyneins purification from *Chlamydomonas* axonemes, 15–18
cytoplasmic dyneins purification, 18
flexibility of tail revealed by negative stain electron microscopy, 22–23
gallery of electron micrographs of inner-arm dynein subspecies, 21f
negative staining electron microscopy, 18–22
single particle analysis procedures, 19f

African protozoan (*Trypanosoma*), 339–340

African trypanosomes, 422

Aggresomes, 292–293

AICD. *See* Alcohol-induced ciliary dysfunction (AICD)

Airway diseases, 404

Alcohol
and *Chlamydomonas* flagella
AICD, 409–410
alcohol alters phosphorylation of specific ciliary proteins, 410–412
alcohol impacts ODA, 410
exposure, 404
and mucociliary function
desensitization of dynein activation, 408–409
kinase–phosphatase relationships in AICD, 409
linkage of CBF to ODAs in mammalian airway cilia, 405–408
mucociliary clearance, 404–405

Alcohol use disorders (AUDs), 403–404

Index

Bent linker, 41
BGEM. *See* Brain Gene Expression Map (BGEM)
Bicaudal-D2 (BICD2), 47–48, 254–255
BicD, 144–146
Bicd2 knockout mouse, 254–255
BICD2 N-terminal fragment (BICD2-N), 254–255
BICD2. *See* Bicaudal-D2 (BICD2)
BicD^{Pa66} mutant, 145
BicDR1, 146
Bidirectional cargo transport
   bidirectional transport behavior, 156f
   dynein properties relevant, 160–161
   effects of membrane fluidity, 162–164, 163f
   experimental and computational work to date, 154–155
   kinesins in bidirectional transport, 158–160
   models of bidirectional transport, 155–158
   roles of MAPs and tubulin PTMs, 161–162
"Biochemical amount" of homogeneous dynein molecules, 3
Biochemical isolation of cytoplasmic dynein, 100
Biochemical purification, 90
   axonemal dyneins, 90–99
   cytoplasmic dynein, 99–103
   storage of dynein, 103
Biological General Repository for Interaction Datasets (BioGRID), 441–442
Biological materials, issues for electron microscopy observation on, 12–13
Biophysical function of dynein in vivo, 139–143
Biophysical properties of dynein in vivo
   biophysical function of dynein in vivo, 139–143
   motility and regulation of dynein in vitro, 137–139
   regulation of dynein motility in vivo, 143–147
   in vivo forces for dynein, 143t
Biophysical studies, 74
Bipolar migration dynein, 270
BIV. *See* Bovine immunodeficiency virus (BIV)
BLAST algorithms, 319–322
Bones, 359–360

Boundary conditions, 197, 197t
Boundary value problem, coefficient equations for, 210–211
Bovine immunodeficiency virus (BIV), 215–216
BOXER, 20
Brain
   brain-imaging studies, 365–366
   development, 271, 274
   malformations, 368–369
   ventricles, 327–328
Brain Gene Expression Map (BGEM), 438–441
Brain-derived neurotrophic factor (BDNF), 291
Breathalyzer test, 405
Bridging model, 74–75
Bronchial circulation, 405
Bronchiectasis, 404
bst2 gene, 223

C

C-terminal domain, 43–45
Caenorhabditis elegans (*C. elegans*), 57–58
Cajal–Retzius cells, 263–264
cAMP, 318–319
cAMP-dependent protein kinase A (PKA), 406
Canonical cGMP-dependent kinase (PKG), 406
CAP-Gly domain. *See* Cytoskeletal-associated protein, glycine-rich domain (CAP-Gly domain)
Carbon-filmed grid, 14–15
Cargo, 72–73
   beads, 117
   cargo binding, dynein motor activation by, 47–48
   directionality exhibits memory, 141
   dynein, 289–290
   stalled, 160–161
Cartilage, 359–360
Catch-bond behavior, 160–161
CB-5083, 186
CBF. *See* Ciliary beat frequency (CBF)
CCD. *See* Charge-coupled device (CCD)
CCDC103, 333–334
CCDC114. See Coiled-coil domain-containing protein 114 (*CCDC114*)
Cdk1. *See* Cyclin-dependent kinase-1 (Cdk1)
cDNA clones analysis, 6

Index

Index

Index

Printed in the United States
By Bookmasters